# Mass Spectrometry in Biomolecular Sciences

# NATO ASI Series

## Advanced Science Institutes Series

*A Series presenting the results of activities sponsored by the NATO Science Committee, which aims at the dissemination of advanced scientific and technological knowledge, with a view to strengthening links between scientific communities.*

The Series is published by an international board of publishers in conjunction with the NATO Scientific Affairs Division

| | |
|---|---|
| **A  Life Sciences** | Plenum Publishing Corporation |
| **B  Physics** | London and New York |
| | |
| **C  Mathematical and Physical Sciences** | Kluwer Academic Publishers |
| **D  Behavioural and Social Sciences** | Dordrecht, Boston and London |
| **E  Applied Sciences** | |
| | |
| **F  Computer and Systems Sciences** | Springer-Verlag |
| **G  Ecological Sciences** | Berlin, Heidelberg, New York, London, |
| **H  Cell Biology** | Paris and Tokyo |
| **I   Global Environmental Change** | |

### PARTNERSHIP SUB-SERIES

| | |
|---|---|
| **1.  Disarmament Technologies** | Kluwer Academic Publishers |
| **2.  Environment** | Springer-Verlag / Kluwer Academic Publishers |
| **3.  High Technology** | Kluwer Academic Publishers |
| **4.  Science and Technology Policy** | Kluwer Academic Publishers |
| **5.  Computer Networking** | Kluwer Academic Publishers |

*The Partnership Sub-Series incorporates activities undertaken in collaboration with NATO's Cooperation Partners, the countries of the CIS and Central and Eastern Europe, in Priority Areas of concern to those countries.*

### NATO-PCO-DATA BASE

The electronic index to the NATO ASI Series provides full bibliographical references (with keywords and/or abstracts) to more than 50000 contributions from international scientists published in all sections of the NATO ASI Series.
Access to the NATO-PCO-DATA BASE is possible in two ways:

– via online FILE 128 (NATO-PCO-DATA BASE) hosted by ESRIN,
Via Galileo Galilei, I-00044 Frascati, Italy.

– via CD-ROM "NATO-PCO-DATA BASE" with user-friendly retrieval software in English, French and German (© WTV GmbH and DATAWARE Technologies Inc. 1989).

The CD-ROM can be ordered through any member of the Board of Publishers or through NATO-PCO, Overijse, Belgium.

**Series C: Mathematical and Physical Sciences – Vol. 475**

# Mass Spectrometry in Biomolecular Sciences

edited by

## Richard M. Caprioli

University of Texas,
Medical School,
Houston, Texas, U.S.A.

## Antonio Malorni

SESMA-C.N.R.,
Napoli, Italy

and

## Giovanni Sindona

Università della Calabria,
Dipartimento di Chimica,
Arcavacata di Rende, Italy

**Kluwer Academic Publishers**

Dordrecht / Boston / London

Published in cooperation with NATO Scientific Affairs Division

Proceedings of the NATO Advanced Study Institute on
Mass Spectrometry in Biomolecular Sciences
Lacco Ameno, Ischia, Italy
June 23–July 5, 1993

**Library of Congress Cataloging-in-Publication Data**

ISBN-13: 978-94-010-6581-8

---

Published by Kluwer Academic Publishers,
P.O. Box 17, 3300 AA Dordrecht, The Netherlands.

Kluwer Academic Publishers incorporates the publishing programmes of
D. Reidel, Martinus Nijhoff, Dr W. Junk and MTP Press.

Sold and distributed in the U.S.A. and Canada
by Kluwer Academic Publishers,
101 Philip Drive, Norwell, MA 02061, U.S.A.

In all other countries, sold and distributed
by Kluwer Academic Publishers Group,
P.O. Box 322, 3300 AH Dordrecht, The Netherlands.

*Printed on acid-free paper*

---

# CONTENTS

Preface............................................................................................................ ix

Contributing Authors.................................................................................... xi

Participants.................................................................................................. xv

**Part I: Methods in Mass Spectrometry**

1.  Ion Formation by Fast Atom Bombardment Mass Spectrometry.............................3
    *G. Sindona*

2.  Matrix-Assisted Laser Desorption-Ionization (MALDI) Mass Spectrometry of Biological
    Molecules..............................................................................................33
    *M. Karas, U. Bahr*

3.  Matrix-Assisted Laser Desorption Mass Spectrometry on a Magnetic Sector Instrument
    Fitted with an Array Detector.................................................................51
    *R. S. Bordoli, K. Howes, R.G. Vickers, R.H. Bateman, and D.J. Harvey*

4.  Electrospray Ionization Mass Spectrometry: Protein Structure.....................61
    *T. B. Farmer and R. M. Caprioli*

5.  Applications of Quantum Chemistry in Mass Spectrometry............................89
    *K. Vékey*

6.  Proton Affinities and Intrinsic Basicities of Alanine and Glycine Studied by Means of
    AM1 and PM3 Methods...........................................................................103
    *T. Marino, N. Russo, and M. Toscano*

**Part II: Instrumentation for Mass Analysis and Detection**

7.  Energy-Isochronous Time-of-Flight Mass Spectrometers.............................111
    *H. Wollnik*

8.  The Detection of High Mass-to-Charge Biological Ions by Fourier Transform Mass
    Spectrometry: Issues and Routes for Instrument Improvement....................147
    *C.L. Holliman, D. L. Rempel and M. L. Gross*

9.  Quadrupole Mass Filters, Quadrupole Ion Traps and Fourier Transform Ion Cyclotron
    Resonance Spectrometers.......................................................................177
    *L. Olimpieri and P. Traldi*

10. Energy Shifts in Collisional Activation.................................................201
    *H.J. Cooper and P. J. Derrick*

11.    Hybrid Instruments of Unusual Geometry...............................................................261
       *D. Fravetto and P. Traldi*

**Part III: Application to Biomolecules**

12.    Continuous-Flow Fast Atom Bombardment Mass Spectrometry: Applications to Dynamic
       Systems...................................................................................................................273
       *R.M. Caprioli*

13.    Hybrid Tandem Mass Spectrometry of Peptides...........................................................299
       *S.J. Gaskell*

14.    Mass Spectrometry in the Study of Advanced Glycation End Products.........................317
       *A. Lapolla, D. Fedele, S. Catinella, P. Traldi*

14.    Nucleic Acids: Overview and Analytical Strategies.....................................................351
       *P.F. Crain*

15.    Monitoring of Human Exposure to Xenobiotics; Identification and Quantification of
       Cancer Initiators in Vivo...........................................................................................381
       *M. Törnqvist*

16.    The Role of Mass Spectrometry in Biomonitoring Exposure to Carcinogens.................397
       *N.Sannolo, V. Carbone, P. Ferranti, I. Fiume, D. Santoro and A. Malorni*

17.    Characterization of Ceramide Mixtures by Fast Atom Bombardment and Tandem Mass
       Spectrometry.............................................................................................................417
       *F.M. Rubino, S. Sonnino*

18.    Bioorganic Studies of a New Photoreceptor Structure.................................................429
       *M. Orlando and M. L. Gross*

**Part IV: Coupling of Chromatography to Mass Spectrometry**

19.    Gas Chromatography and Mass Spectrometry Coupling................................................437
       *J. Abián and E. Gelpi*

20.    Recent Developments in Combined Supercritical Fluid Chromatography/Mass
       Spectrometry (SFC/MS)..............................................................................................461
       *P. Arpino, F. Sadoun, H. Virelizier*

21.    GC-MS Analysis of Volatile Compounds of Mesir.......................................................475
       *Y. Hisil, N. Bagdatlioglu, S. Ötles*

22.    The Application of Mass Spectrometry to Food and Nutrition Research.......................483
       *R. Self, F.A. Mellon, B.A. McGaw, A.G. Calder, G.E. Lobley and E. Milne*

Index...........................................................................................................................517

# PREFACE

The important role of mass spectrometry as an analytical tool in the life sciences is now well recognized. The remarkable development of both instrumentation and methodology allows the structural characterization of biological molecules of all classes; peptides and proteins, carbohydrates, nucleic acids, lipids, drugs and their metabolites, and other xenobiotics. The present capabilities of mass spectrometry are truly impressive, encompassing the analysis of biopolymers of molecular weights of hundreds of thousands and measurements of minute quantities of compound at the attomole level. It is clear that mass spectrometry has had a big impact on many different areas of biological research and quality control in bioprocesses.

The goal of the NATO ASI dedicated to "Mass Spectrometry in the Biomolecular Sciences" is to provide young scientists with an intensive short course on the latest instrumental advancements, developments in appropriate analytical methodology and the opportunity to discuss these topics with experts in the field who are themselves leaders in some of these developments. This NATO ASI was launched in 1988 and the first institute, held in southern Italy (Cetraro) in 1990, was enthusiastically greeted by both junior and senior scientists. In many academic institution, the participation of students in international schools, such as this NATO ASI, is considered an important part of the training of Ph.D. students and young scientists because it aids in the development of their ability to communicate and exchange information. For this reason, senior scientists involved in the first meeting have agreed that an advanced course in mass spectrometric applications in biological sciences should be organized on a regular basis, each two or three years. The impressive advances in the science of mass spectrometry in the three years following the first ASI formed the backdrop for the second meeting on this subject held at the Hotel Augusto in Lacco Ameno, on the island of Ischia in Southern Italy, June 23-to July 5, 1993. In a congenial and casual atmosphere, 103 scientists discussed the developments of the methodology with reference to important applications in biology, bio-medicine, environmental sciences and food chemistry.

This book provides an account on the recent developments of both the technology and applications related to the topics presented at the ASI in 1993. It is divided into four sections. Section I, "Methods in Mass Spectrometry", introduces and reviews some of the newer ionization methods widely employed in the analysis of biological molecules, such as fast atom bombardment, electrospray, and matrix-assisted laser desorption ionization. Further, an introduction to the application of quantum chemistry in mass spectrometry is provided. Section II, "Instrumentation for Mass Analysis and Detection", describes the major instrumentation involved in mass analysis including magnetic, time-of-flight, quadrupole, ion trap and fourier transform mass spectrometers. A discussion of MS/MS applications to large molecules and the principles of low and high energy collisions for structure characterization are also presented. Section III, "Applications to Biomolecules", presents the scope of the technology in biology, medicine, and environmental areas, with specific examples for a selected number of topics. These topics include the use of continuous flow interfaces for the *in vivo* detection of analytes, evaluation of the risks of human exposure to carcinogens, and the analysis of peptides and proteins, lipids and nucleic acids derived from biological sources. Section IV, "Coupling of Chromatography to Mass Spectrometry", presents recent achievements in the on-line combination of separation techniques such as gas chromatography, liquid chromatography, and supercritical fluid chromatography. In addition, applications in the environmental sciences, and in human nutrition, are presented.

The editors acknowledge the special efforts of the other fourteen senior plenary lectures, P. Arpino (France), A. Burlingame (USA), P. Crain (USA), P. Derrick (UK), S. Facchetti (Italy), S.

vii

Gaskell (UK), E. Gelpi (Spain), M. Gross (USA), M. Karas (Germany), H. Morris (UK), R. Self (UK), P. Traldi (Italy), M. Tornqvist (Sweden) and H. Wollnik (Germany) and of the senior keynote lectures, A. Abbondandolo (Italy), G. Falcone (Italy), N. Russo (Italy), T. Theophanides (Greece) and K. Vekey (Hungary). Most have generously contributed to this book.

Special thanks go to the NATO scientific affairs division, the Italian National Research Council (CNR) and Fisons-VG (UK) who provided most of the funds for the institute. Thanks are also due to the French Society for Mass Spectrometry (SFSM), the Portuguese JNCT, Turkish TUBITAK, Greek Ministry of Industry, Energy and Technology and the U.S. National Science Foundation (NSF) for providing travel supports to some junior participants at the ASI.

R. M. Caprioli
A. Malorni
G. Sindona
May, 1995

# CONTRIBUTING AUTHORS

J. Abián      C.I.D. - C.S.I.C., Dept. of Neurochemistry, Jiordi Girona 18-26, 08034-Barcelona, Spain

P. Arpino      Laboratoire de Chimie Analytique, Institut National Agronomique, 16 rue Claude Bernard, 75231 Paris-05, France

N. Bagdatlioglu      Department of Food Engineering, University of Ege, Bornova, Ismir 35100, Turkey

U. Bahr      Institute for Medical Physics and Biophysics, University of Münster, Robert-Koch-Str. 31, 48149 Münster F.R. Germany

R. H. Bateman      Fisons Intruments, Organic Analysis, Analytical MS, Floats Road, Wythenshawe, Manchester, M23 9LE, UK

R. S. Bordoli      Fisons Intruments, Organic Analysis, Analytical MS, Floats Road, Wythenshawe, Manchester, M23 9LE, UK

A. G. Calder      Rowett Research Institute, Greenburn Road,Bucksburn, Aberdeen, AB2 9SB, UK

R. M. Caprioli      Analytical Chemistry Center, Department of Biochemistry and Molecular Biology, University of Texas Medical School, P.O. Box 20708, Houston, Tx 77225

V. Carbone      Servizio di Spettrometria di Massa del CNR c/o Facoltà di Medicina dell'Università di Napoli Federico II, Via Pansini 5, I - 80131 Napoli, Italy

S. Catinella      CNR Research Area, Corso Stati uniti, 4 - 35100 Padua, Italy

H. J. Cooper      Institute of Mass Spectrometry and Department of Chemistry, University of Warwick, Coventry CV4 7Al UK

P. F. Crain      Department of Medicinal Chemistry, University of Utah, Salt Lake City, UT 84112 USA

P. J. Derrick      Institute of Mass Spectrometry and Department of Chemistry, University of Warwick, Coventry CV4 7Al UK

T. B. Farmer      Analytical Chemistry Center, Department of Biochemistry and Molecular Biology, University of Texas Medical School, P.O. Box 20708, Houston, Tx 77225

D. Fedele — Institute of Internal Medicine, Division of Metabolic Disorder, Via Giustiniani 2 - 35100 Padua, Italy

P. Ferranti — Servizio di Spettrometria di Massa del CNR c/o Facoltà di Medicina dell'Università di Napoli Federico II, Via Pansini 5, I - 80131 Napoli, Italy

I. Fiume — Servizio di Spettrometria di Massa del CNR c/o Facoltà di Medicina dell'Università di Napoli Federico II, Via Pansini 5, I - 80131 Napoli, Italy

D. Fravetto — CNR, Area della Ricerca, Corso Stati Uniti 4, I-35020 Padova, Italy

S. J. Gaskell — Department of Chemistry, University of Manchester, Institute of science and Technology, P.O. Box 88, Manchester M60 1QD, UK

E. Gelpi — C.I.D. - C.S.I.C., Dept. of Neurochemistry, Jiordi Girona 18-26, 08034-Barcelona, Spain

M. L. Gross — MidWest Center for Mass Spectrometry, Department of Chemistry, University of Nebraska-Lincoln, Lincoln, Nebraska 68588

D. J. Harvey — Glycobiology Institute, Department of Biochemistry, University of Oxford, South Parks Road, Oxford, OX1 3QU, UK

Y. Hisil — Department of Food Engineering, University of Ege, Bornova, Ismir 35100, Turkey

C. L. Holliman — Midwest Center for Mass Spectrometry, Department of Chemistry, University of Nebraska-Lincoln, Nebraska, 68588-0362 USA

K. Howes — Fisons Intruments, Organic Analysis, Analytical MS, Floats Road, Wythenshawe, Manchester, M23 9LE, UK

M. Karas — Institute for Medical Physics and Biophysics, University of Münster, Robert-Koch-Str. 31, 48149 Münster F.R. Germany

A. Lapolla — Institute of Internal Medicine, Division of Metabolic Disorder, Via Giustiniani 2 - 35100 Padua, Italy

G. E. Lobley — Rowett Research Institute, Greenburn Road,Bucksburn, Aberdeen, AB2 9SB, UK

A. Malorni — Servizio di Spettrometria di Massa del CNR c/o Facoltà di Medicina dell'Università di Napoli Federico II, Via Pansini 5, I - 80131 Napoli, Italy

T. Marino — Dipartimento di Chimica, Università della Calabria, I-87030 Arcavacata di Rende (CS), Italy

B. A. McGaw — School of Applied Sciences, The Robert Gordon University, Aberdeen, AB1 1HG, UK

F. A. Mellon — Institute of Food Research, Norwich Laboratory, Norwich Research Park, Colney Lane, NR4 7UA, UK

E. Milne — Rowett Research Institute, Greenburn Road,Bucksburn, Aberdeen, AB2 9SB, UK

L. Olimpieri — CNR, Research Area, Corso Stati Uniti 4, 35100 Padova, Italy

M. Orlando — MidWest Center for Mass Spectrometry, Department of Chemistry, University of Nebraska-Lincoln, Lincoln, Nebraska 68588

S. Ötles — Department of Food Engineering, University of Ege, Bornova, Ismir 35100, Turkey

D. L. Rempel — MidWest Center for Mass Spectrometry, Department of Chemistry, University of Nebraska-Lincoln, Lincoln, Nebraska 68588

F. M. Rubino — ITBA - CNR, Via Ampere 56, I - 20131 Milano, Italy

N. Russo — Dipartimento di Chimica, Università della Calabria, I-87030 Arcavacata di Rende (CS), Italy

F. Sadoun — Commissariat à l'Energie Atomique, Centre d'Etude de Saclay SPEA/SAIS, 91190 Gif-sur- Yvette, France

N. Sannolo — Sezione di Medicina Occupazionale e Igiene Industriale, Dipartimento di Biochimica e Biofisica, Seconda Università di Napoli, Piazza Miraglia 3 I-80138 Napoli, Italy

D. Santoro — Servizio di Spettrometria di Massa del CNR c/o Facoltà di Medicina dell'Università di Napoli Federico II, Via Pansini 5, I - 80131 Napoli, Italy

R. Self — School of Chemical Sciences, University of East Anglia, Norwich, NR4 7TJ UK

G. Sindona — Dipartimento di Chimica, Università della Calabria, I-87030 Arcavacata di Rende (CS), Italy

S. Sonnino — Study Center for the Functional Biochemistry of Brain Lipids; Department of Medical Chemistry and Biochemistry, The School of Medicine, University of Milan, Italy

M. Törnqvist — Department of Environmental Chemistry, Wallenberg Laboratory, Stockholm University, S-106 91 Stockholm Sweden

M. Toscano — Dipartimento di Chimica, Università della Calabria, I-87030 Arcavacata di Rende (CS), Italy

P. Traldi — CNR Research Area, Corso Stati uniti, 4 - 35100 Padua, Italy

K. Vèkey — Central Research Institute for Chemistry, Hungarian Academy of Sciences, H-1025 Budapest, Pusztaszeri ut 59-67, Hungary

R. G. Vickers — Fisons Intruments, Organic Analysis, Analytical MS, Floats Road, Wythenshawe, Manchester, M23 9LE, UK

H. Virelizier — Commissariat à l'Energie Atomique, Centre d'Etude de Saclay SPEA/SAIS, 91190 Gif-sur- Yvette, France

H. Wollnik — Physikalisches Institut, University Gissen, 35392 Gissen, Heinrich-Buff-Ring 16, Germany

# PARTICIPANTS IN THE INSTITUTE

## BELGIUM

Laurent LECLERQ
Lab. of Mass Spectrometry
University of Liege
B 6 Salt Tilman
4000 Liege

## CANADA

Mike MORRIS
Inst. for Marine Biosciences, NCR
1411 Oxford st., Halifax
Nova Scotia B3H 3ZI

## DENMARK

Kim ALVING
Department of Chemistry
University of Odense
Campusvej 55,
5230 Odense

Kim NORMANN
H.C. Oersted Institute
University of Copenhagen
Universitetsparken 5
2100 Copenhagen

Jens S. ANDERSEN
Department of Molecular Biology
University of Odense
Campusvej 55
5230 Odense

Ole VORM
Dept. of Molecular Biology
University of Odense
Campuvej 55
5230 Odense

Kim V. LARSEN
H.C. Oersted Institute
University of Copenhagen
Universitetsparken 5
2100 Copenhagen

Henrik RAHBEK-NIELSEN
Dept. of Molecular Biology
University of Odense
Campusvej 55
5230 Odense

Ejvind MORTZ
Dept. of Molecular Biology
University of Odense
Campusvej 55
5230 Odense

xiv

# FRANCE

Patrick ARPINO
Institut National Agronomique
16 rue C. Bernard
75231 Paris Cedex 05

Eric CHAPON
Lab. de Chimie Organique Structurale
University P. M. Curie
Batiment F, 7e Etage, 4 Place Jussev
75005 Paris

Adam HACHIMI
Lab. de Spectrometrie de Masse
University of Metz
1,B.D, Arago
57078 Metz Cedex 3

Noelle POTIER
Faculte de Chimie
University of Strasbourg
1, rue B. Pascal
67084 Strasbourg

Benedicte REMAUD
Lab. de Chimie Organique Structurale
University P. M. Curie
Batiment F, 7e etage, 4 Place Jussiev
75005 Paris

Veronique MOREAUX
Institut National Agronomique
16 rue Claude Bernard
75231 Paris 05

# GERMANY

Ute BAHR
Institut fuer Medizinische Physik und
Biophysik
Westfalische Wilhelms Universitaet
Robert Koch str. 31-4400 Munster

Roman Guido BECKER
Physikalische Institut
University of Giessen
Heinrich-Buff-Ring 16
6300 Giessen

Klaus DREISEWERD
Institut fuer Medizinische Physik und
Biophysik
Westfalische Wilhelms Universitaet
Robert Koch str. 31-4400 Munster

Michael KARAS
Istitut. fuer Medizinische Physik und
Biophysik
Westfalische Wilhelms Universitaet
Robert-Koch str. 31
4400 Muenster

Martin SCHURENBERG
Institut fuer Medizinische Physik und
Biophysik
Westfalische Wilhelms Universitaet
Robert Koch str. 31
4400-Munster

Kerstin STRUPAT
Institut fuer Medizinische Physik und
Biophysik
Westfalische Wilhelms Universitaet
Robert Koch str. 31
4400 Munster

Hermann WOLLNIK
Physikalisches Institut
Universitaet Gissen
Heinrich-Buff Ring 16
6300 Giessen

Marek A. WOLTER
J.W. Goethe Univesity
Niederurselar Hank 50
Frannkfurt am Mainz

# GREECE

Jane ANASTASSOPOLOU
Dept of Chemical Engineering
National Technical University
15773 Zografou
Athens

Konstantinos ATHANASSOPOULOS
Dept. of Chemistry
University of Patras
26110 Patras

Nikos BREKULAKIS
National Technical University
15773-Zografou
Athens

Nikos FOTOPOULOS
National Technical University
15773-Zografou
Athens

Theo THEOPHANIDES
Dept. of Chemical Engineering
National Technical University
15773 Zografou
Athens

Vassilios TSIKARIS
Dept. Of Chemistry
University of Ioannina
45110 Ioannina

# HUNGARY

Gabriella POCSFALVI
University of Debrecen
4010 Debrecen P.F. 20

Karoly VEKEY
Hungarian Academy of Sciences
Central Res. Institute
Pusztaszeri u. 59-67
Budapest

Pal SZABO
University of Debrecen
P.O. Box 70
4010 Debrecen

# ITALY

Angelo ABBONDADOLO
CSTA Mutagenesis
Ist. Nazionale per la Ricerca sul Cancro
Viale Benedetto XV 10
16123 Genova

Antonella BERTAZZO
Mass Spectrometry Service
CNR, Corso Stai Uniti 4
35020 Padova

Virgina CARBONI
SESMA, CNR
Via Pansini 5
80131 Napoli

Vinicio CARLONI
Ist. Clinica Medica
Università di Firenze
Viale Morgagni 85
50134 Firenze

Silvia CATINELLA
Mass Spectrometry Service
CNR, Corso Stati Uniti 4
35020 Padova

Gianluca DAMONTE
Ist. Chimica Biologica
Università di Genova
Viale Benedetto XV 1
16132 Genova

Sergio FACCHETTI
Commissione delle Comunità Europee
Centro Comune di Ricerca
21020 ISPRA (VA)

Giovanni FALCONE
Dip. di Fisica
Università della Calabria
87030 Arcavacata di Rende

Donata FAVRETTO
Mass Spectrometry Service
CNR, Corso Stati Uniti 4
35020 Padova

Pasquale FERRANTI
SESMA, CNR
Via Pansini 5-80131 Napoli

Immacolata FIUME
SESMA, CNR
Via Pansini 5
80131 Napoli

Giuseppe GIORDANO
Department of Pediatrics
University of Padova
Via Giustiniani 3
35128 Padova

Gianluca GIORGI
Centro Anal. Determinazioni Strutturali
Università di Siena
via P. Mattioli 10
Siena

Francesca LUCERI
Centro di Spettrometria di Massa
Università di Firenze
Viale Morgagni 65
50134 Firenze

Fulvio MAGNI
Ist. Scientifico S. Raffaele
Via Olgettina 60
20123 Milano

Antonio MALORNI
SESMA-CNR
II Policlinico
Via Pansini 5
80131 Napoli

Anna NAPOLI
Dip. di Chimica
Università della Calabria
87030 Arcavacata di Rende

Maria F. RASETTI
Ist. di Scienze Farmacologiche
Università di Milano
Via Balzaretti 9
20133 Milano

Federico M. RUBINO
ITBA, CNR
Via Ampere 56
20131 Milano

Margherita RUOPPOLO
SESMA, CNR
Via Pansini 5
80131 Napoli

Nino RUSSO
Dip. di Chmica
Università della Calabria
87030 Arcavacata di Rende

Marina SCANDOLA
Dir. Farmacocinetica
Glaxo Ricerche
Via Fleming 4-37100 Verona

Rosa SICILIANO
SESMA, CNR
Via Pansini 5
80131 Napoli

Piero TRALDI
C.N.R., Area di Ricerca di Padova
Corso Stati Uniti 4
35100 Padova

Giovanni SINDONA
Dip. di Chimica
Università della Calabria
87030 Arcavacata di Rende

Francesca ZAPPACOSTA
SESMA CNR
Via Pansini 5
80131 Napoli

Cristina SOTTANI
Lepetit Research Center
Via R. Lepetit 34
21040 Gerenzano (MI)

## THE NETHERLANDS

K. W. LI
Faculty of Biology
Free University Amsterdam
De Bolelaan 1087
1081 HV Amsterdam

Peter G. von YSACKER
Lab. of Industrial Analysis
Eindhoven University of Technology
P.O. Box 513
5600 Eindhoven

Jaroslav SLOBODNIK
Dept. Analytical Chemistry
Free University
De Bolelaan 1083
1081 HV Amsterdam

## POLAND

Witold DANIKIEWICZ
Institute of Organic Chemistry
Polish Academy of Sciences
ul. Kasprzaka 44/52
01-224 Warszawa

Marek SOCHACKI
Centre of Mol. and Macromol. Studies
Polish Academy of Sciences
ul. Sienkiewicza 112
90-363 Lodz

Leszek KONOPSKI
Inst. of Industrial Organic Chemistry
Annopol 6
03-236 Warszawa

## PORTUGAL

Carlos M. FERREIRA DE SOUSA
BORGES
Centro de Espectrometriade Massa
Complexo I, University of Lisboa
Av. Rovisco Pais-1096 Lisboa

Luis HENRIQUES
Dept. de Quimica
ESACB Instituto Politecnico
Quinta Senhora de Mercules
6001 Castelo Branco

Maria Estela JARDIM
Centro Espectrometria de Massa
University of Lisboa
Av. Rovisco Pais
1096 Lisboa

Maria Alzira ALMOSTER FERREIRA
Dept. Of Chemistry
University of Lisboa Campo Grande
1770 Lisboa

## SPAIN

Joaquin ABIAN
Dept. of Neurochemistry
CID-CSIC, Jiordi Girona 18-26
08034 Barcelona

Roser CHALER
Dept. of Environmental Chemistry
CID-CSIC, Jordi Girona 18-26
08034 Barcelona

Emilio GELPI
CSIC - CID
Jordi-Girona 18-26
08304 Barcelona

Silvia LACORTE i BRUGNERA
Dept. of Environmental Chemistry
CID-CSIC, Jordi Girona 18-26
08034 Barcelona

## SWEDEN

Lillemor APLUND
Lab. for Analytical Chemistry
Ist. of Applied Environ Research
Stockholm University
106 91 Stockholm

Emma BERGMARK
Dept. of Radiobiology
Stockholm University
106 91 Stockholm

Henrik KYLIN
Dept. of Environ Chemistry Stockholm
University
106 91 Stockholm

Margareta TORNQVIST
Department of Radiobiology
Stockholm University
10691 Stockholm

Vlado ZORCEC
Dept. of Radiobiology
Stockholm University
106 91 Stockholm

## SWITZERLAND

Marc F. SUTER
EAWAG,
8600 Duebendorf

# TURKEY

Meysum IBRAHIM
Dept. of Chemistry
Middle East Technical University
06531 Ankara

Semih OTLES
Dept. of Food Engineering
Ege University
35100 Bornova- Ismir

Tamerkan OZGEN
Dept. of Chemistry
Ege University
35100 Bornova- Ismir

# UNITED KINGDOM

Su CHEN
Dept. of Chemistry
University of Warwick
Conventry CV4 7AL

Helen COOPER
Dept. of Chemistry
University of Warwick
Coventry CV4 7 AL

Peter J. DERRICK
Department of Chemistry
University of Warwick
Coventry CV4 7AL

Richard P. EVERSHED
Environmental Analytical Section
School of Chemistrsy
University of Bristol
Cantock's Close
Bristol BS8 1TS

Simon S. GASKELL
Dept of Chemistry
UMIST, P.O. Box 88
Manchester M60 1QD

Jonathan HAYWOOD
Dept. of Chemistry
University of Warwick
Conventry CV4 7 AL

Howard MORRIS
Dept of Biochemistry
Imperial College
London SW7 2AZ

David J. REYNOLDS
Dept. of Chemistry
University of Warwick
Conventry CV4 7AL

Sunil SARDA
Biochemistry Department
Imperial College
London SW7 2AZ
London

Ron SELF
Pytchley House
Upgate, Poringland
Norwich NR 14 75H

Joty SHARMA
Biochemistry Department
Imperial College
London SW7 2AZ
London

Elain STIMPSON
Biochemistry Department
Imperial College
London SW7 2AZ

xx

# USA

Mark S. BOLGAR
Bristol Myers
Squibb Pharmaceutical Res. Inst
P.O. Box 4000
Princeton NJ 08543-4000

A. L. BURLINGAME
Mass Spectrometry Facility
Dept of Pharmaceutical Chemistry
University of California
San Francisco CA 94143-0446

Richard M. CAPRIOLI
Analytical Chemistry Center
Medical School
University of Texas
Houston, TX 77225

Ron CERNY
Dept. of Chemistry
University of Nebraska
Lincoln NE 68588

Karl R. CLAUSER
Dept. of Pharmaceutical Chemistry
University of California
San Francisco CA 94143-0446

Dallas CONNOR
Dept. of Pharmaceutical Chemistry
University of California
San Francisco CA 94143-0446

Pamela F. CRAIN
Dept. of Medicinal Chemistry
University of Utah
311 A Skaggs Hall
Salt Lake City, UT 84112

Mario ORLANDO
Department of Chemistry
Columbia University
New York, NY 10019

Joseph DALLAGE
Dept. of Medicinal Chemistry
University of Utah
112 Skaggs Hall
Salt Lake City UT 84112

Terry B. FARMER
University of Texas
Medical School
P.O. Box 20708
Houston TX 77225

Michael L. GROSS
Dept. of Chemistry
University of Nebraska
Lincoln NE 68588

Steven GYGI
Dept. of Medicinal School
University of Utah
112 Skaggs Hall
Salt Lake City UT 84112

Chris L. HOLLIMAN
Dept. of Chemistry
University of Nebraska
Lincoln NE 68588

Michel KELLY
Dept. of Chemistry and Biochemistry
University of Mariland
Baltimore MD 21228

Patrick LIMBACK
Dept. of Medicinal Chemistry
University of Utah
311A Skaggs Hall
Salt Lake City UT 84124

Jeff PATRICK
Chemistry Department
Purdue University
West Lafayette IN 47907

Don REMPEL
Dept. of Chemistry
University of Nebraska
Lincoln NE 68558

Matthew SLAWSON
Northwest Toxicology Lab.
1141 East 3900 South
Salt Lake City UT 84124

Patricia WHEELAN
National Jewish Center for Immunology
and Respiratory Medicine
1400 Jackson
Denver CO 80204

# Part I
# Methods in Mass Spectrometry

# ION FORMATION BY FAST ATOM BOMBARDMENT MASS SPECTROMETRY

GIOVANNI SINDONA
*Dipartimento di Chimica*
*Università della Calabria*
*I-87030 ARCAVACATA DI RENDE (CS)*

ABSTRACT. Fast atom bombardment (FAB) is a mass spectrometric technique which allows the ionization and desorption of polar and thermally fragile molecules directly from their solutions in viscous matrices by means of impinging atoms or ions of 8+20 kilo electronvolt translational energy. The principles of ion formation and desorption will be presented with reference to the applications in the determination of the structure of biological molecules

## 1. The desorption ionization techniques

The direct formation of ionic species from samples in the condensed phase has broadened the perspectives of analytical applications of mass spectrometry (MS) allowing the characterization of involatile and thermally fragile biomolecules. The volatility barrier was overcome in the 1970s by means of high field desorption (FD) of ions from analytes deposited on activated tungsten wires [1,2], and by the applications, in the biomolecular field, of secondary ion mass spectrometry (SIMS) [3] and of the newly developed $^{252}$Cf plasma desorption mass spectrometry (PDMS) [4]. FDMS represented for some years the only possible approach to ion formation from biomolecules with molecular weight in the range of 1-2 KDa and can still provide useful information for low polar compounds such as valinomycin [5] or Carotenoids [6]. The major drawbacks of FDMS are represented both by the formation of unreproducible fragment ions on the emitter surface [7,8] preventing, in the case of unknowns, a straightforward structural assignment, and by the experimental difficulties (emitter-activation, control of the anode temperature, etc.) which do not allow routine operation. PDMS has been widely applied in the biomolecular sciences [9] and has allowed for the first time the use of mass spectrometry in the field of large molecules with molecular weight in the range of some tens of KDa [10]. A breakthrough in the analysis of biomolecules of few kDa was represented by the introduction of fast atom bombardment (FABMS) [11] at the beginning of 1980s. The success of this approach is shown by the number of scientific contributions in the fields of inorganic, organic, bio-organic and bio-medical applications which have appeared in the literature in the last twelve years . The late 1980s saw the appearance of matrix assisted laser desorption (MALDI) [12] and of electrospray ionization (ESI) [13] methods which have completed the armory of the mass spectrometric laboratory by extending MS applications to MDa molecules [14]. All these ionization techniques show dissimilarities in the hardware and in the source of energy employed

3

*R. M. Caprioli et al. (eds.), Mass Spectrometry in Biomolecular Sciences, 3–32.*
© 1996 *Kluwer Academic Publishers.*

for the production of the charged species, yet they provide very similar analytical results when applied to the same analyte [15]. A striking example is provided by the PDMS [16] and SIMS [17] spectra of protected diribonucleoside monophosphates, where the major peaks can be assigned to the quasi-molecular ions and to characteristic fragments originating either from the phosphate backbones or from the protecting groups. The common feature of all these methods, i.e. the formation of ions without previous evaporation of the sample, allow them to be classified as desorption ionization (DI) methods [18]. Aspects connected with the use of FAB will be discussed in the following paragraphs, while some of the other techniques will be covered in other chapters of this book.

## 2. Fast Atom Bombardment Mass Spectrometry

Secondary ions can be produced by bombarding analytes dissolved or suspended in viscous matrices with neutral (Ar, Xe) or ionic ($Cs^+$) beams of translational energy of the order of Kilo electronvolts (KeV) (Fig. 1) [11].

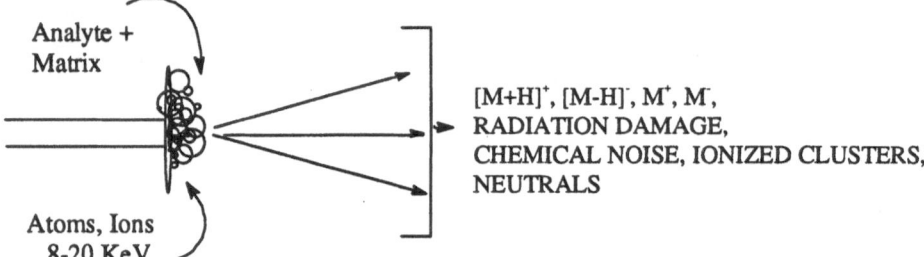

Figure 1. Secondary ion emission by FAB

The production of keV atoms is achieved by commercial guns which are easily adapted to high voltage ion sources: this has favored the rapid diffusion of the technique. The real improvement of the method over the already existing desorption ionization techniques, at the time when FAB was introduced, was due to the use of liquid matrices [19]. The advantages of exposing solutions, or suspensions, rather than solid layers to particle bombardment are represented by the longevity of the sample, which allows averaging of spectra and easy application of MS/MS techniques [20, 21] and by the reduction of the radiation damage, i.e. those processes which cause random degradation of the analyte. The use of matrices, as a means for reducing the strong interactions between molecules in the solid phase, has become important with other DI techniques, such as SIMS [22] and PDMS [23], and has opened new frontiers in the laser desorption (LD) methods with the introduction of MALDI technique [12] which allows the production of ionized biomolecules in the range of hundreds of KDa.

## 2.1 DESORPTION MECHANISM

A detailed mechanism, whereby complex biomolecules are released as charged species from the condensed to the gas-phase by particle bombardment, will probably not ever be described in a quantitative way, i. e., it seems unlikely that the physical and chemical aspects of the phenomenon can be associated to general equations. This, however, is not a limitation of FAB or of the other DI methods only, but is typical of all phenomenological sciences dealing with molecules. It is, in fact, impossible to predict by theoretical calculations the structural feature of a MS spectrum of an electron-ionized molecule, even though electron ionization (EI) is a well understood technique. The number of experiments performed in the last two decades provide, however, know-how which enable the use of these powerful and innovative methodologies in analytical chemistry and biochemistry. Aspects related to the kinetics and energetics of sputtering in FAB will be briefly presented with reference also to comprehensive accounts on the mechanism in molecular SIMS available in the literature [18, 24].

In FAB or related experiments, the kinetic energy carried by the incident particle is first transferred to the internal degree of freedom of the surface system and then used for the formation and sputtering of the ionized species. A Simple knock-on mechanism, based on pure binary elastic collisions, does not provide an adequate model for explaining the formation of secondary ions. The experimentally-determined kinetic energy distributions of the species at $m/z$ 133, formed by DI of PEG using primary beams of 2 and 4keV, is characterized by two distinct maxima ($E_p$, figure 2) and an energy tail extending up to few eV [25].

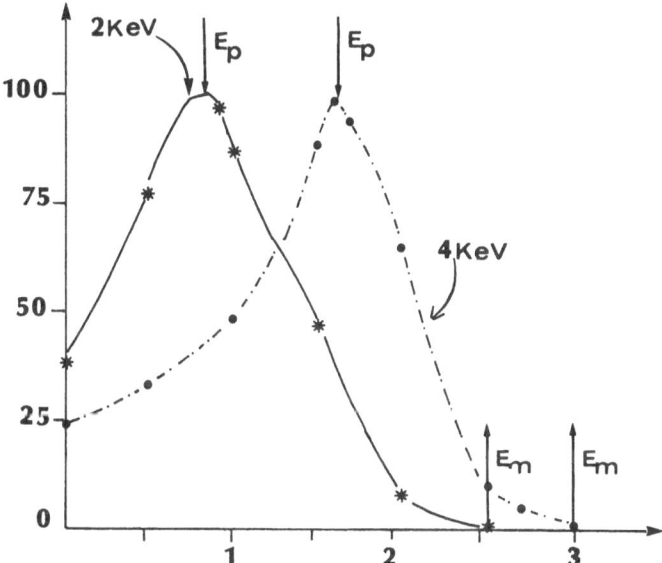

**Figure 2.** Energy distribution of $m/z$ 133 ions generated from PEG by Argon ions with different initial traslational energy

The shift in the peak maxima should not be observed if the ejection process were described by binary-elastic collisional-sputtering, moreover the minima ($E_M$) should have been in the range of KeV since their values are related to the incident energy by equation 1 [26] where U and $E_i$ are

$$E_M = \gamma (1-\gamma) E_i - U \qquad (1)$$

the surface potential barrier and the incident energy, respectively, which are both in the order of KeV, and $\gamma$ is related to the mass of the incident and ejected particles. In the case of the experiment examined here, eq. 1 can be written in the form of equation. 2

$$E_M = 0.2 \, (E_i - U) \qquad (2)$$

It is evident that if a knock-on mechanism is valid the maximum kinetic energy of the sputtered species should be in the order of KeV and not of eV as experimentally observed. If the secondary ion emission were described by models which could involve a Maxwell-type distribution the equation 3 can be written [26].

$$J(E) \propto E(E+U)^{-1/2} \exp [-(E + U)7kT] \qquad (3)$$

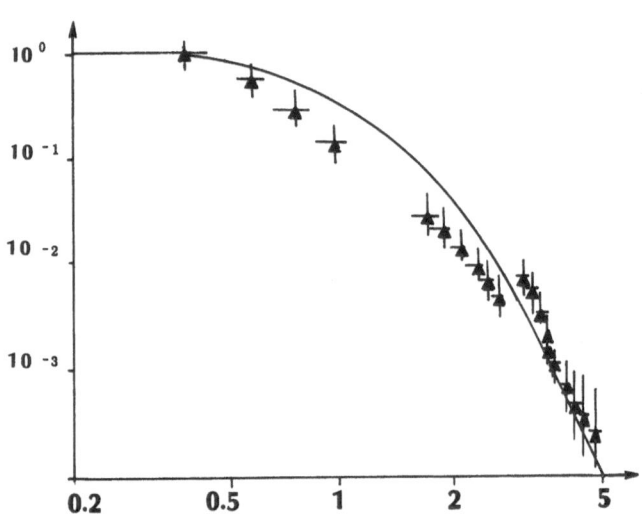

Figure 3. Energy distribution of sodiated glycerol ions

Even though the experimentally determined [27] kinetic energy distribution of sodiated glycerol ions does not fit perfectly the theoretical curve obtained by applying the equation 3 (figure 3,) it may be considered that a local equilibrium can be reached. The physical aspects of a FAB experiment can be summarized as follow: The recoiling species, set in motion by the incident particles, completely loose the memory of the projectile whose kinetic energy is released along the primary particle tracks; in times estimated in the order of $10^{-12}$ s, a near-thermal equilibrium is approached by the system before the ejection of most of the ions and clusters which occurs in $10^{-10} - 10^{-9}$ s. The data, obtained by monitoring the composition of the released gaseous species from chemical or enzymatic equilibria taking place on a FAB probe, do not differ from those provided by conventional experiments in solution [29,30], thus proving that equilibrium conditions can be reached [31]. The dynamic region where chemistry occurs, through ion-molecule reactions which cause the formation of all type of ions found in the mass spectrum, is known as *selvedge* [32].This qualitative model gives insights into the sputtering mechanism of organic molecules which are useful in the interpretation of the FAB spectra.

## 2.2 THE ROLE OF THE MATRIX

The examined analytes have to posses good solubility in the liquid matrices. Other parameters should be taken into account such as its vapor pressure, viscosity, pKa and redox potential which are related to the physical and chemical aspects of the FAB method [19]. Glycerol is a liquid of general use because it fulfills most of the requirements of a matrix, including its good affinity to polar molecules. Many other matrices have been employed such as thioglycerol, aminoglycerol, 3-nitrobenzylalcohol, diethanolamine, sulfolane, "magic bullet", etc. [19]. A useful compilation of ready-to-use FAB spectra of the most common matrices for FAB has been published as appendix to the book recently edited by McCloskey [33].

The importance of the matrix for the longevity of the sample was recognized from the first applications of the method. It was also observed that the spectra of mixtures may not correspond to the composition of the analyte, i.e. the ionic species of some components can be undetected or underestimated. A correlation between the structure of the surface exposed to atom bombardment and the mass spectral response was presented as early as 1983 by Barber and co-workers [34] . The relative abundances of the (M-H)⁻ species of stearic acid ($m/z$ 283) with respect to the (G₃-H)⁻ cluster of the glycerol (G) at $m/z$ 275 was monitored as a function of the sodium stearate concentration, hence as a function of the surface tension of the solutions, at different concentration of analyte, exposed to atom bombardment (fig. 4A)

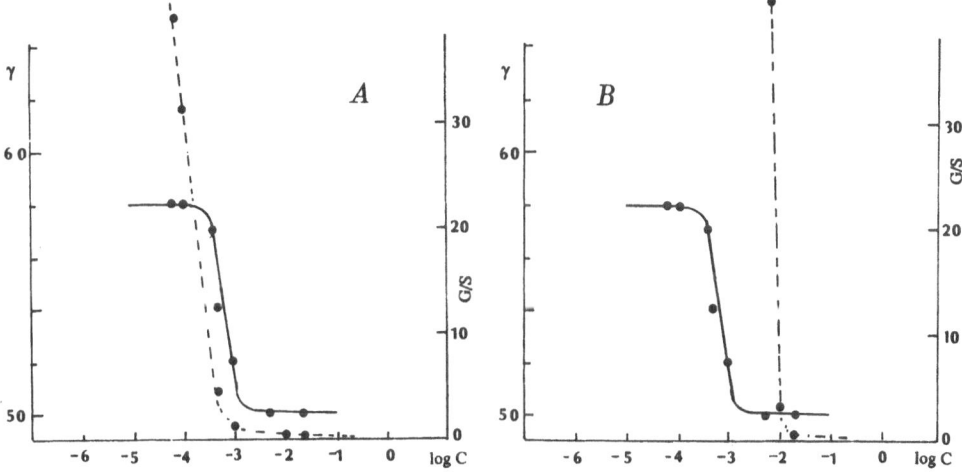

Figure 4. Mass spectral response as a function of surface coverage effect. G/S = $m/z$ 275/283 (A) and $m/z$ 277/285 (B)

It was found that the ratio of the ion abundances $m/z$ 275/283, referred to as glycerol-to-sample ration (G/S), was inversely proportional to sample concentration (dashed line), and directly proportional to the degree of surface coverage by solute molecules, estimated from the variation of the surface tension γ (solid line) which goes from the sub-monolayer to the mono-layer region, both unaffected by sample dilution (dγ/dlogC =0), through a linear dependence of γ versus C corresponding to the formation of analyte monolayers. Therefore, in the sub-monolayer region, when the surface is covered mainly by matrix molecules, the G/S ratio tends to infinite; when at least one mono-layer of analyte is formed, the mass spectrum displays abundant ionic species originating from the sample and no glycerol clusters ( G/S ratio goes to zero). The same experiment was performed in the positive ionization mode, following the relative intensity of the $(M+H)^+$ at $m/z$ 285 of the same compound and that of the glycerol trimer $(G_3+H)^+$ at $m/z$ 277 (fig. 4B). In this condition the G/S value tends to zero when the composition of the surface approaches the micelles region. Therefore, when the micelles region is reached, no glycerol cluster formation is observed in either negative and positive ionization mode. These experiments which associate the surface coverage with mass spectral response, have given rise to two fundamental questions regarding the mechanism of ion formation by DI methods, i.e.: (i) the role of the *preformed ions* and (ii) the origin of the charged sputtered particles. The different sensitivity observed in the detection of $(M-H)^-$ and $(M+H)^+$ species when the carboxylate, a *preformed ion*, is sampled as sodium salt, led the authors to suggest [33] that no chemical reactions were involved in the sputtering of the $(M-H)^-$ species. Assuming, however, that a FAB spectrum reflects the composition of the equilibrium reached by the solvated species, and that the gas-phase data are consistent with those obtained in solution [29,30], and considering a similar solvent effect for glycerol and water, the different mass spectral response can be easily correlated with the analytical concentration of stearic acid and its conjugated base in that environment. The experiments previously discussed could lead to the conclusion that, at least when glycerol is used, the mass spectral response has to be correlated with the surface activity of the analyte. This effect was also investigated in relation to the different distribution of structurally similar compounds between the surface and the bulk [35], which might account for the observation that the relative yields of molecular ion clusters from mixtures do not always match the analytical composition of the solution exposed to particle bombardment (*suppression effect*). Figure 5a shows the FAB spectrum of an equimolar (0,001 M) glycerol solution of $C_6$-$C_{16}$ alkyl trimethylammonium bromides obtained by irradiating only a small section of the sample droplet.

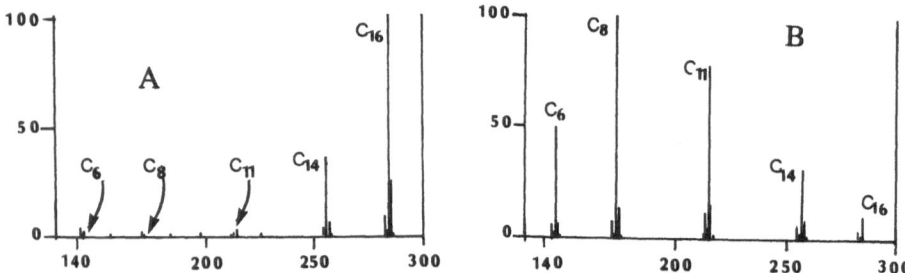

Figure 5. FAB (+) spectrum of equimolar glycerol solutions of $C_6$_$C_{16}$ alkyltrimethylammonium bromides. (A) without surfactant; (B) with surfactant. A small section of the sample droplet is irradiated.

The species $C_{14}$ and $C_{16}$ are the only $[M + H]^+$ ions present in the spectrum, in fact, the clusters labeled as $C_6$ and $C_8$ are mainly due to fragmentation of higher mass precursors. The addition to the same mixture of ammonium salts of an excess of tetramethylammonium stearate gave rise to the positive FAB spectrum of figure 5b. Therefore, in the absence of a surface active additive (the stearate), the shorter chain ammonium salts seem to be well solvated inside the bulk of the solution exposed to bombardment, whereas the top layers should be mainly formed by the longer chain homologues. The added ammonium stearate is a surfactant that reaches more efficiently the top layers of the solution of the ammonium salts examined here partly because it is used in large excess (c= 0,02M) with respect to the other components of the mixture (c= 0,001 M). It has also the property of not interfering with the analytes in the positive FAB spectra, at least in the region of their pseudomolecular ions, since the yield of positive ions from a carboxylate is far less than that of positive ions from alkyl ammonium species. The observation of the shorter chain ammonium cations in the spectrum of fig 5b has been interpreted [35] as due to the surfactant properties of the additive which brings, by ion pair formation, all the cations close to the surface.

Although these experiments provide insights into the understanding of the FAB mechanism of ion formation, the method of enhancing the surface activity of a given component of a mixture by surfactant addition cannot be generalized. Very recently, the use of surfactants to selectively modify the relative response of mixtures has been reinvestigated [36]. It was pointed out that many classes of compounds give better response in m-nitrobenzyl alcohol, a low surface tension matrix, than in glycerol, regardless of their surface activity in this environment. The absolute intensity of the signals of a FAB spectrum remains constant in a given period of time thus suggesting that ion formation and their desorption occur in dynamic equilibrium conditions. The structure of the surface, during the experiment, may differ, therefore, from that in its static state, i.e. when the system is not exposed to particle bombardment. A better understanding of the suppression effect, in the analysis of alkylammonium salt mixtures, was achieved by means of further experiments with a pulsed FAB source and a sector instrument equipped with array detection [37].

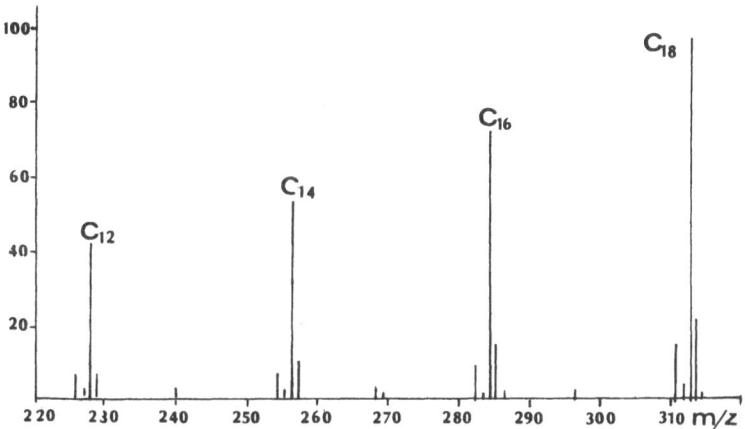

Figure 6. FAB (+) spectrum of equimolar glycerol solutions of $C_{12}$-$C_{18}$ alkyltrimethylammonium bromides. The entire droplet is irradiated

It was found that when the entire droplet of equimolar mixture of $C_{12}$ and $C_{18}$ species dissolved in glycerol is exposed to bombardment the spectrum shows a predominance of the less surface active compounds (figure 6). However, in agreement with early findings, If only a small area of the solution is irradiated, the higher surface-active homologue is more-or-less the only detectable species (Fig. 5A [35]). This differential response is probably due to the Marangoni effect which is in operation when only a small fraction of the sample is irradiated. In this case there is a surface area which can act as a reservoir to refurbish sample in the depletion zone, as a result of the interfacial agitation due to local variations of interfacial tension, and in this condition the more surface active compound, present in the top layer, moves to fill the gap in the damaged area. When the entire sample area is bombarded, a situation which is common to routine FAB operation, a dynamic equilibrium condition is reached whereby the analytes can reach the top layers at the same rate at which they are desorbed. In this condition the shorter alkylammonium cations, which are more mobile, predominate in the spectra. These experiments do not provide definitive answers to the suppression effect and to the non-linearity in the mobility of the species in the bulk, in fact, assuming, as previously mentioned, that the lower $m/z$ peaks are not produced by fragmentation of the higher mass species, the spectrum obtained in the dynamic condition does not reflect either the composition or the surface activity of the analytes. The time dependence of the FAB response provides, however, some very interesting observations which throw more light on the complex mechanism of ion formation. When a 2:1 mixture of $C_{16}$ and $C_{18}$ alkylammonium bromide is analyzed in glycerol, by exposing the entire droplet to atom bombardment, the top layer is initially composed exclusively of the higher-mass higher-surface-active compound (Fig 7 A). When the dynamic equilibrium conditions are reached (fig 7 B) the relative abundance of the species shown in the spectrum approaches the real composition of the analyte mixture, but never gives the 2:1 ratio. It is very interesting to note that if the ion gun is switched off and the system is allowed to relax for 10 sec. and then exposed back to particle bombardment, the spectrum of figure 7A is again obtained.

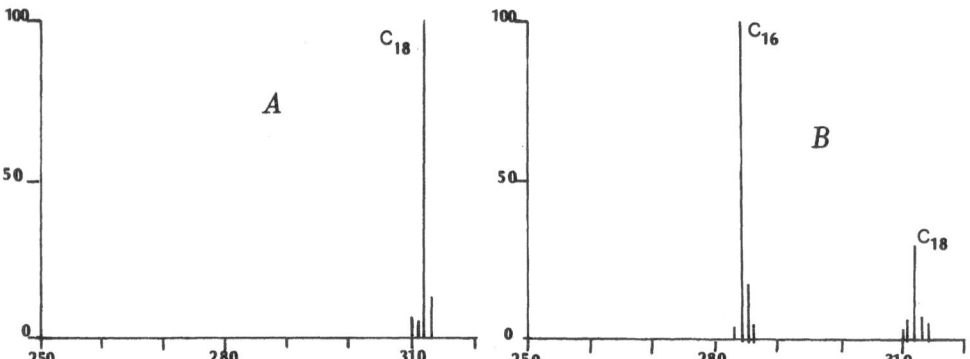

Figure 7. Time dependence of the pulsed FAB (+) spectra of equimolar glycerol solutions of $C_{16}$ and $C_{18}$ alkyltrimethylammoiun bromides.

The composition of the surface under particle bombardment differs from that of the unperturbed system thus allowing the detection of ions formed from molecules initially present in the bulk. The effect of surface coverage in relation to the structural feature of the FAB spectra was investigated by means of a mixture of three alkylammonium salts slightly differing in surface tension ($\gamma$) properties and molecular weight [36]. When the absolute concentration in glycerol of each of the three components was lower than that needed to form a monolayer ($d\gamma/dC \neq 0$, fig. 4) the mass spectrum, taken at the first exposure of the solution to particle bombardment, reflects the relative surface activity of the analytes thus showing three molecular ions of similar intensity. When the absolute concentration of the three species is above that needed to obtain monolayers of each component the initial FAB spectrum displays only ions formed from the more surface active compound. The spectra taken in dynamic conditions reflect both the relative mobility and surfactant properties of the analytes. In particular one of the three component was more-or-less undetected in the spectra, regardless of its initial concentration.

From an analytical point of view, these results indicate that a FAB spectrum is reproducible since the dynamic conditions are reached in less then one second. It seems also clear that the spectrum reflects the composition of the surface, whose "structure" depends on the fluido-dynamic properties of the solution in the adopted experimental conditions. Further insights into the mechanism of surface renewal and on the longevity of the analyte during a FAB experiment were obtained by monitoring the mass spectral response of deuterated glycerol as a function of time [38]. Layers of different labeled glycerol molecules were deposited on a probe-tip, by first applying fully deuterated glycerol ($d_8$), and then allowing H/D exchange of the labile deuterons (those linked to the hydroxyl moieties) to take place for 15 seconds in a room atmosphere of 40% humidity. Some of the top layers are therefore made up of $d_5$ glycerol molecules, while the bulk should contain $d_8$ homologues. The FAB spectra of the positive ions obtained from a section of the droplet ( a beam limiting slit was used) show a time dependent decreasing of the relative intensity of the peak at $m/z$ 98, corresponding to the $(M + H)^+$ ions of glycerol-$d_5$, and an increasing of that of the peak at $m/z$ 102, due to glycerol-$d_8$ (38) (figure 8).

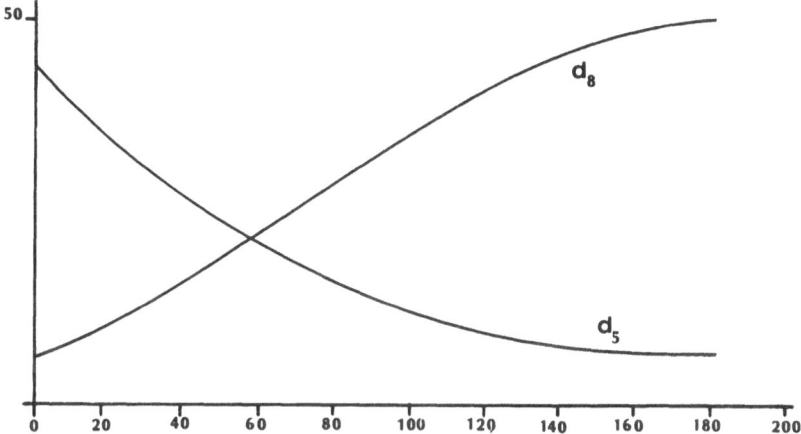

Figure 8. Time dependence of $d_5$ ($m/z$ 98) and $d_8$ ($m/z$ 102) [M+H]$^+$ ions from deuterated glycerol. A small section of the sample droplet is irradiated

When the entire droplet is irradiated, a much faster decay of the initial relative intensity of $m/z$ 98 with respect to the other molecular ion species is observed. This different behavior can now be interpreted on the ground of the recently reported experiments discussed above [36]. The higher intensity of protonated glycerol molecules ($m/z$ 98) over the deuterated ones ($m/z$ 102), when a small portion of the droplet is irradiated, reflects the initial composition of the top layers and the longevity of $m/z$ 98 could be due to a side-filling effect which brings liquid flow from the surroundings to the irradiated area. Assuming that the mobility and the surfactant properties of the different labeled glycerol molecules is similar, the irradiation of the entire droplet causes the rapid achievement of dynamic conditions which affects the composition of the surface. This and the other experiments discussed so far seems to indicate, in any case, that the candidates for the desorption process are those present in the top layers, which are removed cleanly by particle bombardment each at a time. Further studies on the mechanism of surface renewing in a FAB experiment supported the proposal that the diffusion of molecules from the bulk to the top layer is not needed to explain the longevity of sample as a consequence of the formation of a clean surface after each sputtering event [39]. All the experiments previously described, and many others available from the literature on the mechanism of ion formation by FAB deal with the use of glycerol matrix. The latter, in fact, possesses the physico-chemical requirements to be considered as a matrix of general use. The correlation of the mass spectral response with the surface coverage effect may be different when other matrices are used.

A direct consequence of the mechanism of ion formation is the production of stable aggregates where a number of neutral species interact with a charged particle to give cluster ions of various composition . Clustering can provide indirect evidence on the composition of the surface when different matrices are used. Nucleic acid molecules are keen to give in hydrophobic environment horizontal interaction through the formation of multiply hydrogen bonded species (figure 9)

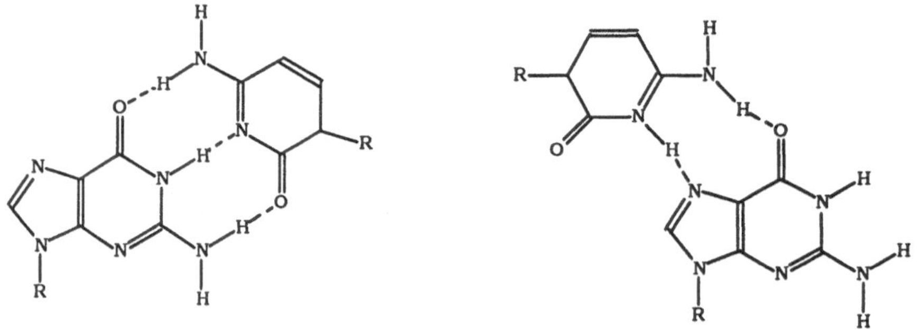

Figure 9. Watson-Crick and Hoogsteen horizontal interactions of deoxyguanosine and deoxycytidine

Equimolar mixtures of deoxyguanosine (G) and deoxycytidine (C) in thioglycerol/water give rise to the FAB spectrum of figure 10 where the formation of $(G_mC_n +H)^+$ species, with $(m+n) \leq 7$ can be observed. When glycerol is used the relative abundances of clusters with $(m+n) \geq 2$ is negligible, whereas proton bound species formed by three nucleosides $[(m+n) = 3]$ are easily detectable in m-nitrobenzyl alcohol [40].These results can be correlated with a different degree of solvation of the species exposed to particle bombardment, in particular in the measurements related to figure 10, water is likely to be lost by evaporation before exposure of the mixture to particle bombardment thus causing an enrichment of the surface with analyte molecules.It is possible to envisage that some regular arrangements of the nucleosides through Watson-Crick and Hoogsteen interaction [41] is reached before the sputtering event, as proved by nuclear magnetic resonance experiments on G, C mixtures which form polyaggregates in polar aprotic solvents [42]. A direct proof that the formation of clusters is not due to gas-phase recombination comes from experiments performed with a split-probe-tip [43]. Mixtures of guanine (Guo) and cytosine (Cyt) give rise to, among others, the ions reported in table 1 (column 2) when the nucleobases are mixed together in a conventional probe tip. In a peculiar experiment, the solutions of the two molecules are deposited on the two halves of a properly designed probe-tip where a gap of ca. 200–µm. was drilled, across the middle. The two solutions do not mix while the probe is positioned to allow a maximum overlap of the atom beam with the two halves. The main result of this measurement is that no peaks due to proton bound dimers formation are present (table 1, column 3).

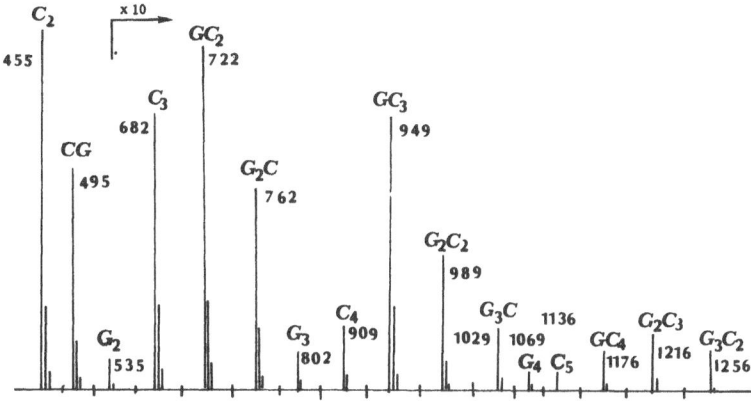

**Figure10. Partial FAB (+) spectrum of an equimolar mixture of deoxyguanosine (G) and deoxycytidine (C) in thioglycerol/water.**

**Table 1. Cluster formation from nucleoside mixtures by FAB**

| ions | Mixtures of Cyt and Guo % rel | split probe tip % rel. |
|---|---|---|
| $[Cyt+H]^+$ | 100 | 100 |
| $[Guo+H]^+$ | 75 | 45 |
| $[Cyt+Guo+H]^+$ | 34 | <1 |

The formation of clusters, as a consequence of the composition of the solution exposed to particle bombardment, could represent a limitation of the method in the case of the characterization of unknowns. This effect is particularly important when salts are present in the solution exposed to bombardment. [44] Clustering of the matrix and of the analyte with cations (positive spectra) or anions (negative spectra) could prevent the formation of $[M+H]^+$ or $[M-H]^-$ species. A very limiting case is represented by the effect of a tenfold molar excess of sodium chloride added to a solution of cytidine in glycerol. The positive FAB spectrum thus obtained was entirely formed of glycerol-sodiated cluster peaks [45]. After two cycles of a proposed desalting procedure the analyte gave the expected FAB spectrum of cytidine. Many methods are available for the pre-treatment of those biological or chemical samples rich in salt content [44]. In most of the applications of FAB, however, the presence of traces of alkali metal salts can provide additional information on the composition of the analyte if the molecular ion peak is accompanied by $[M+Na]^+$, $[M+K]^+$ species, formed by alkali ion attachment in the selvedge. Clustering of the reactants exposed to particle bombardment can be deliberately induced to evaluate physical parameters of those charged species, which can not be produced by other experimental methods. Proton-bound dimers or multimers, where the proton joins together the same or differently structured ligands, are clusters which can be easily released into the gas-phase by positive FAB. This property has allowed the extension of the kinetic method to involatile biomolecules for the determination of the proton affinities (PA) of their gaseous species [46-49]. In a peculiar application, clusters formed by deoxynucleosides or their nucleobases, amines of known PA's and a proton are produced by mixing the analytes in glycerol, the unimolecular dissociations of the proton-bound dimer thus formed are used to assess the unknown PA of the nucleoside or nucleobase [49]. This kinetic approach has also important applications in the peptide research area [50-52]. Therefore an apparent limitation of the method of ion production by FAB, i.e. the formation of clusters, turns out to be very useful in the determination of intrinsic parameters of biomolecules in a non-interacting environment. The efficiency of the method is illustrated by the possibility of correlating structural changes with PA variation. The PA's of 2'-deoxythymidine (1), 2'-deoxyuridine (2) and uridine (3), determined as previously discussed, are 224.9, 224.3 and 223.3 Kcal/mol, respectively. The methyl group in the position 5 of the nucleobases enhances the basicity of the nucleoside thus accounting for the observed $\Delta$PA of 0.6 Kcal/mol. The antagonist effect of the 2'-OH and 5-methyl group is reflected by $\Delta$PA of 1.6 Kcal/mol experimentally determined [53] for 1 and 3

## 2.3 DERIVATIZATION

The efficiency of the ionization process, in terms of sensitivity in the detection of the molecular species and enhancement of gas-phase fragmentation, can be improved by pre- or *in situ* derivatisation of the analyte. This procedure seems in contrast with the peculiarity of D.I. methods, introduced for m.s. analysis of involatile and polar molecules, and is only occasionally used. However, data have been presented which show that the sensitivity in the detection of nucleoside can be enhanced by trimethylsilylation of the analytes [54]. Similar results have been obtained in the analysis of peptides by pre-derivatisation with pentafluorobenzoyl fluoride [55]. In this case it was suggested that the increasing in sensitivity in the positive ionization mode was due to the enhancement of the surface active properties of the derivatised peptides, whereas in the negative ionization mode the higher yields of [M-H]⁻ ions were due to the effect of the pentafluorobenzoyl group which increases the acidity of the analyte. Most of the methods used to improve the mass spectral response consider *in situ* derivatisation of the analytes, often by addition of acids and bases aiming at obtaining precharged species . Variation of the pH of the medium has been extensively exploited in relation to the possible improvement of the mass spectral response [44, 56-57]. Quantitative studies of this effect, however, do not allow any generalization [58, 59]. A double effect is probably associated, for instance, to the use of p-toluensulfonic acid in the analysis of dinucleosides polyphosphates : its low pKa ensures the replacement of alkali ions by protons at the phosphate moieties, whereas its surfactant property allows a better coverage of the top layers exposed to bombardment [60].

## 3. Energetics of FAB ionization method

Even though only a minor fraction ($10^{-3}$- $10^{-4}$) of the sputtered material corresponds to ionized species the sensitivity of the modern mass spectrometers allow the use of the methodology, routinely, at the sub-nanomolar level. Atoms (Xe) of 8÷10 or ions ($Cs^+$) of 10÷20 KeV kinetic energy are usually used for most of the analytical applications of FAB.

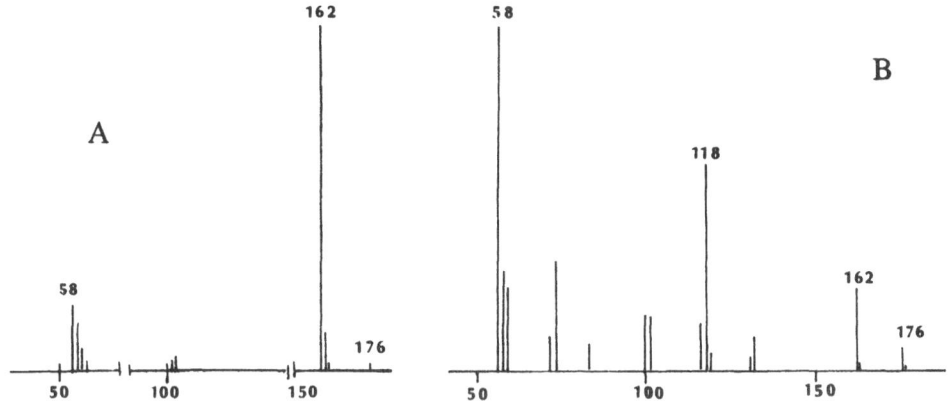

Figure 11. FAB(+) spectrum of L-carnitine inner salt (Kr, 8 KeV). (A) from glycerol, (B) without matrix

The energy of the primary particles ( ions or neutrals) can be correlated with the sensitivity in the detection of the molecular ion species and with the extent of their gas-phase fragmentation. The sensitivity in the detection of tetraphenylphosphonium bromide deposited on different metallic substrates, in typical SIMS experiments, increases by a factor of approximately twenty when the kinetic energy of the bombarding $Ar^+$ particles goes from 0.7 to 5.0 KeV [61]. On the other hand, in the same interval, the extent of fragmentation is inversely proportional to the energy of the incident ions . A five-fold decrease was observed in the ratio of the intact cation to the main fragment ion [61], as a consequence of the different internal energy content of the molecular ions. It can be assumed that the more energetic projectiles set in motion a larger volume of recoiling species thus causing a distribution of the incident energy in a greater number of molecules. Therefore, the higher the number of ionized species released from the surface the lower their internal energy content, hence the extent of fragmentation. Another possible explanation considers that low energetic beams cause the formation, in the selvedge, of smaller clusters whose energy dissipation by neutral losses is less efficient. The energetic of ion formation by D.I. is greatly affected by the presence of solid or liquid matrices. The admixture of the analyte with ammonium chloride, in the case of static SIMS, results in a decrease in the extent of fragmentation [62]. Liquid matrices, in FAB, have also the effect of preserving the analyte from chemical degradation prior to its desorption into the gas phase. The positive FAB spectrum of carnitine dissolved in glycerol (figure 11A) displays very abundant $(M+H)^+$ ions at $m/z$ 162, a weak transmethylation peak at $m/z$ 176 and a fragment at $m/z$ 58 derived from gas-phase unimolecular dissociations of the parent species [63] In the absence of matrix (fig. 11B), the spectrum showed, as expected (see above), extensive fragmentation of the analyte, and the appearance of a peak at $m/z$ 118, due to formal $CO_2$ neutral loss, which was not found in the gas-phase dissociations of the quasi-molecular ions (MIKE and CID experiments), and which could be due to high critical energy processes undergone in the condensed phase. A rationale for the formation of $m/z$ 118 and 176 can be provided by extending to the energized condense phase the classic tools of synthetic organic chemistry. The relevant product ions displayed by a FAB spectrum are usually formed by gas-phase processes and the mechanism of their formation can be inferred by applying the basic principles of gas-phase ion chemistry in relation to the internal energy distribution of the reacting species. It has already been mentioned that the kinetic energy of the secondary ions sputtered from liquid matrices is in the range of few eV, whereas the impinging primary particles usually possess kinetic energy in the range of KeV and this effect rules out a knock-on mechanism of ion formation. Furthermore, experimental evidence suggests [18-24] that in the excited "selvedge region" multiple collisions take place which randomize to some extent the internal energy content of the species before they are released into the gas phase. Therefore, those ions surviving intact the ionization region have undergone a number of collisions that, even if not sufficient to establish a Maxwell-Boltzmann distribution, cause a partial equilibration of the internal energy content in their translational and vibrational modes. A Classic thermochemical approach was used [64, 65] to determine the internal energy distribution of gaseous ions produced by FAB. In a series of experiments the critical energy for the dissociation of some benzylpyridinium cations was determined from the standard heats of formation $(\Delta H_f)$ of reactants and final products. It was found [64] that the experimental data fit with an internal energy distribution of the reacting species which extends over ca 340 kJ/mol (ca 80 kcal/mol,) and has a half-width of 150 kJ mol (36 Kcal/mol). By similar thermochemical consideration it was determined [65] that the average internal energy of secondary ions was ca 110 kJ mol . Therefore most of the gaseous species reacting in the mass spectrometer possess an average internal energy of ca 30 kcal mol which is not sufficient to bring about dissociations of the major part of the covalent bonds present in the analytes. This data, however, does not exclude that high energy species can be formed during the sputtering event, but ensures that fragments formed by gas-phase dissociations, extremely useful for analytical applications, do not derive from dissociative reaction paths of high critical energy.

## 4. Interpretation of FAB mass spectra

The chemistry of the excited analyte/matrix system in the interfacial region depends on the reactivity of the substrates in that particular environment. Multiple collisions occur which bring about the formation of ions by hard and soft acid-base (HSAB) [66] equilibria and redox processes. The HSAB principle applied to bimolecular processes allows the classification of all the ionic reactions available in the excited condensed phase, including simple proton transfer, substitution reactions and clustering. Redox processes may be driven by the availability in the selvedge of low-energy electrons (0-3 eV) that can be captured by low-lying lowest unoccupied molecular orbitals (LUMOs) of the analytes [67,68], other mechanisms are in operation when molecular radical cations ($M^{+\cdot}$) are formed. The reactivity of many classes of compounds can be predicted nowadays on the ground of the experimental data accumulated so far. It must be considered, however, that all type of reactions previously mentioned may be available even when very simple compounds are analysed. The formation of the species reported in table 2 will be discussed in the following section with reference to the behavior of some broad families of compounds analyzed so-far by FAB mass spectrometry.

## 4.1 PRECHARGED SUBSTRATES

When the analytes are sampled as salts, the releasing of intact cations (C) and anions (A) depends on their charge state. (1:1) electrolytes, i.e. those having the composition ($C^+$,$A^-$), give rise to very abundant peaks in the positive (+) and negative (-) ionization mode, respectively, as reported for some phosphonium and ammonium halides [69] or for some inorganic salts [70]. In this case the charged species are already present in the matrix, their releasing into the gas phase, however, cannot be due to simple sputtering processes according to the "preformed ion" model. Desolvation of analyte/matrix clusters takes place in the selvedge, through multiple collisions, some solvated species are often released into the gas phase as shown by the appearance in the high-mass region of the spectrum of analyte/matrix cluster-peaks.

**Table 2.** Peaks found in a FAB spectrum

| type | origin | analytical information |
|---|---|---|
| chemical noise | Low abundant species at every $m/z$ value formed by high energy and a-specific degradations of analytes and matrices | NONE |
| Pseudo-molecular ions | $[M+H]^+$ and $[M-H]^-$ even electron species formed by Hard-Acid-Base interactions | composition of the analyte |
| Radical molecular ions | $M^+$ and $M^-$ odd electron species formed by redox processes | composition of the analyte |
| Cluster ions | Complex species usually formed by non-covalent interactions between neutrals and charged particles. | Composition of the analyte. Chemistry in the selvedge. Fundamental studies |
| Radiation damage | Fragments formed by surface specific fragmentation of analyte and matrix by Soft-Acid-Base interactions and redox processes | Chemistry in the selvedge. Reactivity of analyte/matrix system. Analyte structure |
| Product ions | even, and less usually, odd electron ions mainly formed by gas-phase unimolecular dissociations | Analyte structure |

The observation of $(M+H)^+$ and $(M-H)^-$ species from *inner salts*, i.e. from zwitterionic compounds, is correlated with the availability of reaction channels leading to the neutralization of one of the two net charges present in the analyte. The FAB (+) spectrum of carnitine inner salt in glycerol ( fig. 11A) displays abundant $(M+H)^+$ at $m/z$ 162. In the negative ionization mode and in the same experimental conditions, the relative yield of the correspondent $(M-H)^-$ species is negligible, whereas the base peak of the spectrum is represented by $(M-CH_3)^-$ ions (fig. 12). Moreover the FAB (-) spectrum shows, among others, peaks at $m/z$ 252 due to ionization of neutral carnitine by coordination of (G-H)⁻ species. It can be assumed that the excellent response of the methodology in the positive ionization mode depends on the facile hard acid-base equilibrium between carnitine and glycerol which accounts for the neutralization of the carboxylate moiety In this case the reacting system can be considered equivalent to a (1:1) electrolyte (scheme 1). The neutralization of the positive charge on the ammonium moiety can occur only through carbon-nitrogen bond fission giving rise to $(M-CH_3)^-$ anions, a substitution reaction requiring a soft interaction of the analyte with a nucleophile which could be either the carboxylate moiety of another carnitine molecule or the glycerol solvent. As a matter of fact the spectrum of the positive ions displayed the peak at $m/z$ 176 corresponding to a transmethylated species with nearly 1:10 ratio to the $(M+H)^+$ ions [63]. This value becomes 3:1 when carnitine is exposed neat to ion bombardment ,in typical SIMS experiments [71].

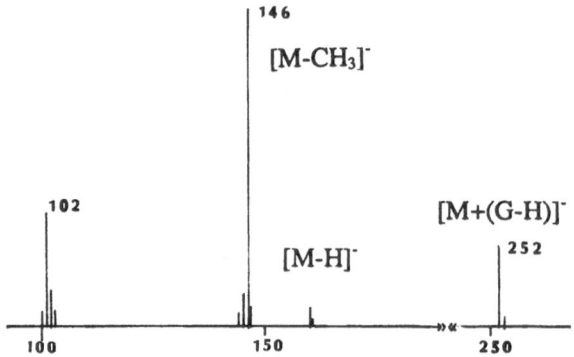

Figure 12. FAB (-) spectra of L-carnitine inner salt (Xe 8 KeV, glycerol)

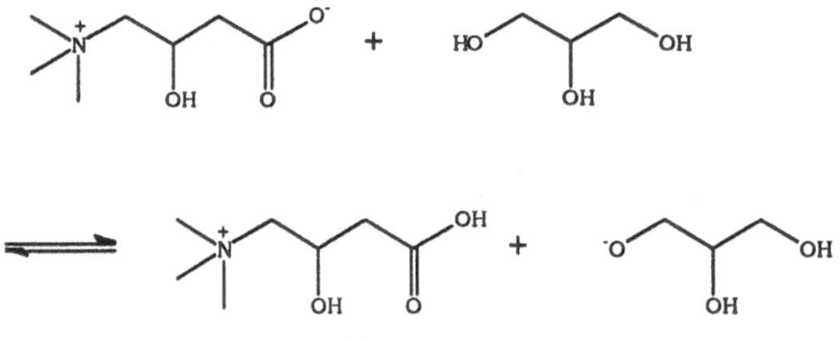

Scheme 1.

The formation of (M-H)⁻ quasi-molecular ions at $m/z$ 160 requires an hard acid-base process which brings about the formation of a polycharged species containing two negative and one positive charges. The high critical energy for the abstraction of a proton from carnitine and the low sputtering yield of multiple charged species by FAB, account for the formation of $m/z$ 160 with 6-7% relative yield, only [63].The appearance in the FAB (-) spectrum (fig 12) of the species [M+(G-H)]⁻ at $m/z$ 252 represents an alternative pathway for the ionization of the intact molecule, i.e. by attachment of anions present in the environment, through soft acid-base interactions. The SIMS and FAB spectra of a very low molecular weight, but extremely polar, compound such as carnitine provide, therefore, a wealth of information on the chemistry of zwitterionic analytes in the energized condensed phase.

The energetics of ion sputtering from liquid matrices prevent the releasing into the gas phase of intact $C^{n+}$ and $A^{p-}$ ions from (n:p) electrolytes, i.e. those having the composition ($C^{n+}$, $A^{p-}$). In the (+) ionization mode it is very likely that cations can be observed which have the composition ($C^{n+}$, $A^{(p-1)-}$) as in the case of some ruthenium(II) polypyridine complexes [72]. The FAB (+) spectrum of [Ru(ppy)₂(2,5-dpp)](PF₆)₂ (4), a (2:1) electrolyte, in m-nitrobenzyl alcohol (NBA) displayed, in fact, very abundant monocharged cations at $m/z$ 793 due to the releasing of a single PF₆⁻ counterion and extremely weak doubly charged species corresponding to cation moiety. The ionization of the intact neutral salt (m.w. 937.63) took place, by analogy with the zwitterionic carnitine molecule, through the coordination of the (M-PF₆)⁺ cation ($m/z$ 793), present in the mixture, giving rise to a peak at $m/z$ 1731, whose MIKE spectrum displayed a very abundant product ion at $m/z$ 793. A similar behavior was exhibited by other isomers such as 5 and analogues [72]. The LUMO of the Ruthenium complexes previously mentioned is centered on the π* orbital of the dpp ligand ( $E_{1/2}$ = - 1.03 V) they can undergo, therefore, very easily redox processes. The FAB (-) of 4 displayed a very abundant peak at $m/z$ 938 corresponding to a radical anion having the composition of the intact salt. This is a typical case of a one-electron capture reaction taken by a substrate which can interact with the free electrons produced during the exposure of the analyte/matrix system to particle bombardment. The other significant species present in the (-) spectrum was the peak at $m/z$ 1083, formed by a soft interaction of the intact salt with a PF₆⁻ anion found in the selvedge. Ionization of neutrals by electron capture or anions attachment is a widely encountered process which ought to be taken into account in the interpretation of a FAB (-) spectrum.

4                                    5

FAB (-) represents the methodology of choice in the analysis of phosphodiester molecules of biological importance such as nucleotides and short oligonucleotides [73] which are generally isolated as salts, due to the low pKa of the corresponding acids which is close to one unit. In the case of (1:1) electrolytes formed by the phosphodiester and alkali metal or ammonium cations, exhaustive information can be obtained on the composition and structure of the anion. In this case, once more, no hard acid-base equilibria are involved in the formation of the charged particles.

6

7

DMT: dimethoxytritil

The FAB (-) spectrum of the sodium salt of the lypophilic derivative of thymidine monophosphate (TMP, 6), a synthetic intermediate in the production of a drug delivery system [74], displayed, in fact, the base peak at $m/z$ 847 corresponding to the phosphodiester anion and few diagnostic fragments at $m/z$ 545 and 321 due to the releasing of the DMT protecting group and to the formation of a phosphomonoester fragment. Very limited information was obtained in the positive ionization mode. In this condition the formation of a $(M+H)^+$ quasi-molecular ion requires the neutralization of the phosphate group and the attachment of an extra proton, moreover the protonated species thus formed are very unstable because of the presence of the extremely acid-labile DMT group. The FAB(+) spectrum displayed, in fact, a very abundant peak at $m/z$ 303 due to the dimethoxytrytil (DMT) cation . A peculiar behavior was exhibited by dithymidylic acid (TpT), triethylammonium salt (7), a (1:1) electrolyte, analyzed, in glycerol, both in the positive and negative ionization mode [75]. The (+) spectrum was dominated, as expected, by the presence of a peak at $m/z$ 102 due to the cation. The molecular ion region displayed, among others, a peak at $m/z$ 648 corresponding to intact salt ionized by proton attachment (fig 13a). Ammonium phosphates and carboxylates are extremely soluble in low-polar organic solvent, such as chloroform or dichloromethane, because they can be described as intimate ion-pairs [76], and this molecular arrangement can be still preserved in glycerol thus accounting for the detection of the intact protonated salt. The phosphodiester anion (TpT-H)⁻ at

$m/z$ 545, the base peak of the (-) spectrum (fig. 13b), can be recorded with high sensitivity and its gas-phase dissociations provide a number of diagnostic fragment ions very useful for structural elucidation. The sensitivity in the detection of dinucleotides phosphodiesters by (-) FAB is enhanced when alkali metal are replaced by ammonium counterions [77]. It can be assumed, therefore, that the formation of ion-pairs increases the surfactant properties of the analytes, at least when glycerol is used as a matrix.

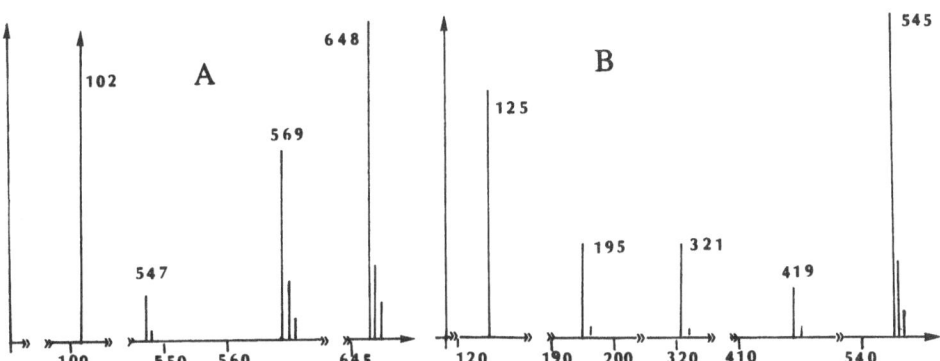

**Figure 13. FAB spectra of dithymidilic acid triethylammonium salt (Xe 8KeV, glycerol).**
(A) positive ions; (B) negative ions

Oligonucleotide strands can be considered as ($C^+$, $A^{n-}$) electrolytes. When they are sampled as triethylammonium salts in glycerol, the FAB (-) spectrum usually displays abundant peaks due to the monocharged anions $A^-$ (fig. 14). Their formation is driven by hard acid-base reactions which neutralize (n-1) negative charges of the polyphosphate oligomers [21].

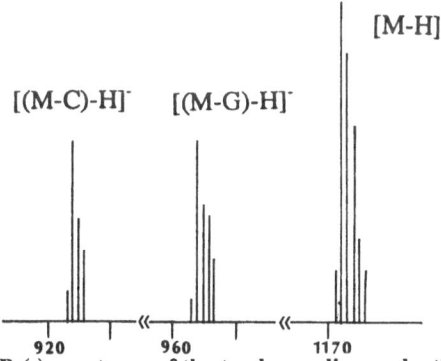

**Figure 14. Partial FAB (-) spectrum of the tradeoxyoligonucleotide GCGC (Xe, 8 KeV, glycerol) sampled as triethylammonium salt.**

Polycharged anions are usually absent or account for a very low percentage of the total ion current [69]. A close inspection of the distribution of peaks within the molecular anion cluster have revealed that the ionization of DNA strands produces also (M-2H)$^-$ , M$^-$ radicals and

(M+H)⁻ anions [78]. The (M-H)⁻ species, however, always provide the major contribution to the shape of the peak envelope. The formation of odd and even electron ions from these polycharged precursors provides a picture of all the possible chemical processes available in FAB ionization of organic salts. It can be concluded that organic and inorganic salts are amenable to analysis by FAB. (1:1) electrolytes usually provide good (+) and (-) spectra showing the intact cations or anions, respectively. The probability of detecting polycharged species is normally very low, therefore the mass spectral response for $(C^{n+}, A^{p-})$ species depends on the chemistry undergone by the analytes in the selvedge as a function of the properties of the analyte/matrix system. However, it can be excluded that, even in very simple cases, the production of ions derives by simple knock-on of the "preformed" species existing in solution.

## 4.2. NEUTRAL POLAR SUBSTRATES

Fusapyrone 9 is an antifungal metabolite isolated from rice kernels [79]. The structural feature of the molecule, i.e. its molecular weight in the range of one kDa and the presence of polar groups makes it amenable to FAB analysis.

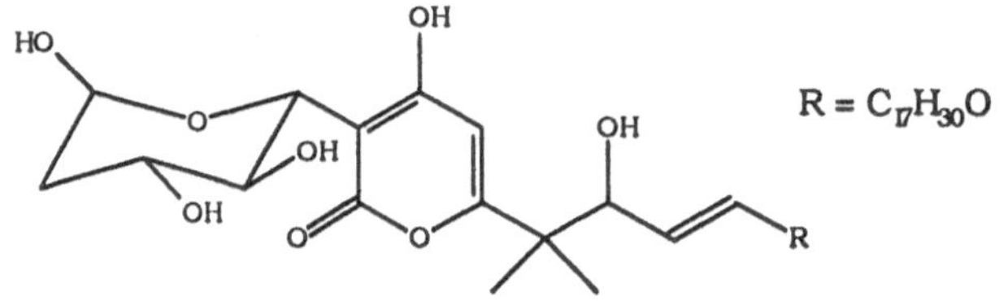

$$R = C_{17}H_{30}O$$

9

The (+) spectrum obtained from NBA showed (M+Na)⁺ and (M+H)⁺ peaks at $m/z$ 629 (22%) and 607 (46%), respectively, formed by interactions of the analyte polar groups (hard nucleophiles) with Na⁺ and H⁺ (hard electrophiles) [80]. The (M-H)⁻ species at $m/z$ 605 was the base peak of the spectrum of the negative ions, formed by hard acid-base reaction.

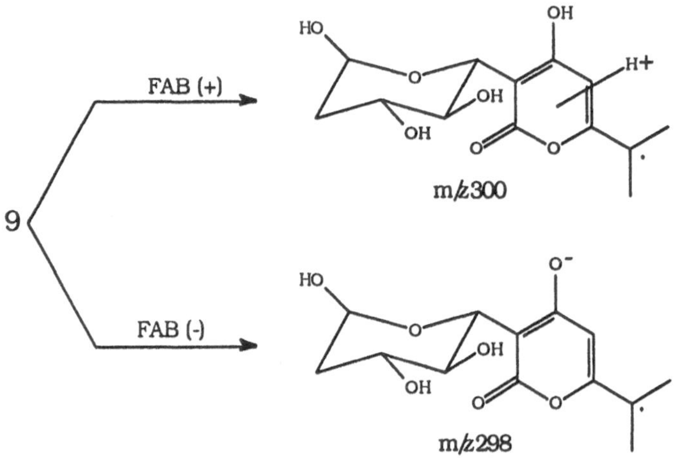

FAB (+)

$m/z$ 300

FAB (-)

$m/z$ 298

9

Scheme 2

Mass spectrometry provides, therefore, a quick and simple solution to the understanding of the molecular composition of unknowns isolated from natural sources. The presence of sodiated and protonated peaks can be considered a typical structural feature of (+) FAB spectra of substances extracted from complex natural matrices; the detection of (M-H)⁻ pseudo molecular ions complement the information obtained in the positive ionization mode and allows the characterization of the analyte.The complete elucidation of the structure of fusapyrone 9 has required, as expected, the extensive use of both $^{13}$C and $^{1}$H nmr spectroscopy; FAB, however, was unique in providing, at the nanomolar scale and quickly, the first and appropriate piece of information, i.e., the molecular composition. The application of MS/MS methods to 9 and some other related compounds of the same family have enabled the identification of peculiar gas-phase reaction channels (scheme 2) which could be useful for the structural assignment of similar molecules which might be isolated in the future [81].

As previously seen a neutral molecule can be easily ionized by alkali metal ion attachment. This process can be deliberately induced either for the understanding of gas-phase interactions of metal ions and neutrals or as an analytical tool for structure determination. Simple di and tripeptides mixed with glycerol saturated with alkali metal ion hydroxides and exposed to 25 KeV Cs⁺ beams, release into the gas phase monocharged anions (10) formed by the doubly ionized substrate which chelates the monocharged cation [82]. Their formation results from hard acid-base interaction in the excited selvedge region.

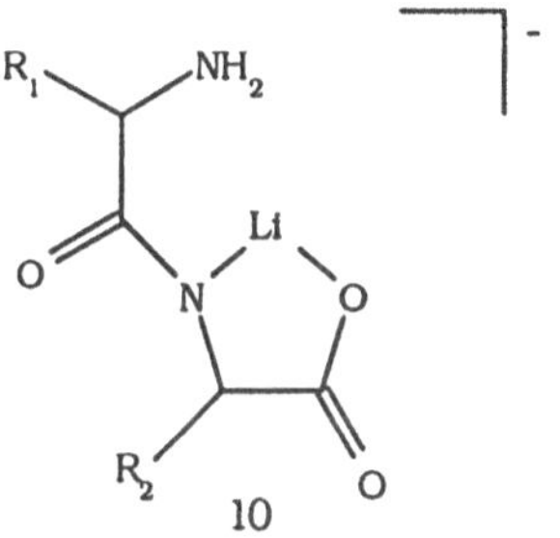

Neutral substrates can undergo one-electron oxidation or reduction processes in the selvedge. Depending on the structure of the analyte, the positive and negative radical ions thus formed can survive the excited region where they are formed and can be released into the gas phase. Halogenated nucleosides, analyzed by FAB, in glycerol, in both positive and negative ionization mode, displayed peaks due to (M+H)⁺ and (M-H)⁻ species, respectively, together with quasimolecular cations and anions where the halogen (X) was replaced by one hydrogen atom. The occurrence of this reaction in the interfacial excited region was proved by the appearance of appropriate analyte clusters with the matrix and by the observation of H/D exchange when deuterated glycerol was used [83]. The H/X exchange in the series 5'-Iodo, 5'-Fluoro deoxyuridine was in the range 5.3 to 33%, in the positive, and slightly lower in the negative ionization modes.

$$R\text{-}X + e^- \rightarrow [R\text{-}X]^{\cdot-} \rightarrow R^{\cdot} + X^- \rightarrow R\text{-}H$$

**Scheme 3**

The mechanism whereby the dehalogenation process takes place may require an initial one-electron capture process, followed by the formation of neutral radicals of the analyte, which can than exchange hydrogen, or deuterium, atoms with the matrix, according to scheme 3 [67]. Dehalogenation processes have been observed for other halogenated molecules such as the antibiotic vancomycin exposed in glycerol to particle bombardment [67]. It has been observed, however, that the replacement of a chlorine by a hydrogen atom does not occur when the same analyte is sampled in NBA matrix [84] In this case the matrix acts as an electron scavenger thus preventing the analyte from the one-electron capture processes. The occurrence of single electron transfer processes is inferred, in the examined case, by the peculiar chemistry undergone by the ionized analytes in the selvedge, since the initially formed radical anions do not survive the ionization region and are not shown in the FAB (-) spectra. Deoxynucleosides can be analyzed routinely by FAB and they usually provide abundant $(M+H)^+$ and $(M-H)^-$ species [73, 85-86]. N-4-benzoyl-deoxycytidine, a very simple derivative of the correspondent natural nucleoside, yielded in the negative ionization mode $(M-H)^-$ and $M^-$ molecular ion species of similar abundances [87]. The electron capture process, in this case, provides radical anions of sufficient stability to be released intact into the gas-phase (fig 15)

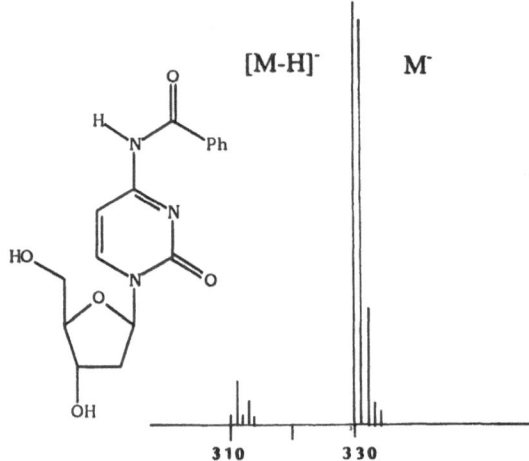

Figure 15. Partial FAB (-) spectrum of *N*-4-benzoyldeoxycytidine (Xe 9.5 KeV, glycerol)

Single electron transfer processes drive the formation of reduced species within the selvedge, often by formal hydrogenation of the analytes [88-90]. The evaluation of the reduction processes as a function of both matrix and analytes reduction potential was performed with some organic dyes of known redox potential [91]. It was shown that the typical FAB matrices can be arranged in the following order of decreasing reductive power:

Glycerol > Sulfolane > Thioglycerol > NBA

It follows that the extent of single electron transfer processes can be controlled by the proper choice of the matrix, as previously seen for the antibiotic vancomycin, or by adding to matrices, of high reductive power, appropriate electron scavengers such as p-benzoquinone [92]. The role of the matrix can be clearly seen by matching the FAB (+) spectra of methylene blue, an organic dye, taken in different experimental conditions. The analyte, a (1:1) electrolyte, yields in glycerol [91] molecular ion clusters formed by the $C^+$ (35% only), $(C+H)^+$ and $(C+H_2)^+$ cations. In the absence of solvent the spectrum did not show any pseudomolecular ion due to reduction processes.

Unlike radical anions, radical cations are not usually observed in a FAB spectrum. This has been attributed to the energetics of organic ion formation from liquid matrices [67], which is a consequence of both the ionization energy of a typical organic molecule (in the range of 7÷10 eV) and of the internal energy distribution of the gaseous ions (in the range of 0÷3 eV). High energy processes are available in the selvedge and gaseous ions with internal energy higher than 3 eV might be sputtered as a function of the structure of the analytes and of the chemistry undergone by this species prior to the sputtering event. Polycyclic aromatic hydrocarbons (PAH), of ionization energy in the range 6.98÷8.15 eV, yield, by (+) FAB, in sulfolane matrix, abundant $M^+$ radical cations [94]. The latter are not observed when NBA is used as matrix. According to the scale of the reductive power of matrices, reported above, it can be excluded that the removal of one electron from the substrates takes place through charge transfer complexation [95] between analyte and matrix, since the presence of NBA should favor the process. In this particular case, therefore, it can be considered that the life-times of the resonance-stabilized PAH radical cations, formed in the selvedge allow their releasing into the gas-phase.

Charge transfer complexation [95] represents another means of radical cation formation by FAB.

$$\text{Donor} + \text{Acceptor} = (\text{DA}) \rightarrow D^{+\cdot} + A^{-\cdot}$$

This mechanism has been invoked to account for the formation of substantial amounts of $M^{+\cdot}$ and $M^{-\cdot}$ from chlorophyll $b$ [96], a molecule with high-lying HOMO and low-lying LUMO.

11 - 17

**Table 3. Formation of molecular ions from compounds 11-17 by FAB (+)**

| Compound | $R^1$ | $R^2$ | $R^3$ | $R^4$ | NBA | Glycerol |
|----------|-------|-------|-------|-------|-----|----------|
| 11 | Tos | Me | Me | OMe | $M^{+\cdot}$ | $(M+H)^+$ |
| 12 | Tos | i-Pr | Me | OMe | $M^{+\cdot}$ | $(M+H)^+$ |
| 13 | Tos | Et | Me | OMe | $M^{+\cdot}$ | $(M+H)^+$ |
| 14 | Tos | Et | i-But | OMe | $M^{+\cdot}$ | $(M+H)^+$ |
| 15 | Tos | H | s-But | OMe | $M^{+\cdot}$ | $(M+H)^+$ |
| 16 | Tos | H | i-Pr | OBzl | $M^{+\cdot}$ | $(M+H)^+$ |
| 17 | Tos | Me | i-Pr | OBzl | $M^{+\cdot}$ | $(M+H)^+$ |

Tosyl protected amino acid esters are useful intermediates in the synthesis of their N-alkylated derivatives [97]. The FAB (+) spectra of **11-17** displayed $(M+H)^+$ pseudomolecular ions when analyzed in glycerol whereas in NBA they afforded $M^+$ radical cations only. The formation of the latter were suppressed when copper sulfate was added to the NBA matrix.

The charge transfer complexation mechanism now accounts for the production of radical species. It can be assumed, in fact, that the one-electron transfer takes place within complexes formed by the analyte, the donor, and m-nitrobenzyl alcohol, the acceptor. The process is not allowed when glycerol, a bad acceptor species is used.

## 4.3 BIOLOGICAL SAMPLES BEARING LABILE PROTECTING GROUPS

The chemical synthesis of biologically important molecules requires the use of protecting groups, which, by definition, are organic functionalities removable in mild conditions [98]. FAB represents an ionization method which can provide, even in the case of labile molecules structural information.

|   18   |   19   |   20   |

FAB has been successfully applied in the field of natural and synthetic peptides, whose principles of reactivity and sequencing will described elsewhere in this book. They represent typical polar molecules possessing hard and soft acid-and basic sites at the N and C terminal moieties, in the polyamide backbone and in the side chain groups. Very abundant $(M+H)^+$ and $(M-H)^-$ quasimolecular ions can be, therefore, formed and desorbed in the gas phase by particle bombardment [99]. The presence of acid or base labile groups on partially or fully protected peptides affects their chemistry in the selvedge and in the gas-phase. In very simple cases, i.e. when the oligomers contain a single protecting group, it can be removed before the m.s. experiment [100]. On the other hand, when an analytical check is needed on synthetic segments representing building blocks for further elongation of the oligomer, it is important to achieve exhaustive information on the intact substrate. t-Butoxycarbonyl (Boc, 18) benzyloxycarbonyl (Z, 19) and N-9-Fluorenylmethoxycarbonyl (Fmoc, 20) are urethane groups widely employed in the protection of the N-terminal residue of a peptide chain.

The (+) FAB spectra of a number of Boc-peptides [100] show peaks due to consecutive elimination of isobutene and carbodioxyde neutrals giving rise to N-terminal free species which are identical to those generated by chemical removal of the t-butyloxycarbonyl group. Three different series of N-terminal sequence ions will occur in the spectrum, as a consequence of the complete and partial degradation of the urethane group which produces, besides $(M+H)^+$ ions, the $[(M-C_4H_8)+H]^+$ and $[(M-Boc)+H]^+$ precursors.

The Z- protecting group of aminoacid and peptide methyl esters is less labile than Boc- towards acid treatment in solutions. The dipeptide Z-(S)-Phe-(S)-Phe-OMe gives rise by FAB to abundant $(M+H)^+$ species at $m/z$ 461 either in glycerol or NBA solution. A minor peak at $m/z$ 327, corresponding to the N-terminal free molecule, was present in the NBA spectra. The ratio

327/461 was time dependent and after nearly 10 minutes of exposure to atom bombardment $m/z$ 327 became the base peak of the spectrum with concomitant disappearance of the $(M+H)^+$. This behavior is in agreement with the "buffering" power of the matrix which protect the analyte from radiation. The occurrence of this surface process is indirectly proved by the availability of gaseous reaction paths leading to the degradation of the same urethane moiety. The $(M+H)^+$ species at $m/z$ 461 undergo unimolecularly (MIKE spectra) N-terminal "deprotection" which, unlike the process occurring in the selvedge, proceeds by loss of carbon dioxide which is driven by the migration of the benzyl group to the terminal and amidic nitrogen atoms [101].

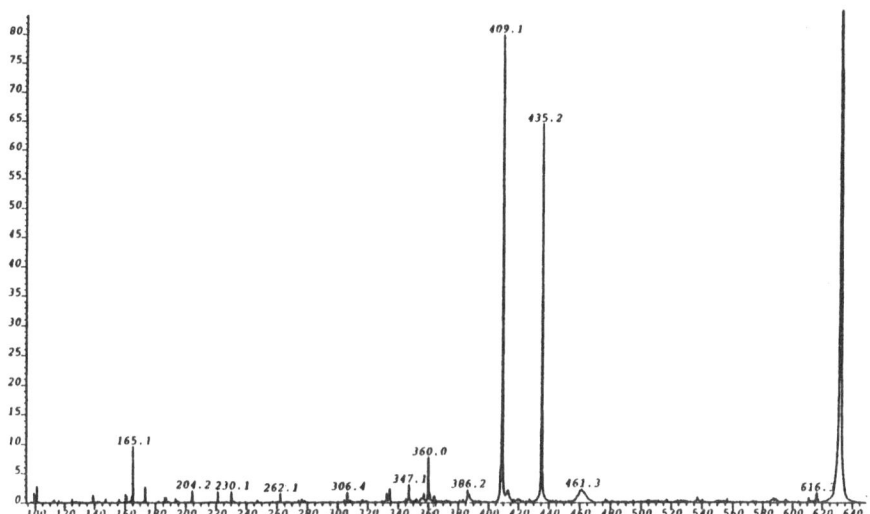

Figure 16. FAB-MS/MS spectra of the [M-H]⁻ at $m/z$ 631 of the Fmoc protected sequence 25-28 of Porcine Calcitonin

The Fmoc protecting group is routinely used in solid phase synthesis of peptides. Protocols are available which allow the isolation of fully protected peptides which can be subsequently used for block synthesis of larger oligomers. FAB (-) provides an unique tool for a satisfactory and quick structural assignment. The (M-H)⁻ species undergo gas-phase rearrangements to N-terminal free and N- terminal isocyanate peptide ions (fig. 16), from which exhaustive sequence information can be obtained [102].

Partially and fully protected oligonucleotides are amenable to FAB analysis especially in the negative ionization mode [73,86]. In some particular cases, however, the information on the composition of the analyte may be missed. The base peak of the FAB (+) spectrum of 3'-$O$-(5'-dimetoxytritylthymidine) $O$-2,4-dichlophenyl $S$-methyl phosphate (21) was given by the dimethoxytrityl (DMT) cation at $m/z$ 303. The highest $m/z$ detectable peaks in the FAB (-) spectra were 783 and 653 in glycerol/thioglycerol and glycerol, respectively. Both species do not correspond to the quasi-molecular anion of the analyte and are formed by demethylation of the thiomethyl group ($m/z$ 783) or by solvolysis of phosphate dichlorophenoxy group ($m/z$ 653) occurring in the selvedge as a function of the analyte/matrix composition. The application of the

HSAB principle provide, once more, a rationale for the observed behavior. In the (+) ionization mode the protonation of the substrate by hard acid-base equilibria leads to very unstable species which preferentially release the DMT cations. In the (-) ionization mode, soft acid-base equilibria predominates over the deprotonation processes as a consequence of the lability of this peculiar phosphate moiety.

21

## 5. Conclusions

Cations, anions, charged radicals and neutrals are sputtered from liquid matrices by FAB. In the previous sections it has been proposed that even electron ions can be formed by hard and soft acid-base equilibria which do not involve any odd electron (radical) intermediate. An alternative explanation takes into account the possibility that the initial step of ion formation by FAB is represented by single electron transfer processes giving rise to radical ions, which then undergo interactions with neutral and charged species, present in the selvedge, to give the sputtered ions whic are displayed by the mass spectrum [103]. The origin and the structure of ions produced by FAB of neat glycerol, still the most commonly used matrix, have been recently evaluated in details [104]. On the grounds of the many experimental data obtained, it was proposed that the effect of the penetrating projectile is the formation at the interfacial region of cavities containing free electrons, matrix fragment ions and matrix molecules. Therefore the analyte/matrix chemistry in the selvedge cannot be exclusively described either by radical or by ionic processes, only. Considering that the intimate mechanisms whereby charged species are formed and sputtered by FAB are difficult to prove in details, a rationale for ion formations should rely on the basic concepts of chemical reactivity which can help in the understanding of the interfacial processes induced by particle bombardment [105].

# 6. References

1    H. D. Beckey, *Principles of Field Ionization and Field Desorption Mass Spectrometry*, Pergamon Press, Oxford (1977)
2    H. D. Beckey, K. Levsen, F. W. Rollgen and H. R. Schulten, *Surf. Science*, **70**, 325 (1978)
3    A. Benninghoven, D. Jaspers and W. Sichtermann, *Appl. Phys.*, **11**, 35 (1976)
4    R. D. Macfarlane and D. F. Togerson, *Science*, **191**, 920 (1976)
5    M. M. Sheil and P. J. Derrick, *Org. Mass Spectrom.*, **27**, 1000 (1992)
6    S. Takaichi, *Org. Mass Spectrom.*, **28**, 785 (1993)
7    H. Budzikiewicz and M. Linscheid, *Biomed. Mass Spectrom.*, **4**, 103 (1977)
8    M. Linscheid and A. L. Burlingame, *Org. Mass Spectrom.*, **18**, 245 (1983)
9    R. D. Macfarlane, in *Methods in Enzymology, Volume 193 Mass Spectrometry*, ed. by J. A. McCloskey, pp. 263-280, Academic Press, San Diego (1990)
10   G. Jonsson, A. Hedin, P. Hakansson, B. U. M. Sundqvist, H. Bennich and P. Roepstorff, *Rapid. Commun. Mass Spectrom.*, **3**, 190 (1989)
11   M. Barber, R. S. Bordoli, R. D. Sedgwick and A. N. Tyler, *J. Chem. Soc. Chem. Commun.*, 325 (1981)
12   M. Karas, D. Bachmann and F. Hillenkamp, *Anal. Chem.*, **57**, 2935 (1985)
13   J. B. Fenn, *J. Am. Soc. Mass Spectrom.*, **4**, 524 (1993)
14   T. Nohmi and J. B. Fenn, *J. Am. Chem. Soc.*, **114**, 3241 (1992)
15   H. Fales, E. A. Sokoloski, L. K. Pannell, P. Quan-long, D. L. Klayman, A. J. Lin, A. Brossi, J. A. Kelly, *Anal. Chem.*, **62** (1990).
16   C. J. McNeal, K. K. Ogilvie, N. Y. Theriault and M. J. Nemer, *J. Am. Chem. Soc.*, **104**, 972 (1982)
17   W. Ens, K. G. Standing, J. B. Westmore, K. K. Ogilvie and M. J. Nemer, *Anal. Chem.*, **54**, 960 (1982)
18   S. J. Pachuta and R. G. Cooks, *Chem. Rev.*, **87**, 647 669 (1987)
19   E. De Pauw, *Mass Spec. Review*, **5**, 191-212 (1986)
20   G. Sindona, N, Uccella and K. Weclawek, *J. Chem. Res. (S)*, 184 (1982)
21   M. Panico, G. Sindona and N. Uccella, *J. Am. Chem. Soc.*, **105**, 5607 (1983)
22   A. Ba-Isa, K. L. Busch, R. G. Cooks, A. Vincze and A. Granoth, *Tetrahedron*, **39**, 591 (1983)
23   C. J. McNeal and R. D. Macfarlane, *J. Am. Chem. Soc.*, **108**, 2132 (1986)
24   J. Sunner, *Org. Mass Spectrom.*, **28**, 805-823 (1993)
25   L. Kerner and S. P. Markey, *Int. J. Mass Spectrom. Ion Processes*, **59**, 157 (1984)
26   G. Falcone, S. Sroubek, G. Sindona and N. Uccella, *Int. J. Mass Spectrom. Ion Processes*, **83**, 223 (1988)
27   G. J. Q. Van der Peyl, W. J. Van der Zande, R. Hoogerbrugge and P. G. Kistemaker, *Int. J. Mass Spectrom. Ion Processes*, **67**, 147 (1985)
28   J. Sunner, A. Morales and P. Kebarle, *Anal. Chem.*, **60**, 98 (1988)
29   R. M. Caprioli, in *Specialist Periodical Reports Mass Spectrometry*, ed. by M. Rose, Vol. 8, pp 185-209, The Royal Society of Chemistry, London (1986)
30   R. M. Caprioli, *Mass Spectrom. Rev.*, **6**, 237-287 (1987)
31   F. Greco, A. Liguori, G. Sindona and N. Uccella, *J. Am. Chem. Soc.*, **112**, 9092 (1990)

30

32  F. Honda, G. M. Lancaster, Y. Fukuda and J. W. Rabalais, *J. Chem Phys.*, **69**, 4931 (1978)

33  C. E. Costello, in *Methods in Enzymology, Volume 193 Mass Spectrometry*, ed. by J. A. McCloskey, pp. 875-882, Academic Press, San Diego (1990)

34  M. Barber, R. S. Bordoli, G. J. Elliot, R. D. Sedgwick and A. N. Tyler, *J. Chem. Soc., Faraday Trans. 1*, **79**, 1249 (1983)

35  W. V. Ligon, Jr. and S. B. Dorn, *Int. J. Mass Spectrom. Ion Processes*, **61**, 113 (1984)

36  A. N. Tyler, L. K. Romo and R. B. Cody, *Int. J. Mass Spectrom. Ion Processes*, **122**, 25 (1992)

37  A. N. Tyler, L. K. Romo, M. H. Frey, B. D. Musselman, J. Tamura and R. B. Cody, *J. Am. Soc.Mass Spectrom.*, **3**, 637 (1992)

38  W. V. Ligon, Jr., *Int. J. Mass Spectrom. Ion Phys.*, **52**, 183 (1983)

39  S. S. Wong, F. W. Rollgen, I. Manz and M. Przybylski, *Biomed. Mass Spectrom.*, **12**, 43 (1985)

40  F. Greco, G. Sindona e N. Uccella, *Chim. Ind. (Milan)*, **71**, sm46 (1991)

41  W. Saenger, *Principles of Nucleic Acid Structure*, Springer-Verlag, New York (1984)

42  N. G. Williams, L. D. Williams and R. B. Shaw, *J.Am. Chem. Soc.*, **111**, 7205 (1989)

43  D. P. Michaud, J. N. Kyrianos, T. F. Brennan and P. Vouros, *Anal. Chem.*, **62**, 1069 (1990)

44  C. Fenselau and R. J. Cotter, *Chem. Rev.*, **87**, 501 (1987)

45  D-C. Moon and J. A. Kelley, *Biomed. Environ. Mass Spectrom.*, **17**, 229 (1988)

46  R. G. Cooks and T. L. Kruger, *J. Am. Chem. Soc.*, **99**, 1279 (1977)

47  S. A. McLuckey, D. Cameron and R. G. Cokks, *J. Am. Chem. Soc.*, **103**, 1313 (1981)

48  G. Bojesen, *J. Am. Chem. Soc.*, **109**, 5557 (1987)

49  F. Greco, A. Liguori, G. Sindona and N. Uccella, *J. Am. Chem. Soc.*, **112**, 9092 (1990)

50  G. S. Gorman, J. P. Spier, C. A. Turner and I. J. Amster, *J. Am. Chem. Soc.*, **114**, 3986 (1992)

51  Z. Wu and C. Fenselau, *Rapid Commun. Mass Spectrom.*, **6**, 403 (1992)

52  X. Cheng, Z. Wu and C. Fenselau, *J. Am. Chem. Soc.*, **115**, 4844 (1993)

53  A. Napoli, A. Liguori and G. Sindona, *Rapid Commun. Mass Spectrom.*, **8**, 89 (1994)

54  D. Slowikowski and K. H. Schram, *Biomed. Environ. Mass Spectrom.*, **13**, 263 (1986)

55  J. P. Kiplinger, L. Contillo, W.L. Hendrick and A. Grodski, *Rapid. Commun. Mass Spectrom.*, **6**, 747 (1992)

56  S. A. Martin, C. E. Costello and K. Biemann, *Anal. Chem.*, **54**, 2362 (1982)

57  A. Malorni, G. Marino and A. Milone, *Biomed. Environ. Mass Spectrom.*, **13**, 477 (1986)

58  W. V. Ligon Jr. and S. B. Dorn, *J. Am. Chem. Soc.*, **110**, 6684 (1988)

59  J. Shiea and J. Sunner, *Org. Mass Spectrom.*, **26**, 38 (1991)

60  H. Moser and G. W. Wood, *Biomed. Environ. Mass Spectrom.*, 15, 547 (1988)

61  B. E. Winger, O. W. Hand and R. G. Cooks, *Int. J. Mass Spectrom. Ion Processes*, **84**, 89 (1988)

62  K. L. Busch, B.-H. Hsu, Y.-X. Xie and R. G. Cooks, *Anal. Chem.*, **55**, 1157 (1983)

63  A. Liguori, G. Sindona and N. Uccella, *J. Am. Chem. Soc.*, **108**, 7488 (1986)

64  D. H. Williams and S. Naylor, *J. Chem. Soc. Chem. Commun.*, 1408 (1987)

65  F. Derwa, E. De Pauw and P. Natalis, *Org. Mass Spectrom.*, **26**, 117 (1991)

66  T. L. Ho, *Chem. Rev.*, **75**, 1 (1975)

67  D. H. Williams, A. F. Findeis, S. Naylor, B.W. Gibson, *J. Am. Chem. Soc.*, **109**, 1980 (1987)

68  R. L. Cerny, M. L. Gross, *Anal. Chem.*, **57**, 1160 (1985)

69  D. H. Williams, C. Bradley, G. Bojesen, S. Santikarn and L. C. E. Taylor, *J. Am. Chem. Soc.*, **103**, 5700 (1981)

70  C. Javanaud and J. Eagles, *Org. Mass Spectrom.*, **18**, 93 (1983)

71  S. E. Unger, R. J. Day and R. G. Cooks, *Int. J. Mass Spectrom. Ion Phys.*, **39**, 231 (1981)

72  G. Denti, S. Serroni, G. Sindona and N. Uccella, *J. Am. Soc. Mass Spectrom.*, **4**, 306 (1993)

73  G. Sindona in *Mass Spectrometry in the Biological Sciences: A Tutorial*, ed. by M. L. Gross, pp. 383-405, Kluwer A. Press, Dordrecht (NL) ,(1990)

74  A. Liguori, A. Procopio and G. Sindona, *207th A.C.S. Meeting*, San Diego (USA), March 13-17, 1994

75  G. Sindona, N. Uccella and K. Weclawek, *J. Chem. Res.(S)*, 184 (1982)

76  S. Winstein, E. Clippinger, A. H. Fainberg, R. Heck and G. C. Robinson, *J. Am. Chem. Soc.*, **78**, 328 (1956)

77  L. Grotjahn, H. Blöcker and R. Frank, *Biomed. Mass Spectrom*, **12**, 514 (1985)

78  J. A. Laramèe, D. Arbogast and M. L. Deinzer, *Anal. Chem.*, **61**, 2154 (1989)

79  A. Evidente, L. Corti, C. Altomare, A. Bottalico, G. Sindona, A. L. Segre and A. Logrieco, *Natural Toxins*, **2**, 4 (1994)

80  F. A. Carey and R. J. Sundberg, *Advanced Organic Chemistry*, 3rd edition, Plenum Press, New York, 1990, part A, p. 230.

81  A. Evidente, A. Napoli and G. Sindona, *13th Int. Mass Spectrometry Conference*, Budapest, 29 August-2 September, 1994

82  P. Hu and M. L. Gross, *J. Am. Soc. Mass Spectrom.*, **5**, 137 (1994)

83  S. S. Sethi, C. C. Nelson, J. A. McCloskey, *Anal. Chem.*, **56**, 1975 (1984)

84  A. N. R. Nedderman and D. H. Williams, *Biological Mass Spectrom.*, **20**, 289 (1991)

85  K. H. Schram, in *Biomedical Applications of Mass Spectrometry*, C. II. Suelter and J. T. Watson eds., John Wiley & Sons, New York, 1990, pp. 203-287

86  P. F. Crain, *Mass Spectrom. Rev.*, **8**, 505-554 (1989)

87  G. Sindona, results from our laboratory

88  O. W. Hand, L. D. Detter, S. A. Lammert, R. G. Cooks and R. A. Walton, *J. Am. Chem. Soc.*, **111**, 5577 (1989)

89  J. L. Aubagnac, R. M. Claramunt and D. Sanz, *Org. Mass Spectrom.*, **25**, 293 (1990)

90  J. L. Aubagnc, I. Gilles, R. Astier, G. Gosselin, J. L. Imbach and M.C. Bergogne, *Rap. Commun. Mass Spectrom.*, **7**, 41 (1993)

91  J. N. Kyranos and P. Vourous, *Biomed.Environ. Mass Spectrom.*, **19**, 628 (1990)

92  A. Agnello and E. De Pauw, *Org. Mass Spectrom.*, **26**, 175 (1991)

93  S. M. Scheifers, S. Verma and R. G. Cooks, *Anal.Chem.*, **55**, 2260 (1983)

94  C. Dass, *J. Am. Soc. Mass Spectrom.*, **1**, 405 (1990)

95  E. De Pauw, *Anal. Chem.*, **55**, 2196 (1983)

96  R. G. Brereton, M. B. Bazzaz, S. Santikarn and D. H. Williams, *Tetrahedron Lett.*, **24**, 5775 (1983)

97    D. Papaioannou, C. Athanassopoulos, V. Magafa, N. Karamakos, G. Stavropoulos, A. Napoli, G. Sindona, D. W. Aksnes and G. W. Francis, *Acta Chem. Scand.*, **48**, 324 (1994)

98    T. W. Greene and P. G. M. Wuts, *Protecting Groups in Organic Synthesis*, John Wiley & Sons, New York, 1991

99    K. Biemann, in *Methods in Enzymology, Volume 193 Mass Spectrometry*, ed. by J. A. McCloskey, pp. 351-374, Academic Press, San Diego (1990)

100   E. R.Bathelt and W. Heerma, *Biomed. Environ. Mass Spectrom.*, **14**, 53 (1987)

101   E. Mammoliti, G. Sindona and N. Uccella, *Org. Mass Spectrom.*, **27**, 495 (1992)

102   K. Barlos, D. Gatos, A. Napoli and G. Sindona, *42nd ASMS Conference on Mass Spectrometry and Allied Topics*, Chicago (USA), May 29 - June 3 (1994)

103   E. Clayton and A. J. Wakefield, *J. Chem. Soc., Chem. Commun.*, 969 (1984)

104   K. A. Caldwell and M. L. Gross, *J. Am. Soc. Mass Spectrom.*, **5**, 72 (1994)

105   L., D. Detter, O. W. Hand, R. G. Cooks and R. A. Walton, *Mass Spec. Rev.*

# MATRIX-ASSISTED LASER DESORPTION-IONIZATION (MALDI) MASS SPECTROMETRY OF BIOLOGICAL MOLECULES

M.KARAS, U.BAHR
*Institute for Medical Physics and Biophysics*
*University of Münster*
*Robert-Koch-Str. 31, 48149 Münster*
*F.R. Germany*

ABSTRACT. Matrix-assisted laser desorption/ionization mass spectrometry (MALDI-MS) is an analytical technique for fast and precise mass determination of biological molecules. Intact molecular ions are produced by short pulsed laser irradiation of the biomolecules which are embedded in a matrix consisting of small highly absorbing organic molecules. Mass analysis is carried out in a linear or reflector time-of-flight mass spectrometer. The accessible mass range is 500 000 Da, a mass accuracy of up to 0.01% can be reached, the sample amounts required are 1 pmol or less. Proteins, glycoproteins, oligonucleotides and oligosaccharides can be analyzed.

## 1. Introduction

The development of new ionization techniques during the two past decades has enabled important achievements in bioorganic mass spectrometry for the analysis of biopolymers such as proteins and carbohydrates. The general problem to be overcome is to convert the polar, thermally labile biomolecules into intact isolated ions in the gas phase. The so-called desorption/ ionization techniques use different physical approaches to achieve this goal; field desorption [1] applies a high electric field to the sample; in fast atom bombardment [2] and $^{252}$Cf plasma desorption [3] the sample is bombarded by highly energetic ions or atoms; thermospray ionization [4] and electrospray ionization [5] form ions directly from small, charged droplets. Laser desorption [6,7] and the newly developed version of this method, matrix-assisted laser desorption/ionization (MALDI) [8] make use of short, intense pulses of laser light to induce the formation of intact gaseous ions.

## 2. Development of MALDI-MS

Already in the 1970s, first results to use laser light as mass spectrometric ionization method for organic molecules have been reported [9-13]. $CO_2$-lasers with a wavelength of 10,6 $\mu$m in the infrared (IR) and lasers emitting in the far ultraviolet (UV) proved to be most useful. In order to transfer the energy twithin a very short time and to avoid thermal decomposition of labile organic molecules lasers with pulse widths on the nanosecond timescale, such as

*R. M. Caprioli et al. (eds.), Mass Spectrometry in Biomolecular Sciences, 33–49.*
© *1996 Kluwer Academic Publishers.*

Q-swiched Nd-YAG, excimer or TEA-$CO_2$-laser were used. The pulsed desorption of ions favours the combination of the laser desorption ion source with a time-of-flight (TOF) mass analyzer [14,15] or a fourier transform ion cyclotron resonance (FT-ICR) mass analyzer [16,17], both make it possible to record complete mass spectra for each laser shot.

The initial experiments on laser desorption of organic ions all showed a restriction of the analysis to biocompounds with molecular masses below ca. 2000 Da. For UV-laser desorption a "soft" desorption of molecular ions was only found for compounds exhibiting a strong resonance absorption at the incident laser wavelength. For larger molecules - even if they contain a chromophoric group - the obtained energy/volume was believed to be insufficient at moderate irradiances (laser power density), higher irradiances resulted in fragmentation (photodissociation) rather than in the desorption of intact molecular ions. Samples which cannot be resonantly excited at the laser wavelength need very high irradiances (laser power/area, $W/cm^2$) for ion production which inevitably destroy large organic molecules. These limitations, which prohibited a general use of laser desorption for organic mass spectrometry, promoted the idea to use a matrix for the analysis of non-absorbing and/or high-molecular-mass biopolymers [18]. The matrix consists of a strongly UV-absorbing organic compound known to show a soft laser desorption for the neat compound and the high mass molecules are finely dispersed therein. Thereby, the latter can be desorbed and ionized irrespective of their individual absorption characteristics [19,20].

## 3. Ion Formation in Matrix-Assisted Laser Desorption/Ionization

The matrix was introduced to serve the following functions :
- Its strong absorption at the wavelength of the incident laser light makes it possible to transfer the energy of the laser light to the usually solid sample in a controllable way. For the adequate irradiances used in the range of $10^6$ to $10^7$ $W/cm^2$, laser pulse lenghts of some ns and a matrix absorptivity of some thousand $1 \, mol^{-1} \, cm^{-1}$ typically one photon per matrix molecule in the uppermost layer is absorbed. The calculation is based on linear absorption and validity of Beer's Law. This also results in an exponentially decaying in-depth excitation with typical penetration depths ($I=I_0/e$) of ca. 100 nm. This energy is sufficient to induce the ablation of a sample volume setting free intact matrix and analyte neutrals and ions.
- The analyte is diluted in the matrix forming a solid sovent which is present in the matrix-sample mixture at a very high molar excess. This prevents analyte molecules from aggregation and reduces strong intermolecular interaction.
- Analyte ionization is strongly promoted by photochemical reactions. Proton transfer from active matrix species, initiated by photoionization resulting in both intact molecule and fragment ions, is supposed to form the major ionization mechanism for analytes which are able to form stable (de)protonated ions.

An increasing number of research investigations is available today illuminating the desorption-ionization process. Neutral and ion desorption exhibits a strong dependance on the applied laser irradiance [18-21]. Once a distinct value is reached (threshold irradiance), ion yield is strongly increasing with the irradiance, typically with the fifth or sixth power. Best results are obtained in a small irradiance range at and slightly above the threshold value.

Matrix and analyte ions show a strongly forwarded emission (normal to the target surface) and have supersonic initial velocities irrespective of their mass [22,23].

Despite their high initial velocities, ion accelerated into the mass spectrometer show deficit (too low) energies increasing with increasing irradiance; this can be rationized by collisions of ions with neutrals in the initially very dense material cloud and/or by ion formation by chemical reactions taking place in the acceleration region [24]. The general notion that the ions formed and observed in the mass spectra are essentially intact molecule ions has to be replaced by a more refined picture. Depending on the matrix and physical conditions in the ion source molecule ions may show strong metastable dissociation [25,26]. Whereas small peptide ions undergo backbone cleavages resulting in structurally indicative fragments ions, show larger proteins mainly the loss of small neutrals such as ammonia, water and carbon dioxide [27]. Incorporation of analyte molecules in matrix crystals formed upon evaporation of the solvent seems to be an essential prerequisite for a successful MALDI analysis. Formation of a solid solution of analyte in matrix crystals has been proven for two matrices, manely 2,5-dihydroxybenzoic acid [28] and sinapinic acid [29]; in the latter case a preferential built-in of a test protein onto a hydrophobic crystal surface was documented. This incorporation is also the basis for the observed high tolerance of some high-quality matrices against some ubiquitous contaminants such as inorganic salts and buffers by preferential incorporation of proteins and exclusion of salts.

Though not unequivocally proven, there is good evidence that the matrix plays an active role in the ionization of analyte ions. The inspection of the LD mass spectra of the matrix compounds themselves revealed some very unexpected features such as high intensity radical cat- and anions accompanying even-electron molecule and fragment ions. This was taken as an indication that a good matrix is not able to form stable ions and that matrix ions function as gas phase acids or bases resulting in (de)protonated analyte ion species [30].

## 4. Matrices and Preparation

Even though considerable progress has been obtained in the understanding of the MALDI process, it is not possible yet to select a matrix theoretically. The absorption at the UV-laser wavelength is unfortunately not a very restrictive criterion and that is why all of the matrices in use today have been found empirically just by trial and error. Compatibility of solvents required for matrix and analyte, absence of chemical reactivity and vacuum stability are further obvious properties to be fulfilled by a matrix.

Even though many compounds work as matrix in principle, only very few can be regarded as high-quality matrices for UV-MALDI. These are :
- Sinapinic acid and some related cinnamic acid derivatives [31,32]
- 2,5-Dihydroxybenzoic acid [28].
- α-Cyano-4-hydroxycinnamic acid [33]
- 3-Hydroxypicolinic acid [34].
These widely-used matrices all fulfill some additional criteria. They allow for a fast, easy and reproducible preparation; they show a pronounced tolerance towards several ubiquitous contaminants and reagents; they only yield a low-intensity adduct ion signal, i.e. an ion signal formed by the addition of usually a matrix fragment to an analyte molecule.

3-Hydroxypicolinic acid does not have the latter property for proteins as analytes, but has enabled progress for oligonucleotide MALDI; this indicates that different matrices for different classes of analytes may be required.

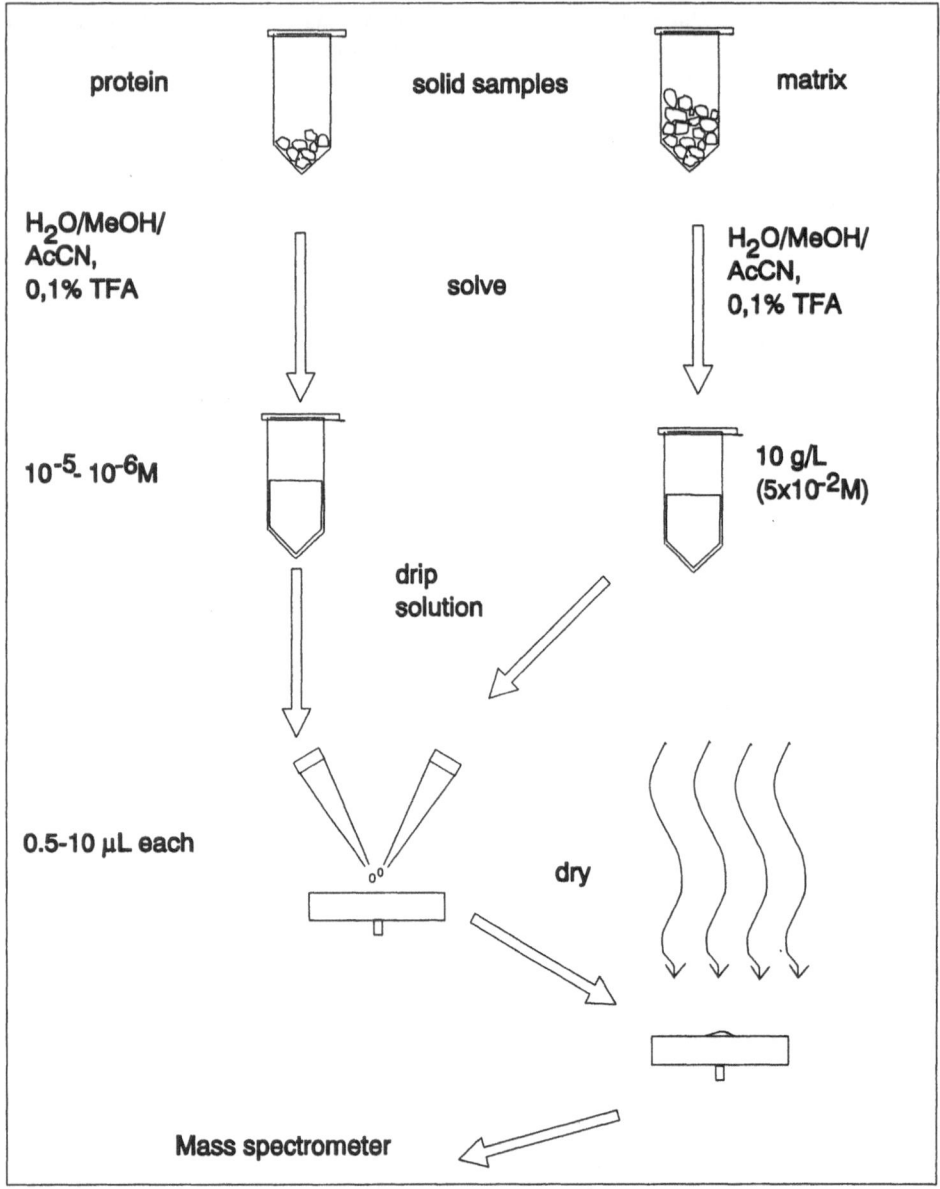

Fig.1: Sample preparation procedure for MALDI mass spectrometry

The preparation of samples for MALDI analysis is simple and fast. (Fig. 1). For protein analysis a 5-10 g/L solution of the matrix material is prepared in either pure water or a mixture of water and organic solvent (acetonitrile, ethanol); a mixture of water acidified by trifluoroacetic acid (0.1%) and acetonitrile (2:1) is a well-suited solvent. $10^{-5} - 10^{-7}$ M solutions of the analyte are prepared in the same solvent as the matrix. Small amounts of both solutions (between 0,5 - 10 $\mu$l) are then mixed together on the metal sample support (usually stainless steel) to give a final analyte concentration of 0.005 - 0.05 $\mu$g/$\mu$L in the mixture. The solvent is evaporated and the sample tranferred to the vacuum chamber of the mass spectrometer. Depending on the matrix used finely dispersed small crystallites or extended crystalline areas at the rim of the droplet can be seen by a microscopic observation. The most intense ion signals are usually associated with these well-developed crystalline regions of the sample [28,29].

## 5. Instrumentation

### 5.1. LASER DESORPTION ION SOURCE

Ion sources for MALDI [35-38] use pulsed lasers with pulse durations of 1 - 200 ns. Most commonly used are UV-lasers such as nitrogen ($N_2$)-lasers emitting at 337 nm or Nd-YAG lasers, whose emission wavelength of 1064 nm has been transferred to 355 nm or 266 nm by

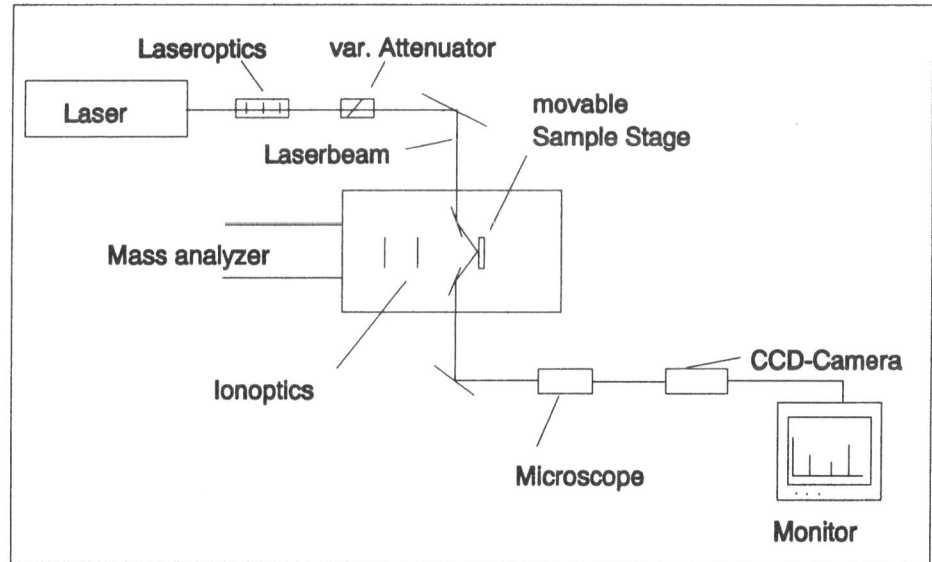

Fig.2: Schematic drawing of a laser ion source

frequency tripling or quadrupling using nonlinear optical crystals. Moreover, excimer lasers (193, 248, 308 and 351 nm), frequency doubled, excimer pumped dye lasers (220 - 300 nm) have been used. In the visible wavelength region dye lasers and in the infrared TEA-CO$_2$ at 10,6 $\mu$m and Er-YAG at 2,94 $\mu$m have been used. For the required irradiances in the range of 10$^6$ to 10$^7$ W/cm$^2$ the laser beams are focussed to spot sizes between 30 and 500 $\mu$m by suitable optical lenses. The angle of incidence of the laser beam on the sample surface varies between 15 - 70°. The irradiance at the sample surface is a critical parameter, the minimum (threshold) irradiance to produce ions is well defined and best results are obtained for a laser irradiance no more than ca 20% above the threshold. Thus the intensity of the laser beam on the sample has to be carefully adjusted either by neutral density filters, angle-dependant reflection attenuaters or polarizers. The position of the laser focus on the sample surface can be changed either by moving the sample towards a fixed laser spot or by steering the laser beam.

Optical control of the sample is a valuable means to yield optimal MALDI results. In Fig. 2 a laser ion source with movable sample stage and a video-microscope for obervation is shown schematically.

## 5.2. MASS ANALYZER

MALDI of large molecules is usually coupled to TOF mass analysis, although several applications have been performed on FT-ICR [39,40] or magnetic sector analyzers [27,41], recently first results with quadrupole ion trap mass spectrometers have been reported [42-44].

In TOF analyzers the mass to charge ratio of an ion is determined by measuring its flight time. After acceleration of the ions in the ion source to a fixed kinetic energy, they pass a field free drift tube with a velocity proportional to $(m_i/z_i)^{-1/2}$ ($m_i/z_i$ is the mass-to-charge ratio of a particular ion species). Due to their mass-dependant velocities, ions are separated during their

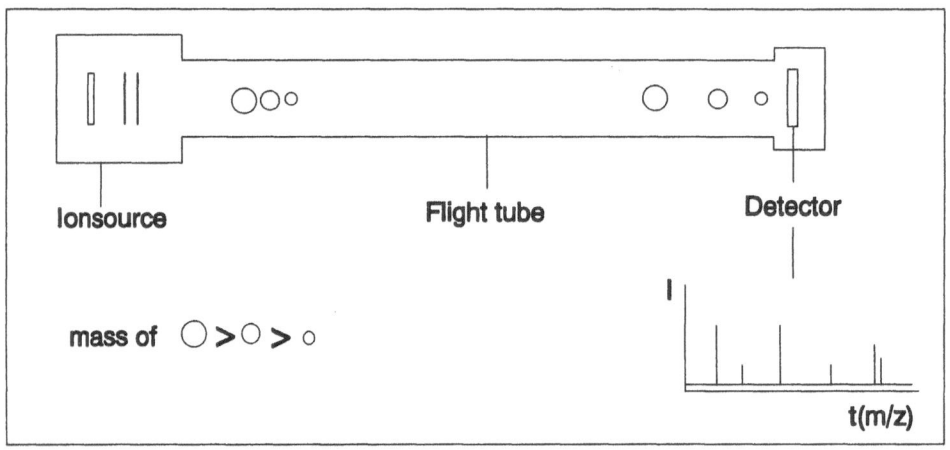

Fig. 3: Schematic drawing of ion separation in a linear time-of-flight instrument.

flight. A detector at the end of the flight tube produces a signal for each ion species. Typical flight times are between a few microseconds and several 100 $\mu$s.

Fig. 3 shows schematically ion separation in a linear time-of-flight instrument. The flight path lengths are typically between 0.5 and 3 m, the accelaration voltages lie between 1 - 30 kV. The mass resolution $m/\Delta m$ in linear TOF instruments is limited to 200 - 800 (hwfm) due to the initial energy distribution ions aquire during desorption. To enhance mass resolution a reflectron TOF mass analyzer is used (see Fig. 4). The ions

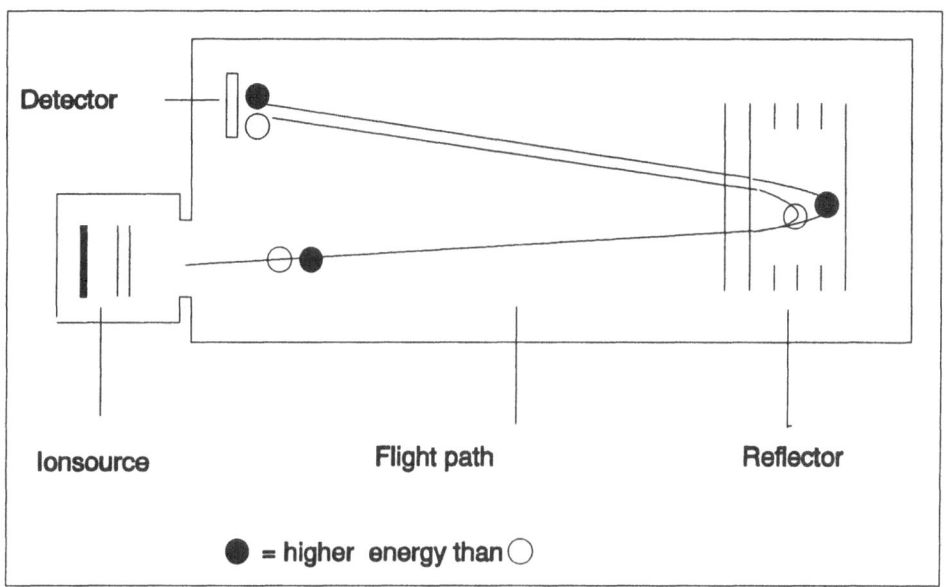

**Detector**

**Ionsource**        **Flight path**        **Reflector**

● = higher energy than ○

Fig. 4: Schematic drawing of ion separation in a reflector time-of-flight instrument.

are decelerated in the ion mirror and turn around at different locations in the reflecting electric potential gradient. Ions of higher kinetic energy spend a longer time in the ion mirror than ions of lower kinetic energy. The arrival time spread can thus be largely corrected for at the detector and an increase in mass resolution up to 6000 can be obtained.

## 5.3. ION DETECTION AND DATA COLLECTION

Ion detection in MALDI-MS is done by either conventional discrete-dynode secondary electron multipliers or microchannelplates. The high mass ions produce either electrons or low-mass ions at the conversion dynode of the multiplier. These particles are than used to start the multiplication cascade in an electron multilier. In MALDI instruments where a low ion acceleration energy is used postacceleration of high mass ions is needed in order to compensate for the lower detection efficiency at lower ion velocities. This is achieved by a separated dynode held at a potential of typically 20 kV placed in front of the multplier.

The signal of the detector is either amplified with a fast linear amplifier or directly digitized by a digital oscilloscope (transient recorder) with a sampling rate of 100 Msample/s. The data are then transferred to a PC for spectrum evaluation. For an accurate mass determination, typically 10 - 50 spectra, each from a single laser shot, are summed to improve the signal-to noise ratio.

## 6. MALDI-mass analysis of proteins

Most applications on MALDI-MS have been done for proteins and no limitations of the application caused by primary, secondary or tertiary structure has yet been discovered. Proteins with different solution-phase properties, including proteins that are insoluble in aqueous solutions can be analyzed as well as glycoproteins containing a large amount of carbohydrate.

### 6.1. CHARACTERISTICS OF MALDI ANALYSIS OF PROTEINS

Fig. 5 shows as typical MALDI spectrum the spectrum of the protease Subtilisin Carlsberg with a molecular weigth of 27 288. As usually observed, the molecular ion

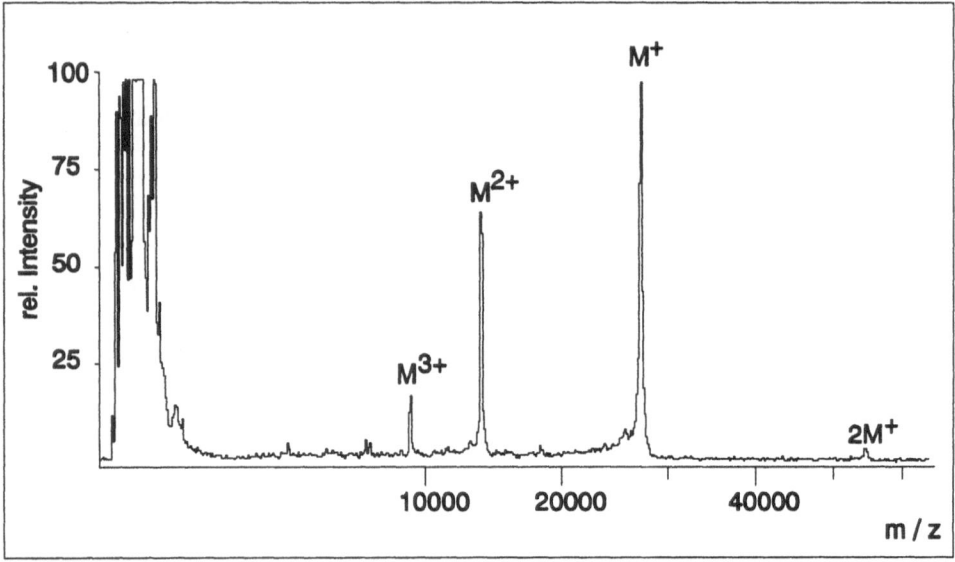

Fig.5: MALDI spectrum of Subtilisin Carlsberg

peak is the most prominent signal accompanied by doubly and triply charged molecular ion signals as well as a dimer. The relative abundances of the ion signals depend on the matrix and the concentration of the analyt. In the mass range below 500 Da ion signals from the matrix appear. Both positive and negative ion spectra with comparable signal intensities can be obtained. Positive ions appear as protonated, negative ions as deprotonated species.

The technique is relatively immune to many non-proteinaceous impurities. Salts or buffers in concentrations normally used for biochemical procedures often do not have to be removed before analysis. With sinapinic acid as matrix inorganic salts like NaCl and denaturation agents like urea or guanidinium hydrochloride up to 2 mM in the protein solution and buffering agents like ammonium bicarbonate, ammonium acetate, citrate, glycine, hepes, tris up to 200 mM do not strongly affect the analysis [31,45]. The same holds for other matrices. Up to 10% SDS (sodium dodecyl sulfate) in the protein solution is tolerable with DHB as matrix [28]. Complex mixtures of different proteins can be measured.

Mass determination accuracy as high as 0.01% for proteins below 30 000 Da can be achieved [45,46]. The mass calibration has to be done using well-defined reference compounds. For linear TOF instruments, an internal calibration, e.g. adding of the calibrant proteins to the investigsted sample, is essential to obtain this level of accuracy; for reflectron instruments external calibration by analysis of the reference compounds in a second analysis is usually sufficient. Above 30 000 Da a decrease in mass accuracy is observed which is supposed to be due to non-resolved adducts between analyte and matrix.

Typical sample amounts used for MALDI analysis are about 1 pmol. In several cases 1-10 fmol have been used for analysis [28,47]. The amount of material consumed for a spectrum is much less than the total amount loaded onto the sample support.

The mass range accessible for TOF mass analysis is in principle unlimited. Among the highest masses detected so far in MALDI analysis is the dimer of a monoclonal antibody with a molecular weight of 294 960 Da (see Fig.6).

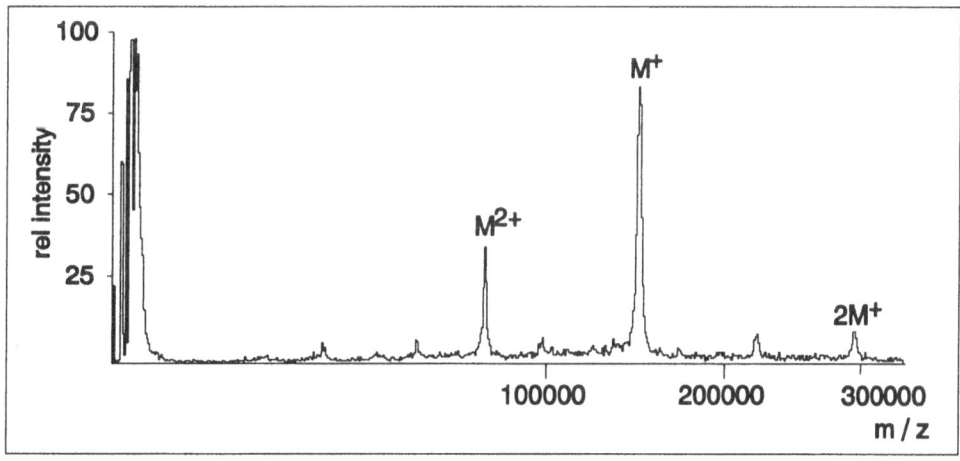

Fig.6: MALDI mass spectrum of a monoclonal antibody

The time for sample preparation and for data acquisition is only some minutes, more time-comsuming are the calibrations generally needed to obtain high mass accuricies.

## 6.2. APPLICATIONS IN PROTEIN ANALYSIS

*6.2.1.Characterization of intact proteins.* MALDI-MS is used to characterize intact proteins via determination of the molecular weight where a quick control of integrity and purity of approtein preparation is needed. A MALDI mass spectrum can give a first indication to posttranslational modifications like phosphorylation or glycosylation.

*6.2.2.Characterization of proteins by inspection of cleavage peptides.* Characterization of proteins by MALDI-MS can also be done via molecular weight determination of peptide fragments. Peptide fragments produced by enzymatic or chemical degradation of a protein can either be analyzed in the unfractionated mixture or after separation with HPLC or gel electrophoresis.

Fig. 7 shows the MALDI spectrum of cytochrome C after enzymatic digest [48]. To 0.5 $\mu$g Cytochrome C 0.25 $\mu$g Trypsin was added and incubated at pH 8 and 25 °C over night. 1 $\mu$l of the reacting mixture was mixed with DHB as matrix and loaded onto the sample target for MALDI analysis. Trypsin releases lysine and arginine from the C-terminus of the peptide, 22 fragments are expected, except the C-terminal amino acids (Lys$^{101}$-Glu$^{104}$) all fragments have been identified as [M+H]$^{+}$, in several cases also as [M+Na]$^{+}$ and [M+K]$^{+}$-ions. In the mass range below 550 Da peaks marked with an asterix stem from the matrix, the metal target or known impurities. Table 1 shows all identified fragment masses together with the calculated masses. The accuracy of mass determination is between 0.01-0.1%.

*6.2.3. Combination of SDS-gel electrophoresis and MALDI-MS.* Sodium dodecyl sulfate-polyacrylamide gel electrophoresis (SDS-PAGE) and high-resolution two-dimensional polyacrylamide electrophoresis (2-DE) are the most widely used methods for separation of complex protein mixtures and molecular weight determination of proteins. Unfortunately, the mass accuracy achieved with this method is only 5-10%. For more precise mass determination, the separated proteins can be removed from the gel and measured by MALDI-MS [49,50]. MALDI-MS directly from the gel is under investigation in a current research project. Good results have been obtained for proteins electrotransferred onto polyvinylidene diflouride (PVDF) or polyamide membranes (electroblotted), a method often used for further protein chemical analysis like sequencing or amino acid composition analysis. The piece of membrane containing the protein of interest was cut out, soaked in matrix solution and transferred to the mass spectrometer for MALDI analysis [51,52]. Fig. 8 shows a spectrum of carbonic anhydrase gained in this manner. For laser desorption an Er-Yag-Laser (2.94 $\mu$m) and succinic acid as matrix were used. The spectrum has nearly the same quality than a MALDI spectrum than from a normal preparation. A problem arises from staining of the proteins, as multiple attachment of dye molecules to the analyte can lead to peak broadening. Staining by colloidal pigments is well-suited for MALDI analysis.

*6.2.4.Protein Ladder Sequencing.* A new approach to protein sequence analysis by MALDI MS has been reported [53]: It consists of two steps, a controlled generation of sequence-defining peptide fragments, each differing from the next by one amino acid and the subsequent MALDI-MS analysis of this fragment mixture. Each amino acid in the sequence

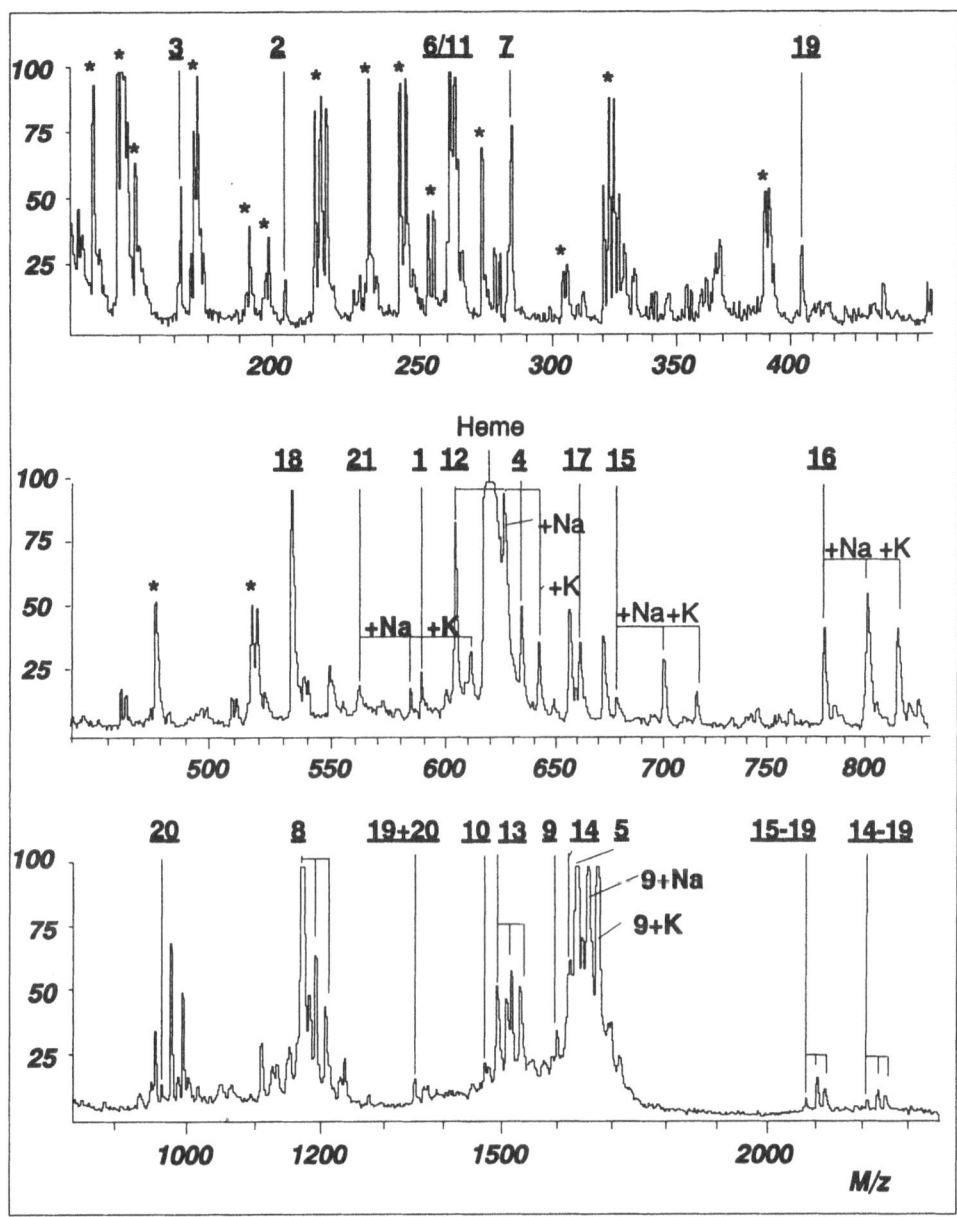

Fig.7: MALDI mass spectrum of cytochrome C fragments from degradation with trypsin
(from [48] with permission of Finnigan MAT, Bremen)

| 1 | Gly$^1$ - Lys$^5$ + Acetyl | 588.68 | 589.2 |
|---|---|---|---|
| 2 | Gly$^6$ - Lys$^7$ | 203.24 | 204.1 |
| 3 | Lys$^8$ | 146.19 | 147.1 |
| 4 | Ile$^9$ - Lys$^{13}$ | 633.79 | 634.3 |
| 5 | Cys$^{14}$ - Lys$^{22}$ + Heme | 1635.0 | 1635.2 |
| 6 | Gly$^{23}$ - Lys$^{25}$ | 260.29 | 260.9 |
| 7 | His$^{26}$ - Lys$^{27}$ | 283.33 | 284.2 |
| 8 | Thr$^{28}$ - Arg$^{38}$ | 1168.33 | 1169.4 |
| 9 | Lys$^{39}$ - Lys$^{53}$ | 1598.57 | 1599.5 |
| 10 | Thr$^{40}$ - Lys$^{53}$ | 1470.57 | 1471.4 |
| 11 | Asn$^{54}$ - Lys$^{55}$ | 260.30 | 260.9 |
| 12 | Gly$^{56}$ - Lys$^{60}$ | 603.72 | 604.3 |
| 13 | Glu$^{61}$ - Lys$^{72}$ | 1495.71 | 1494.9 |
| 14 | Glu$^{61}$ - Lys$^{73}$ | 1623.89 | 1624.0 |
| 15 | Tyr$^{74}$ - Lys$^{79}$ | 677.8 | 678.3 |
| 16 | Met$^{80}$ - Lys$^{86}$ | 779.05 | 779.4 |
| 17 | Lys$^{87}$ - Arg$^{91}$ | 660.77 | 661.4 |
| 18 | Lys$^{88}$ - Arg$^{91}$ | 532.6 | 533.4 |
| 19 | Thr$^{89}$ - Arg$^{91}$ | 404.42 | 405.2 |
| 20 | Glu$^{92}$ - Lys$^{99}$ | 964.12 | 964.8 |
| 21 | Lys$^{100}$ - Glu$^{104}$ | 561.4 | 562.4 |
| 22 | Lys$^{101}$ - Glu$^{104}$ | 433.3 | - |

Table 1: Measured and calculated masses of the fragments of cytochrome C

was identified from the mass differences between adjacent peaks. Peptides with molecular weights up to 3500 Da have been sequenced in this manner with high accuracy. The method was also used to directly locate a phosphoserine residue in a phospholipide [53]. For protein ladder sequencing some pmol of staring material are sufficient, a very high sample throughput at very low cost per cycle is achieved

*6.2.5. Post source decay (PSD) fragment analysis.* Another promising approach for peptide sequencing has been developed using analysis of metastable fragments in a reflector TOF mass

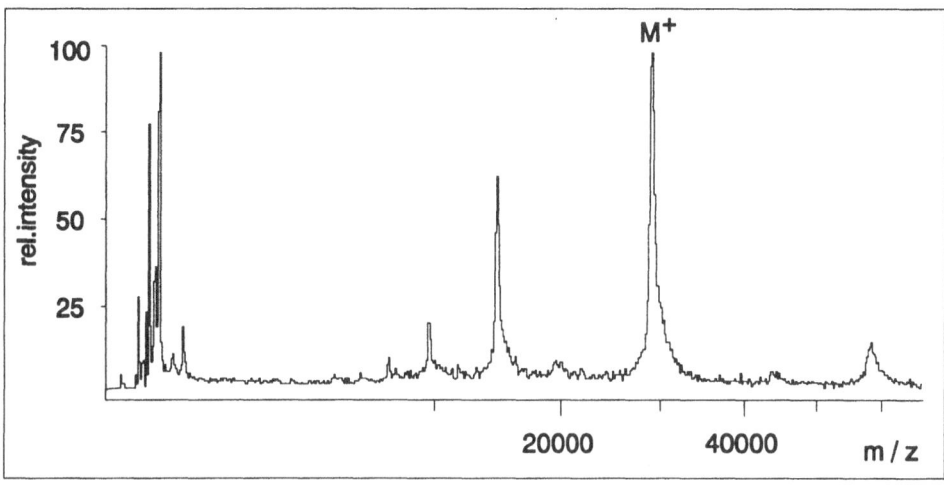

Fig. 8: MALDI mass spectrum of carbonic anhydrase directly from a PVDF membrane

spectrometer [25,54]. Whereas prompt fragmentation is a minor process under in MALDI, metastable fragmentation can be observed for small peptides to a large extent. Depending on various instrumental parameters (gas pressure, composition of residual gas, matrix) both uni- and bimolecular decay (i.e. by collisions) of laser-desorbed peptides takes place in the field-free drift tube. To detect all fragments the potential of the ion reflector has to be tuned according to the kinetic energy of the ions. Thus a series of spectrum segments can be obtained from which a complete fragment ion spectrum can be reconstructed. Several small peptides have been analyzed with this method showing the complete series of sequence fragments in the mass spectrum, obtained in several minutes.

*6.2.6. Identification of proteins.* A rapid and sometime sufficient approach for identifiying proteins is the comparison of masses determined by MALDI-MS from protein digests (fingerprint) and calculated masses in a database [55,56]. Using this method in a study of 40 proteins it has been shown that often as few as three or four experimentally determined peptide masses are sufficient to uniquely identify the sample when screened agained a fragment data base [57].

## 7. MALDI MASS ANALYSIS OF NUCLEIC ACIDS

Whereas MALDI-MS has found broad application in protein analysis, investigations on nucleic acids are limited. The highest mass detected for nucleic acids so far is about 40000 Da from a rRNA [58]. Oligo(deoxy)ribonucleotides and mixtures thereof [34, 59-65] have been analyzed as well as small RNA samples [59,66]. Ultraviolet as well as infrared lasers have been used with different matrices[58]. Good results have been obtained with an Er-Yag laser at 2,94 $\mu$m wavelength and succhinic acid as matrix [58] and with a $N_2$-laser using 3-hydroxy-picolinic acid as matrix [34,63]. Oligo(deoxy)ribonucleotides yield stronger signals and better mass resolution for negative than for positive ions. Only for higher molecular weight tRNAs and rRNAs the signal intensity is higher for the positive ions.

Fig. 9 shows the mass spectrum of negative ions of an oligodeoxynucleotide 5'-d[CGG AAT TCG GAA GTG GGC CGC CTT CAG]-3 received from UV-MALDI with 3-hydroxypicolinic acid as matrix.

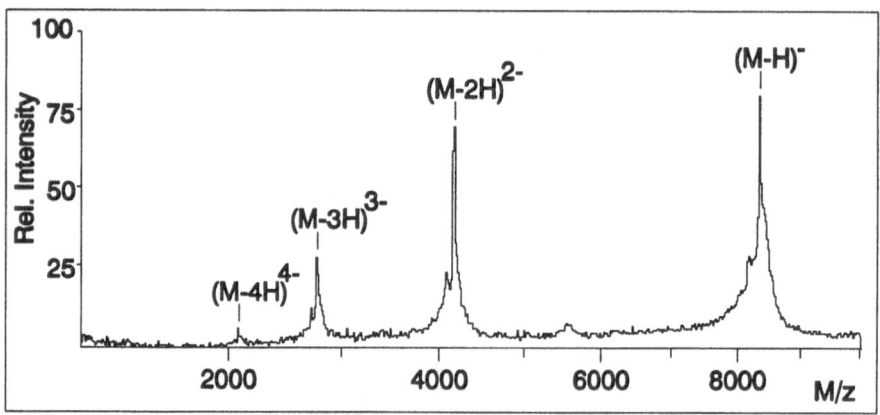

Fig.9: Negatice ion MALDI mass spectrum of a oligodeoxynucleotide 27 mer
       5'-d[CGG AAT TCG   GAA GTG  GGC CTT CAG]-3'

One problem in nucleotide analysis is the formation of metal ion attachment arising mainly from alkali salt impurities in the matrix solution or on the target. This often leads to a broad pseudomolecular ion distribution due to multiple cation attachment which can not be resolved any longer for larger molecules. A simple means to avoid this is to exchange the metal cations against ammonia ions using suitable cation exchange polymer bead [58]. This can be done by adding the beads to the sample solution or directly to the sample droplet on the target. No negative influence on the desorption process could be observed.

## 8. MALDI MASS ANALYSIS OF CARBOHYDRATES

Underivatized oligosaccharides have also been successfully analyzed by MALDI-MS using different matrices [67-69]. The oligomer distribution of dextrins and dextrans with molecular weights up to 15 000 Da has been reported. Only 10 fmol of sample was needed to obtain spectra from native and permethylated glycosphingolipids and gangliosides [70,71]. In contrast to proteins, carbohydrates show under MALDI conditions only alkai-attached molecular ions. It seems to be more difficult to ionize larger oligosaccharides, the mass range has yet not exceeded 15 000 Da. Here more work is necessary to expoit the potential for MALDI analysis of these compounds.

Fig.10 shows the MALDI spectrum of a glycopeptide obtained from a pronase degradation of a surfacelayer glycoprotein of *Thermoanaerobacter thermohydrosulforicus* strains L 111-69 [72].

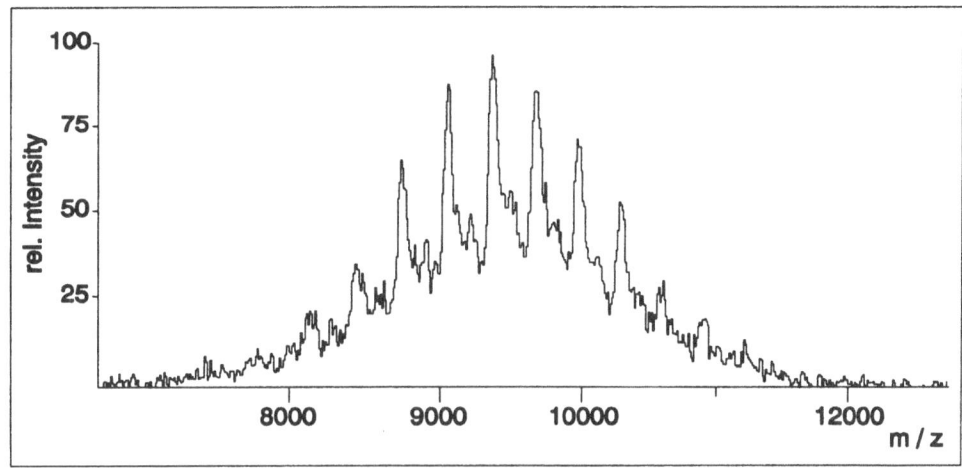

Fig.10: MALDI mass spectrum of an O-glycopeptide o-methyl-Rha-(Man-Rha)
$_n$-Man-Rha-Rha- Rha-Gal-Tyr,Pro,Val

The O-glycopeptide has the structure o-methyl-Rha-(Man-Rha)$_n$-Man-Rha-Rha-Rha-Gal-Tyr, Pro,Val. The spectrum shows a series of $[M+2Na-H]^+$ ion peaks with the most abundant peak at m/z 9364 (calculated value 9360.4) representing the the compound with 26 repetetive disaccharide units. The adjacent peaks have mass differences of 308 Da reflecting various numbers of disaccharide units.

## 9. REFERENCES

1.    Beckey HD (1977) Principles of field desorption mass spectrometry. Pergamon Press, Oxford
2.    Barber M, Bordoli RS, Sedgwick RS, Tyler AN (1981) J ChemSoc Chem Commun:325
3.    Macfarlane RD, Torgerson FD (1976) Science 109:920
4.    Blakely CR, Carmody JJ, Vestal ML (1980) Anal Chem 52: 1636
5.    Whitehouse CM, Dreyer RN, Yamashita M, Fenn JB (1985) Anal Chem 57: 675
6.    Hillenkamp F, Ehring H (1992) In: Gross ML (ed) Mass spectrometry in the biological sciences: a tutorial. Kluwer Academic Publishers, Dordrecht, Boston, London. pp 165-179
7.    Cotter JR (1987) J Anal Chim Acta 195:45-59
8.    Hillenkamp F, Karas M, Beavis RC, Chait BT (1991) Anal Chem 63:1196A-1203A
9.    Vastola FJ, Mumma, RO, Pirone AJ (1970) J Org Mass Spectrom 3:101
10.   Posthumus MA, Kistemaker PG, Meuzelaar HLC, deBrauw MC (1978) Anal Chem 50:985
11.   Stoll R, Röllgen FW (1979) Org Mass Spectrom 14:642-645
12.   Cotter RJ (1981) Anal Chem 53: 1306-1307
13.   Hardin ED, Vestal ML (1981)Anal Chem 53: 1492-1497
14.   Hillenkamp F, Unsöld E, Kaufmann R, Nitsche R (1975) Appl Phys 8: 341-348
15.   Van Breemen RB, Snow M, Cotter RJ (1983) Int J Mass Spectrom Ion Phys 49:35-50

16. Weller RR, MacMahon TJ, Freiser BS (1990) In:Lubman D: (Ed) Lasers and Mass Spectrometry, Oxford University, New York, pp 249-270
17. Köster C, Kahr MS, Castoro JA, Wilkins CL (1992) Mass Spectrom Reviews 11:495-512
18. Karas M, Bachmann D, Bahr U, Hillenkamp F (1987) Int J Mass Spectrom Ion Proc 78:53-68
19. Karas M, Hillenkamp F (1988) Anal Chem 60: 2299-2301
20. Karas M, Bahr U, Hillenkamp F (1989) Int J Mass Spectrom Ion Proc 92:231-242
21. Ens W, Mao Y, Mayer F, Standing KG (1991) Rapid Commun Mass Spectrom 5: 117
22. Beavis RC, Chait BT (1991) Chem Phys Lett 181:479
23. Huth-Fehre T, Becher CH (1991) Rapid Commun Mass Spectrom 5:198
24. Zhao J, Ens W, Standing KG, Verentschikov A (1992) Rapid Commun Mass Spectrom6:671
25. Spengler B, Kirsch D, Kaufmann R, Jaeger E (1992) Rapid Commun Mass Spectrom 6:105
26. Spengler B, Kaufmann R (1992) Analusis 20:91
27. Annan RS, Köchling HJ, Hill JA, Biemann K (1992) Rapid Commun Mass Spectrom 6:298-302
28. Strupat K, Karas M, Hillenkamp F (1991) Int J Mass Spectrom Ion Proc. 111:89-101
29. Beavis RC, Bridson JN (1992) J Phys D, Appl Phys 26:442
30. Ehring H, Karas, M, Hillenkamp F (1992) Org Mass Spectrom 27: 472
31. Beavis RC, Chait BT (1989) Rapid Commun Mass Spectrom 3:436
32. Beavis RC, Chait BT (1989) Rapid Commun Mass Spectrom 3:432
33. Beavis RC, Chaudhary T, Chait BT (1992) Org Mass Spectrom 27:156
34. Wu KJ, Steding A, Becker CH (1993) Rapid Commun Mass Spectrom 7: 142
35. Feigl P, Schueler B, Hillenkamp F (1983) Int J Mass Spectrom Ion Phys 47:15
36. Overberg A, Karas M, Bahr U, Kaufmann R, Hillenkamp F (1990) Rapid Commun Mass Spectrom 4:293
37. Overberg A, Karas M, Hillenkamp F (1991) Rapid Commun Mass Spectrom 5:128
38. Spengler B, Cotter RJ (1990) Anal Chem 62: 7932.
39. Buchanan MV, Hettich RL (1993) Anal Chem 65: 245A-259A
40. Castoro JA, Köster C, Wilkins CL (1993) Anal Chem 65: 784-788
41. Hill JA, Annan RS, Biemann K (1991) Rapid Commun Mass Spectrom 5:395-399
42. Schwartz JC, Bier ME (1993) Rapid Commun Mass Spectrom 7: 27
43. Chambers DM, Goeringer DE, McLuckey SA, Glish GL (1993) Anal Chem 65:14-20
44. Jonscher K, Currie G, McCormack AL, Yates JR (1993) Rapid Commun Mass Spectrom 7:20-26
45. Beavis RC, Chait BT (1990) Proc Natl Acad Sci 87:6873
46. Beavis RC, Chait BT (1990) Anal Chem 62:1836
47. Karas M, Ingendoh A, Bahr U, Hillenkamp F (1989) Biomed Environm Mass Spectrom 18: 841-843
48. Loo AJ, Ingendoh A, Bahr U (1993) Finnigan MAT Vision 2000,Application Data Sheet No 2
49. Wang YK, Liao P-C, Allison J, Gage DA, Andrews PC, Lubman DM, Hanash SM,

Strahler JR (1993) J Biol Chem 268: 14269
50. Chait BT, Kent SBH (1992) Science 275: 1885
51. Eckerskorn C, Strupat K, Karas M, Hillenkamp F, Lottspeich F (1992) Electrophoresis 13:664-665
52. Strupat K, Karas M, Hillenkamp F, Eckerskorn C, Lottspeich F (1994) Anal. Chem. in press
53. Chait BT, Wang R, Beavis RC, Kent SBH (1993) Science 262:89
54. Kaufmann R, Spengler B, Lützenkirchen F (1993) Rapid Commun Mass Spectrom 7:902
55. Henzel WJ, Billeci TM, Stults JT, Wong SC, Grimley C, Watanabe C (1993) Proc Natl Acad Sci USA 90: 5011
56. Mann M, Hojrup P, Roepstorff P (1993) Biol Mass Spectrom 22:338
57. Pappin DJC, Hojrup P, Bleasby AJ (1993) Current Biol 3:327
58. Nordhoff E, Ingendoh A, Cramer R, Overberg A, Stahl B, Karas M, Hillenkamp F, Crain PF (1992) Rapid Commun Mass Spectrom 6: 771-776
59. Karas M, Bahr U, Gießmann U (1991) Mass Spectrom Rev 10: 335-357
60. Börnsen KO, Schär M, Widmer HM (1990) Chimica 44:412
61. Huth-Fehre T, Gosine JN, Wu KJ, Becker CH (1991) Rapid Commun Mass Spectrom 5:378
62. Parr GR, Fitzgerald MC, Smith LM (1992) Rapid Commun Mass Spectrom 6:369
63. Nordhoff E, Cramer R, Karas M, Hillenkamp F, Kirpekar F, Kristiansen K, Roepsdorff P (1993) Nucl Acid Res 15: 3347
64. Keough T, Baker TR, Dobsen RLM, Lacey MP, Riley TA, Hasselfield JA, Hesselberth RE (1993) Rapid Commun Mass Spectrom 7:195-200
65. Tang K, Allman SL, Jones RB, Chen CH, Araghi S (1993) Rapid Commun Mass Spectrom 7:63-66
66. Hillenkamp F, Karas M, Ingendoh A, Stahl B (1990) in: Burlingame AL, Mc Closkey JA (eds) Biological mass Spectrometry, Elsevier, Amsterdam, pp49-60
67. Mock KK, Davey M, Cotrell JS (1991) Biochem Biophys Res Commun 177:644
68. Stahl B, Steup M, Karas M, Hillenkamp F (1991) Anal Chem 63: 1463
69. Harvey DJ (1993) Rapid Commun Mass Spectrom 7:614-619
70. Egge H, Peter-Katalinic J, Karas M, Stahl B (1991) Pure Appl Chem 63: 491-498
71. Juhasz P, Costello CE (1992) J Am Soc Mass Spectrom 3: 785-796
72. Bock K, Schuster-Kolbe J, Altmann E, Allmaier G, Stahl B, Christian R, Sleytr UB, Messner P (1994) J Biol Chem, in press

# MATRIX-ASSISTED LASER DESORPTION MASS SPECTROMETRY ON A MAGNETIC SECTOR INSTRUMENT FITTED WITH AN ARRAY DETECTOR

R.S. BORDOLI, K. HOWES, R.G. VICKERS AND R.H. BATEMAN
*Fisons Instruments, Organic Analysis, Analytical MS, Floats Road, Wythenshawe, Manchester, M23 9LE, UK.*

and

D.J. HARVEY,
*Glycobiology Institute, Department of Biochemistry, University of Oxford, South Parks Road, Oxford, OX1 3QU, UK.*

ABSTRACT. Matrix assisted laser desorption ionisation (MALDI) has been carried out on a high performance double focusing mass spectrometer fitted with an integrating focal plane array detector. High sensitivity (low fmol detection) was obtained with peptides such as renin substrate. Underivatised complex oligosaccharides were also detectable at the 200 fmol level. The advantages of the array detector for improved resolution (5000 FWHM) and the benefits in complex oligosaccharide and glycolipid analyses relative to MALDI-TOF systems are shown. This improved resolution enabled ambiguities from the TOF data to be interpreted.

## 1. Introduction

Matrix assisted laser desorption ionisation (MALDI) (1) produces intermittent bursts of ions that are associated with each pulse of the laser beam. Consequently, MALDI is unsuitable for scanning-type mass spectrometers and is more readily associated with time-of-flight (TOF) analysers which detect the entire mass spectrum simultaneously and additionally can exploit the high mass capability of the MALDI technique with good sensitivity. However, the mass resolution offered by TOF instruments is low and can cause problems with mixture analysis when the mass difference between components is small.

The use of an integrating focal plane array detector appears to offer a solution to the combination of laser desorption with a magnetic mass spectrometer (2) for molecules with masses under about 10 kDa. Additionally, it might be anticipated that the better mass resolution afforded by the magnetic sector and array combination will have advantages over the TOF instruments, where, even with a reflectron, mass resolution is typically not greater than 2000 (FWHM).

This work investigates the advantages of using an integrating focal plane diode array detector, on a magnetic sector mass spectrometer for MALD ionisation compared to linear TOF systems.

51

*R. M. Caprioli et al. (eds.), Mass Spectrometry in Biomolecular Sciences, 51–60.*
© 1996 *Kluwer Academic Publishers.*

## 2. Experimental

### 2.1 MASS SPECTROMETER

The mass spectrometer used in this work was a VG AutoSpec-FPD (Fisons Instruments, Organic Analysis, Analytical MS, Manchester, UK). This instrument has an $E_1BE_2$ geometry (Fig. 1.) and a mass range of 5000 Dalton at 8 keV ion energy. An inhomogeneous field ESA mounted between B and $E_2$ allows the beam to be diverted onto a 50 mm chevron microchannel plate (MCP) focal plane array detector, optically coupled to a 2048 channel photo diode array. The dispersion of the inhomogeneous ESA and array detector can be varied to give a mass ratio from 1.2:1 at 2000 resolution (FWHM) to 1.07:1 at 5000 resolution (FWHM). To bring the array detector system into operation requires only that the appropriate voltages are applied to the inhomogeneous ESA, MCP assembly and phosphor. Similarly when the array detector is not required, the inhomogeneous ESA plates are grounded and the ion beam passes through $E_2$ to the point detector. In this way, switching between the array and point detectors is quite rapid. Furthermore, the facility to be able to add a quadrupole, orthogonal acceleration TOF or double focusing magnetic mass analyser after the point detector to form an MS/MS instrument is unimpaired.

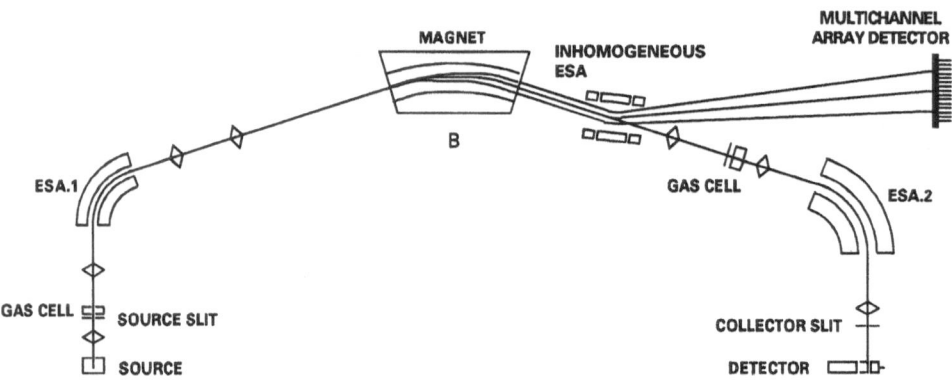

**Fig. 1.** Schematic diagram of the ion optics of the VG AutoSpec-FPD mass spectrometer.

A pulsed UV (337 nm wavelength) nitrogen laser (VSL-337i, Laser Science Inc., Cambridge, MA, USA) was mounted horizontally above the ion source housing. The laser beam was focused by a lens and reflected by an adjustable mirror, through a sapphire window into the ion source housing. The pulse width was 4 nsec (FWHM) with a repetition rate variable up to 20 Hz and an energy of 180 µJ/pulse measured directly using a flux meter. A graduated filter and adjustable iris were available to adjust the spot size and reduce the laser power if necessary.

Laser desorption and ionisation of samples occurred in a standard AutoSpec Fast Atom Bombardment (FAB) ion source using the normal sample stage consisting of a thin stainless steel strip (6 mm long x 1.5 mm wide). A multiple sample stage to allow several samples to be simultaneously loaded into the ion source on the side entry probe has been designed. The initial location of the laser spot was easily achieved by placing a small piece of paper on the sample probe and observing the laser-induced fluorescence through the viewport. Two micrometers attached to the mirror mount could be adjusted to position the laser spot at the top of the sample stage. One micrometer would then move the spot vertically along the length of the stage, whilst the other moved the spot horizontally across the stage. During acquisition it was possible to move the laser up and down in parallel tracks along the sample. In the instrument tuning mode it was possible to manually fire single laser shots or continuously fire with a repetition rate of 1 - 20Hz.

## 2.2. CALIBRATION

The array detector was mass calibrated by stepping a single mass peak (usually $m/z$ 132.9 from CsI in FAB mode) across the detector by adjusting the ion acceleration and ESA voltages appropriately whilst the magnetic field was kept constant. These data were used to construct a calibration curve of relative mass vs channel number. Typically 15 steps were adequate to define this relationship however the software can accommodate a greater or lesser number if required.

Since the same source can be used for both FAB and MALDI, calibration was easily achieved by using the Cs primary ion source and inserting a sample of CsI into the ion source on the sample probe.

During data acquisition, the centre mass (i.e. relative mass = 1.000) of the array for each exposure was defined by reading the Hall probe field sensor in the magnetic field and converting this to mass using a previously recorded Hall probe calibration of the magnetic field. Thus the mass at any point on the array detector is the product of the centre mass for that exposure and the relative mass at that point.

## 2.3. DATA ACQUISITION

Typically exposures of 1 second duration were taken with a laser pulse repetition rate of 20 Hz so that each exposure contained data from 20 pulses. The laser was synchronised with the acquisition cycle so that between exposures the laser was not firing. The system could either repetitively record a single section of the mass spectrum of the same mass range as the detector, or record multiple exposures adjacent to one another to build up larger mass ranges. In this latter case adjacent exposures overlapped slightly so that no data were lost. To improve further the quality of the data several cycles could be averaged together.

Real time displays of the chromatogram of the ion current in each exposure and the mass spectra could both be viewed to monitor the quality of the data. The micrometer drives on the adjustable mirror were used to steer the beam around the sample surface to find the optimum production of signal. With some samples a complete spectrum could be acquired from a single spot, however, with other samples it seemed necessary to move the laser spot more frequently.

## 2.4. SAMPLE PREPARATION

Peptides and glycolipids were from Sigma Chemical Co. (Poole, Dorset, UK). Reference *N*-linked oligosaccharides were obtained from Oxford Glycosystems (Abingdon, Oxon. UK). Ovalbumin oligosaccharides were obtained from hen ovalbumin by hydrazinolysis. Oligosaccharide samples (about 100 pmol) were mixed on the laser target with a saturated solution of 2,5-dihydroxybenzoic acid (2,5-DHB, 1 µl) (3 to 5) in 70% acetonitrile and 30% water, allowed to dry and were then recrystallised from ethanol (5). Peptides were mixed with α-cyano-4-hydroxycinnamic acid (6) dissolved in 70% acetonitrile and 30% water. Glycolipids were examined in both of the above matrices and esculetin (6,7-dihydroxycoumarin) with sample loadings comparable to those used with the oligosaccharides. However, the matrix solutions were made up in 100% acetonitrile to prevent the glycolipids from precipitating. The sample preparation strategy is essentially the same as that adopted in MALDI-TOF experiments.

## 3. Results and Discussion

In the MALDI mode it was not necessary, as in MALDI-TOF mass spectrometry, to operate the laser close to the ion appearance (threshold) power to obtain good resolution and mass accuracy. This is because the double focusing magnetic instrument better accommodates the energy spread of the ions and the temporal spread is of no consequence, thus the power could be increased to maximum to attain higher ion beam currents. There were two consequences of this:

The first was that the sample consumption was higher than with TOF instruments, which necessitated moving the laser spot more frequently although this was not a problem since good quality spectra were always obtained from a single sample loading (once the preparation was optimised). In spite of the higher sample consumption, sensitivity was still good. For example, serial dilutions gave a detection limit of 5 fmol (loaded onto the target) for renin substrate and 200 fmol (5:1 S/N ratio) for an underivatised sample of a typical *N*-linked biantennary oligosaccharide $[(Gal)_2(GlcNAc)_4(Man)_3(Fucose)]$ ($[M + Na]^+ = 1809.6$ Da).

The second consequence of the higher laser power was the appearance of fragment ions such as those from the triantennary oligosaccharide shown in Fig. 2. which are normally absent from the TOF spectrum. Although the data are not presented here, we have observed that under these conditions the water loss peak 18 Da below the molecular ion produced from peptides is larger than normally observed in MALDI-TOF spectra.

The resolution and signal to noise ratio, even under low resolution conditions, were considerably higher than that from TOF instruments, as illustrated by the spectra of *N*-

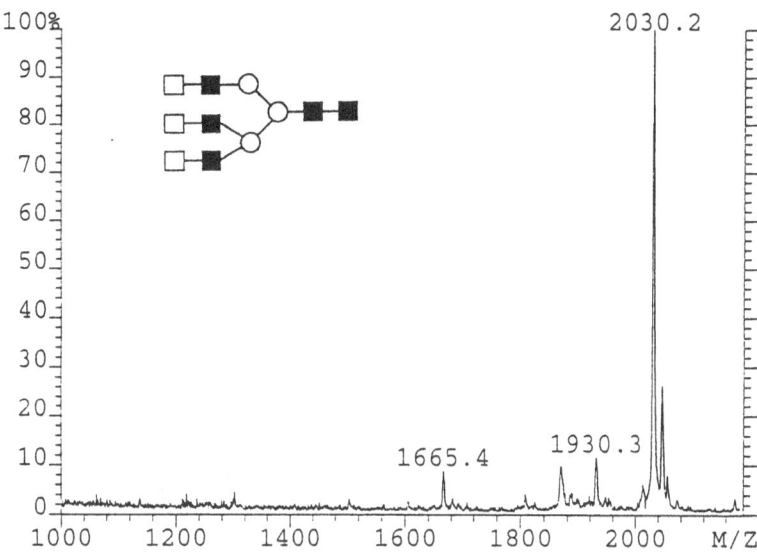

**Fig. 2.** Low resolution MALDI spectrum of the triantennary oligosaccharide (Gal)$_3$(GlcNAc)$_5$(Man)$_3$ (α-cyano-4-hydroxycinnamic acid matrix) . □ = Galactose, O = Mannose, ■ = *N*-acetylglucosamine

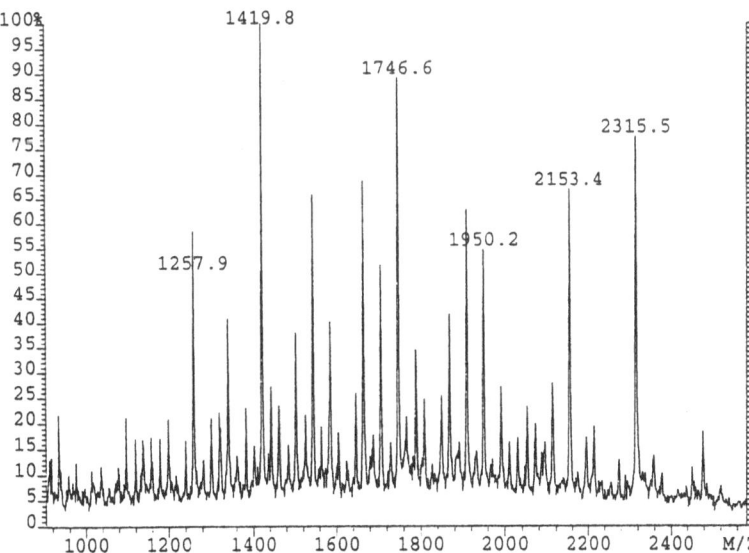

**Fig. 3.** Low resolution MALDI spectrum of *N*-linked oligosaccharides released from ovalbumin (2,5-dihydroxybenzoic acid matrix).

linked oligosaccharides from ovalbumin (Fig. 3). Mass accuracy was typically within 0.5 Da of the accepted values, which was sufficient to assign a composition in terms of the isobaric monosaccharide content below about 1500 Da, but above this figure the possible combinations became sufficiently numerous that mass differences (e.g. 2 Da) between possible structures necessitated operation at higher resolution conditions.

Fig. 4. shows separation of the isotopic cluster from the biantennary oligosaccharide $(Gal)_2(GlcNAc)_4(Man)_3(Fucose)]$ ($[M + Na]^+$ 1809.64 Da) under high resolution conditions to illustrate that this is easily achievable with the array detection system.

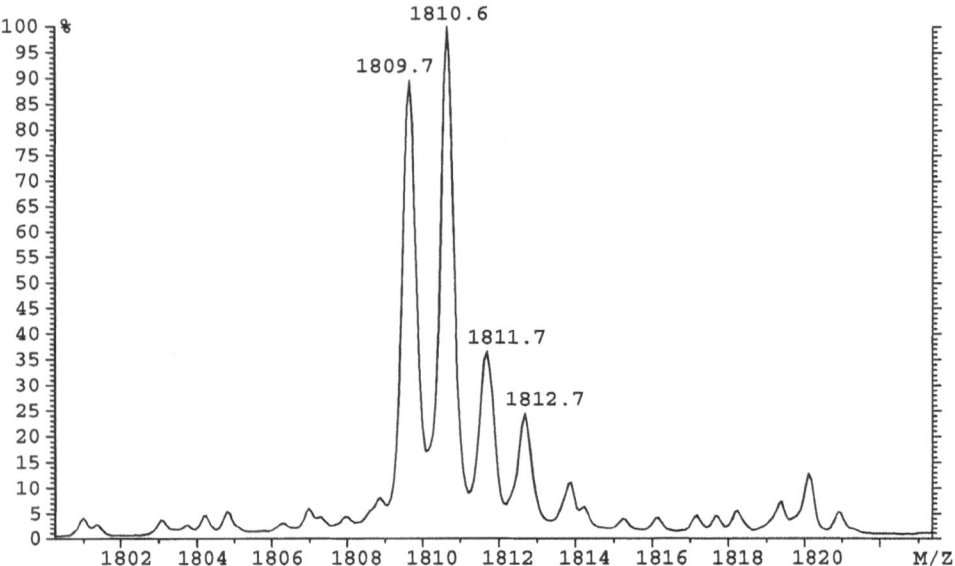

**Fig. 4.** Molecular ion region ($[M + Na]^+$) from the MALDI spectrum of the biantennary oligosaccharide $(Gal)_2(GlcNAc)_4(Man)_3(Fucose)$ (2,5-dihydroxybenzoic acid matrix) recorded with the array detector at 5000 resolution (FWHM).

High resolution was used to confirm the absence of the high-mannose oligosaccharide $(Man)_9(GlcNAc)_2$ (monoisotopic mass for $[M + Na]^+$ = 1905.64 Da) from the peak in the ovalbumin spectrum at $m/z$ 1909. This peak shown at high resolution in Fig. 5. was produced by an oligosaccharide of composition $(Hex)_4(HexNAc)_6$ (monoisotopic mass 1907.67 Da). This differentiation would not be possible with a lower resolution TOF mass spectrometer.

Another area in which the higher resolution of the sector-array instrument proved to be invaluable was in the examination of glyco- and phospho-lipids. These compounds are obtained from biological sources as mixtures containing acyl and sometimes sphingosine bases differing in mass by only two Daltons as the result of varying degrees of unsaturation. No difficulty was experienced in resolving these constituents as illustrated in Fig. 6.

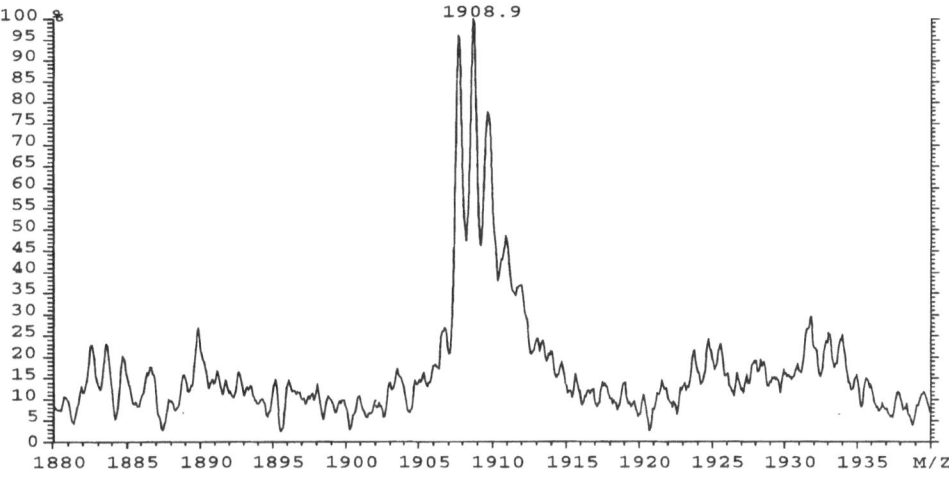

**Fig. 5**. High resolution MALDI data from the peak at $m/z$ 1909 in Fig. 3. showing the absence of $(Man)_9(GlcNAc)_2$ in $N$-linked oligosaccharides released from ovalbumin.

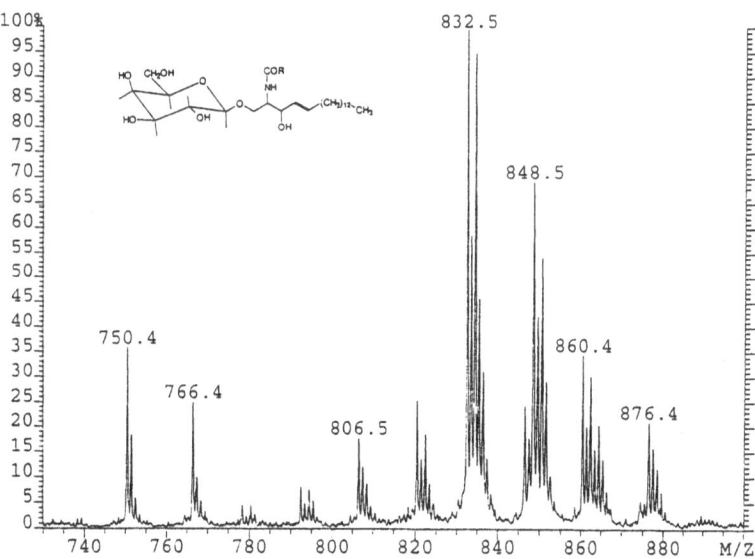

**Fig. 6.** MALDI spectrum of a mixture of galactocerebrosides (types I and II) from bovine brain. Molecular species differing in the degree of unsaturation can be clearly seen.

Furthermore, when examined by TOF mass spectrometry, the spectra of gangliosides were found to contain very broad, unresolved, peaks thought to be due to fragmentation as the result of decarboxylation and loss of sialic acid (Fig. 7) (7 and unpublished observations). When examined with the sector-array instrument, these peaks were completely resolved as shown in Fig. 8.

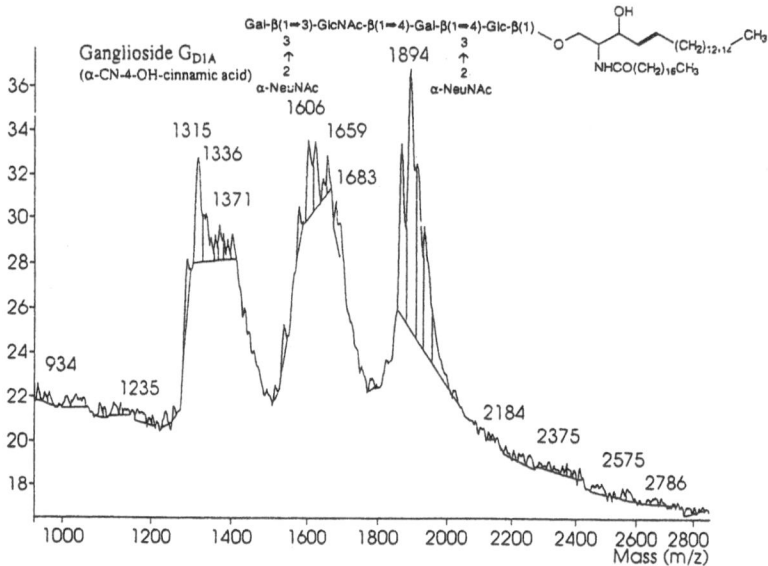

**Fig. 7.** Upper mass region of the MALDI spectrum (α-cyano-4-hydroxycinnamic acid matrix) of disialoganglioside $G_{D1a}$ recorded with a TOF mass spectrometer. The broad peaks from the fragment ions are typical of spectra recorded on linear TOF systems (see also ref. 7). Peak broadening is further enhanced by the formation of sodium and potassium salts by the constituent sialic acids.

## 4. Conclusions

The array detector enabled MALDI spectra to be acquired on a magnetic sector instrument and with the advantage of greatly increased resolution and S:N ratio to that provided by a TOF mass spectrometer. In this study the high mass applications have not been explored, however, it is anticipated that TOF instruments may be more suitable for these analyses since no prior knowledge of the molecular weight range of the sample is required.

Detection limits for the MALDI-array system were 5 fmol for renin substrate ($MH^+$ ion) and 200 fmol for a biantennary $N$-linked oligosaccharide ($[M + Na]^+$ ion). It appears that MALDI produces spectra from small amounts of sample with substantially higher signal to background than would be expected to be obtained by FAB. A second advantage is that by synchronising the laser pulses with acquisition, the sample is being depleted only whilst the signal is being acquired.

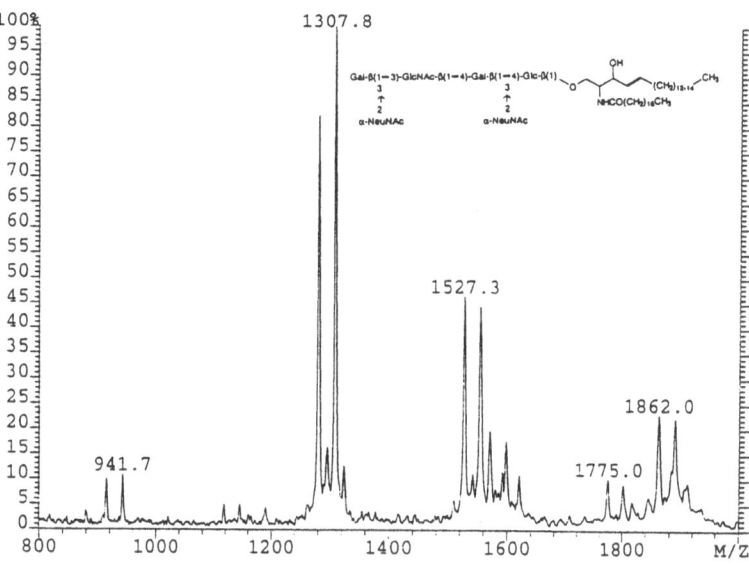

**Fig. 8.** The spectrum of the same compound as in Fig. 7. recorded with the sector-array mass spectrometer showing clear resolution of the molecular ion cluster and fragment ions.

High resolution mass measurements of peaks in complex oligosaccharide mixtures has enabled full structural assignments in terms of isobaric monosaccharide composition to be deduced. This, however, is not to be confused with accurate mass measurement. Linear TOF measurements lack the necessary resolution to be able to separate the peaks of interest in such mixtures to allow even nominal mass measurement.

**References.**

1. M. Karas and F. Hillenkamp, *Anal. Chem.* **60**, 2299 (1988).

2. R.S. Annan, H.J. Kochling, J.A. Hill and K. Biemann, *Rapid Commun. Mass Spectrom.* **6**, 298 (1992).

3. K. Strupat, M. Karas and F. Hillenkamp, *Int. J. Mass Spectrom Ion Processes*, **111**, 89 (1991).

4. B. Stahl, M. Steup, M. Karas and F. Hillenkamp, *Anal. Chem.*, **63**, 1463 (1991).

5. D.J. Harvey, *Rapid Commun. Mass Spectrom.* **7,** 514 (1993).

6. R.C. Beavis, T. Chaudhary and B.T. Chait, *Org. Mass Spectrom.*, **27**, 156 (1992).

7. P. Juhasz and C. E. Costello, *J. Am. Soc. Mass Spectrom.*, **3**, 785 (1992).

# ELECTROSPRAY IONIZATION MASS SPECTROMETRY: PROTEIN STRUCTURE

TERRY B. FARMER and RICHARD M. CAPRIOLI
*Analytical Chemistry Center*
*Department of Biochemistry and Molecular Biology, University of Texas Medical School*
*P.O. Box 20708, Houston, Tx 77225*

ABSTRACT. Electrospray ionization mass spectrometry (ESI MS) has proved extremely useful in the analysis of proteins in addition to the determination of molecular weight. In conjunction with peptide mapping and tandem MS, it has been used to elucidate the primary structure of peptides and proteins. ESI MS analysis alone is not sufficient for examination of secondary structure, but comparison of deuterium exchange experiments with information gained by other analytical techniques provides valuable insight. Protein tertiary structure can be examined by changing solvent temperature, pH, and organic solvent content and by changing instrumental conditions to examine shifts in the number and distribution of charge states in the resulting spectra. Metal-protein interactions have also been successfully analyzed by ESI MS to determine the specificity and stoichiometry of these interactions. Quaternary structure has been resolved for small peptides at high concentrations and for larger proteins utilizing extended m/z mass analyzers.

## 1. Introduction

In the last few years, matrix-assisted laser desorption mass spectrometry (MALD MS) and electrospray ionization mass spectrometry (ESI MS) have revolutionized the study of large biomolecules. The ease of use and accuracy of these techniques for molecular weight determination has led to an acceleration of research in the field of proteins and nucleic acids using mass spectrometry. Recent results suggest that these techniques may also provide a means to study inter- and intramolecular interactions between biomolecules.

This chapter will focus on the use of ESI MS for the analysis of the structure of proteins with examples illustrating work at the primary, secondary, tertiary, and quaternary level, as well as the interaction of proteins with ligands. The goal of this chapter is to present a brief review of the kinds of information that can be obtained and how these experiments are performed. This chapter is not intended to be a comprehensive review, but an introduction for the investigator who is not expert in mass spectrometry but who would like to plan experiments that are amenable to mass spectrometric analysis. Nucleic acids and separation methods combined with ESI MS are covered in other chapters of this book and will not be discussed here.

*R. M. Caprioli et al. (eds.), Mass Spectrometry in Biomolecular Sciences, 61–88.*
© *1996 Kluwer Academic Publishers.*

Dole et al. [1] studied pneumatically generated ions by gas phase mobility, experiments that were the forerunner of modern ESI techniques. In the early 1980's, Fenn and coworkers [2-4] further developed fundamental aspects of the technique and introduced an electrospray source coupled to a quadrupole analyzer. Spectra of polyethylene glycol oligomers showed the generation of a series of peaks that corresponded to successive numbers of charges on the molecular species giving a bell-shaped distribution of multiply-charged ions.

Three processes can be envisioned in the formation of gas phase ions from a liquid sample in the ESI process: production of charged droplets by dispersion and disintegration of the liquid sample at room temperature, reduction in the size of these charged droplets by solvent evaporation, and production of gas-phase ions [5]. The mechanism of the production and/or transfer of ions from the liquid to the gas phase is not yet completely understood. There are several theories that attempt to explain the generation of the charged particles and their distribution. Briefly, in the "ion evaporation" model, solvent evaporation leads to a decrease in the radius of the charged droplet with an increase of the surface field strength through a process of successive decompositions to form smaller droplets and, when the charge density exceeds that allowed for the droplet diameter, ion evaporation occurs. In a second model electrohydrodynamic decomposition of the droplet above the Rayleigh limit is thought to produce small droplets from which ions are formed through desolvation by solvent evaporation, charge condensation, and coulomb repulsion. Thus, the first model describes the emission of gas phase ions from a droplet because of high electric field strengths and the second the production of very small droplets containing a single ion that eventually completely desolvates. Detailed discussions of these models can be found in several recent publications [5-8].

## 2. Determination of Molecular Weight

The monoisotopic molecular weight of a protein can be calculated from its chemical formula by summing the exact atomic masses of the most abundant isotopes ($C = 12.0000$, $H = 1.0078$, $N = 14.0030$, $O = 15.9949$, etc.). The average molecular weight can be calculated in a similar manner using the average atomic weights of the elements ($C = 12.011$, $H = 1.008$, $N = 14.007$, $O = 15.999$, etc.). Currently one of the major uses of ESI MS is the determination of molecular weights of proteins. For low resolution instruments, the average m/z (mass-to-charge ratio) of each of the multiply-charged ions can be used to calculate the average molecular weight of the analyte. High resolution instruments have the added advantage of being able to resolve the individual isotopic peaks in the spectrum and the m/z values of the monoisotopic multiply-charged peaks can be used to calculate the monoisotopic molecular weight. Figure 1 shows the low resolution positive ion ESI mass spectrum of horse skeletal muscle apomyoglobin with the characteristic multiply-charged ion distribution. From the measured m/z values of the ions, the charge of the peaks can be calculated using the following equation:

$$n_2 = \frac{M_1 - M_a}{M_2 - M_1}$$

where $M_2$ and $M_1$ are the m/z of adjacent multiply-charged peaks, $n_2$ is the charge state of $M_2$, $M_2 > M_1$, and $M_a$ is the m/z of the adduct ion (e.g. $(M+H)^+$, $(M+Na)^+$). In the case of the apomyoglobin spectrum where hydrogen (H) is the adduct and $M_1 = 808.22$ and $M_2 = 848.70$, then $n_2 = 20$. Thus, m/z 848.7 has a charge state of $+20$ and is the $[M+20H]^{+20}$ ion. The molecular weight of the protein can be then calculated from the following equation:

$$M_r = n_2(M_2 - M_a)$$

where $M_r$ is the average molecular weight. For the spectrum of apomyoglobin, this example calculation gives $M_r = 16954$. The theoretical molecular weight for horse skeletal muscle apomyoglobin is 16951.49

[9] and thus the measured value is 0.015% (147 ppm) high. It is noted that, for this example, the calculation was made from only two multiply-charged peaks. Computer algorithms are available to perform this calculation or "deconvolute" the mass spectrum to obtain a molecular weight that is the average of many individual calculations over the multiply-charged ion distribution increasing the precision of the measurement [10]. Using the mass spectrum shown in Figure 1, the computer-processed deconvoluted spectrum (inset in Figure 1) gives a molecular mass of 16952.2, a value which is +0.004% or 42 ppm from the theoretical value.

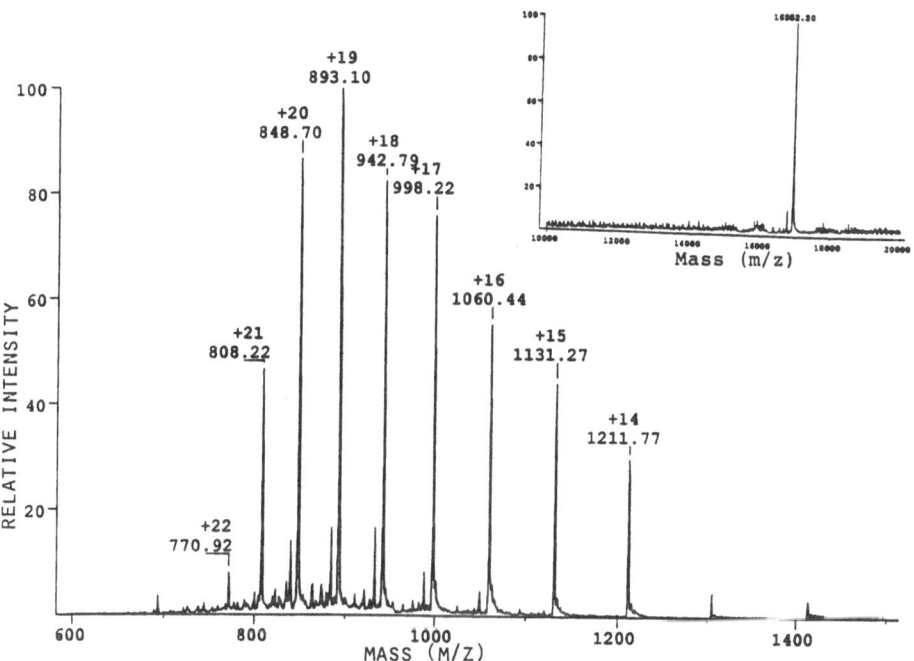

Figure 1. ESI mass spectrum obtained from horse heart apomyoglobin (1 pmol/µl)in 50% methanol with 5% acetic acid by continuous infusion. The inset shows the deconvoluted spectrum obtained from the multiply-charged ion distribution.

In general, molecular weight values obtained by ESI MS have an accuracy of ≤0.01%, although this depends on many experimental parameters. This accuracy is remarkable when compared with the more traditional methods of molecular weight determination such as gel electrophoresis or gel chromatography that have typical accuracies of 1-10%. In addition, the amount of sample used for ESI analysis is in the 0.5-50 picomole range and time of analysis is usually less than 30 minutes.

A large number of proteins have been characterized by electrospray mass spectrometry [11-15].

Molecular weight determinations have been used to identify unknowns, verify an amino acid sequence, identify post-translational modifications, and identify degraded or incomplete recombinant proteins. In addition, it has been shown that under certain conditions, structural features of the protein and protein-protein and protein-ligand interactions can be studied, as discussed later in this chapter. However, ESI MS is not without difficulties. For example, ion formation is sensitive to salt and even moderate salt concentrations (100 $\mu$M) in samples can substantially interfere with the analysis of the compound of interest. Also, several investigators have reported that different solvents, different pH values, and various alterations in instrumental parameters have led to changes in the distributions of charges and the degree of fragmentation observed in ESI spectra. These aspects of the technique will be further discussed below.

## 3. Primary Structure

The primary structure of a protein is the sequence of amino acid residues in the polypeptide chain. The amino acid sequence is most often determined by chemical methods such as the Edman technique, although for many years mass spectrometry has been used in conjunction with enzymatic and chemical digestion to determine the sequence of peptides. Previously, one of the drawbacks of mass spectrometry has been the limited size of the polypeptide that could be analyzed, i.e., the maximum molecular weight range of the mass spectrometer.

### 3.1. PEPTIDE MAPPING

Peptide mapping is a powerful method for protein analysis that involves chemical or enzymatic digestion followed by molecular weight determinations of the resulting fragments. It is used for verification of the primary structure of proteins, identification of sites of post-translational modifications, and identification of mutations. Fast atom bombardment mass spectrometry (FABMS) and plasma desorption mass spectrometry (PDMS) have been extensively used to analyze the molecular weights of the peptides in the protease digests.

The tryptic digest of apolipoprotein AI and subsequent direct analysis of the digest by ESI MS (Figure 2) [16] can be used to show the feasibility of ESI MS for peptide mapping. In general, ions from all the tryptic fragments were observed with the exception of several small peptides of m/z < 400 and fragment $T_{18}$. However, the $T_{17} + T_{18}$ fragment was seen in the spectrum, confirming the presence of $T_{18}$. The overlap of ion peaks is due to incomplete resolution of peptides that are separated by only 1-2 m/z units. Most of the tryptic fragments identified in the spectrum were doubly-charged ions, a consequence of the presence of at least two basic sites on tryptic peptide fragments. Table 1 lists the fragments with their designations in the original protein, i.e. $T_1$, $T_2$, etc., the sequences, the calculated molecular weights and the observed (measured) m/z values.

C-terminal sequencing can be accomplished both by enzymatic and chemical methods. Carboxypeptidases are exopeptidases that selectively cleave the C-terminal residues from polypeptides with free carboxy groups. The amino acid residues released by carboxypeptidases are obscured in ESI MS spectra especially at low sample levels because of the salts in buffers and other low molecular weight compounds that produce ions. However, the truncated polypeptide species can be identified. In one report, an on-line reactor was constructed to monitor the enzymatic digestion of several proteins using carboxypeptidase P and the methodology was demonstrated by sequencing glucagon, a 29 amino acid peptide [17]. Larger proteins, such as cytochrome C, apomyoglobin and carbonic anhydrase, were tested without first being denatured, but only apomyoglobin was partially sequenced. Although not effective with native proteins, exopeptidase digestion would be useful for small proteins and peptides.

Partial hydrolysis of a large protein and isolation of the C-terminal peptide can be used to characterize recombinant proteins to assure that the correct protein has been made. Digestion by carboxypeptidase A and Y was used to determine the C-terminal sequence of interferon-$\gamma$ [18]. Carboxypeptidase A digestion showed two major products which corresponded to the complete sequence and the sequence

Table 1. Sequences, calculated molecular weights[a], and observed m/z values (from Figure 2) of tryptic peptides from human apolipoprotein AI. (Reproduced with permission from Reference 15.)

| Fragment | Sequence | Calculated molecular weight | Observed m/z values (M+H)[+] | (M+2H)[2+] | (M+3H)[3+] |
|---|---|---|---|---|---|
| $T_1$ (1-10) | DEPPQSPWDR | 1226.3 | | 614.1 | |
| $T_2$ (11-12) | VK | 245.3 | 246.8 | | |
| $T_3$ (13-23) | DLAITYTVDVLK | 1235.4 | | 618.7 | |
| $T_4$ (24-27) | DSGR | 433.4 | | 217.0[b] | |
| $T_5$ (28-40) | DYVSQFEGSALGK | 1400.5 | | 701.1 | |
| $T_6$ (41-45) | QLNLK | 614.7 | | 308.9 | |
| $T_7$ (46-59) | LLDNWDSVTSTFSK | 1612.7 | | 807.1 | |
| $T_8$ (60-61) | LR | 287.4 | 287.9 | | |
| $T_9$ (62-77) | EQLGPVTQEFWDNLEK | 1933.1 | | 967.4 | |
| $T_{10}$ (78-83) | ETEGLR | 703.7 | | 352.9 | |
| $T_{11}$ (84-88) | QEMSK | 621.7 | | 312.1 | |
| $T_{12}$ (89-94) | DLEEVK | 731.8 | | 367.1 | |
| $T_{13}$ (95-96) | AK | 217.3 | 219.0[b] | | |
| $T_{14}$ (97-106) | VQPYLDDFQK | 1252.4 | | 627.1 | |
| $T_{15}$ (107) | K | 146.2 | | [c] | |
| $T_{16}$ (108-116) | WQEEMELYR | 1283.4 | | 642.6 | |
| $T_{17}$ (117-118) | QK | 274.3 | | [c] | |
| $T_{18}$ (119-123) | VEPLR | 612.7 | | [c] | |
| $T_{19}$ (124-131) | AELQEGAR | 872.9 | | 437.3 | |
| $T_{20}$ (132-133) | QK | 274.3 | | [c] | |
| $T_{21}$ (134-140) | LHELQEK | 896.0 | | 449.2 | 300.0 |
| $T_{22}$ (141-149) | LSPLGEEMR | 1031.2 | | 516.6 | |
| $T_{23}$ (150-151) | DR | 289.3 | 291.2 | | |
| $T_{24}$ (152-153) | AR | 245.3 | 246.8 | | |
| $T_{25}$ (154-160) | ARVDALR | 781.9 | | 391.8 | 262.6 |
| $T_{26}$ (160-171) | THLAPYSDELR | 1302.4 | | 651.5 | 435.7 |
| $T_{27}$ (172-173) | QR | 303.3 | [d] | | |
| $T_{28}$ (174-177) | LAAR | 430.5 | | 217.0[b] | |
| $T_{29}$ (178-182) | LEALK | 572.7 | | 287.9 | |
| $T_{30}$ (183-188) | EWGGAR | 603.6 | | 303.2[d] | |
| $T_{31}$ (189-195) | LAEYHAK | 830.9 | | 416.6 | 278.6 |
| $T_{32}$ (196-206) | ATEHLSTLSEK | 1215.3 | | 608.5 | 406.3 |
| $T_{33}$ (207-208) | AK | 217.3 | 219.0[b] | | |
| $T_{34}$ (209-215) | PALEDLR | 813.91 | | | |
| $T_{35}$ (216-226) | QGLLPVLESFK | 1230.5 | | 616.1 | |
| $T_{36}$ (227-238) | VSFLSALEEYTK | 1386.6 | | 694.2 | |
| $T_{37}$ (239) | K | 146.2 | | [c] | |
| $T_{38}$ (240-243) | LHTQ | 474.5 | 475.4 | | |
| $T_{14}$ + $T_{15}$ | | 1380.6 | | 691.2 | 461.5 |
| $T_{15}$ + $T_{16}$ | | 1411.5 | | 706.6 | 471.3 |
| $T_{17}$ + $T_{18}$ | | 869.0 | | 435.8 | [e] |
| $T_{20}$ + $T_{21}$ | | 1152.3 | | 577.0 | 385.3 |
| $T_{29}$ + $T_{30}$ | | 1158.3 | | 580.2 | 387.2 |
| $T_{33}$ + $T_{34}$ | | 1013.2 | | 507.0 | 338.9 |
| $T_{36}$ + $T_{37}$ | | 1314.7 | | 758.4 | [e] |

a   Average molecular weight is calculated from the natural isotopic abundance.

b   Peak corresponding to the (M+2H)[2+] ions from $T_4$ and $T_{28}$ were not resolved.

c   Protonated ions from these peptides were not observed. However, ions corresponding to larger incompletely cleaved peptides containing these residues were observed.

d   The (M+H)[+] ion of peptide $T_{27}$ could not be resolved from the (M+2H)[2+] ion of peptide $T_{30}$.

e   The peaks were not resolved completely.

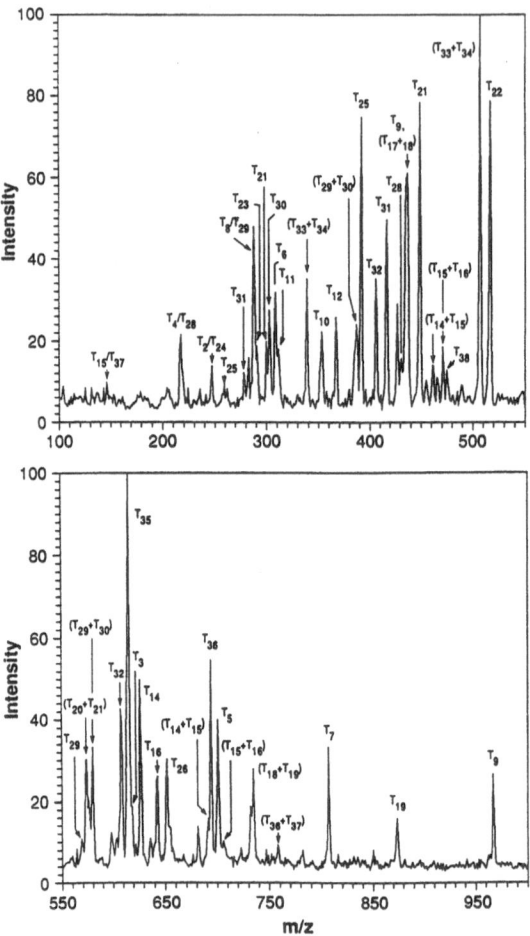

Figure 2. ESI mass spectrum of the unfractionated tryptic digest of human apolipoprotein AI in 50/50 methanol/water with 5% acetic acid. $T_i$ designates molecular ion species of tryptic peptides and these are listed in Table 1. (Reproduced with permission from Reference 15.)

lacking the C-terminal glutamine. Carboxypeptidase Y digestion identified a minor component lacking C-terminal glutamine and serine.

Chemical hydrolysis using high concentration organic acids allows C-terminal sequencing with some highly specific cleavages. Hydrolysis with vapors of 90% pentafluoropropionic acid or heptafluorobutyric acid with subsequent analysis by ESI MS showed a series of successive C-terminal ions [19], as exemplified by the sequencing of magainin 1, a 23 amino acid peptide, as shown in Figure 3. Specific cleavages were observed on the C-terminal side of aspartic acid and the N-terminal side of serine (fragment 1-7). Eight peptides from C-terminal cleavages were observed in the spectrum.

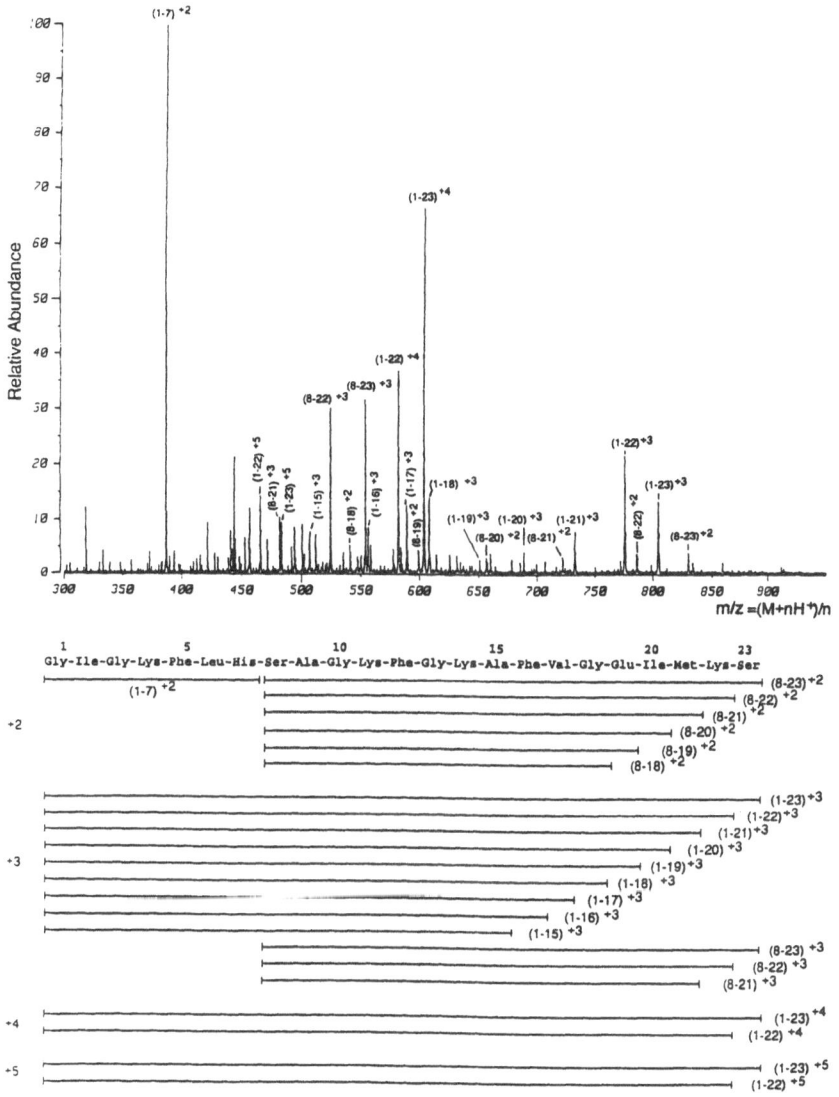

Figure 3. ESI mass spectrum of 125 pmol of the peptide magainin hydrolyzed with 90% $F_5C_2COOH$ in the vapor phase with 50 mg solid dithiothreitol for 2 hours. The dried hydrolysate was dissolved in 25 $\mu$l 50% methanol with 2% acetic acid solution prior to mass spectrometric analysis. (Reproduced with permission from Reference 19.)

### 3.2. TANDEM MS

Tandem MS has been used quite effectively for the sequence analysis of peptides below 3000 Da. Specific enzyme digests of larger proteins followed by FAB mapping and MS/MS of peptide mixtures have become well-documented procedures. However, the inability to handle larger polypeptides and proteins has been a serious limitation. Recent work with ESI MS has indicated that the analysis of larger polypeptides is possible. ESI and MS/MS were used to sequence individual multiply-charged ions of RNase A, as shown in Figure 4 [20]. The majority of fragment ions seen in the spectra are b and y fragments (Roepstorff and Folhman nomenclature [21]) and are due to the cleavage of the CO-NH bond of the peptide backbone. Figure 4A and 4B are the CID spectra of the $(M+12H)^{+12}$ and the $(M+13H)^{+13}$ ions respectively. Figure 4C and 4D are the CID spectra of the same two ions after reduction of disulfide bonds, showing an increase in intensity of product ions upon reduction. The generation of two or more complementary product ion pairs is often observed and can aid in the interpretation of the spectrum. Thus, in Figure 4D, $y_8$ and $b^2_{116}$ are complementary ions for the $[M+13H]^{+13}$ charge state. However, some knowledge of the protein sequence is necessary to assign sequence specific fragmentations for the CID spectra of the intact protein.

Hydrogen-deuterium exchange and ESI and tandem MS have been used in conjunction with computer algorithms for interpreting peptide mass spectra to increase the reliability of sequence determination [22]. The mass shift between undeuterated and deuterated peptides is a direct measure of the number of exchangeable protons. This places restraints upon the number of different amino acid compositions that are possible for a molecular weight. CID of the deuterated peptide helps narrow the possible sequence assignments and aids the computer algorithm in ranking the sequence possibilities.

Dissociation of variants of albumin proteins demonstrated the utility of tandem MS for larger molecular

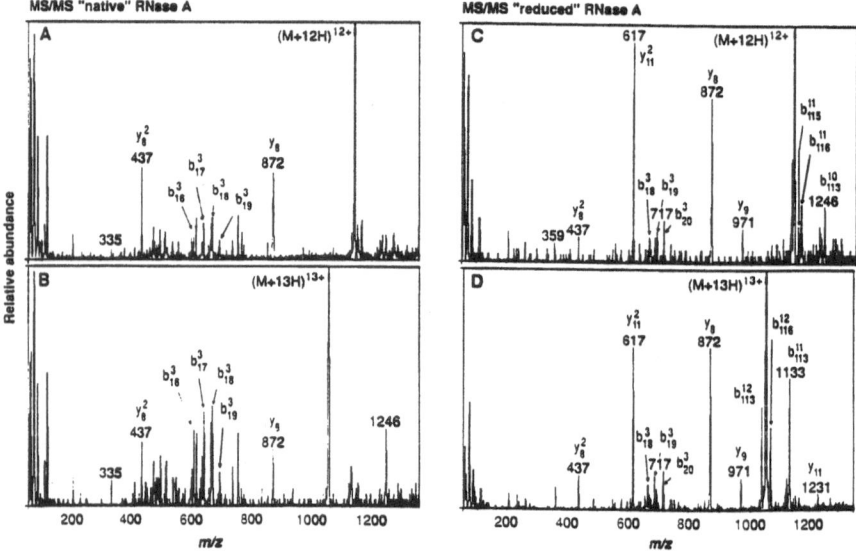

Figure 4. MS/MS CID spectra from A) $[M+12H]^{+12}$ and B) $[M+13H]^{+13}$ peaks of native RNase A and C) $[M+12H]^{+12}$ and D) $[M+13H]^{+13}$ peaks for the reduced form of RNase A. The parent ions undergo collisions with with the energies held constant at approximately 1820 eV. The superscript denotes the charge state with the absence of the superscript indicating a singly-charged ion. (Reproduced with permission from Reference 20.)

weight proteins [23]. Tandem MS of albumins permitted verification of residues which lie between positions 18 and 28 from the $NH_2$ terminus. Conventional Edman sequencing had shown one difference in the initial 30 amino acids between goat and sheep albumin, the presence of His rather than Asn at position 18. ESI MS followed by tandem MS analysis, however, gave identical spectra indicating the presence of Asn at position 18 for both goat and sheep albumins which was confirmed by tryptic digestion with subsequent HPLC analysis and by ESI MS. Mass spectra of this protein from dog, sheep and horse showed that nearly all the more abundant product ion peaks detected have been assigned as multiply charged $b_n$ ions arising from fragmentation within the first 30 residues from the amino terminus.

## 4. Secondary Structure

Secondary structure is the locally ordered three dimensional structure of an amino acid chain which includes $\alpha$-helices, $\beta$-sheets, turns, and random coils. Noncovalent bonds such as Van der Waal's forces, hydrogen bonds, and ionic interactions, help create and stabilize this structure. $\alpha$-Helices are stabilized by hydrogen bonds of the peptide backbone which can be disrupted by changes in pH, solvent and ionic strength. $\beta$-Sheets participate in hydrogen bonding as well, but these are generally interstrand not intrastrand. In a typical protein, 60% of the residues participate in secondary structure as determined by x-ray crystallography. Solution spectroscopy is generally used to determine the presence of $\alpha$-helices and $\beta$-sheets in a protein [24]. Although thus far ESI MS has not been able to distinguish between $\alpha$-helices and $\beta$-sheets, it can help elucidate conformational changes within a protein that occur when compared with data from other techniques.

### 4.1. DEUTERIUM EXCHANGE

Hydrogen-deuterium exchange has been used to characterize the structure of molecules, gain information on chemical and biological reactions, and aid in the interpretation of mass spectra. Amino acid residues typically contain 1-5 exchangeable hydrogens. When proteins are dissolved in deuterated solvents, the exchange of deuteriums occurs for all hydrogen atoms attached to oxygen, nitrogen, and sulfur which are accessible to the solvent. The mass shift between deuterated and undeuterated forms is a direct measure of the number of exchanged hydrogen atoms.

ESI MS spectra of tuna cytochrome C in low pH solution were obtained using normal and deuterated solvents [25]. Like bovine cytochrome C, there are two different distributions of peaks as shown in Figure 5. In $D_2O$, the maxima are believed to relate to two different conformations with molecular weights of 12102 (tight conformation) and 12139 (loose conformation)(Figure 5b) as compared to 11986 in $H_2O$ (Figure 5a). The number of exchangeable protons measured from the spectrum was 116 (tight) and 153 (loose). With the aid of a molecular modeling program, these data were used to postulate that 115 protons were accessible to solvent in the folded protein. Loss of the N- and C-terminal $\alpha$-helical structures lead to 151 exchangeable protons on exposure to the solvent. These results are in good agreement with other reports that cytochrome C loses much of its $\alpha$-helical character under acidic conditions at low ionic strength. The differences between the calculated and the experimental number of exchanged protons may be due to the static structure that is generated in the modeling program, whereas a protein in solution is flexible.

Well-characterized peptide models of growth hormone analogs have been used to study the relationship between primary and secondary structure [26]. Changing the solvent from water to trifluoroethanol or methanol induces the formation of $\alpha$-helices from random coil. Both circular dichroism (CD) and nuclear magnetic resonance (NMR) data show that formation of the helix is primarily dependent on solvent composition. Deuterium exchange was used in conjunction with ESI MS to determine $\alpha$-helical content and this result was then compared to CD data. The exchange of one deuterium for one hydrogen atom results in the increase of one mass unit per exchange site. Peptides that exchanged more slowly were those that formed the helical structure as confirmed by CD measurements. This procedure allows

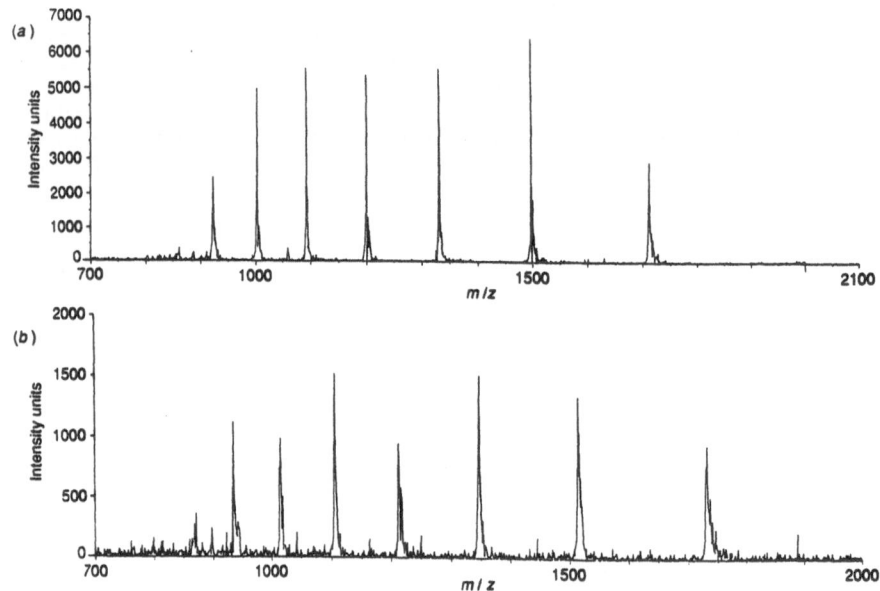

Figure 5. ESI mass spectra of tuna cytochrome C (0.6 mg/ml) obtained in a) 60% $H_2O$/35% $CH_3OH$/5% $CH_3COOH$ and b) 60% $D_2O$/35% $CH_3OD$/5% $CH_3COOD$. (Reproduced with permission from Reference 25.)

measurement of the presence of helical content, but does not identify the location of the helix.

Protein folding intermediates are important in understanding structure and biological activity, but these intermediates often have very short lifetimes, some on the order of milliseconds. Pulse labelling and H-D exchange followed by analysis of the folded protein allows the determination of the number of hydrogen atoms that were exposed at a particular time in the folding pathway. Analysis of the folding behavior of hen egg white lysozyme by ESI MS measured the populations of protein molecules distinguished by different numbers of amides that are inaccessible to the deuterated solvent at various times during the folding process [27]. Different fractions fold at different rates and ESI MS, in conjunction with pulse-labelling, allowed the identification of refolded proteins by molecular weight. The H-D labeling distribution determined by ESI MS, with additional data obtained by NMR, has been used to delineate three species in the folding pathway.

## 5. Tertiary Structure

Tertiary structure refers to the three-dimensional structure of a single chain polypeptide describing not only local (secondary) structure, but also the spatial location of residues. This includes the gross shape of a protein and, for example, the location of a residue on the interior or exterior of a compact globular protein. Forces that stabilize tertiary structure are Van der Waal's forces (1-2 kcal/mol), hydrogen bonding (3-7 kcal/mol), hydrophobic interactions (3-5 kcal/mol, representing free energy required to unfold a nonpolar side chain from the protein interior into aqueous surroundings), electrostatic interactions (3-7 kcal/mol), and disulfide bonds (covalent, 50 kcal/mol). As discussed below, the presence of some of these interactions can be observed using ESI MS.

Noncovalent interactions can be disrupted by the addition of organic solvents (hydrophobic interactions), change in pH (hydrogen bonds, Van der Waal's forces, electrostatic interactions), addition of denaturing agents (hydrophobic interactions), and thermal changes (hydrophobic interactions, hydrogen bonds). Disulfide bonds can be disrupted by reducing agents such as dithiothreitol (DTT). These studies involving the folding and unfolding of the protein molecule are of importance because the behavior of the protein under these conditions and its interactions with ligands offers clues to the mechanism of its biological activity.

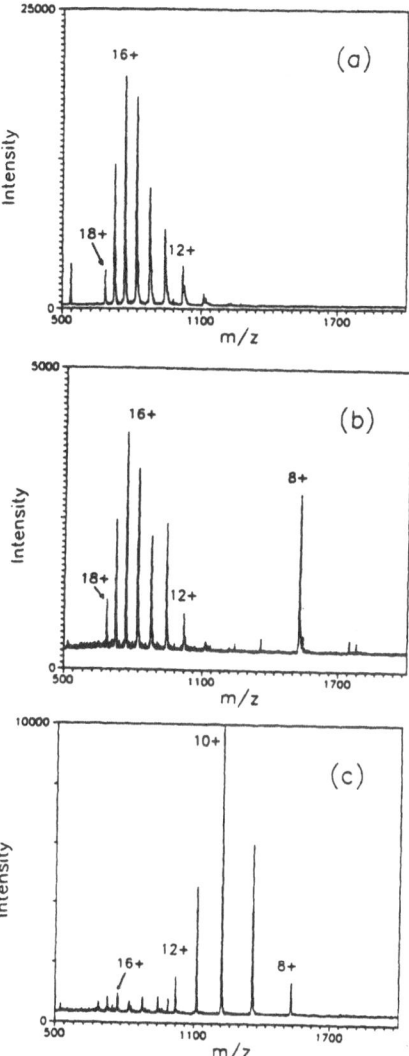

Figure 6. ESI mass spectra of bovine cytochrome C ($1 \times 10^{-5}$ M) with different acid concentrations in aqueous solutions a) 4% acetic acid, pH 2.6,  b) 0.2% acetic acid, pH 3,  c) no acid, pH 5.2. (Reproduced with permission from Reference 28.)

### 5.1. pH CHANGE

A tightly folded protein in the native state has fewer sites available for proton attachment than a protein in the extended, denatured state. ESI MS makes use of this because it is able to measure gas phase charge state distribution, which in some cases reflects liquid state charge distribution. For example, the change in charge states of cytochrome C at different pH values has been demonstrated [28] and is shown for one case in Figure 6. Three charge states were observed: native where +10 is the most prevalent m/z ion (Figure 6c), denatured where +16 is most prevalent m/z ion (Figure 6a), and an intermediate state where +8 is the most prevalent m/z ion (Figure 6b). These three states have been observed using other methods and clearly demonstrate that ESI MS, at least in this case, allows preservation of conditions present in solution.

Other proteins were investigated at acidic and basic pH with positive and negative ESI [29]. Equine myoglobin showed similar distributions at acid and basic pH in the positive and negative ion mode. As another example, native lysozyme gave only a slight shift in the charge distribution in the positive ion spectrum with a change from pH 3 (charge states range from +6 to +10) to pH 10 (charge states range from +6 to +9). No negative ion spectra were obtained possibly due to the lack of anionic sites available for deprotonation in the folded molecule. Reduction of the disulfide bond for lysozyme caused an increase in the charge state distribution at pH 3 for positive ion ESI (charge states range from +6 to +14) and also for the negative ion spectrum (charge states range from -7 to -11), reflecting the more unfolded state of the protein. Reduction at basic pH causes the protein to precipitate so that no spectra of lysozyme were obtained at high pH.

### 5.2 THERMAL DENATURATION

Thermal denaturation has been examined in studies of protein unfolding using ESI MS to analyze eight different proteins that were heated at low pH [30]. Cytochrome C and lysozyme produced changes in charge distribution similar to those seen with the addition of acid. ß-Lactoglobulin and α-chymotrypsinogen showed less significant changes in charge distribution, while other proteins (myoglobin, insulin, and alcohol dehydrogenase) were unaffected by the increasing temperature. The latter cases were attributed to the protein already being completely denatured at pH 3. The maximum temperature needed to denature proteins is a measure of the hydrogen bonding and hydrophobic interactions. At low pH, there is less thermal energy needed to unfold molecules because molecular interactions are already weakened.

Denaturation as a function of temperature and pH was investigated for bovine ubiquitin and cytochrome C [31] using ESI MS. To perform these experiments, an apparatus for controlling and measuring temperature just prior to spraying was constructed. Bovine ubiquitin is a tightly-folded protein resistant to denaturation due to a hydrophobic core with a large portion of molecule involved in H-bonding. Increasing temperature from 25°-75°C produced a change from a single charge distribution (+7 predominant peak) to two distributions (+7, +11). Increasing the temperature further to 93°C completed the transition to the higher charge state distribution (+11). This process has been corroborated by NMR studies, showing that initially the protein is in a tightly folded conformation and as it begins to unfold, the charge state becomes bimodal over a temperature range of 70-95°C with half-denaturation occurring at a temperature of 85°C [32]. Studies of cytochrome C using ESI MS show more complex distributions between 25° and 40°C and between 67° and 75°C at pH 3.6. These transitions also have been suggested from other experimental data obtained with CD and NMR.

Reversibility of heat induced denaturation was studied by ESI MS for bovine cytochrome C [31]. An initial temperature at 25°C gave a spectrum with +8 as the predominant multiply-charged ion (Figure 7a). Increasing the temperature to 90°C for 4 minutes and cooling to 25°C (Figure 7c) resulted in a spectrum showing the same distribution as that of the protein sprayed at 88°C (Figure 7b). Renaturation did not occur during the 25 min cooling period. However, denaturation and renaturation are entirely dependent

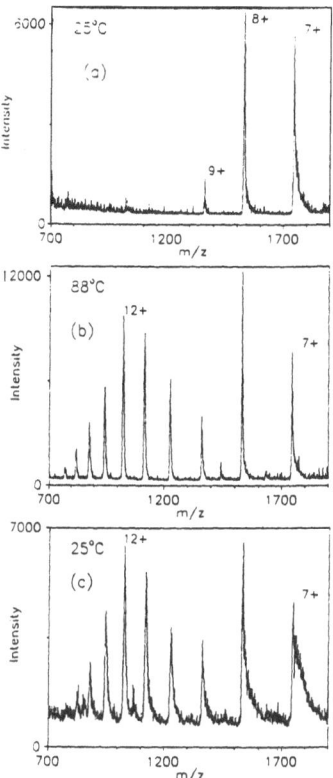

Figure 7. ESI mass spectra of bovine cytochrome C in 5 mM ammonium acetate buffer, pH 6.6, with spray needle solution temperature of a) 25°C and b) 88°C. c) Spectrum obtained after heating the protein solution at 90°C for 4 minutes in a test tube with subsequent cooling to 25°C prior to electrospray analysis. (Reproduced with permission from Reference 31.)

on the nature of the protein itself. For example, bovine ubiquitin showed complete renaturation under these conditions.

Heating a protein sample often produces a mass spectrum that has a different change-state distribution from that obtained at room temperature, a phenomenon that can be of practical importance when using instruments having relatively low m/z range. For example, proteins containing many disulfide bridges are more difficult to denature, and those containing relatively few basic groups may have too few charges to be recorded in the m/z range of the instrument. An example of this is wheat germ agglutinin which has 12 basic groups and a large number of disulfide bridges. The mass spectra at low and high temperature in a denaturing solvent is shown in Figure 8. The protein is stable at 25°C, pH 2.4 with 50% methanol, but no ions are observed in the spectrum up to the instrumental limit of m/z 2000. Upon heating to 70°C, the protein is more highly protonated with the +9 charge state as the dominant ion in the recorded spectrum. The mass spectra of thermolysin and subtilisin Carlsberg were also obtained at elevated temperatures [31] and gave similar results, i.e., a shift to higher charge states (lower m/z) upon heating in a denaturing solvent.

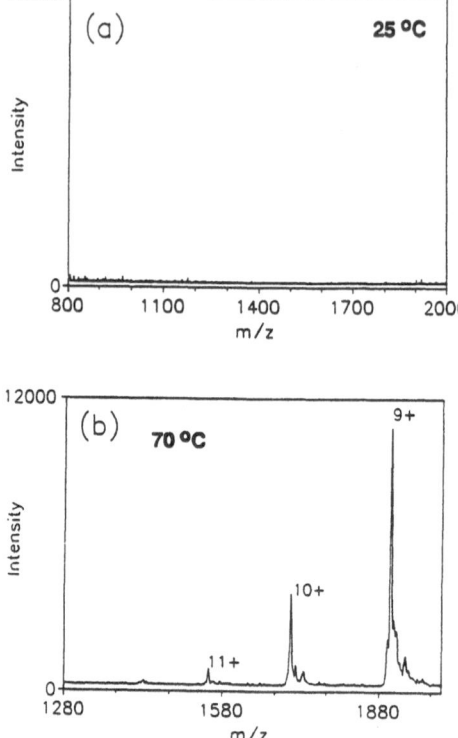

Figure 8. ESI mass spectra of wheat germ agglutinin ($1 \times 10^5$ M) in 50% methanol with 5% acetic acid obtained at a) 25°C and b) 70°C. (Reproduced with permission from Reference 31.)

## 5.3. ADDITION OF ORGANIC SOLVENTS

Solvent-induced conformational changes in proteins were studied with ubiquitin [33] employing different amounts of acetonitrile (18%, 12%, 0%). Shifts in the m/z of ions produced by ESI were observed as shown in Figure 9. A comparison of the effect of different solvents in changing the charge distribution of bovine ubiquitin showed that acetonitrile and isopropanol were the most effective in denaturing the protein to give a higher charge state. At 0% acetonitrile (Figure 9c) some of the lower m/z distribution is visible in the spectrum. Acetonitrile was used as a sheath liquid and it acts as an organic solvent on the sprayed solution causing some lower m/z distribution to be apparent. When water was used as a sheath liquid (Figure 9d), the lower charge state distribution disappeared and only the m/z distribution reflecting the native form of the protein was seen in the spectrum.

In one study [34], isopropanol appeared to be more effective than methanol in denaturing myoglobin. However, it has been suggested that the most important factor in this regard is the decrease in the dielectric constant of water through addition of solvent. Indeed, this was the case in a study of the measurements of the helical content of ubiquitin [35]. Table 2 shows the dielectric constants for solvents

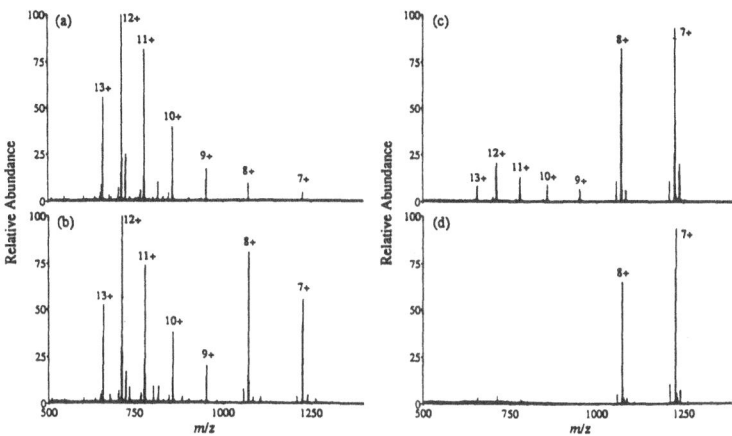

Figure 9. ESI mass spectra of bovine ubiquitin in a) 18:77:5, b) 12:83:5, and c) 0:95:5 acetonitrile + water + acetic acid with acetonitrile as the sheath liquid. d) Same sample conditions as c., but water was used as the sheath liquid. (Reproduced with permission from Reference 33.)

Table 2. Organic Solvents used for ESI MS.

| Solvent | Dielectric Constant | % needed for denaturation of bovine ubiquitin [35] |
|---|---|---|
| Methanol | 32.70 | 36-42 |
| Acetonitrile | 37.50 | 12-18 |
| Isopropanol | 19.92 | 12-18 |
| Acetone | 20.70 | 18-24 |
| Water | 78.30 | ----- |

used in ESI MS as well as the experimentally derived mixture proportions needed for denaturation of bovine ubiquitin [31,36].

Differences in the exchange of labile protons have been used to identify folding intermediates for solvent induced conformational changes of cytochrome C [37] and RNase A [38] when analyzed by NMR. Only recently has mass spectrometry been used in such experiments and ESI MS was employed to study conformational changes of bovine ubiquitin using H-D exchange [39]. In $D_2O:CH_3OOD$ (99:1) CD and NMR data suggested a tight conformation and, after the addition of $CH_3OOD$, large conformational changes associated with unfolding were observed. Ubiquitin has a total of 144 protons of which 73% exchanged after 90 minutes for the native conformation, and 96% exchanged after 60 minutes for the unfolded form. The maximum charge state of +8 observed in the ESI spectrum of the folded protein indicates that, on average, only 7 or 8 basic sites appear to be protonated. This is in agreement with NMR data that showed that the N-terminus and four lysine residues are involved in hydrogen bonding through salt bridges. Unfolded ubiquitin, on the other hand, has a maximum of +13 charges as determined by ESI MS. The location of the exchanged deuteriums cannot be determined from the molecular ion distribution, but analysis by tandem MS and chemical or enzymatic digestion can provide such information.

The same trend was seen with tuna cytochrome C in deuterated solvents [25]. In a 35% organic/5% acetic acid solution, the ESI spectrum shows two distinct distributions with 18 residues protonated out of a total of 25 basic groups, i.e., seven basic groups appear to be shielded. In deuterated solvents, the protons of the protein were calculated to be 116 unexchanged and 153 exchanged. A molecular modeling program was used to calculate the ionization state of each residue at the pH employed and it was concluded that 115 exchangeable protons were available out of a total of 209, providing good agreement with the experimental data.

## 5.4. DISULFIDE BONDS

Reduction of disulfide bonds in proteins results in a change in the ESI spectrum from a higher m/z distribution for the native, unreduced state to a lower m/z distribution for the reduced protein. This is attributed to the unfolding of the protein and subsequent protonation of groups that were previously shielded. For example, native lysozyme, which contains four disulfide bonds, shows a multiply-charged ion of +13 to be the most abundant at pH 3 [29]. After reduction with DTT there is a decrease in the m/z value of the maximum charge state as well as an increase in the number of charge states as shown in Figure 10.

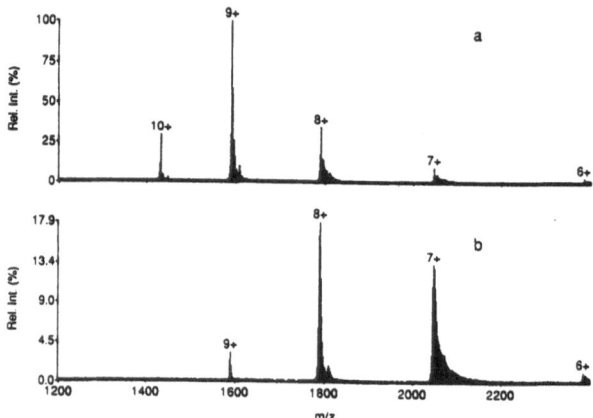

Figure 10. ESI mass spectra of hen egg white lysozyme at pH 3 in 50% methanol in A) native form and b) disulfide reduced form. (Reproduced with permission from Reference 29.)

Another study examined native lysozyme and bovine serum albumin in 5% acetic acid [40]. For lysozyme the initial distribution showed +10 to +14 charge states in the spectrum and upon addition of DTT the maximum charge state seen in the spectrum increased to +20. The increase in the number of charge states was also seen for bovine serum albumin. It was noted, however, that for the larger proteins (≥10,000 daltons) the number of charge states seen is less than the number of basic residues present in the protein. The authors note that this may be due to incomplete reduction or to a situation where there is too high a charge concentration on a small region of the droplet surface and taxing the electrospray process. The spectrum of RNase A, a protein having four disulfide bonds, from m/z 500-1300 shows [M+15H]$^{+15}$ to be the highest charge state seen with the +12 multiply-charged species being the most prominent peak in the spectrum. After reduction with DTT, the +23 charge state is the highest detected with the +16 ion being the predominant peak in the spectrum. The folding of RNase A has been well studied by a variety of techniques and the increase in charge state is interpreted as a consequence of a more extended conformation after reduction [20].

## 5.5. HEME-GLOBIN INTERACTION

Some native proteins are associated with low molecular weight cofactors or prosthetic groups that can be either noncovalently or covalently bound. Denaturation or other changes in the conformation of the protein weakens non-covalent interactions. One of the well-studied interactions is that of globin proteins with heme. Myoglobin and hemoglobin have noncovalently bound heme groups which are bound in a hydrophobic pocket of the protein by Van der Waal's forces. Acid denaturation of horse and whale myoglobin occurs between pH 4.5-3.5 with subsequent loss of the heme. Figure 11a shows the ESI spectrum at pH 3.35 with only one charge state distribution corresponding to apomyoglobin. In the spectrum of this protein at pH 3.9, shown in Figure 11b, two charge distributions are observed corresponding to the holoprotein (protein + heme) and apoprotein [41].

Studies of folding and unfolding of myoglobin using ESI MS were reported [42]. The loss of the heme was observed upon acidification and refolding with reattachment of the heme was initiated by addition

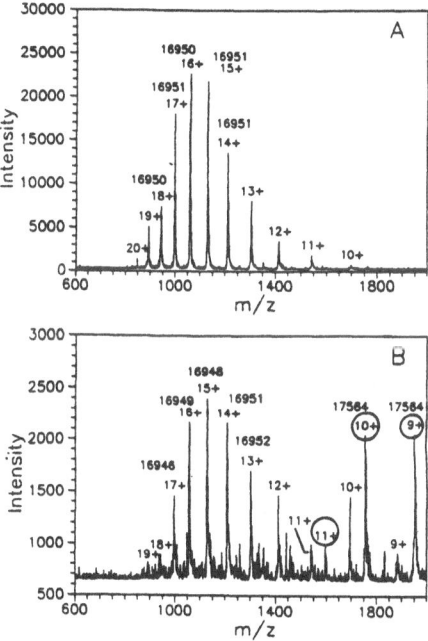

Figure 11. ESI mass spectra of equine skeletal muscle myoglobin (20-40 μM) obtained at a) pH 3.35 and b) pH 3.9. The circled charge states designate peaks corresponding to the intact heme-globin complex in myoglobin and the most intense peaks are also labeled with molecular masses derived from measured m/z values. (Reproduced with permission from Reference 41.)

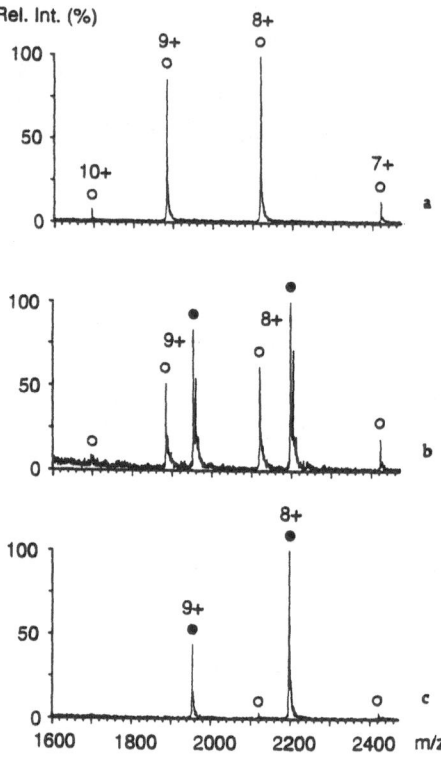

Figure 12. Partial ESI mass spectra of equine myoglobin. Ammonium hydroxide was added to the denatured protein solution in 10% acetic acid to give a) pH 5, b) pH 6, and c) pH 8. (•) indicates peaks due to noncovalent reattachment of the heme group. (Satellite peaks on the right side of the peaks in b are due to acetate adducts.) (Reproduced with permission from Reference 42.)

of ammonium hydroxide to pH 8. At pH 5 (Figure 12a), a native charge-state distribution was observed. Upon increasing the pH to 6, a charge-state distribution corresponding to myoglobin with the heme bound appeared (Figure 12b) and heme binding seemed to be complete at pH 8 (Figure 12c). To determine if the role of heme in folding and unfolding is active or passive, one experiment was done in which heme was extracted from myoglobin with solvent after acidification. Upon addition of ammonium hydroxide to adjust the pH, there was a shift in the charge distribution similar to that of the holoprotein. Although these data do not provide proof of the final structure of the refolded protein, it does suggest that the presence of the heme was not critical for this folding pathway.

For native myoglobin, +9 is the highest charge state seen in the spectrum with only minor changes in the charge distribution profile occurring between pH 5-10. Denatured myoglobin can have +25 as the highest charge state in the spectrum. There are 32 basic amino groups and 21 acidic groups in myoglobin and all but the two histidines involved in heme binding are found on the surface as determined by x-ray crystallography. It is possible that the +9 charge state reflects the difference between the 30 positive charges and the 21 negative charges while the other charge-pairs are involved in salt bridges.

## 5.6. METAL-PROTEIN INTERACTIONS

Transition metals such as iron, manganese, cobalt, and nickel are important in biological regulation since they bind and stabilize protein structure and also sometimes play a functional role. Transition metals are not found in large amounts in the cell and are bound to other molecules because they are toxic. Some proteins require noncovalently bound alkali divalent cations (e.g. magnesium and calcium) for activity and these are present in relatively high concentrations in the cell. Such interactions are of great interest to biochemists: if a macromolecule binds metal, can it be removed, and what is the binding specificity and stoichiometry? ESI MS has been shown to be useful in such studies because it can provide information on the effect of pH, changes in temperature, ionic strength, and solvent dielectric constants for metal-protein interactions.

Cys-metal binding was studied by synthesizing estrogen receptor DNA binding domain, a 71 amino acid protein containing two clusters of 4 Cys. At acidic pH, the ESI mass spectrum of the apopeptide showed that the predominant ion was the $+10$ charge state [43]. Upon the addition of $Zn^{++}$, ions were observed corresponding to peptide with 0, 1, and 2 bound $Zn^{++}$, as shown in Figure 13, where the predominant ion is again the $+10$ charge state. With fully or partially oxidized Cys no change in the charge state distribution occurred and no specific peptide/metal stoichiometry (1 peptide to 2 metal cations) was seen if zinc was added to the peptide. The binding of $Cu^{++}$ by this same protein was examined as well [44], because $Cu^{++}$ inhibits its activity. Upon binding of 4 $Cu^{++}$ atoms, the dominant charge state changed to $+8$, as compared to $+10$ for the apopeptide. $Cu^{++}$ binding stoichiometry was also verified by atomic absorption measurements. Because of the change in charge state distribution and the difference in the number of copper and zinc atoms bound to this protein, a metal-induced conformational change that affects activity has been postulated [45].

A portion of the histidine rich glycoprotein with the sequence $(GHHPH)_5G$ has been synthesized to investigate copper binding [46]. The molecular weight of the apoprotein was determined by ESI MS. The holopeptide showed the formation of four new peaks that differed by an average of 64.3 Da corresponding to the binding of up to 5 $Cu^{++}$ ions. Decreasing the pH decreased the number of $Cu^{++}$ ions bound to 1-2 at the highest acid concentration. Spectrophotometric titration of histidine rich glycoprotein in water with $CuSO_4$ showed a maximum of 5.1 $Cu^{++}$ ions bound per mole of peptide and

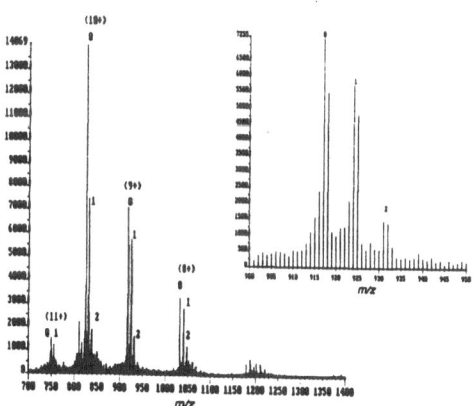

Figure 13. ESI mass spectra of the 71-residue estrogen receptor DNA binding domain peptide (8 nmol/ml) occupied with bound zinc(II) atoms. The electrospray solution contained 80 $\mu$M $ZnSO_4$ in water. Peaks representing peptides with 0, 1, or 2 bound Zn atoms are indicated. The inset shows the $+9$ charge state data in greater detail. (Reproduced with permission from Reference 43.)

TABLE 3. Estimated and calculated molecular masses for Ncp7, its N-terminal peptide, Ncp7 (1-35), and its C-terminal peptide, Ncp7 (29-55), fragments. Reproduced with permission from Reference 49.

| Sample | Peak (m/z) | Absolute intensity | Positive charge | Mass measured (Da) | Final measured mass (Da) | Standard deviation | Calculated mass (Da) |
|---|---|---|---|---|---|---|---|
| Ncp7 | 645.5 | 390 | 10 | 6444.92 | 6444.22 | 0.40 | 6444.53 |
| | 717.0 | 1470 | 9 | 6443.93 | | | |
| | 806.5 | 1380 | 8 | 6443.94 | | | |
| | 921.6 | 670 | 7 | 6444.15 | | | |
| | 1075.0 | 230 | 6 | 6443.95 | | | |
| | 1289.9 | 80 | 5 | 6444.46 | | | |
| Ncp7 + 2Zn²⁺, 4H⁺ complex | 731.0 | 85 | 9 | 6569.93 | 6567.75 | 0.29 | 6568.53 |
| | 822.3 | 285 | 8 | 6570.34 | | | |
| | 939.7 | 1015 | 7 | 6570.84 | | | |
| | 1096.1 | 1215 | 6 | 6570.55 | | | |
| | 1315.2 | 470 | 5 | 6570.96 | | | |
| Ncp7-(1-35)-peptide | 514.7 | 620 | 8 | 4109.54 | 4109.19 | 0.26 | 4109.86 |
| | 588.0 | 5000 | 7 | 4108.94 | | | |
| | 685.9 | 5160 | 6 | 4109.35 | | | |
| | 822.8 | 1580 | 5 | 4108.96 | | | |
| | 1028.3 | 550 | 4 | 4109.17 | | | |
| Ncp7-(1-35)-peptide + Zn²⁺, 2H⁺ complex | 597.0 | 600 | 7 | 4171.94 | 4171.91 | 0.10 | 4171.86 |
| | 696.3 | 5420 | 6 | 4171.75 | | | |
| | 835.4 | 3130 | 5 | 4171.96 | | | |
| | 1044.0 | 990 | 4 | 4171.97 | | | |
| Ncp7-(29-55)-peptide | 630.5 | 3020 | 5 | 3147.46 | 3147.07 | 0.36 | 3147.66 |
| | 787.7 | 4580 | 4 | 3146.77 | | | |
| | 1050.0 | 1340 | 3 | 3146.98 | | | |
| Ncp7-(29-55)-peptide + Zn²⁺, 2H⁺ complex | 643.0 | 580 | 5 | 3209.96 | 3209.54 | 0.44 | 3209.66 |
| | 803.4 | 3680 | 4 | 3209.57 | | | |
| | 1070.7 | 850 | 3 | 3209.08 | | | |

quantitative chromatography as well as MALD MS verified stoichiometry and binding. The specificity of $Cu^{++}$ binding was examined by adding $Mn^{++}$. No peptide bound $Mn^{++}$ was seen in the ESI spectrum, and thus the binding for $Cu^{++}$ is thought to be specific and not a gas-phase event.

Metallothioneins bind up to seven atoms of divalent ($d^{10}$) transition metals such as copper, zinc, and cadmium. Removal of the metal ions can be accomplished by chelation or acidification. Apometallothionein shows a +6 charge state as the dominant ion in the ESI spectrum, while that of metallothionein shows seven $Cd^{++}$ ions bound. NMR and X-ray data show that the protein is folded when fully bound with metal. This and other data obtained by ESI indicated that the conformation of the protein is retained during ESI MS analysis [47].

Synthetic nucleocapsid protein (Ncp-7) of HIV-1 has two $Zn^{++}$ binding domains containing His which bind two equivalents of $Zn^{++}$ tightly and stoichiometrically to induce formation of folded domains similar to zinc fingers. At pH 5.0, His residues are protonated and can no longer bind zinc ions as coordinating ligands. The ESI MS spectrum of the apoprotein shows the +9 charge state to be the base peak for Ncp-7 without the addition of zinc [48]. Increasing the zinc concentration changes the charge state so that the +6 state is the base peak with 0, 1, and 2 atoms of zinc bound. A small amount of 3 zinc atoms bound suggested a third binding site. Two shorter fragments of Ncp-7 were synthesized, each containing one zinc binding domain: Ncp-7 (1-35) and Ncp-7 (29-55) (see Table 3). Ncp-7 (1-35) without zinc had the +5 and +6 charge states as dominant peaks in the spectrum which shifted to +5 with the addition of $ZnSO_4$. The stoichiometry of binding was 1:1. Tandem MS of the +6 state showed little formation of the apopeptide molecular ion species, indicating the metal-Ncp-7(1-35) complex is very stable, and that the third potential binding site was on the Ncp-7(29-55) fragment. Ncp-7(29-55) without zinc showed the +4 charge state as the dominant peak in the spectrum which did not change with the addition of even higher zinc concentrations, suggesting a lower metal binding affinity for this fragment. An additional series of peaks that seemed to result from the complexing of a second $Zn^{++}$ was seen in the spectrum of NCp-7(29-55). Substitution of $Ca^{+2}$ for the second $Zn^{+2}$ was readily accomplished indicating nonspecific binding with probable attachment of the cations to the carboxyl group of the C-terminus or another acidic residue.

## 6. Quaternary Structure

The quaternary structure of a protein is the arrangement of the covalent or noncovalent subunits of the protein complex. The simplest quaternary structure is composed of two identical subunits with the same primary and tertiary structure. The techniques often used to determine the number of subunits found in a protein complex are x-ray crystallography and electron microscopy. Hydrodynamic data of an intact protein can provide the molecular weight with accuracies in the range of 1-10%. Upon denaturation with SDS or reducing agents, the molecular weight of the subunits and the stoichiometry can be determined.

Oligomeric proteins with more than one type of subunit often show stepwise association and dissociation. For example, the hemoglobin tetramer dissociates as follows: $\alpha_2\beta_2 \rightarrow 2\ \alpha\beta \rightarrow 2\alpha + 2\beta$. Van der Waal's forces predominate in interactions between subunits with a single Van der Waal's contact equal to about 100-500 calories. Generally there are over 100 contacts to form a subunit interface so Van der Waal's forces contribute 10-50 kcal to subunit association. The typical association constant, $K_D$, is equal to approximately $10^{-8}$ to $10^{-16}$ moles/l, which corresponds to -11 to -22 kcal/mol at 25°C. Loss of entropy occurs upon subunit association that corresponds to 20-30 kcal of free energy at room temperature. Therefore, a negative free energy ($-\Delta G°$) of -30 to -50 kcal/mol is needed to form these complexes with most of this energy coming from hydrophobic interactions [24].

Factors that differentiate between specific and nonspecific associations have been considered using ESI MS [49]. The most dominant peak in the spectrum should show a reasonable stoichiometry with little or no aggregation visible in spectrum. Noncovalent interactions are dissociable under harsher interface conditions and MS/MS and these can be used to distinguished covalent interactions. Complex dissociation by solvent modification (change in pH, temperature, solvent, etc.) should also produce a change in the

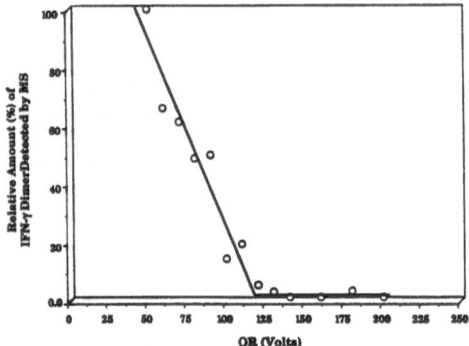

Figure 14. Effect of changing the voltage at the atmosphere to vacuum transition on the gas-phase IFN-γ noncovalent complex showing percentage of IFN-γ dimer detected at different orifice voltages. (Reproduced with permission from Reference 50.)

spectrum. ESI should be sensitive to modifications of components in the complex, i.e., variants of one component should produce a change in the relative intensity of ions and perhaps the overall charge state of the complex.

The appearance of multimers in the ESI spectra of proteins at relatively high concentration can be observed. The spectrum of glucagon at 100 $\mu$M concentration shows formation of the dimer $(2M+7H^+)^{+7}$. The CID spectrum of the dimer species at low energy (800 eV) show $(M+4H^+)^{+4}$ and $(M+3H)^{+3}$, complementary ions confirming the dimer peak. Similar results showing formation of dimer were obtained for bovine insulin B-chain (oxidized) and A-chain [49]. Gentle conditions, both in solution

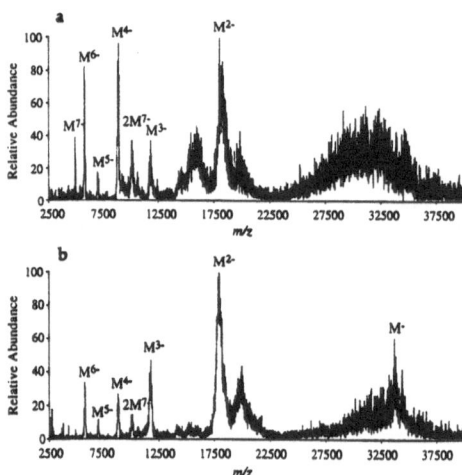

Figure 15. Negative ion ESI mass spectra of porcine pepsin in 5% acetic acid with countercurrent $N_2$ flow at a) room temperature, and b) heated to 80°C. (Reproduced with permission from Reference 51.)

83

and at the electrospray interface, are necessary to observe these multimers. Because of the high concentrations of proteins used, these multimers are not considered to have physiological significance.

Protein-protein interaction of interferon-γ (IFN-γ) has been reported using ESI MS [50]. IFN-γ is a homodimer with monomer molecular weight of approximately 17 kDa. Analysis of IFN-γ under normal conditions (50% methanol/0.1% TFA at pH 2.5) resulted in a spectrum with an ion distribution from +8 to +22 charge states and a molecular mass corresponding to the monomer. At pH 6.7 in distilled water, this distribution included charge states from +8 to +19. The pI of IFN-γ is 9 and when the ESI spectrum was obtained at pH 9 in 30% NH₄OH, the resulting spectrum showed evidence of dimer peaks that correspond to the dimeric form of IFN-γ. One of the characteristics of the spectrum of a homodimer is that although the even number charge states for the dimer overlap those of the monomer, the odd charge states for the dimer are distinct in the spectrum. Because random aggregation has been shown to occur for proteins such as cytochrome C due to concentration effects, dilution experiments for IFN-γ were performed with and without cytochrome C present in the solution. IFN-γ dimer ions were also detected at low concentrations while no cytochrome C dimers or IFN-γ/cytochrome C peaks were observed, indicating that nonspecific interaction between the proteins were not significant. Dissociation of the IFN-γ dimers was also investigated by changing the voltage at atmosphere-to-vacuum transition. This has the effect of imparting more energy into the ions which eventually would tend to dissociate them and even lead to some fragmentation. Dissociation was initiated at 60v with complete dissociation at ≥ 130 V as shown in Figure 14. Thus, little energy was required to induce gas-phase dissociation.

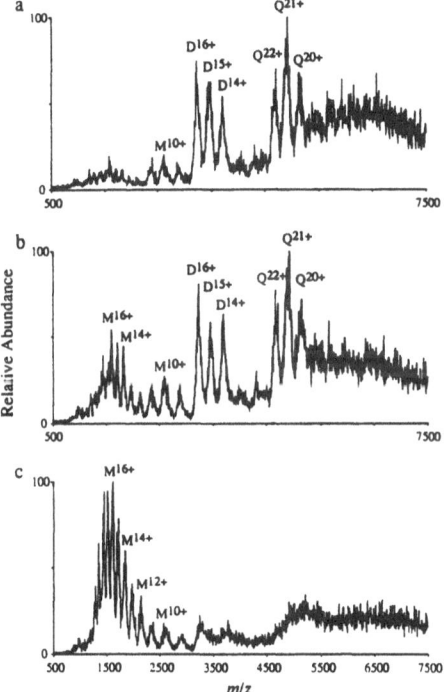

Figure 16. ESI mass spectra of concanavalin A in aqueous 10 mM NH₄OAc (pH 6.7) using an extended m/z range quadropole mass spectrometer with a) capillary temperature of 160°C, b) capillary temperature of 170°C, and c) capillary temperature of 185°C. (Reproduced with permission from Reference 52.)

Investigation of protein-protein interactions using an extended m/z range instrument was investigated [51] because of the low number of multiply-charged ions in some molecular complexes. The results were quite interesting in that high mass-to-charge ratio ions (> 4000) were observed for bovine cytochrome C and porcine pepsin using a quadropole mass spectrometer with a m/z 45,000 range. Cytochrome C in an aqueous solution gives a charge state distribution from + 12 to +2, and protein aggregates were observed between m/z 2000 and 6000 that correspond to dimers, trimers, tetramers, and pentamers. Addition of 10 mM ammonium acetate to the solution significantly decreased the extent of aggregation, as expected for noncovalent complexes.

Porcine pepsin (35 kDa), a highly acidic protein, was analyzed in 1% acetic acid (pH 2.5) in order to observe high mass-to-charge ratio negative ions. The charge state distribution was -7 to -2, substantially lower than the maximum charge state of -42 obtained in basic solution. Dimer and trimer aggregate peaks were also observed along with broad ion distributions at m/z 16,000 and another from m/z 25,000 to 40,000. This is shown in Figure 15 with the broad ion distributions probably the result of molecular ion dissociation products, higher order aggregates, or incomplete desolvation. By increasing temperature on the countercurrent gas ($N_2$), these peaks were reduced supporting the conclusion that they are probably due to incomplete desolvation [51].

Concanavalin A (Con A) is a homotetramer with a monomer molecular weight of 25.5 kDa. It has a pH dependent solution equilibrium between dimer and tetramer. The ESI spectrum of Con A in 10 mM $NH_4OAc$ (pH 6.7) was obtained on an extended m/z range quadropole mass spectrometer with both dimer and tetramer visible in the spectrum shown in Figure 16 [52]. Only dimeric and tetrameric peaks were observed with no trimer or pentamer peaks, suggesting a specific interaction. Changing the inlet capillary temperature produced dissociation to the monomeric species, indicating the interaction is noncovalent. These data strongly support the proposal of association of this protein.

## 7. Conclusions

Chemistry, biochemistry, and molecular biology have benefitted greatly from the development of ESI mass spectrometry as a useful analytical tool, not only for proteins and their interactions with other molecules and ligands, but also glycoproteins, lipoproteins, nucleic acids, and other classes of molecules. ESI has been successfully used both in the positive and negative ion mode to obtain structural information on biomolecules.

ESI MS is particularly useful for the analysis of the primary structure of proteins. Both enzymatic and chemical digestion give sufficient information to sequence small proteins and peptides. Enzymatic digestion mixtures often can be analyzed directly without purification providing care is taken to eliminate specific salts and other compounds (e.g., phosphate, SDS) and keep other compounds at low concentrations. However, in such direct analyses, generally not all fragments are observed. Tandem MS can be used effectively with ESI MS to obtain sequence data and, for large molecules, is often able to generate useful information on terminal sequences.

Secondary structure is more difficult to analyze by ESI MS alone, but useful information on the helical structure of proteins and protein folding pathways can be obtained in conjunction with other analytical techniques to yield insight into biological activity. Most of the information has been obtained by using H-D exchange and proposals have been made for using enzymatic digestion or tandem MS to locate the site of this exchange.

Tertiary structure can be observed in a variety of ways using ESI MS to measure intramolecular interactions together with changes in solvent, pH, temperature, H-D exchange, and addition of reducing agents. Interaction of proteins with ligands can also be observed as exemplified by the work done with heme and globins and metal-protein interactions. These measurements can give information on protein conformation, binding, and the charge states of bound metals.

The determination of the quaternary structure by ESI MS has made significant progress over the last two years. It has been shown that gentle experimental conditions (low acid, low organic solvent, correct pH levels, and low energy conditions) are necessary to record intact multimeric complexes. The advent of high m/z range analyzers appears to be an important event for quaternary structure determination, particularly for molecules with high molecular weight subunits, allowing the generation of multimers with fewer charge states. The concentrations needed to observe these interactions by ESI MS are still quite high (often 1 $\mu g/\mu l$), but the subunit interactions, in many cases, have been shown to be specific and are not due solely to aggregation.

ESI MS is one of the most useful mass spectrometric techniques for protein analysis that has yet been developed. It is a rapid and facile method for molecular weight analysis that has applications which continue to expand rapidly, showing utility in many fields of research. Sensitivity continues to increase allowing research at trace levels (femtomoles) to obtain *in vivo* information such as the determination of endogenous metabolites.

## 8. References

1.  M. Dole, L.L. Mack, and R.L. Hines, "Molecular Beams of Macroions", *J. Chem. Phys.* **49**(3), pp. 2240-2249 (1968).

2.  M. Yamashita and J.B. Fenn, "Electrospray Ion Source. Another Variation on the Free-Jet Theme", *J. Phys. Chem.* **88**, pp. 4451-4459 (1984).

3.  M. Yamashita and J.B. Fenn, "Negative Ion Production with the Electrospray Ion Source", *J. Phys. Chem.* **88**, pp. 4671-4675 (1984).

4.  J.B. Fenn, M. Mann, C.K. Meng, S.F. Wong, and C.M. Whitehouse, "Electrospray ionization for mass spectrometry of large biomolecules", *Science* **246**, pp. 64-71 (1989).

5.  Paul Kebarle and Liang Tang, "From Ions in Solution to Ions in the Gas Phase", *Anal. Chem.* **64**(22), pp. 972a-986a (1993).

6.  F.W. Röllgen, U. Lüttgens, Th. Dülcks, and U. Giessmann, "On the Release of Ions From Charged Droplets in Electrospray Mass Spectrometry", *Proceedings of the 41st ASMS Conference on Mass Spectrometry*, May 31-June 3, 1993, San Francisco, pp. 1a-b (1993).

7.  J.B. Fenn, "Ion Formation from Charged Droplets: Roles of Geometry, Energy, and Time", *JASMS* **4**, pp. 524-535 (1993).

8.  B.A. Thomson and J.V. Iribarne, "Field induced ion evaporation from liquid surfaces at atmospheric pressure", *J. Chem. Phys.* **71** (11), pp. 4451-4463 (1979).

9.  J. Zaia, R.S. Annan and K. Biemann, "The correct molecular weight of myoglobin, a common calibrant for mass spectrometry", *RCMS* **6**(1), pp. 32-36 (1992).

10. M. Mann, C.K. Meng, J.B. Fenn, "Interpreting Mass Spectra of Multiply Charged Ions", *Anal. Chem.* **61**, pp. 1702-1708 (1989).

11. R.D. Smith, J.A. Loo, C.G. Edmonds, C.J. Barinaga and H.R. Udseth, "New Developments in Biochemical Mass Spectrometry: Electrospray Ionization", *Anal. Chem.* **62** (9), pp. 882-889 (1990).

12. J.A. Loo, C.G. Edmonds, R.D. Smith, M.P. Lacey, and T. Keough, "Comparison of Electrospray Ionization and Plasma Desorption Mass Spectra of Peptides and Proteins", *Biomed. Environ. Mass Spectrom.* **19**, pp. 286-294 (1990).

13. A.V. Dorslaer, F. Bitsch, B. Green, S. Jarvis, P. Lepage, R. Bischoff, H.V.J. Kolbe, and C. Roitsch, "Application of Electrospray Mass Spectrometry to the Characterization of Recombinant Proteins up to 44 kDa", *Biomed. Environ. Mass Spectrom.* **19**, pp. 692-704 (1990).

14. R. Feng and Y. Konishi, "Analysis of antibodies and other large glycoproteins in the mass range of 150,000-200,000 by electrospray ionization mass spectrometry", *Anal. Chem.* **64**(18), pp. 2090-2095 (1992).

15. S.K. Chowdhury and B.T. Chait, "Analysis of mixtures of closely related forms of bovine trypsin by electrospray ionization mass spectrometry: use of charge state distributions to resolve ions of the different forms", *Biochem. Biophys. Res. Commun.* **173**, pp. 927-931 (1990).

16. S.K. Chowdhury, V. Katta, and B.T. Chait, "Electrospray Ionization Mass Spectrometric Peptide Mapping: A Rapid, Sensitive Technique for Protein Structure Analysis", *Biochem. Biophys. Res. Commun.* **167**(2), pp. 686-692, 1990.

17. K.J. Rosnack and J.G. Stroh, " C-Terminal Sequencing of Peptides Using Electrospray Ionization Mass Spectrometry", *RCMS* **6**, pp. 637-640 (1992).

18. F. Maquin, B.M. Schoot, and P.G. Devaux and B. N. Green, "Molecular Weight Determination of Recombinant Interleukin 2 and Interferon Gamma by Electrospray Ionization Mass Spectrometry", *RCMS* **5**, pp. 299-302 (1991).

19. A. Tsugita, K. Takamoto, M. Kamo, and H. Iwadate, "C-terminal Sequencing of Protein: A novel partial acid hydrolysis and analysis by mass spectrometry", *Eur. J. Biochem.* **206**(3), pp. 691-696 (1992).

20. J.A. Loo, C.G. Edmonds, and R.D. Smith, "Primary Sequence Information from Intact Proteins by ESI Tandem Mass Spectrometry", *Science* **248**, pp. 201-204 (1990).

21. P. Roepstorff and J. Fohlman, "Proposal for a Common Nomenclature for Sequence Ions in Mass Spectra of Peptides", *BMS* **11**, p. 601 (1984).

22. N.F. Sepetov, O.L. Issakova, M. Lebl, K. Swiderek, D.C. Stahl, and T.D. Lee, "The Use of Hydrogen-Deuterium Exchange to Facilitate Peptide Sequencing by Electrospray Tandem Mass Spectrometry", *RCMS* **7**, pp. 58-62 (1993).

23. J.A. Loo, C.G. Edmonds, and R.D. Smith, "Tandem Mass Spectrometry of Very Large Molecules: Serum Albumin Sequence Information from Multiply Charged Ions Formed by Electrospray Ionization", *Anal. Chem.* **63** (21) pp. 2488-2499 (1991).

24. P.R. Cantor and C.R. Schimmel, Biophysical Chemistry, W. Freeman and Co., San Francisco (1980).

25. C. Brown, P. Camilleri, N.J. Haskins, and M. Saunders, "Probing Protein Conformation by a Combination of Electrospray Mass Spectrometry and Molecular Modelling", *J. Chem. Soc.*

*Commun.* **10**, pp. 761-6 (1992).

26.     C.L. Stevenson, R.J. Anderegg, and R.T. Borchardt, " Probing the Helical Content of Growth Hormone-Releasing Factor Analogs Using Electrospray Ionization Mass Spectrometry", *JASMS* **4**, pp. 646-651 (1993).

27.     A. Miranker, C.V. Robinson, S.E. Radford, R.T. Aplin, and C.M. Dobson, "Detection of Transient Protein Folding Populations by Mass Spectrometry", *Science* **262**, pp. 896-900 (1993).

28.     S.K. Chowdhury, V. Katta, and B.T. Chait, "Probing Conformational Changes in Proteins by Mass Spectrometry", *JACS* **112**, pp. 9012-3 (1990).

29.     M.A. Kelly, M.M. Vestling, and C. Fenselau, "Electrospray Analysis of Proteins: A Comparison of Positive-ion and Negative-ion Mass Spectra at High and Low pH", *OMS* **27**, pp. 1143-1147 (1992).

30.     J.C.Y. Le Blanc and D. Beuchemin, "Thermal Denaturation of some Proteins and its Effect on their Electrospray Mass Spectra", *OMS* **26**, pp. 831-839 (1991).

31.     U.A. Mirza, S.L. Cohen, and B.T. Chait, "Heat-Induced Conformational Changes in Proteins Studied by Electrospray Ionization Mass Spectrometry", *Anal. Chem.* **65**, pp.1-6 (1993).

32.     P.D. Cary, D.S. King, C. Crane-Robinson, E.M. Bradbury, A. Rabbani, G.H. Goodwin, and E.W. Johns, "Structural Studies on two High-Mobility-Group Proteins from Calf Thymus, HMG-14 and HMG-20(ubiquitin), and Their Interaction With DNA", *Eur. J. Biochem.* **112**, pp. 577-586 (1980).

33.     J.A. Loo, R.R. Ogorzalek Loo, H.R. Udseth, C.G. Edmonds, and R.D. Smith, "Solvent-induced Conformational Changes of Polypeptides Probed by Electrospray-ionization Mass Spectrometry", *RCM* **5**(3), pp. 101-105 (1991).

34.     T.T. Herskovits, B. Gadegbeku, and H. Jaillet, "On the Structural Stability and Solvent Denaturation of Proteins I. Denaturation by the Alcohols and Glycols", *J. Biol. Chem.* **245**, pp. 2588-2598 (1970).

35.     K.D. Wilkinson and A.N. Meyer, "Alcohol-Induced Conformational Changes of Ubiquitin", *Arch. Biochem. Biophys.* **250**, pp. 390-401 (1986).

36.     J.A. Dean, ed., Lange's Handbook of Chemistry, 12th edition, McGraw-Hill Book Co., New York, pp. 10-103 to 10-117 (1979).

37.     H. Roder, G.A. Elove, and S.W. Englander, "Structural characterization of folding intermediates in cytochrome c by H-exchange labelling and proton NMR", *Nature* **335** (6192), pp. 700-704 (1988).

38.     J.B. Udgaonkar and R.L. Baldwin, "NMR Evidence for an early framework intermediate on the folding pathway of Ribonuclease A", *Nature* **335** (6192), pp. 694-699 (1988).

39.     V. Katta and B.T. Chait, "Conformational Changes in Proteins Probed by Hydrogen-exchange

Electrospray-ionization Mass Spectrometry", *RCMS* **5**, pp. 214-217 (1991).

40. J.A. Loo, C.G. Edmonds, H.R. Udseth, and R.D. Smith, "Effect of Reducing Disulfide-Containing Proteins on Electrospray Ionization Mass Spectra", *Anal. Chem.* **62**, pp. 693-698 (1990).

41. V. Katta and B.T. Chait, "Observation of the Heme-Globin Complex in Native Myoglobin by Electrospray-Ionization Mass Spectrometry", *JACS* **113**, pp. 8534-5 (1991).

42. R. Feng and Y. Konishi, " Stepwise Refolding of Acid-Denatured Myoglobin: Evidence from Electrospray Mass Spectrometry", *JASMS* **4**, pp. 638-645, 1993.

43. M.H. Allen and T.W. Hutchens, " Electrospray-ionization Mass Spectrometry for the Detection of Discrete Peptide/Metal-ion Complexes Involving Multiple Cysteine (Sulfur) Ligands", *RCMS* **6**, pp. 308-312 (1992).

44. T.W. Hutchens, M.H. Allen, C.M. Li, and T-T. Yip, "Occcupancy of a $C_2$-$C_2$ type 'Zinc-finger' protein domain by copper: Direct observation by electrospray ionization mass spectrometry", *FEBS Lett.* **309**(2), pp. 170-174 (1992).

45. T.W. Hutchens and M.H. Allen, "Differences in the Conformational State of a Zinc-finger DNA-binding Protein Domain Occupied by Zinc and Copper Revealed by Electrospray Ionization Mass Spectrometry", *RCMS* **6**, pp. 469-473 (1992).

46. T.W. Hutchens, R.W. Nelson, M.H. Allen, C.M. Li, and T-T. Yip, "Peptide-Metal Ion Interactions in Solution: Detection by Laser Desorption Time-of-flight Mass Spectrometry and Electrospray Ionization Mass Spectrometry", *BMS* **21**, pp. 151-159 (1992).

47. X. Yu, M. Wojciechowski, and C. Fenselau, "Assessment of Metals in Reconstituted Metallothioneins by Electrospray Mass Spectrometry", *Anal. Chem.* **65**, pp. 1355-1359 (1993).

48. A. Surovoy, D. Waidelich, and G. Jung, "Nucleocapsid protein of HIV-1 and its $Zn^{2+}$ complex formation analysis with electrospray mass spectrometry", *FEBS Lett.* **311**(3), pp. 259-262 (1992).

49. R.D. Smith, K.J. Light-Wahl, B.E. Winger, and J.A. Loo, "Preservation of Non-covalent Associations in Electrospray Ionization Mass Spectrometry: Multiply Charged Polypeptide and Protein Dimers", *OMS* **27**, pp. 811-821 (1992).

50. E.C. Huang, B.N. Pramanik, A.T. Tsarbopolous, P. Reichert, A.K. Ganguly, P.P. Trotta, T.L. Nagabhushan, and T.R. Covey, "Application of Electrospray Mass Spectrometry in Probing Protein-Protein and Protein-Ligand Noncovalent Interactions", *JASMS* **4**, pp. 624-630 (1993).

51. B.E. Winger, K.J. Light-Wahl, R.R. Orgorzalek Loo, H.R. Udseth, and R.D. Smith, "Observation and Implications of High Mass-to-Charge Ratio Ions from Electrospray Ionization Mass Spectrometry", *JASMS* **4**, pp. 536-545 (1993).

52. R.D. Smith and K.J. Light-Wahl, " The Observation of Noncovalent Interactions in Solution by Electrospray Ionization-Mass Spectrometry: Promise, Pitfalls, and Prognosis", *BMS* **22**, pp. 493-501 (1993).

# APPLICATIONS OF QUANTUM CHEMISTRY IN MASS SPECTROMETRY

KÁROLY VÉKEY
*Central Research Institute for Chemistry*
*Hungarian Academy of Sciences*
*H-1025 Budapest, Pusztaszeri ut 59-67*
*Hungary*

ABSTRACT. Quantum chemistry is used with increasing frequency for the evaluation and explanation of mass spectrometric characteristics. Here a short overview is given on the semiempirical and *ab initio* theoretical methods, their advantages and limitations. Beside energetics, various other parameters (geometry, bond order, energy partitioning, spin and charge density, free valence) can be used to characterise ion structures, reactions and possibly to predict fragmentation behaviour, and these are discussed in some detail. Three illustrative examples are given: description of a simple bond cleavage, discussion of differences between isomers and a calculation of kinetic energy release in charge separation processes.

## 1. Introduction

A system in quantum mechanics is described by a wave function, from which every property of the system can be calculated. The application of quantum mechanics to chemical problems is usually called quantum chemistry. The concept of molecular orbital (MO) is used very often, hence the other frequently used terminology: 'MO methods', loosely synonymous to quantum chemistry. Calculations can become very complex, and almost invariably require approximations. Nevertheless, the results in some cases can be more accurate, than experimental measurements.

In mass spectrometric applications the most important information to be obtained from a calculation is the energy of the system (of ground and of transition states). Unfortunately, this is also the most difficult to determine precisely (with an accuracy which is chemically meaningful). Usually better results are obtained if relative energies are considered, i.e. that of a set of isomers. Determination of other structural characteristics, like molecular geometry, is usually less critical. Quantum chemical methods are often divided into two broad categories, to '*ab initio*' and to 'semiempirical' methods. Both yield similar types of information solving more or less similar equations, though their accuracy, reliability and speed is different.

With the advent of fast and easily available computers, quantum chemistry could become a very useful tool helping the evaluation or explanation of experimental results. The usefulness of such theoretical methods is indicated by an increasing number of publications in the field mass spectrometry in which results of quantum chemical calculations are also included. The present paper gives a short overview of this field, and shows a few selected examples. Most publications using molecular orbital methods in mass spectrometry deal with energetics. While this is undoubtedly important, here emphasis is put on the use of other parameters, like bond orders, spin density, etc., which can be advantageously used for a qualitative description of structures and reactions.

*R. M. Caprioli et al. (eds.), Mass Spectrometry in Biomolecular Sciences*, 89–102.

## 2. Theoretical methods

Closed and open shell species are treated by different formalisms, the former by the so-called (restricted) Hartree-Fock (HF or RHF) method, which considers electron pairs. Open shell systems, on the other hand, are usually calculated by the unrestricted HF (UHF) method, which treats electrons of $\alpha$ and $\beta$ spin separately.

'Semiempirical' methods consider only valance electrons and make various 'shortcuts' in the calculations (some integrals, which are small and difficult to calculate are neglected). To compensate for the approximations and to obtain better results a number of empirical parameters are incorporated into the program. The most widely used semiempirical methods are MNDO and AM1[1-5]. MNDO is the 'older' one and is used somewhat more frequently in mass spectrometric applications. Both are well tested, the average error in calculated heats of formations was about 50 kJ/mol both for MNDO and for AM1[5]. Both are, however, relatively unsuccessful determining ionisation energies - these are often over 1 eV off.

It is often claimed that semiempirical methods are unreliable, i.e. it is difficult to know how accurate (or how inaccurate) is the obtained result for an unusual structure. Nevertheless, surprisingly accurate results were obtained recently for very unusual structures like transition states for doubly charged ions[6], ionised silanes[7,8], distonic ions,[8] etc. If one encounters 'unusual' structures in MNDO or AM1 calculations it is always worth (if possible) making some checks with ab initio methods, maybe on analogous, but smaller systems.

The main advantage of semiempirical methods is their speed: they are very fast and can easily be run on personal computers. The speed of calculations is an especially important aspect when relatively large molecules are considered: computation time increases with ca the 4-th power of the molecular size (or, more precisely, with the number of basis functions). Over 5 times larger molecules can be treated by MNDO than by small basis set *ab initio* calculations, the latter not necessarily giving better results.

In the case of *ab initio* methods neither shortcuts in the calculations, nor empirical constants are used. These methods together with some interesting examples were discussed in an excellent recent review[9]. *Ab initio* methods are characterised by the size of the basis set and if (and what level of) electron correlation treatment is used. The simplest (but unfortunately also the least accurate) *ab initio* calculation uses a so-called minimum (usually STO-3G) basis set and electron correlation is not taken into account. Even such a simple *ab initio* calculation is usually considered somewhat more reliable than one using a semiempirical method - though this is often debated.

If possible (computation time allows) it is expedient to use a so-called split basis set, like 3-21G, or 6-31G (for a detailed description of basis sets and their abbreviation see e.g.[9]) - the results will become more reliable. Polarisation functions (indicated by * or **) further increase accuracy (especially for transition states), a basis set indicated as 6-31G** can be considered as a reasonably large and accurate one. Calculations using a 6-31G** set, however, require ca. 100 times more computer time than that with a minimum basis set, so large basis sets can not always be used.

To predict some structural characteristics, like ionisation energies and energetics of loose structures, electron correlation (e.g. at the MP2 level) has to be taken into account. To derive precise energetics, sometimes vibrational zero point energies are also considered, though these seldom change relative energies by more than 5-10 kJ/mol. Using large basis sets, good electron correlation treatment and vibrational zero point energies (e.g. 'G1' or 'G2' level of *ab initio* theory) energies can be predicted within 15 kJ/mol. The drawback is, that such calculations are very time consuming even on large computers, so these can be used only for small molecules (3 - 4 atoms plus hydrogens).

Geometry, and also other structural characteristics like bond orders, spin and charge density etc., are usually less sensitive to the level of theory used, than energetics. To make a compromise between accuracy and computation time, often the geometry is calculated at a 'lower level', while for energy calculations a larger basis set and perhaps electron correlation treatment is used. Irrespective of the level of theory used, the results are always more reliable if relative values are compared, or if e.g. changes in homologous series of molecules are considered.

## 3. Relative energies

In mass spectrometric applications most usually energetics are compared, e.g. relative energies of isomeric systems. There are numerous examples, one interesting case is the comparison of methanol ($CH_3OH^{+\bullet}$) and methyleneoxonium ($^{\bullet}CH_2OH_2^+$) radical cations: The latter, somewhat unconventional distonic ion structure, is more stable by 33 kJ/mol calculated by a high level of theory, in good agreement with experimental results[9].

The most important question concerning relative stability is the accuracy of the calculated results. To increase the accuracy of a calculation there is the practical problem of increasing computer time in the case of small molecules. Larger molecules, on the other hand, may not be calculated by a high level of theory, so either one has to accept lower accuracy, or the problem could not be studied by molecular orbital methods. Of course, high accuracy is always *desirable*, but the accuracy *required* depends a lot on the problem studied.

One 'extreme' case, which requires very high accuracy, is concerning relative gas phase basicity. The protonation reaction $AH^+ + B \rightarrow A + BH^+$ takes place, if the gas phase basicity of B is larger than that of A. Few kJ/mol differences can easily be measured experimentally, and 10-15 kJ/mol difference means that practically only $AH^+$ or only $BH^+$ is observed in the spectra. Theoretical methods, to be of any practical help in such a case, should be very accurate, indeed.

This problem may be partially overcome, i.e. a lower accuracy may be sufficient, when series of compounds are studied, e.g. the effect of substitution on gas phase basicity. In such a case relative energies can be compared, and a *difference* of 10 kJ/mol obtained by a medium quality *ab initio* calculation can be reliable.

In most mass spectrometric experiments internal energies are in the range of several hundred kJ/mol, even excess energies of reactions may be larger than 100 kJ/mol. For this reason if the heats of formation of two structures are only a few tens of kJ/mol apart, both are likely to be accessible under mass spectrometric conditions. In such a case very precise calculation of heats of formation may not always be necessary and lower quality calculations, like a small basis set *ab initio* or a semiempirical methods, will give meaningful results. Faster calculations have another advantage, i.e. more (computer) time could be spent to obtain information on the potential energy surface. This may be more important to predict fragmentation behaviour, than obtaining precise energetics for the ground states of the molecular and few selected fragment ions.

### 4. Potential energy profiles

In mass spectrometry often competitive reactions are studied. The main factor influencing whether a reaction takes place or which reaction is most likely, is mainly determined by the energetics of the reactions. Other factors effecting reaction kinetics, like 'entropy' of activation or the 'tightness' of the transition state, have usually a smaller effect, and are neglected in the following treatment.

The energetics of reactions can be best illustrated on a so-called potential energy profile (e.g. Fig 1): on the horizontal axis the 'reaction co-ordinate' on the vertical axis the energy (total energy or heat of formation) of the system is indicated. The 'reaction co-ordinate' is the bond length in the case of a simple bond cleavage, but may also be a expressed as a linear combination of internal co-ordinates. In a complex rearrangement process the 'reaction co-ordinate' could even be a formal parameter indicating the lowest energy route connecting two structures .

Potential energy profiles were studied in various cases by D.H. Williams, R.D. Bowen and collaborators in the '70-es[10,11]. At that time calculations were not very advanced, so heats of formation of various species and transition states were either experimentally determined (often with large error margins) or estimated. The reaction co-ordinate was nearly always a schematic representation of the reaction. In spite of these gross simplifications use of such potential energy profiles proved very valuable: isomerizations, reaction mechanisms, kinetic energy release in metastable processes were successfully explained.

With the advent of relatively fast and accurate quantum chemical methods potential energy profiles are often calculated. Usually only the minima and the transition states are considered, the curve connecting these is typically less relevant. Sometimes, however, two-dimensional potential energy maps are also calculated. This is a reliable way of determining the lowest energy reaction pathway and the transition state of a reaction. Potential energy maps are especially valuable in the case of rearrangement processes, often giving further information on the reaction mechanism.

Potential energy profiles are very valuable for studying reaction mechanisms, comparing isomers, and are frequently used. The main difficulty of their application is that their determination is very time consuming, especially if high quality calculations are used. It is often expedient to calculate potential energy diagrams or two-dimensional maps by lower quality, but fast, methods and then calculate the characteristic minima and transition states by higher quality calculations.

### 5. Geometry, spin and charge density, free valence, bond order and energy partitioning

Beside energetics - which are very important and are frequently used - various other information can be obtained from quantum chemical calculations. These can be used to characterise ion structures and reaction pathways. Unfortunately, theoretical information is packed in large matrices, which have no direct meaning in 'classical' chemistry. For a qualitative evaluation of the results it is desirable to transform these matrices into 'classical' parameters. Geometry, free valence, spin and charge densities can be successfully used to describe the molecular structure in qualitative terms. Bond orders and energy partitioning can also be used to predict fragmentation processes[12-14].

The geometry is obviously an important and comprehensible way of characterising the structure of a molecule or an ion. The change in the geometry upon ionisation, if significant, may give hints of likely fragmentation processes as well. Spin and charge densities are usually given at each atom of the molecule. The spin density is practically zero at every atom in a closed shell (singlet) species. In a radical (doublet state) the sum of spin densities is unity. The spin density may have positive or negative sign, but has a large value usually at one or a few atoms - the radical is well 'localised'. According to the 'classical' description of fragmentation processes reactions are often 'induced' by the radical site - therefore the spin density is often helpful for explaining or predicting mass spectrometric behaviour. Free valence is qualitatively similar to the spin density and it always has a positive sign, a large value indicating strong radical character.

The charge density, on the other hand, is nearly always 'delocalised': it is usually large on most atoms in a molecule. The sum of charge densities is exactly zero in a neutral, unity in a singly charged ion. Due to the 'inductive' effect atoms of larger electronegativity attract electrons, and become negatively charged. This explains that hydrogens, which have a low electronegativity, have usually a significant positive charge and heteroatoms, like oxygen or nitrogen, are usually negatively charged. It is often expedient to add the charges on the hydrogens to that on the 'heavy' atom to which they are connected. This way it is easier to evaluate the charge distribution on the molecule, and the results will also be less dependent on the basis set used. When studying ions it is often worth subtracting the charge density of the corresponding neutral from that of the ion. This will result in a charge distribution which is easier to evaluate, reflects the effect of ionisation better and is closer to the conventional description of ion structures.

Bond orders indicate bond strengths in a qualitative, and easily understandable way. They represent a very good description of the molecular structure, and have the advantage that calculated values are close to those expected 'classically': The bond order of a single bond is close to one, that of a double bond to two, etc. Bond orders do not depend significantly on the basis set used, 'semiempirical' and *ab initio* methods also result in similar values. Bond orders correlate well with Mulliken population analysis. The latter has, however, become less popular in the last decades due to its large basis set dependence. Bond orders, and the change of bond orders upon ionisation, can be used to predict likely fragmentation reactions. Some characteristic examples will be discussed below.

Partitioning of the energy to diatomic contributions (i.e. characteristic to each bond) is a very recent method[13,14]. Its great advantage is, compared to other parameters, that it is an energetic quantity. It correlates with dissociation energy, though (unfortunately) it is not an approximation of it: energy partitioning values may be 3-4 times larger, than dissociation energies. Energy partitioning values correlate very well with bond orders, and can also be used to predict fragmentation processes.

Geometry, spin and charge density, bond orders and energy partitioning may advantageously be used to describe the molecular structure and predict likely fragmentation processes in a qualitative way. Sometimes they can be used in a quantitative or semi-quantitative way as well, e.g. for the comparison of isomers. Their great practical advantages are that they can be calculated very fast (in fact, much faster than an SCF calculation), unlike energy profiles, and that they are not very sensitive to basis sets, i.e. calculations at a lower level give sufficiently accurate data. Small changes in the geometry also do not effect significantly these parameters. For a quick initial study of bond orders, spin density etc., these may be calculated in the 'vertical' ion, so geometry optimisation of the ion may not be necessary. Of course, the use of bond orders or energy partitioning is not a substitute for energy or energy profile calculations. However, they can give some initial estimates which are likely fragmentation processes, or how does a given structural change effect possible reaction channels. They can also give valuable help simplifying energy profile calculations: e.g. a 'strong' bond, with bond order 1.3, is unlikely to dissociate, so this does not have to be included in a very time consuming potential hyper surface calculation.

## 6. Selected examples

### 6.1. DESCRIPTION OF A SIMPLE DISSOCIATION PROCESS

α-Cleavage to a carbonyl bond is a very simple process, and the primary reason of this discussion is the demonstration of the use of various quantum chemical characteristics discussed above. Methyl-ethyl ketone was selected as the test compound. Calculations were performed by unrestricted Hartree-Fock (UHF) ab initio method, using minimum (STO-3G) basis set[15]. The geometry of the molecular ion was optimised without restrictions. For other points on the potential surface all geometrical parameters, but the reaction co-ordinate, were optimised. In methyl-ethyl ketone there are two possible α cleavages leading to the elimination of a methyl or an ethyl radical, respectively. Of these two processes ethyl elimination has the lower activation energy, and this will be discussed in detail below.

Ionisation changes the structural characteristics of methyl-ethyl ketone. Most importantly, the bond order of the carbonyl bond decreases from 1.99 in the neutral to 1.53 in the ion, indicating that ionisation significantly reduces the double bond character. There is a corresponding lengthening of the bond from 1.22 to 1.30 Å. The radical site will be on the oxygen. Ionisation increases the charge density on oxygen by 0.35 units, on the carbonyl carbon ($C_2$ in Fig. 1) by 0.12 units, the rest is spread over the ion. Even though following ionisation the carbonyl bond becomes much weaker, it remains the strongest bond, and it is unlikely to be cleaved.

The most traditional, and probably also the most informative, description of a fragmentation

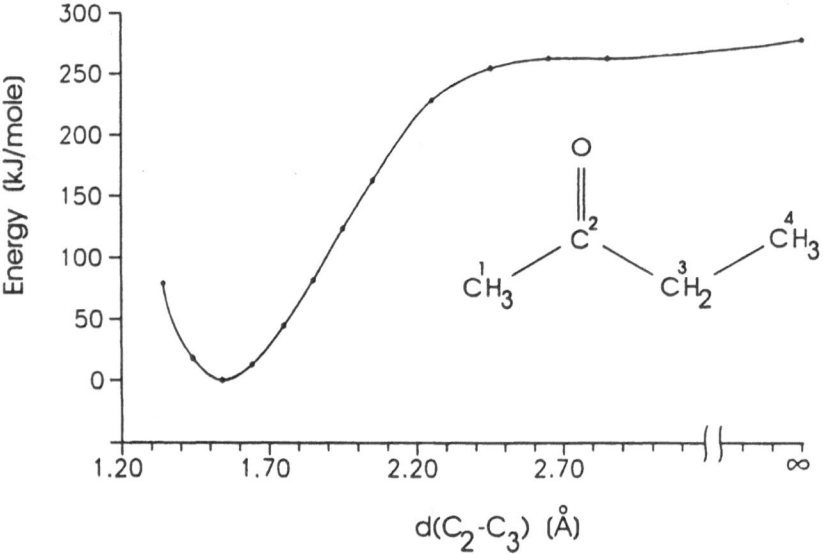

Fig. 1. Potential energy profile of the dissociation reaction $CH_3COC_2H_5^{+\bullet} \rightarrow CH_3CO^+ + C_2H_5^\bullet$ (from Ref 15)

process, this case α cleavage, is the potential energy diagram, shown in Fig. 1. The reaction co-ordinate is the bond length between the 'carbonyl' and the 'ethyl' carbons ($C_2$ and $C_3$ carbon atoms in Fig. 1), and this is shown on the horizontal axis. On the vertical axis the total energy of the $C_4H_8O^{+\bullet}$ system is shown (in kJ/mol), relative to the ground state of the molecular ion. The distance between $C_2$ and $C_3$ is 1.54 Å at equilibrium, very similar to that in the neutral molecule (1.55 Å). Elongation of the carbonyl - ethyl bond results in a monotonous increase in energy, reaching a plateau 260 kJ/mol above the ground state at around 2.7 Å distance. The potential energy curve shows that the reverse reaction has no activation energy: This supports the general expectation that addition of radicals to closed shell species proceeds with no or small activation energy.

The potential energy diagram is very useful for the description of energetics, but it does not tell much about the reaction mechanism. For a qualitative description of a structure or a process bond orders are very useful . These are shown, as a function of the carbonyl-ethyl distance, in Fig. 2.

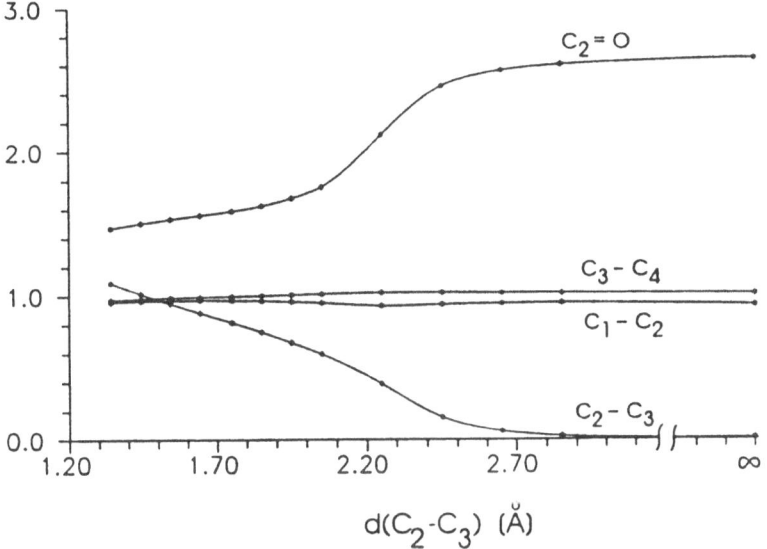

Fig. 2. Change of the bond orders in the course of the dissociation reaction $CH_3COC_2H_5^{+\bullet} \rightarrow CH_3CO^+ + C_2H_5^{\bullet}$ (from Ref 15).

The bond orders of the $C_1$-$C_2$ and the $C_3$-$C_4$ bonds are close to unity, and do not vary during the reaction. The carbonyl-ethyl ($C_2$-$C_3$) bond order is also close to unity at the equilibrium geometry, but decreases steadily as the bond elongates, reaching practically zero value around 2.7 Å distance. The carbonyl bond ($C_2$-O) shows an opposite effect: it becomes stronger (the bond order increases from 1.6 to 2.6) in the course of ethyl elimination. It is interesting that the bond orders in the reaction change relatively little up to 2.0 Å, the major part of the reorganisation of the electron structure (as shown by the bond orders) takes place between 2.0 and 2.5 Å. On the other hand, over 60 % of the dissociation energy is used to elongate the ethyl-carbonyl bond to 2.0 Å .

The structural change in the course of ethyl elimination is also reflected by the position of the radical site. One way of measuring the position or the distribution of the unpaired electron is by the free valence, and as shown in Fig. 3 (top part, dashed lines). At equilibrium geometry the

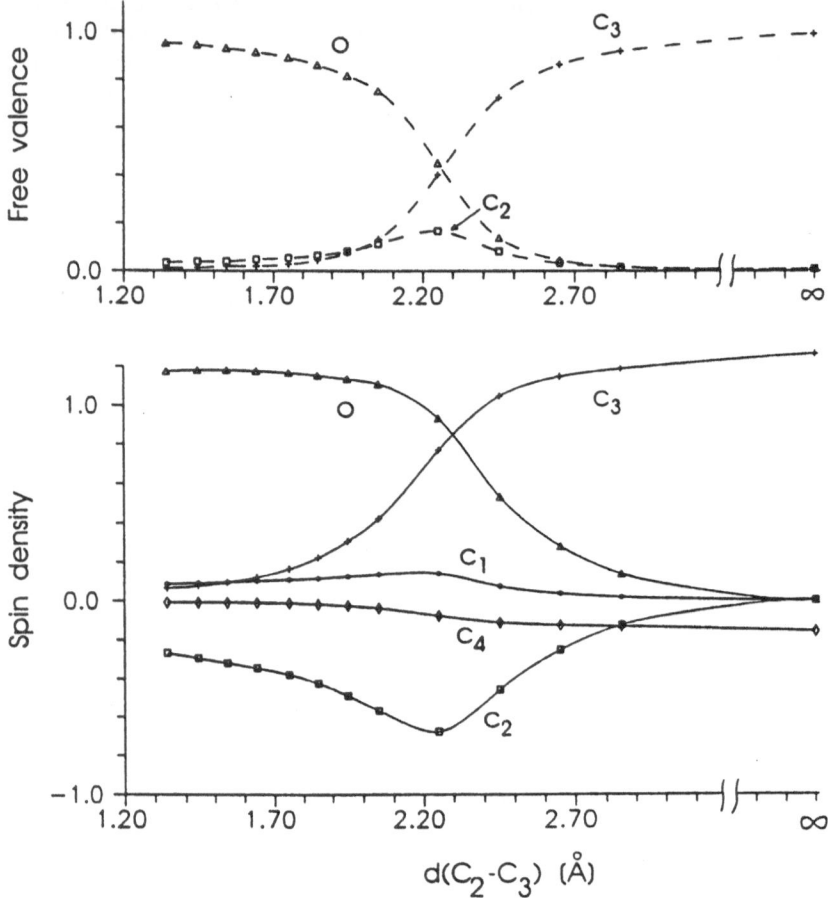

Fig. 3. Change of the spin densities (bottom part, solid lines) and free valence (top part, dashed curves) in the course of the dissociation reaction $CH_3COC_2H_5^{+\bullet} \rightarrow CH_3CO^+ + C_2H_5^\bullet$ (from Ref 15).

radical site is at the oxygen, which is transferred to $C_3$ of the ethyl group during the reaction. The shift of the unpaired electron occurs between 2.0 and 2.5 Å; at similar distances as the major change in the bond orders. Somewhat surprisingly, there is a significant contribution to the radical site at the $C_2$ atom as well.

The spin densities (Fig. 3, lower part, solid curves) on the oxygen and on $C_3$ show a similar change during ethyl elimination as the free valence discussed above. The spin density on $C_2$ at around 2.2 Å bond length is large (similarly to the free valence) but it has, somewhat surprisingly, a negative sign.

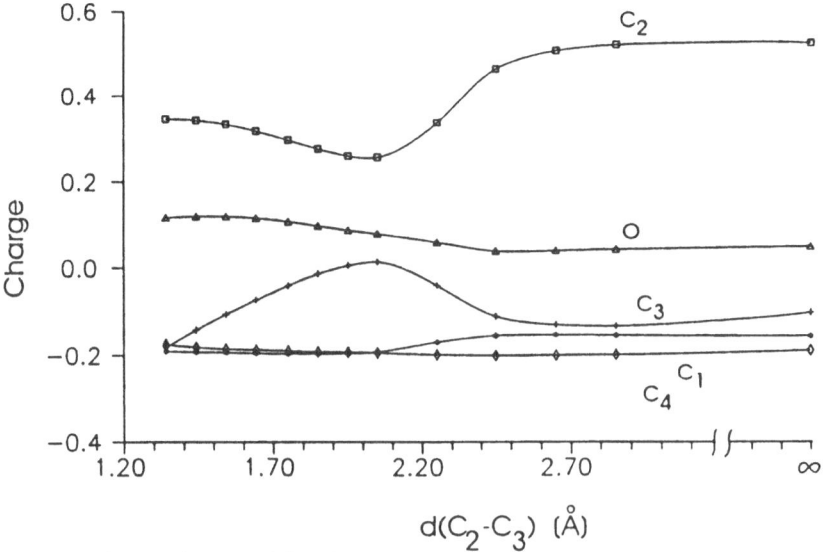

Fig. 4. Change of the charge densities on various atoms in the course of the
dissociation reaction $CH_3COC_2H_5^{+\bullet} \to CH_3CO^+ + C_2H_5^\bullet$ (from Ref 15).

The positive charge, unlike the unpaired electron, is spread over the ion, it is 'delocalised'. A
significant part of the positive charge is carried by the hydrogens in ionised methyl-ethyl ketone
(altogether 0.9 unit), while the charge density on oxygen, which formally carries the charge, is
only 0.1 (Fig 4). The charge on oxygen, $C_1$ and $C_5$ changes only little in the fragmentation, the
major change is on $C_2$ and $C_3$. Increasing the $C_2$ - $C_3$ bond length from equilibrium to 2.0 Å
shifts part of the positive charge from the carbonyl to the ethyl unit. At 2.0 Å this abruptly
changes, and the charge shifts back to the carbonyl, in fact $C_2$ will become much more positive,
than in the equilibrium geometry.

The characteristics derived from molecular orbital calculations may be used for a qualitative
description of $\alpha$-cleavage, utilising the terminology of 'classical' structural chemistry. Ionisation of
methyl-ethyl ketone may be best described by removal of an electron from a non-bonding orbital
of the oxygen, the resulting ion having two canonical forms: In one the double bond is not
effected (formal bond order is 2.0) and both the radical and the charge site will be on the oxygen.
In the other form there is a single bond between oxygen and carbon, the radical site remaining on
the oxygen, but the charge will be on $C_2$. Lengthening the ethyl - carbonyl bond up to 2.0 Å
requires a relatively large amount of energy, but the electronic structure does not change much:
The electron pair between $C_2$ and $C_3$ shifts only slightly towards the carbonyl, decreasing the
bond order and increasing the positive charge on the ethyl group. A further increase in the
distance between the ethyl and the carbonyl units (between 2.0 and 2.5 Å) causes less change of
energy, but a larger change in the electronic structure: The bond between the carbonyl and ethyl
groups breaks completely, the radical site shifts from the oxygen to the ethyl, the positive charge
shifts back to $C_2$, and the oxygen-carbon bond strengthens from a nearly double to a nearly triple
bond.

## 6.2. COMPARISON OF ISOMERS

The second example is a typical application of a theoretical study to explain the mass spectrometric behaviour of isomers. The studied compounds were *cis* and *trans* 1-carbomethoxy-2-trimethylsilylcyclopropane[7], their electron impact mass spectra are shown in Fig. 5. Some characteristic features of the spectra and some differences between the isomers are the following:

(1) The molecular ion is of very low abundance in the *trans* isomer, but it is completely absent in the *cis*;

(2) There are three fragmentation pathways of the molecular ion: Carbon - silicon bond cleavage resulting in either methyl elimination or trimethylsilyl ion formation, and a rearrangement reaction. These three processes yield ions of comparable abundance in the case of the *trans* isomer, but only methyl elimination is abundant in the case of the *cis* one.

(3) The [M-methyl]$^+$ ion fragments further with a rearrangement reaction, involving bond formation between silicon and oxygen, resulting in the [Me$_2$SiOMe]$^+$ ion. This process is, interestingly, much more favourable (daughter/parent abundance ratio) in the *trans*, where the interacting substituents are farther from each other, than in the *cis* isomer.

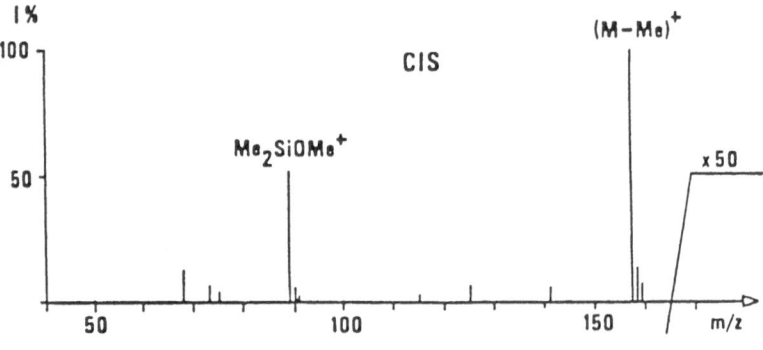

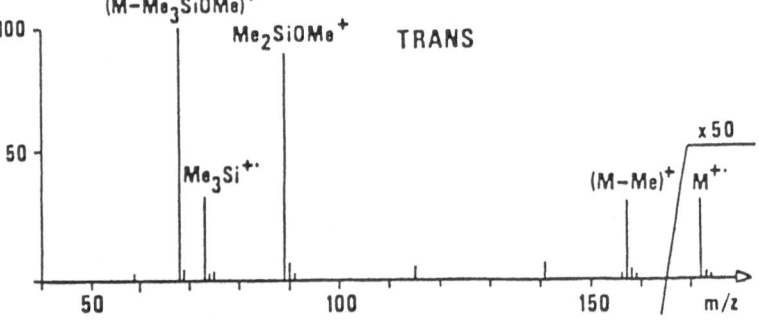

Fig. 5. Electron impact mass spectra of *cis* and *trans* 1-carbomethoxy-2-trimethylsilylcyclopropane, obtained at 15 eV ionisation energy (from Ref 7).

Quantum chemical calculations (using MNDO-UHF formalism) have been used to explain the contrasting behaviour of these isomers. The most characteristic feature of the adiabatic molecular ion of the *trans* isomer is one very weak (bond order: 0.29) and elongated (by 0.25 Å) silicon - methyl bond. 90% of the positive charge is on the silicon containing part of the ion, 80% of the spin density is on the loose methyl group. Dissociation of this methyl radical requires only 56 kJ/mol energy. As the 'vertical' molecular ion is 70 kJ/mol above that of the adiabatic structure. These two energy values explain very well the low abundance of the molecular ion, and the facile methyl loss observed in the spectrum (Fig. 5).

Study of the potential energy surface of the molecular ion yielded another minimum (also ground electronic state), only 12 kJ/mol above the adiabatic structure. This ion is characterised by a weak (bond order 0.56) and loose (elongated by 0.08 Å) bond between silicon and the carbomethoxycyclopropyl unit. Existence of this, energetically easily accessible, structure explains that the SiMe$_3$$^+$ ion can easily be formed from the *trans* molecular ion in competition with the methyl loss discussed above.

Calculations in the case of the *cis* isomer indicate that the molecular ion is unstable, spontaneous methyl loss occurs upon ionisation (i.e. dissociative ionisation takes place). This does explain the lack of molecular ion and that the only primary fragment of large abundance is the [M-methyl]$^+$ ion. The resulting structure (Fig. 6a) is very stable, due to bond formation between silicon and the carbonyl oxygen. The stabilisation energy, compared to the *trans* molecular ion, is very large, 190 kJ/mol. The structure of the carbomethoxy group is completely reorganised: the 'carbonyl' bond order is decreased while the methoxy C-O bond order is increased significantly. There is a corresponding change in the bond lengths as well. The structure of the -COO unit therefore resembles to that of a protonated carboxylic acid.

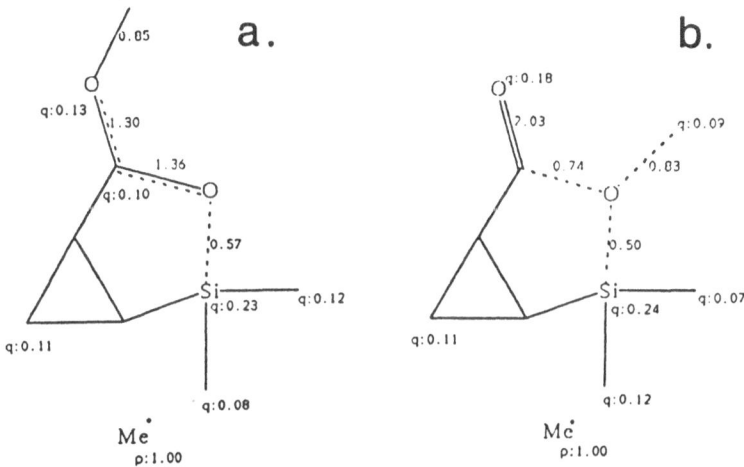

Fig. 6. Structure of the *cis* 1-carbomethoxy-2-trimethylsilylcyclopropane molecular ion. (a) the lowest energy minimum, Si - carbonyl oxygen bond; (b) Si - methoxy oxygen bond. Numbers on the Figure indicate characteristic bond orders (adapted from Ref 7).

Rotation of the carbomethoxy group by ca. 180° results in another structure (Fig. 6b). Stabilisation, in this case, derives from bond formation between silicon and the methoxy oxygen. Stabilisation energy, compared to the trans molecular ion, is 110 kJ/mol. There is a large, 210 kJ/mol, barrier to inter conversion of the two structures, which can be best described as isomers.

Secondary fragmentation of [M-methyl]$^+$ ions, in the case of both isomers, result in [Me$_2$SiOMe]$^+$ ions, as discussed above. The reaction requires breaking the stable five membered ring in the *cis*, whereas there is no such barrier in the *trans* isomer. This explains that the daughter to parent ratio is smaller in the case of the *cis*, than in the *trans* isomer (Fig. 5).

The very large energy differences observed between the isomers justifies the use of semiempirical methods in this example for evaluating mass spectrometric differences. If energy differences between the isomers would have been much smaller, say, ca 10-30 kJ/mol, this would have two consequences: (1) Higher level of calculations should have been used, *ab initio*, with at least a split basis set and (2) it would have been necessary to evaluate, in a more quantitative way, if small energetic differences would really effect the mass spectra.

6.3. DETERMINATION OF KINETIC ENERGY RELEASE

Kinetic energy release (KER) is a very important characteristic of unimolecular reactions. This can be relatively easily measured using sector type mass spectrometers, and mass spectrometry is the only practical way of its experimental determination. Part of the KER originates from the excess energy of a reaction, which is distributed among vibrational degrees of freedom and the reaction co-ordinate. This part of the KER is usually relatively small (rarely exceeds 50 meV) and may be calculated by RRKM calculations.

Higher KER values, especially those exceeding 100 meV, usually have a different origin. When the reverse reaction has an activation energy (RAE) a significant part of it may be converted to KER in addition to that discussed above. By a combined application of detailed molecular orbital and molecular dynamic calculations the partitioning ratio may be determined, but this is feasible only for very small systems. Generally it is little known, how much of the RAE is released as KER, and how much is converted to internal energy.

A special case of large KER is observed in charge separation reactions of doubly charged ions. These give large (few eV) KER. It is generally *assumed* that the reverse activation energy is mainly due to the coulombic repulsion between the two charges, and that *all* of this energy will be converted to KER. From the KER, therefore, the distance between the two charges in the molecule can be estimated. The problems associated with this method are that the basic assumptions have not been substantiated, the unreasonable idea of localised charges is used and it is not always clear, where are the charges located. This problem can be approached in a quantitative way using quantum chemical calculations[6]. In the example below the semiempirical MNDO method were used to study the [C$_6$H$_6$]$^{2+}$ → [C$_4$H$_3$]$^+$ + [CH$_3$]$^+$ reaction.

Previous work has shown that [C$_6$H$_6$]$^{2+}$ ions originating from various precursors isomerize to a common structure before loosing CH$_3^+$, and this is likely to be the [CH$_3$-C$_4$-CH$_3$]$^{2+}$ open chain species. Starting from the equilibrium geometry of this ion the potential energy (heat of formation) along the reaction co-ordinate of CH$_3^+$ loss has been determined, and this is shown in Fig. 7. Part of the potential energy is due to coulombic repulsion of the charges on the two separating particles. This was also calculated (from charge densities and ion geometries along the reaction co-ordinate) and this is also shown on Fig. 7 (E$_r$). The potential and the coulombic repulsion energy curves are very close to each other between the transition state and infinite

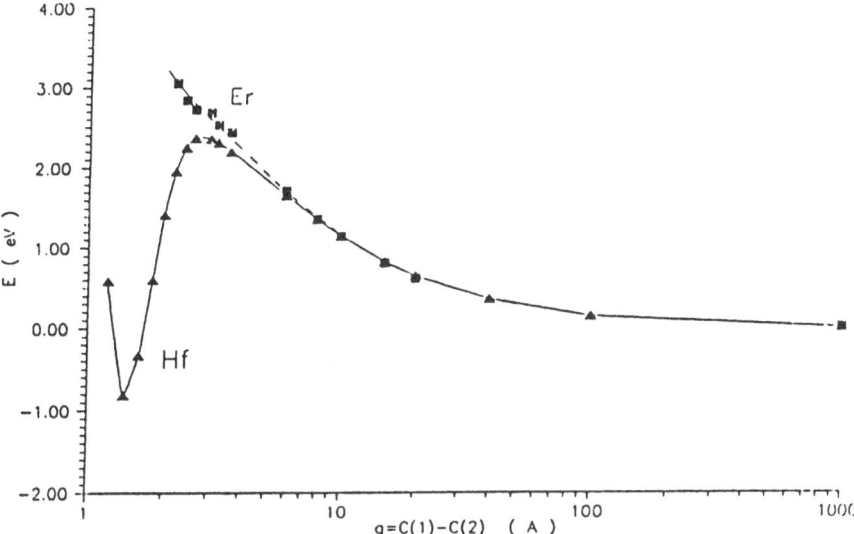

Fig. 7. Heat of formation ($H_f$) of the linear $[CH_3-C_4-CH_3]^{2+}$ ion along the reaction co-ordinate of methyl ion elimination calculated by the MNDO method, relative to the sum of heats of formation of the products. The electrostatic repulsion energy ($E_r$) between the two separating singly charged products is also shown (from Ref 6).

separation. This supports the traditional assumption that the reverse activation energy is mainly due to electric repulsion of the two charges in a charge separation process.

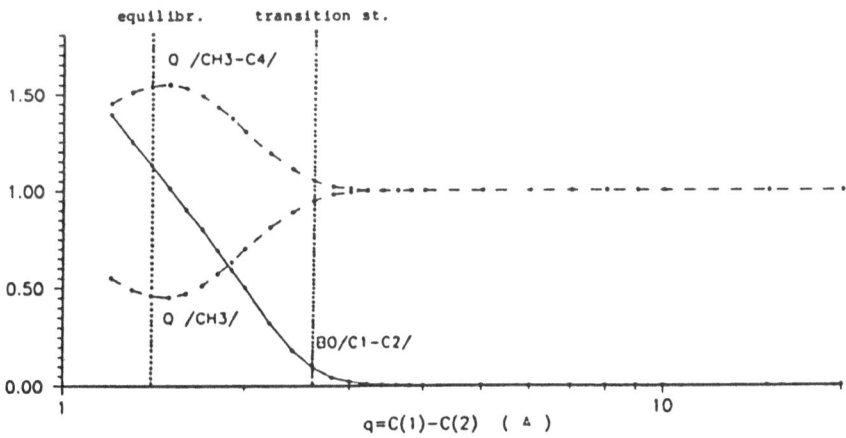

Fig. 8. Charge densities (Q) on the separating $[CH_3-C_4]$ and $[CH_3]$ units and the bond order (BO) between them(from Ref 6).

The charges on, and the bond order between the two separating products ($CH_3$-$C_4^+$....$CH_3^+$) have also been calculated along the reaction co-ordinate, and these are shown in Fig. 8. The bond order between the two parting ions decreases to nearly zero and the charge densities become close to unit values at the transition state. This indicates that the only significant interaction between the $CH_3$-$C_4^+$ and $CH_3^+$ units in the transition state is electric repulsion. This, in turn, suggests that exchange of internal energy between the separating particles is unlikely in the transition state or at longer distances. This strongly suggests that all of the reverse activation energy will be released as kinetic energy of the products. It could also be argued that the KER should be equal to the electrostatic repulsion energy at (or close to) the transition state, which is - as discussed above - very close to the reverse activation energy.

Comparison of experimental and calculated values for benzene show a very satisfying agreement between the observed KER (2.77 eV) with both the calculated reverse activation energy (2.43 eV) and the electrostatic repulsive energy at the transition state (2.81 eV). This suggests that even modest quality calculations may be useful to check the reaction mechanisms or reactive configurations of some doubly charged ions.

## 7. Conclusions

The present paper gave a short overview of the most important quantum chemical methods which are often used in mass spectrometry. Most importantly and most commonly energy values and energy profiles are determined by these calculations. In the present paper various examples are discussed where, beside energetics, less commonly used structural characteristics were also calculated and used for the description of ion structures or reaction mechanisms.

## 8. References

1   M.J.S. Dewar and W. Thiel, J. Am. Chem. Soc., 99 4899 (1977)
2   M.J.S. Dewar and W. Thiel, J. Am. Chem. Soc., 99 4907 (1977)
3   Halim, N. Heinrich, W. Koch, J. Schmidt and G. Frenking, J. Comput. Chem., 7, 93 (1986)
4   D. Higgins, C. Thomson and W. Thiel, J. Comput. Chem., 9, 702 (1988)
5   J.J.P. Stewart, J. Comput. Chem., 10, 221 (1989)
6   K. Vékey and G. Pócsfalvi, Org. Mass Spectrom. 27 1203 (1992)
7   K. Vékey, Á. Somogyi, J. Tamás and G. Pócsfalvi, Org. Mass Spectrom., 27 869 (1992)
8   K. Vékey, B. Paizs, Á. Somogyi, J. D. Knausz and G. Pócsfalvi, Org. Mass Spectrom., in print
9   L. Radom, Org. Mass Spectrom., 26, 359 (1991)
10  D.H. Williams, Phil. Trans. R. Soc. Lond. A293, 117 (1979)
11  R.D. Bowen, D.H. Williams and H. Schwarz, Phil. Trans. R. Soc. Lond. A293, 117 (1979)
12  Á. Somogyi, Á. Gömöri, K. Vékey and J. Tamás, Org. Mass Spectrom., 26 936 (1991)
13  I. Mayer and Á. Gömöry, Int. J. Quant. Chem., Ser. Symp. B, accepted for publication
14  I. Mayer and Á. Gömöry, J. Mol. Struct. Theochem, accepted for publication
15  Á. Gömöry and K. Vékey, in preparation

# PROTON AFFINITIES AND INTRINSIC BASICITIES OF ALANINE AND GLYCINE STUDIED BY MEANS OF AM1 AND PM3 METHODS

T. MARINO, N. RUSSO and M. TOSCANO
*Dipartimento di Chimica*
*Universita' della Calabria,*
*I-87030 Arcavacata di Rende (CS),*
*Italy*

ABSTRACT. The protonation process of alanine and glycine have been studied using the AM1 and PM3 advanced semiempirical methods. For both the systems we found that the N atom is the preferred protonation site. Both methods give proton affinities and intrinsic basicities in agreement with the experimental measurements. The protonation process does not induce changes in the absolute minimum conformation.

## 1. Introduction

The study of physical properties and chemical reactivities of molecules in gas phase is subject of current interest because of the progress in the techniques for the generation of the ions. In addition the knoweledge of fundamental properties as basicity and proton affinity of biomolecular systems is interesting for many points of view including the explanation of proton transfer reactions. Pulsed ion cyclotron resonance (ICR) experiments have been used for obtaining information on the behaviour of large ions in a solvent-free environment [1-9]. In particular the proton affinities (PA) and intrinsic basicities (GB) have been determined for many $\alpha$-amino acids [7,8]. Very recently the study has been extended to the 20 most common $\alpha$-amino acids by using the same experimental technique [9].

Although the experimental measurements give precious data, other significant information (i.e. the more stable protonated form and potential energy surface) can be obtained only by theoretical studies. In this contribution we report the results of a theoretical investigation of the protonation process of glycine and alanine performed by using the AM1 (Austin Method) [10] and PM3 [11] parametrization of advanced semiempirical MNDO ( Modified Neglect Differential Overlap) [12] method.

## 2. Method

All the computations have been performed employing the Mopac code [13]. For the full structural optimizations the analytical gradients procedure has been used [14]. The criterium to terminate the optimization has been increased 100-fold over normal limits of

*R. M. Caprioli et al. (eds.), Mass Spectrometry in Biomolecular Sciences, 103–108.*
© 1996 *Kluwer Academic Publishers.*

MOPAC using the PRECISE option. All structures have been characterized as a stationary points and true minima on the potential energy surfaces have been obtained using the keyword FORCE that also gives the vibrational frequencies.

The thermodynamic quantities have been obtained by means the THERMO option. Considering the process

$$B + H^+ = BH^+ + \Delta H_f$$

the PA has been obtained as follows

$$PA = -\Delta H_f = -[(\Delta H_{BH^+}) - (\Delta H_{H^+} + \Delta H_B)]$$

The GB has been derived from

$$GB = -\log K = (\Delta H_f - T\Delta S) / RT$$

where $\Delta S$ has been obtained from thermochemical computations.

All data are calculated at 350 K. For $H^+$ the experimental heat of formation (367.2 kcal/mol) has been used [15].

## 3. Result and discussion

As a first step of the work we have determined the most stable protonation site. For both the studied molecules two different atoms have been considered as candidates for proton acceptors (see scheme). Results are collected in table 1. As it is shown in this table, in all cases, the most stable protonation is that in which the proton is added to the amino nitrogen. In the case of glycine the Gly-OH$^+$ complex lies at 15.9 and 26.1 kcal/mol over the Gly-NH$^+$ one at AM1 and PM3 level respectively.

The energy differences between Ala-NH$^+$ and Ala-OH$^+$ systems are 18.4 kcal/mol at AM1 level and 28.6 kcal/mol at PM3 one.

The termochemical results for glycine and alanine are reported in table 2 together with the available experimental data [7-9].

| | |
|---|---|
| $CH_3$-CHNH$_2$-COOH (Ala) | $NH_2$-CH$_2$-COOH (Gly) |
| $CH_3$-CHNH$_3^+$-COOH (Ala-NH$^+$) | $NH_3^+$-CH$_2$-COOH (GlyNH$^+$) |
| $CH_3$-CHNH$_2$-COH$^+$OH (Ala-OH$^+$) | $NH_2$-CH$_2$-COH$^+$OH (Gly-OH$^+$) |

**Scheme**

Table 1. Heat of formation and proton affinities calculated for neutral and different protonated complexes. All values are in Kcal/mol.

| System | Method | ΔHf | PA |
|---|---|---|---|
| Gly | AM1 | -101.5 | - |
| | PM3 | -95.8 | - |
| Gly-NH$^+$ | AM1 | 64.1 | 201.6 |
| | PM3 | 69.3 | 202.0 |
| Gly-OH$^+$ | AM1 | 80.0 | 185.7 |
| | PM3 | 95.4 | 175.9 |
| Ala | AM1 | -105.0 | - |
| | PM3 | -101.1 | - |
| Ala-NH$^+$ | AM1 | 56.2 | 206.0 |
| | PM3 | 60.7 | 205.4 |
| Ala-OH$^+$ | AM1 | 74.6 | 187.6 |
| | PM3 | 89.2 | 176.8 |

The AM1 gives a PA value for glycine and alanine of 201.6 and 206.0 kcal/mol respectively. At PM3 level we found a PA value of glycine of 202.0 kcal/mol and 205.4 kcal/mol for alanine. Considering the experimental errors, both PM3 and AM1 PA's values agree well with the experimental evidences. Same agreement has been previously found for the protonation of nucleic acid nucleobases [16] and nucleosides [17]. Nothwithstanding the small differences between the AM1 and PM3 results and experimental values, the methods are able to reproduce the trend in the PA data. In order to obtain the intrinsic basicity we have calculated the entropy variation in the protonation process. At 350 K, the calculated AM1 entropy contribution to the free energy of the considered reaction is 0.8 and 0.3 kcal/mol for glycine and alanine amino acids respectively. The PM3 computations give for glycine 0.2 kcal/mol and for alanine 1.1 kcal/mol. This means that the entropy contributions are always neglegible. On the other hand the extimation of these contributions by using rotational entropy values are of only 0.7 kcal/mol [9]. Using our computed values we have obtained the gas-phase basicity listed in table 2. The agreement between absolute GB values at AM1 and PM3 levels with the experimental data is good. In fact we obtain values that are slightly smaller by about 3 percent. The difference in gas-phase basicity between alanine and glycine found with the AM1 method is 3.8 kcal/mol while the results found from different experimental determinations [7-9] range from 2.8 to 11.6 kcal/mol.

Previous works have demonstred that AM1 is able to predict the conformation of unionized amino acids [18]. In particular for glycine, alanine and serine AM1 structures are very similar to those obtained at ab- initio level employing the 4-21G basis sets.

Figure 1 shows the structures of the calculated PM3 absolute minima for neutral and protonated glycine and alanine amino acids. For both the neutral systems the structure of the absolute minima obtained at PM3 level of computations are similar to those resulting from AM1 and ab-initio 4-21 G ones.

Table 2. Gas-phase basicities (GB) and Proton affinities (PA) of glycine and alanine. All values are in kcal/mol.

| Method | Glycine | | Alanine | |
| | PA | GB (350 K) | PA | GB (350 K) |
|---|---|---|---|---|
| AM1 | 201.6 | 192.5 | 206.0 | 197.8 |
| PM3 | 202.0 | 193.8 | 205.4 | 197.9 |
| Exp.[a] | 204.0-207.4 | 200.7-204.1 | 211.5-214.0 | 212.3-213.2 |
| Exp.[b] | 208.2 | - | 212.2 | - |
| Exp.[c] | 213.0 | 204.6 | 215.8 | 207.4 |

a) from ref. 9; b) from ref. 8; c) from ref. 7

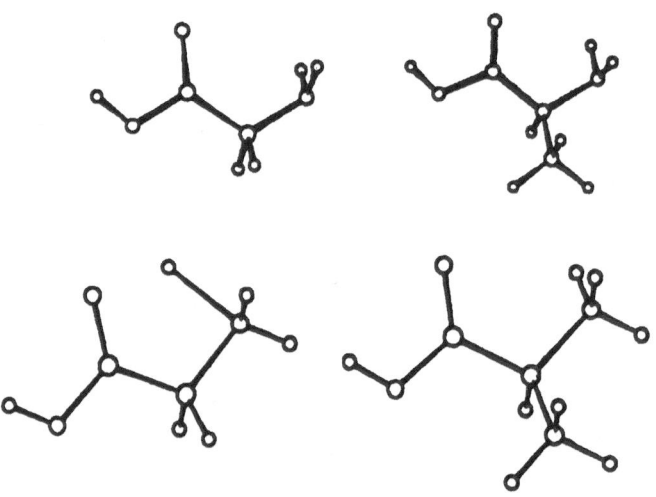

Figure 1. PM3 structure for the absolute energy minimum of unionized and ionized glycine and alanine.

The more stable glycine and alanine conformation are found to be the trans form. For both the systems the oxydril hydrogen point toward the carbonil oxygen.
For both the neutral systems the structure of the absolute minima at PM3 level of computations are similar to those obtained from AM1 one. In fact, both methods give for glycine and alanine a trans structure ($\Psi=\Phi=180°$). The protonation process does not induce changes in the conformational properties. In fact, with the addition of $H^+$ to the amino nitrogen (Gly-$NH^+$, Ala-$NH^+$) the absolute conformation of glycine and alanine skeleton remains trans. The values of torsional angles at AM1 level for Gly-$NH^+$ and Ala-$NH^+$ are $\Psi=180°$ and $\Phi=176°$ respectively. The PM3 computations give the same results.

## 4. Conclusions

On the basis of our preliminary AM1 and PM3 study on the protonation process of glycine and alanine $\alpha$-amino acid we can draw the following conclusions:
- Both AM1 and PM3 method are able to give correct trend on the proton affinity and intrinsic basicity. In particular the AM1 one give absolute values that are in good agreement with the experimental results.
- The used methods give structures that are in agreement with those obtained from more sophisticated and computationally expensive ab-initio methods.
- The protonation process does not induce change in the conformational properties of the unionized glycine and alanine amino acids.
Work is in progress for all other common amino acids and containing alanine dipeptide.

## Aknowledgements

The financial contribution of the Italian Ministero dell' Universita' e della Ricerca Scientifica e Tecnologica (MURST) is highly appreciated.

## References

1. M. Borgarello, R. Houriet, E. D. Raczynka and T. Drapala, J. Org. Chem. 55 (1990) 38.
2. M. Berthelot, M. Decouzon, J-F. Gal, C. Laurence, J-Y. Le Questel, P-C. Maria and J. Tortajada, J. Org. Chem. 56 (1991) 4490.
3. M. Meot-Ner and L. W. Sieck, J. Am. Chem. Soc. 113 (1991) 4448.
4. J. E. Szulejko and T. B. McMahon, J. Am. Chem. Soc. 115 (1993) 7839 and references therein.
5. X. Cheng, Z. Wu and C. Fenselau, J. Am. Chem. Soc. 115 (1993) 4844.
6. G. S. Gorman and I. J. Amster, J. Am. Chem. Soc. 115 (1993) 5729.
7. M. J. Loke and R. T. McIver, J. Am. Chem. Soc. 105(1983) 4227.
8. M. Meot-Ner, E. P. Hunter and F. H. Field, J. Am. Chem. Soc. 101 (1979) 686.
9. G. S. Gorman J. P. Speir, C. A. Turner and I. J. Amster, J. Am. Chem. Soc. 114, (1992) 3986.
10. M. J. S. Dewar, E. G. Zoebisch, E. F. Healy and J. J. P. Stewart, J. Am. Chem. Soc. 107 (1985) 3902.
11. J. J. P. Stewart, J. Comp. Chem., 10 (1989), 209, ibidem 221.

12. M. J. S. Dewar and W. Thiel, J. Am. Chem. Soc. 99 (1977) 4899.
13. J. J. P. Stewart, QCPE 455, Department of Chemistry, Indiana University, Bloomington, Indiana.
14. H. B. Schlegel, J. Comput. Chem., 3 (1982) 214.
15. H. Stull (ed.), JANAF Thermochemical Tables (NBS), US Gov. Print Office (1971).
16. N. Russo and M. Toscano, in "Mass Spectrometry in the Biological Sciences: A Tutorial", M. L. Gross (Ed.), Kluwer, Dordrecht, 1992, P. 311-322.
17. T. Marino, V. Milano, N. Russo and M. Toscano, J. Mol. Struct. (Theochem), submitted.
18. M. Masumara, J. Mol. Struct. (Theochem), 164 (1988) 299.

# Part II
# Instrumentation for Mass
# Analysis and Detection

# Energy–Isochronous Time–of–Flight Mass Spectrometers

## H. Wollnik

2. Physikalisches Institut, University Giessen, 35392 Giessen,
Heinrich–Buff–Ring 16, Germany

## Abstract

Time–of–flight mass spectrometers are discussed for systems in which the ion flight times are independent of the energies of ions and depend only on their mass–to–charge ratios. Such systems can be of the reflector–type geometry as well as consist of magnetic or electrostatic sector fields. Ion trajectories are described as well as some technical details of how ion bunches are recorded.

# 1    Introduction

Ions of different mass–to–charge ratios $(m/q)$ can be separated from each other laterally or longitudinally if they have been accelerated by the same potential $V$, i.e. that their energy–to–charge ratios $K/q$ are equal. In most cases, however, the energy–to–charge ratio $K/q$ varies over some small range since usually the ions are formed at slightly different potentials. In laterally dispersive magnetic sector field analyzers, in which the ion deflections are proportional to the momentum–to–charge ratios $\propto \sqrt{Km}/q$ of the ions, this energy spread causes a lateral beam broadening and thus a reduction of mass resolving power. This can be avoided by use of an additional electrostatic sector field [1,2,3] that deflects ions of different energy–to–charge ratios differently, independent of the masses of the ions [see Fig.1]. Thus the slightly different deflections of ions of slightly different energies can be compensated such that the system has a lateral mass dispersion that is not compromised by the energy spread of the ions.

*R. M. Caprioli et al. (eds.), Mass Spectrometry in Biomolecular Sciences, 111–146.*
© *1996 Kluwer Academic Publishers.*

In longitudinally dispersive systems such a variation of the ions' energy–to–charge ratios $K/q$ causes a spread in the ion velocities $v \propto \sqrt{K/m}$ or ion flight times $\propto 1/v$ and thus a longitudinal beam spread which also results in a reduction of the longitudinal mass resolving power. This can only be avoided if the ions of higher energies reach the final ion detector via a detour as compared to ions of lower energies so that the ion flight times become sole functions of the ions' mass–to–charge ratios [4,5,6]. In other words the ion flight times are energy isochronous.

In time–of–flight (TOF) mass analyzers all initially formed ions are finally recorded independent of their masses, though the ions of different masses arrive after each other. However, the mass spectrum here can and must be recorded in a very short time, perhaps in $100\mu sec$ if all ions were started at the same time[1]. This allows to monitor relatively rapid variations in the amount of provided sample material as would be the case in a GC–TOF combination in which a time–of–flight system would continuously monitor the mass distribution in the effluent of a gas chromatograph or of a capillary zone electrophoresis system.

For all such mass–spectrometric investigations it is well appreciated that a time–of–flight mass analyzer has no limit as to how heavy an ion can be. Besides problems in forming a heavy ion and in recording it, a TOF mass analyzer even achieves higher mass resolving powers though with increased flight times for ions of increased masses.

# 2    Principles of TOF–Mass–Analyzers

Time–of–flight (TOF) mass spectrometers separate ions of different masses because of their different flight times. These flight times depend on the ion flight distance as well as on the ion velocity, $v$ in $[mm/\mu sec]$ [3], i.e. in millimeters per microsecond

$$v = |\mathbf{v}| \approx 13.891332\sqrt{\frac{K}{m}} = 13.891332\sqrt{V\frac{q}{m}} \tag{1}$$

---

[1]Alternatively all ions of mass $m$ and charge $q$ must be entered in times proportional to $\sqrt{m/q}$ [see section 3.2.1]

with the ion mass $m$ in Daltons [u] and the ion energy $K$ in electron volts [eV], where $V$ is the potential by which the q–times charged ions were accelerated.

In a field–free region of length $L$ thus the flight time $T = L/v$ depends on the ions' energies $K = qV$ as well as on the ions' masses $m$. If one sends the more energetic and thus faster ions on properly dimensioned detours as compared to trajectories of reference ions of mass $m_0$ and energy $K_0$, one can make the ions' flight times independent of their energies.

One way to achieve such custom–taylored detours is to let the ions under investigation move in a homogeneous magnetic field of flux density B= |**B**| in which all particles of mass–to–charge ratio (m/q) and energy–to–charge ratio (K/q) move on circles of radii $\rho$ along which the centrifugal force $mv^2/\rho$ is balanced by the centripetal force $q(\mathbf{v} \times \mathbf{B})$. For relativistically slow ions thus the flight path per turn has a length of $2\pi\rho$ with $\rho$ in mm

$$\rho \approx 0.1439711 \frac{\sqrt{mK}}{qB} = 0.1439711 \frac{\sqrt{V(m/q)}}{B} \tag{2}$$

if $K$ is given in electron Volts, $V$ in Volts, $B$ in Tesla and $m$ in Daltons u. Since both $\rho$ and $v$ of Eq.(1) and (2) are proportional to $\sqrt{K} = \sqrt{qV}$ the overall flight time $\overline{T}$ of ions of given $m/q$ is calculated for one turn in $\mu sec$ as:

$$\overline{T} = \frac{2\pi\rho}{v} \approx \frac{0.0651195m}{Bq}, \tag{3}$$

showing that independent of the ions' energies K, the flight times for a full turn or for a specified fraction thereof are proportional to the ions' mass–to–charge ratios $m/q$ only and inversely proportional to the magnetic flux density $B$. This fact has led to the development of the cyclotron accelerator [4] but also to [5] the Ion Cyclotron Resonance Mass Spectrometer (ICR MS)

If ions of mass–to–charge ratios $m_0/q_0$ have flight times $T_0$, ions of $m/q = (m_0/q_0) + \Delta(m/q) = (m_0/q_0)(1 + \delta_m)$ have flight times $T + \Delta T = T(1 + \delta_t)$. Thus the time resolving power $T/\Delta T = 1/\delta_t$ and the mass–to–charge resolving powers $(m/q_0)/\Delta(m/q) = 1/\delta_m$ are equal:

$$\frac{1}{\delta_t} = \frac{\overline{T}}{\Delta\overline{T}} = \frac{1}{\delta_m} = \frac{m_0/q_0}{\Delta(m/q)}. \tag{4}$$

Note here that ion mass differences of $N$ per cent result in flight time differences of $Nq_0/q$ per cent. Highly charged ions $q \gg q_0$ thus become increasingly difficult to separate.

If the ion energy-to-charge ratios $K/q$ vary only slightly around $K_0/q_0$ of reference ions, we may write

$$\frac{K}{q} = \frac{K_0}{q_0} + \Delta\frac{K}{q} = \frac{K_0}{q_0}(1 + \delta_K) \tag{5}$$

with $\delta_K \ll 1$ and postulate only that the ion flight paths are designed such that the overall flight times $T$ are independent of $\delta_K$, but that they can still depend on $K_0/q_0$. Such systems may consist of homogeneous and inhomogeneous magnetic and electrostatic sector fields separated by field-free regions but possibly amended also by magnetic and electrostatic quadrupoles and rotationally symmetric lenses.

## 2.1  Repeller Field TOF Mass Analyzers

In Fig.2 an idealized time-of-flight mass analyzer is shown in which an ion beam is sent into an electrostatic repeller field [6,7] that enforces the mentioned detour of the more energetic ions by their deeper penetration into the field. A q-times charged ion of mass $m$, and velocity $\mathbf{v} = iv_x + jv_z$ penetrates into the repeller field to a depth $\bar{z} = K/(qE) = \overline{V}/E$ where $\overline{V} = K/q$ is the potential by which the ion under investigation had been accelerated in z-direction in the ion source and the acceleration region.

Denoting the distance between the ion source and the entrance to the repeller field (see Fig.2) by $L_0$ one finds the total ion flight times to be twice the flight times $T_t$, the ions need to move from the ion source to the apex of the individual ion trajectories in the repeller field. This total flight time equals $2T_t$ with

$$T_t = \frac{L_0}{v_z} + \frac{\bar{z}}{v_z/2} = \frac{L_0}{v_z} + \frac{2\overline{V}}{Ev_z}. \tag{6}$$

Here the first term describes the flight time in the field-free region in which the ions move in z-direction with the velocity $v_z$, and the second term describes the flight times in the repeller field in which the velocity components in z-direction are on average $v_z/2$ if for the moment the electrostatic field is postulated to be constant. This $v_z$ is found from Eqs.(1,5) for $K = q\overline{V}$ as

$$v_z \approx 13.891332\sqrt{\frac{q\overline{V}}{m}} = v_{z_0}(1 + \frac{\delta_K}{2} - \frac{\delta_K^2}{8} + \ldots)$$

with $v_{z_0} = 13.891332\sqrt{\overline{V}_0/(m/q)}$. Thus Eq.(6) can be written as

$$T_t = T_0 + (T|\delta)\delta_K + (T|\delta\delta)\delta_K^2 + \ldots \tag{7}$$

$$
\begin{aligned}
T_0 &= (L_0 + \frac{2\overline{V}_0}{E})\frac{1}{2v_{z_0}} \implies \frac{2L_0}{13.891332}\sqrt{\frac{m/q}{\overline{V}_0}} \\
(T|\delta) &= (\frac{2\overline{V}_0}{E} - L_0)\frac{1}{2v_{z_0}} \implies 0 \\
(T|\delta\delta) &= (3L_0 - \frac{2\overline{V}_0}{E})\frac{1}{8v_{z_0}} \implies \frac{T_0}{8}.
\end{aligned}
$$

The right–hand terms here are obtained for $L_0 = 2\overline{V}_0/E$ in which case $(T|\delta)$ vanishes [6,7]. Thus, the ion flight times $T_0$ reveal the ion mass distribution quite accurately even if the ion energies and velocities are only approximately equal, i.e. $\delta_K \neq 0$. Note here that the condition $(T|\delta) = 0$ prevails also for ions of different charges $q$, a finding that is quite important for the modern ion sources for multiply charged ions as for instance the electrospray sources.

For slightly different mass–to–charge ratios $(m_0/q_0)+\Delta(m/q) = (m_0/q_0)(1+ \delta_m)$ the ion flight times are $T_0 + \Delta T = T_0(1 + \delta_t) \propto \sqrt{(m_0/q_0)(1 + \delta_m)}$ so that in such TOF mass analyzers the mass–to–charge resolving power is half as large as the time resolving power

$$\frac{1}{2\delta_t} = \frac{\overline{T}}{2\Delta\overline{T}} = \frac{1}{\delta_m} = \frac{m_0/q_0}{\Delta(m/q)}. \tag{8}$$

Note also that ion mass differences of $N$ per cent result in flight time differences of $N/(2q)$ per cent. Higher and higher charged ions thus become increasingly more difficult to separate.

## 2.2   Sector Field TOF Mass Analyzers

Also in deflecting fields the length of the flight path of more energetic ions can be increased relative to the length of the flight path of ions of lower energies. In case of homogenous magnetic fields the ion flight times per turn are independent of the ion velocities. Besides technological problems, however, one should note that a homogeneous magnetic field has focusing properties only in the plane of deflection so that the ion paths are helices which lead the ions more and more away from the magnet mid plane as their path length increases more and more.

If one limits oneself to ions that have at least approximately equal energy-to-charge ratios $K/q = (K_0/q_0)(1 + \delta_K)$ with $\delta_K \ll 1$, one can build high performance TOF mass analyzers [8,9,10] in which the ion flight times are independent of the linear term in $\delta_K$ and depend only on $\delta_K^2, \delta_K^3$.... In such systems one can achieve lateral focusing not only in the plane of deflection but also in the perpendicular plane [3] by using one or preferable several magnetic or electrostatic sector fields.

One such time–of–flight mass analyzer [11,12] has been constructed of four identical sector magnets of deflection angles of $\phi_B = 81°$ each, that deflect the reference ions along radii of $\rho_0 = 1.1m$ and ensure that to first order the ion flight times are independent of the relative energy deviation $\delta_K$. Through this 14 m long TOF mass analyzer the ions pass in about $500nsec$ with calculated flight time deviations of $< 0.01nsec$ full width at half maximum (FWHM). Since the particle detectors, however, have timing errors of $\leq 0.1nsec$, only resolving powers of about $(m_0/q_0)/\Delta(m/q) \approx 3000$ can be reached. The accuracy of the determination of the $m_0/q_0$ value, however, is much better than $1/3000$ since this accuracy is the precision with which the center of gravity of the mass peak in question is determined. This accuracy is proportional to $(m/\Delta_m)\sqrt{N}$ where $N$ is the number of ions that contribute to the mass peak in question. Higher mass resolving powers can be reached only if the ion flight paths are elongated as can be achieved by an enlarged TOF mass analyzer or by building the system as a ring [13] through which the ions must move repeatedly (see section 5).

Note here that ions formed in nuclear fragmentation reactions are always multiply charged, so that there are several flight times, for ions of a given mass (see Fig. 3). Note further that for ions of $m/q$ being for instance 2.5 or 3.0 there are many combinations of m and q so that it usually is advisable not to use such ratios.

# 3 TOF Mass Analyzers for Low–Energy Ions

In case of low–energy ions, one can built energy–isochronous systems for which the term $(T|\delta)$ in Eq.(6) vanishes [6,7,8] by using magnetic or electrostatic sector fields or electrostatic repeller fields. This repeller field region [see Fig.2] can even be improved by introducing an intermediate grid and thus form two regions of different field strengths [14,15,16]. In this case one can achieve $(T|\delta) = (T|\delta\delta) = 0$, so that the first non–vanishing term is proportional to $\delta_K^3$ and thus quite small.

To avoid a deterioration of the optical beam properties by grids one can use grid–free ion repellers or mirrors [17]. To predict the ion–optical properties of such ion mirrors is mathematically difficult, however, to build the systems is relatively simple. A feasible TOF system that incorporates one grid–free ion mirror is shown in Fig.4 where this TOF system is shown connected to a gas chromatograph. Since here an axis of rotational symmetry exists, the beam must be adjusted more precisely than in gridded mirror systems which requires beam deflection plates and at least one additional lens as shown in Fig. 4.

The mass resolving power of any such TOF mass analyzer is limited by the ratio of the total ion flight time and the length of the ion pulse or the jitter of the start and stop pulses. The principally best results can be expected if each ion creates its own start and stop signal [18] by producing secondary electrons when passing through a very thin foil and when impinging on a stop detector (see Fig 5.). The start pulse also can be obtained from the formation process of the ions to be investigated [19,20,21]. Similar good results can be expected if the ion formation process is intimately linked to the ion start pulse [22].

In almost all cases the performance of a TOF mass analyzer improves with an increased overall flight time. A long total flight time on the other hand requires a large system or one that uses the existing path length several times (see also section 4) but it still requires to either produce the ions in a very short time or to bunch them together by static or by dynamic field arrangements [6,17,20,23,24].

## 3.1 Pulsed Ion Production

If ions are formed only during short time intervals the pulse length $\Delta T$ is mass independent but the flight time $T$ is proportional to $\sqrt{m/q}$ [see Eqs.(1,7)] assuming that all ions were accelerated by the same potential difference $V_0$. In this case the mass resolving power improves as the square root of the mass-to-charge ratio of the ions increases. A TOF mass analyzer that achieves $(m_0/q_0)/\Delta(m/q) \approx 1000$ for ions around $m_0/q_0 = 100$, consequently should obtain $(m_0/q_0)/\Delta(m/q) \approx 30000$ for ions around $m_0/q_0 = 100000$. For the investigations of large molecule ions [25], thus a pulsed ion production is most useful and desirable.

One way to form ions during a short time, is to shoot a short pulse of laser light into a relative dense puff of gas or into a gas jet [26] thus forming ions by single- or multi-photon excitation. This allows to investigate the ions of interest with good mass resolving power, though relatively large quantities of material are required.

Bombarding a solid or liquid surface by individual energetic fission products, i.e. Plasma Desorption Mass Spectrometry (PDS)[19], short bunches of ions, i.e. Secondary Ion Mass Spectrometry (SIMS)[20], or short bursts of laser light [21] can form one or several ions of heavy organic molecules in each shot. Though the ionization process in all three cases is not well understood, one can say that energy is coupled into the solid or liquid matrix in which the molecules of interest are individually embedded and that these molecules are simply lifted from the matrix and pulled away by an accelerating field for which purpose they must be ionised either instantly or within a very short time after leaving the surface.

Of the three mentioned methods the pulsed laser technique is the most effective and easiest to handle today. This method has become efficient, however, only after one had learned to modify the matrix such as to effectively absorb the laser light, the so called Matrix Assisted Laser Desorption Ionization (MALDI)[27,28]. At this point one should also note that all three desorption methods [19,20,21] make extremely efficient use of the sample material. In special cases the finally unused material can even be recouped for additional analyzes.

The described methods and here especially MALDI, see ref.[28,29], is capable of ionizing and accelerating ions larger than 300000 Daltons. These molecular ions also have been mass analyzed in time–of–flight mass spectrometers, however, so far the achieved mass–to–charge resolving powers were limited. The reason is that large molecule ions move slowly, so that the impact of such ions on a metal surface does not produce secondary electrons that could quickly be accelerated. What one records are probably only secondary electrons formed by post–accelerated fragment ions that because of their different velocities within the ion recording devices arrive over a longer time interval.

Since the peak widths due to the impaired ion recording are large, so far there was no need to use very sophisticated TOF systems. Thus, up to now mostly straight non–isochronous TOF mass analyzers without repeller fields have been used for the MALDI procedure [28,29] or the PDMS method of ref [19]. In cases of the PDMS method the ion flight times are determined from their start and stop signals by time–to–digital (TOC) converters. In cases in which ions are formed [21,27,28,29] by a single shot of a laser pulse at a low repetition rate, the mass spectrum can be recorded by a transient recorder, that records the varying signal intensity in many channels of perhaps $1nsec$ widths. Here usually one laser shot constitutes a full experiment in which case there is plenty of time to transfer the recorded mass spectrum to a computer for display and manipulation.

## 3.2    Quasi–Continuous Ion Production

To produce ions efficiently it often is advantageous to produce them continuously, store the formed ions for some time and extract them during short intervals only[2]. Assume that after some time of a continuous ion production an ion cloud has been formed between the grids of Fig.6 and that all these ions are at rest if $V_1$, equals $V_2$. A sudden change of $V_2$ then establishes an electric field $E_2 = (V_2 - V_1)/L_2$ over the ion cloud that expells these ions towards the detector plane in Fig.6. Ions that initially are close to grid 2 in this set up are accelerated to higher velocities than ions that initially are close to grid 1. On the other hand the latter are already closer to the detector plane.

---

[2]A very flexible way to build a correspondig system is to store ions in an ion trap, possibly tuned such as to be non mass selective, extract them in a pulsed mode and mass analyze them in a time–of–flight mass analyzer [30]. If this ion trap stores only ions of a limited mass range the system has also MS–MS capabilities.

Thus both groups of ions can arrive at the detector plane simultaneously if all potentials and distances have been chosen properly [6,17].

The overall ion motion in the system of Fig.6 is very similar to the ion motion in the second part of the system of Fig.2, i.e. from the apex of the parabolas to the final ion detector. Thus the resulting flight time $\overline{T}$ can be calculated from Eq.(6) if $E_1 = V_1/L_1$ is equal to $E_2 = (V_2 - V_1)/L_2$ with

$$V_0 = \frac{(V_1 + V_2)}{2} \qquad \delta_K = (V_2 - V_1)\frac{z}{L_2}.$$

For the more involved case $E_2 \neq E_1$ a similar relation can be derived with $(V_0 - V_1)/V_0 = \kappa$ and $E_1/E_2 - 1 = \lambda$:

$$\overline{T} = T_0 + (T|\delta)\delta_K + (T|\delta\delta)\delta_K^2 + (T|\delta\delta\delta)\delta_K^3 + \dots \qquad (9)$$

$$T_0 v_{z_0} = L_0 + \frac{2L_1}{1-\kappa}(1 + \sqrt{\kappa}\lambda)$$

$$2(T|\delta)v_{z_0} = -L_0 + \frac{2L_1}{1-\kappa}(1 + \frac{\lambda}{\sqrt{\kappa}})$$

$$8(T|\delta\delta)v_{z_0} = 3L_0 - \frac{2L_1}{1-\kappa}(1 + \frac{\lambda}{\sqrt{\kappa^3}})$$

$$16(T|\delta\delta\delta)v_{z_0} = -5L_0 + \frac{2L_1}{1-\kappa}(1 + \frac{\lambda}{\sqrt{\kappa^5}})$$

with $v_{z_0} \approx 13.891332\sqrt{V_0/(m/q)}$. To achieve $(T|\delta) = 0$ one must fulfill[3] the relation

$$2L_1/L_0 = (1 - \kappa)/(1 + \frac{\lambda}{\sqrt{\kappa}}) \qquad (10)$$

for instance by adjusting $\lambda$, i.e. $E_1/E_2$. In this case the other three coefficients simplify to:

$$T_0 v_{z_0} = L_0(1 + \kappa) + 2L_1$$

$$8(T|\delta\delta)v_{z_0} = \frac{T_0 v_{z_0} - 2L_0}{\kappa} + 2L_0$$

$$16(T|\delta\delta\delta)v_{z_0} = (2L_0 - T_0 v_{z_0})\frac{1+\kappa}{\kappa^2} - 4L_0$$

---

[3]In real life one should not postulate $(T|\delta) = 0$ but rather $(T|\delta) - (T|\delta\delta\delta)\delta_k^2 = 0$ in which case a part of the aberration $(T|\delta\delta\delta)\delta_k^3$ is compensated by $(T|\delta)\delta_k$.

Note here that $(T|\delta\delta)$ vanishes for $\lambda = 2\kappa\sqrt{\kappa}/(1 - 3\kappa)$. However, it should be noted that for an experiment it is important that $(T|\delta\delta)$ as well as $(T|\delta\delta\delta)$ stay below some limits. Eq.(9) was derived under the assumption that in the initial ion cloud the ions were at rest as far as the motion in z–direction was concerned, though in principle, the x–component of the ion velocity $\tilde{v}$ can have arbitrary values.

If the ions have a velocity with an initial z–component, there is an additional term to the flight time of Eq.(9). An ion that starts at point $A$, i.e. at $z_A$ in Fig.6 and moves towards $B$, arrives at $z_B$, after a time $\Delta\tilde{T} = (z_B - z_A)/(\tilde{v}_z/2)$. From there it starts to move towards the detector plane as any other ion that had been at rest at point $B$ initially. The potential difference $\Delta\tilde{V}$ between $A$ and $B$ is $\Delta\tilde{V} = (z_B - z_A)E_2$ found from the initial $\tilde{v}_z \approx 13.891332\sqrt{\Delta\tilde{V}(q/m)}$ in $mm/\mu sec$ and the initial energy $\tilde{K}_z = q\Delta\tilde{V}$ in eV. Thus one can determine $\Delta\tilde{T}$ in $\mu sec$ as:

$$\Delta\tilde{T} = \frac{2\Delta\tilde{V}}{\tilde{v}_z E_2} \approx \frac{2\sqrt{\Delta\tilde{V}(m/q)}}{13.89 E_2} \approx \frac{2\sqrt{\tilde{K}_z m}}{13.89 q E_2} \approx \frac{\tilde{v}_z(m/q)}{96.48 E_2}. \tag{11}$$

Analogously one can consider an ion that starts from point $A$ with a velocity $\tilde{v}_z$ towards the detector. This ion behaves like one that has started from point $C$ in Fig.6 a time $\Delta\tilde{T}$ earlier.

For an ion of initial velocity $\tilde{v}_z$ or an equivalent energy $q\Delta\tilde{V}$ thus the overall flight time to the detector in Fig.6 is $\overline{T} \pm \Delta\tilde{T}$ as determined from Eqs.(9,11). For a typical initial energy of $\tilde{K} = q\Delta\tilde{V}_z \approx \pm 0.25 eV$ and $E_2 \approx 25 V/mm$ one thus finds from Eq.(11): $\Delta\tilde{T} \approx \pm\sqrt{m}/(13.89133 \cdot 25q)\mu sec$, i.e. $\approx \pm 28.8q$ $nsec$ for a $q$–times charged ion of 100 Daltons. Note here that for a constant $q\Delta\tilde{V}_z$ one finds $\Delta\tilde{T} \propto \sqrt{m}/(qE_2)$ which together with $\overline{T} \propto \sqrt{m/q}$ [see Eq.(10)] yields $\overline{T}/\Delta\tilde{T} \propto E_2/\sqrt{q}$. Note further that differently from the case of a pulsed ion production of section 3.1 here the time–of–flight resolving power $\overline{T}/\Delta\tilde{T}$ and thus [see Eq.(7)] the mass resolving power $m/\Delta m = \overline{T}/(2\Delta\tilde{T})$ is the same for ions of different masses but improves with increasing field strength $E_2$. Since there is always some electronic jitter and pulse broadening, there is a lower limit for $\Delta\tilde{T}$ and consequently an upper limit for the mass resolving power. This is mainly important for the low mass ions for which $\Delta\tilde{T}$ is small.

### 3.2.1 The storage ion source

One way to make use of the considerations of section 3.2, is to build a storage ion source [17,23,24] as shown in Fig.6. In this case one chooses $E_2 \neq 0$ or $V_2 \neq V_1$ only for a short period while for most of the time one has $E_2 = 0$ or $V_2 = V_1$. During this time an electron beam ionizes all gas atoms in this region and simultaneously keeps all ions in the potential well formed by the space charge of the electron beam. During the period when $V_2$ differs from $V_1$, the ions are extracted and bunched to the "detector plane" by proper choice of $E_2$, $L_1$ and $L_0$ [see Fig.6 and Eq.(10)]. Thus ions may be formed for $\approx 100\mu sec$, extracted during $\approx 1\mu sec$ and bunched to form an ion pulse of $\approx 10...50nsec$ duration at the "detector plane".

Of these stored ions often only the heavy–molecule ions are of interest though the light ions are usually more abundant. Thus one can try to eliminate a large percentage of the light ions by applying a high frequency field over the ion cloud that removes the quickly accelerated light ions but leaves the heavy ions untouched or swings them a little back and forth only [31]. Since in the extracted ion beam at the end the light ions that are formed slightly prior to the ion extraction, are still predominant, one often deflects those to the side by a pulsed orthogonal field in order not to overload the ion detector.

Principally, the position where the ion bunching occurs can be changed to an arbitrary distance from the ion source by changing $\lambda$ or $E_2$ appropriately in Eq.(9,10). However, this position can also be considered to be a new ion–source from which all ions start within a time $\pm \Delta \tilde{T}$ as calculated from Eq.(11). The advantage of such an arrangement is that a time–of–flight system as shown in Fig.6 can be followed by a second time–of–flight system as shown in Fig.2 or Fig.4.

In any case the second TOF mass analyzer must handle an ion beam of a relatively large energy spread

$$\Delta K = \pm q E_2 z_{max} \leq q(V_2 - V_1) \tag{12}$$

with $z_{max} \leq L_2/2$. This energy spread – that quite desirably should be small – increases with $E_2$ while the flight time spread of Eq.(12), which also should be small, decreases with $E_2$. Thus, compromises are unavoidable.

### 3.2.2 The "orthogonal" TOF mass analyzer

Another way to make use of the above considerations is to build the TOF mass analyzing system as shown in Fig.7 and make use of the fact that $\tilde{v}_x$ can also be non zero. Such an arrangement [32,33,34] allows to use an external ion source that forms a continuous ion beam and sends it in x–direction into the region between the grids 1 and 2 shown in Fig.7. This approach has the advantage that an otherwise tested, efficient and good DC ion source can be used that had been designed with no compromises with regard to a bunched operation.

In this case the angle of divergence $\alpha_0 \approx \pm\tilde{v}_z/\tilde{v}_x$ of the incoming beam determines the maximal $\pm\tilde{v}_z$ for a given $\tilde{v}_x$ and thus the maximal $\Delta\tilde{T}$ according to Eq.(11). This angle $\alpha_0$ and consequently the maximal $\tilde{v}_z$ and $\Delta\tilde{T}$ usually can be changed by placing a lens of focal length F between the real ion source and the bunching device of Fig.7. This makes the ion beam wider in z–direction but –and this is important– more parallel.

Detrimental in the design of this "orthogonal" TOF mass analyzer is that the flight time $\overline{T} \approx L_0/v_z$ of the ions measured between the time of the extraction, i.e. the time when the bunching pulse is applied to the grids 2 and/or 1 in Fig.7, and the ion arrival at the ion detector, is usually larger than the time $\tilde{T} \approx L_x/\tilde{v}_x$ necessary to fill the ion source except for very heavy ions since $L_0/v_z \ll L_x/\tilde{v}_x$. For some period thus ions will move in x–direction beyond the grid arrangement shown in Fig.7 and consequently are lost for any investigation.

### 3.2.3 TOF–mass analyzers for the investigation of fast changing mass distributions

In some cases the sample material is only available for a limited time, a time that may barely be sufficient for the recording of a full mass spectrum in a scanning mass spectrometer. This situation exists for instance if the effluent of a gas chromatograph must be analyzed [35,36] as is indicated in Fig.4. If some portion of these molecules would be ionized and mass analyzed in a scanning mass spectrometer, for instance a quadrupole mass filter [37,38], one decade of the obtained mass spectrum would usually be recorded in some large fraction of a second, while the GC peak has a width of about one second. Thus it is obvious that the number of ions that enter the spectrometer varies with time. Consequently the intensity distribution of the mass spectrum becomes skewed.

In a time–of–flight mass spectrometer on the other hand the situaton is different. Here all ions are accelerated by for instance $1000V$ so that all q–times charged ions have energies of $1000qeV$ and thus move with velocities of $\approx 14\sqrt{Kq/m} \approx 22mm/\mu sec$ if $q = 3$ and $m$ is 1200 Daltons. The total ion flight time in such a $1m$ long system thus is about $80\mu sec$ if an ion mirror is employed so that the system length is used twice. Over the short time of $80\mu sec$, however, the intensity in a typical 1 second GC–peak [see Fig.8] is practically constant. This would even be the case if the GC–peak would be reduced to 0.1 seconds or less [35].

Also, TOF–mass analyzers record all ions that are formed in the ion source at any time if a storage ion source is used as described in section 3.2.1. This should be compared to the performance of a scanning mass spectrometer that at one instant records only perhaps 1/3000 of these ions, i.e. those of one mass since all others are eliminated. Thus, a TOF–mass analyzer is very well suited to obtain mass spectra of GC–effluents with high detection efficiency.[4] This property is most important if the mass spectrum of the substance of a GC–peak is unknown since in a slow scanning mass spectrometer the timing could be such that the most intense mass peak is recorded when the GC–peak intensity is low so that it still could go by undetected.

Using a $50\mu m$ capillary of 2 m length in the system of ref.[39] we have produced GC–peaks of about 100 msec width [see Fig.9] by recording TOF mass spectra all the time and adding the intensity recorded for all masses from 30 to 150 Daltons. This fast adding as well as storing the result with modest electronic effort perhaps every 20 msec required some effort [39]. Integrating Transient Recorders [ITR] that can achieve this task have been built [39,40]. However, though there were great improvements in fast electronic circuitry and even more so in fast computers over the last years, this part of the technology of TOF mass analyzers still has room for improvements.

---

[4]In ref.[39] a mass spectrum with a signal to noise ratio of about 1:40 was obtained from a few pg of hexachlorobenzene.

Also fast chemical reactions can be studied advantageously by a TOF mass analyzer. An example for such an investigations is illustrated in Fig.10. There a puff of gas is brought to a surface at which the molecules may undergo some surface reaction. To study this reaction one must obtain a full mass spectrum in a time short as compared to the molecule reaction time. Additionally such an investigation must be very sensitive since the number of molecules in the gas puff must be so small that only a small fraction of a molecule monolayer can develop on the surface in question. Otherwise not the decomposition of the molecules on the surface would be analyzed but rather the molecule decomposition on a molecule–covered surface. In ref.[41] such a specialized TOF mass analyzer is described by which the decomposition of acetic acid molecules on a heated palladium surface was investigated (see Fig.10) with a surface coverage of less than $1/1000$ of a monomolecular layer.

# 4 Multipass TOF Mass Analyzers

In most TOF mass analyzers the mass resolving power is limited by the time needed to form or to record a short ion bunch. Thus – as stated in sections 3 and 4 – increasing the length of the flight path also increases the mass resolving power. Most efficiently this is done by using the same physical path repeatedly.

## 4.1 Ring TOF Mass Analyzers

One way to make an ion fly along the same path several times, is to form the flight path to be a ring. In this case one designs the system optimally such that

1. the beam diameters are identical after each turn for a beam of multi-energy ions, and

2. the flight time per turn is independent of the relative energy spread $\delta_K = \Delta K/K_0$

If a small cloud of ions moves in such a storage ring, the passing charges can produce signals in some pick–up electrodes with the recorded frequency being a measure for the ion mass very similar as in a Fourier–Transform mass analyzer [8,42,43,44]. Also one can eject the ions from the ring and determine their flight time after an additional flight distance plus n–turns [42] in the ring. For low energy ions such a ring could consist of 8 identical toroidal electrostatic sector fields [see Fig.11].

## 4.2 Folded TOF Mass Analyzers

Another way to increase the ion flight path is to arrange several TOF systems in series [45,46]. Here the optical quality of the ion mirrors is important since the ion beam passes through a relatively large number of such mirrors.

Alternatively one can use ion mirrors that are operated in a pulsed mode for instance like indicated in Fig 12. In such a system one could bounce ions back and forth repeatedly [46] between the mirrors I and II. Mirror I here must be switched off when the ion bunch to be analyzed enters the system, but it must be powered on, when the fastest ions of the bunch return after one passage through the system. On the other hand, mirror II must be switched off after a certain time in order to let the ion beam reach the final ion detector after so and so many reflections. In such a system the time separation between neighbouring mass ions increases with every passage through the system and so does the mass resolving power, provided the full width at half maximum (FWHM) of the ion bunches stays about constant as in ref.[46].

# 5 Fast Ion Detection

Though one can easily appreciate the advantages of obtaining mass spectra quickly, there are also difficulties caused by the need for fast analog electronics and fast computers to record short ion bunches. Usually this is achieved by letting the ions impinge on channel plates or on fast open dynode multipliers[5]. In special cases converter electrodes are foreseen [47] on which the impinging ions form secondary electrons that are then guided to and amplified by double channel plates. Such converter foils are advantageously

---

[5]Such double channel plates can be built as chevron devices with no space between them. Usually, however, they are separated by fractions of a mm and the electrons are accelerated over this space by a few 100 Volts as is the case also between the last channel plate and the electron collector electrode [see Figs.3,12]

covered by for instance Cs or technologically easier CsI to improve the yield of secondary electrons [48]. For the combination of a converter electrode with channel plates, furthermore, the ion detection efficiency is usually increased by about a factor of two as compared to ions that impinge directly onto the first channel plate. This effect results from the fact that only ions that impinge on the channel plate surface create effectively secondary electrons that may be sucked into the amplifying channels, while ions that enter into the channels themselves release secondary electrons only in some depth in the channel. Thus only a fraction of the channel is available for amplification and the overall amplification factor is reduced for these electrons

For each intial ion at the end several $10^6 - 10^7$ electrons are obtained with the electron pulse rising in about $200psec$ and falling off in about $800psec$. Thus signals of $\approx 1nsec$ full width at half maximum (FWHM) need to be processed further and if several ions are contained in one bunch, signals of several $nsec$ FWHM. Commonly all electrons that leave the last channel plate are accelerated to a collector electrode, i.e. a flat conductive surface of about the area of the channel plate, from where the signal must be conducted away through a good coaxial cable. Here it is important to also use well matched vacuum feed throughs in which signal reflections caused by changes in their impedance are minimized.

To couple the signal from the collector electrode to a coaxial cable has been done since some time by conical coupling devices [49] of perhaps 80mm in length. Very good results have been obtained since some years, however, also [50] by simply starting the coaxial cable – usually terminated by a $50\Omega$ resistor – only a few mm away from the collector electrode (see the distance L in Fig.13). In this case the signals are reflected back and forth over this short distance but since the signal velocity is close to the velocity of light the reflected signal overlaps with the primary signal after about $100psec$ or less. Thus the frequencies of these signal distortions are in the $10GHz$ range and consequently are not amplified in the downstream electronic circuitry that usually has frequency limits of some $100MHz$.

Rather important is also that capacitors are installed parallel to the channel plates and at least the last dynodes of an open multiplier. Quite important is, however, also a good capacitor that connects the last channel plate or dynode surface to ground [see Fig. 13] since this capacitor directly reduces high frequency ringing. In some cases it also is advantageous to have the collector electrode not at ground potential and to couple the fast signal through a capacitor into the coaxial cable – usually with no noticeable signal degradation. This can efficiently be done by placing a thin plastic foil between the collector electrode and another plate of about equal size [see Fig.13]. This arrangement can even be improved [51] if this foil is replaced by an insulator that has a high dielectric constant.

For some applications it is of interest to not only record the arrival time of an ion but also its position. The most simple way to do this is to divide the electron collector into many electrodes for instance arranged like a checker board with individual amplifiers. The problem is here, however that neighbouring electrodes may be hit simultaneously. In order to obtain this position information very quickly, also an arrangement like shown in Fig.14 can be useful, in which a grid system is installed behind a stack of channel plates. The center grid then records the time signal while the signals from the two meander like grids yield signals the delay times of which are measures for the $x$ and $y$ positions of the ion under investigation [52].

# 6 Summary

Until 20 years ago time–of–flight mass spectrometers were quite popular. However, the quickly improved performance of sector field mass analyzers and quadrupole mass filters had at that time pushed them off the main stage of mass spectrometric research. In recent years, however, a renaissance of time–of–flight mass analyzers has occurred, which is largely due to the improved system technology, as well as new electronic circuitry and computer technology.

The applications of TOF–mass spectrometry so far utilize mainly

1. its unlimited mass range, which features also a mass resolving power that improves with increasing ion mass in case of a pulsed ion production

2. its fast scanning capability that allows to monitor fast changing substance concentrations

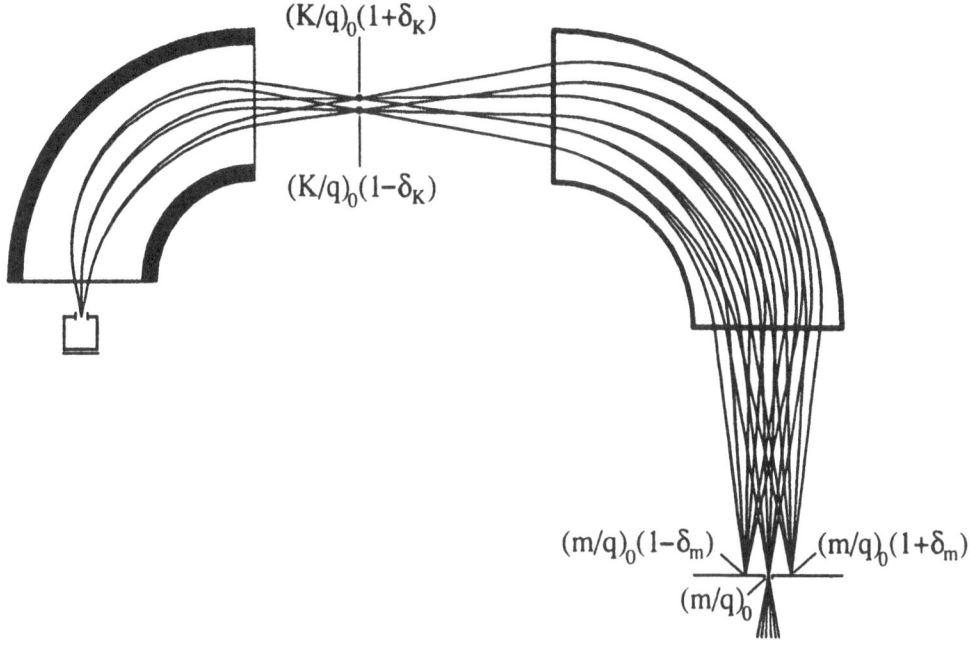

$(K/q)_0(1+\delta_K)$

$(K/q)_0(1-\delta_K)$

$(m/q)_0(1-\delta_m)$  $(m/q)_0(1+\delta_m)$

$(m/q)_0$

Figure 1. An angle– and energy–focusing – often called double focusing – sector field mass analyzer is shown, . In this system ions of equal mass–to–charge ratio $(m_0/q_0)$ and two energy–to–charge ratios $(K/q) = (K_0/q_0)(1 \pm \delta_K)$ are shown as passing through the same exit slit, while ions of different mass–to–charge ratios are eliminated. Note that three mass–to–charge ratios $(m_0/q_0)$ and $(m/q) = (m_0/q_0)(1 \pm \delta_m)$ are indicated. Note also that the width of the beam of one "ion species" should be larger than the corresponding slit opening in order to finally obtain the intensity distribution in this beam when it is scanned across the exit slit.

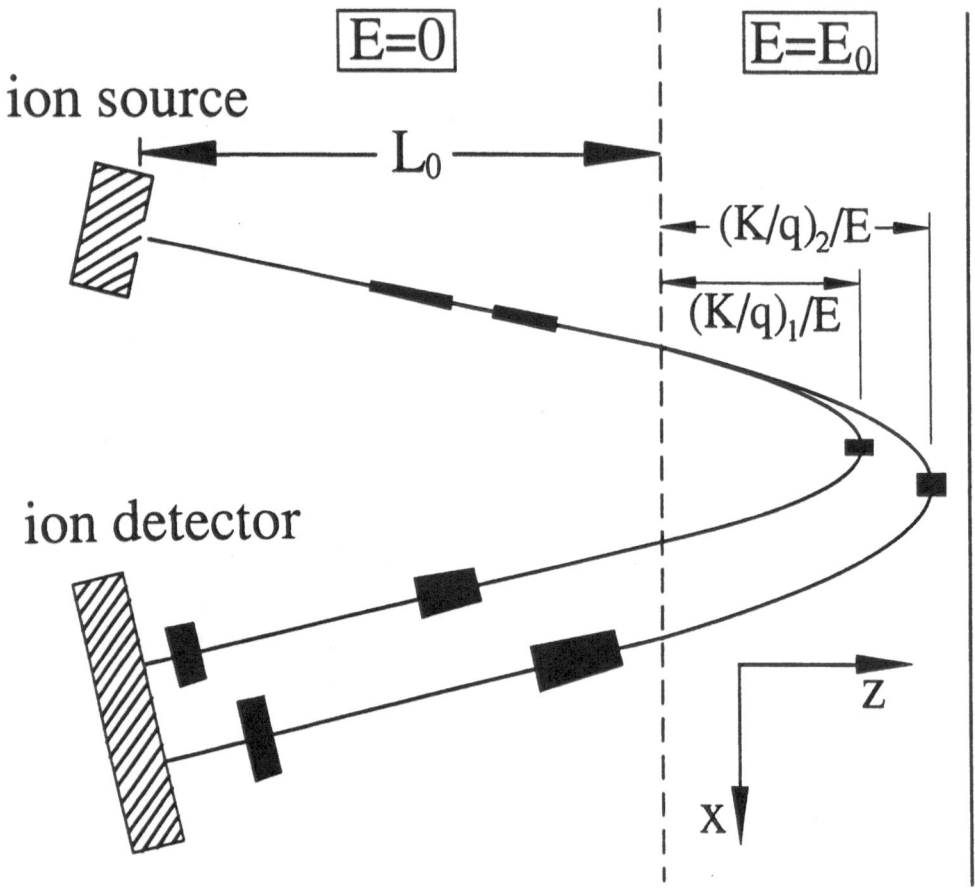

Figure 2. An energy–isochronous time–of–flight mass analyzer is shown that includes an electrostatic repeller field $E = E_0$ into which more energetic and thus faster ions penetrate deeper. This flight path elongation compensates – if properly dimensioned – an increased ion velocity such that the overall flight time becomes independent of energy deviations [7] to first order. Note the indicated positions of two groups of ions of equal masses but different energy–to–charge ratios $(K/q)_1$ and $(K/q)_2$ shown for different times. Note also that the ions must pass through the meshes of a grid that defines the extent of the repeller field but that deteriorates the ion beam.

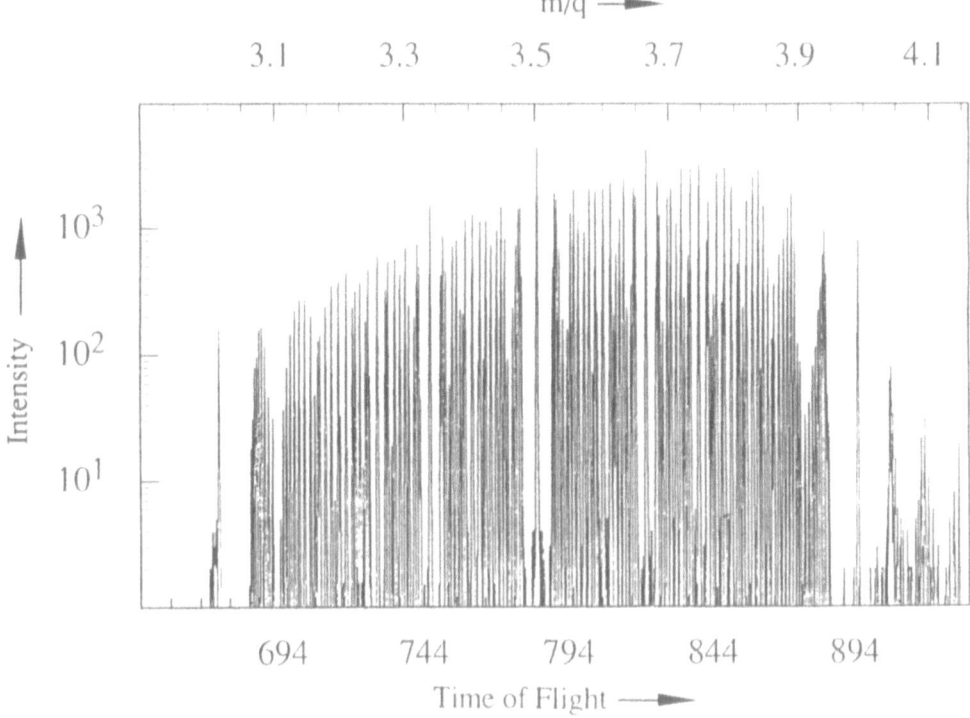

Figure 3. A mass to charge spectrum obtained by the TOF mass analyzer of ref [11,12] is shown, in which the obtained ion intensity is plotted as function of the ion flight time for the investigated proton–induced Th–fragments which have energies of $\approx 1 MeV/u$ and usually spread over about 5 charge states. Note that there are many ions that have the same $m/q$ for instance $m/q = 24/8 = 27/9 = 30/10 = 33/11 = ...$ but that there are gaps in the neigbourhood of $m/q$–values 2.5, 3.0 etc since $q$ can take up integer values only. To distinguish ions of the same $m/q$ value one can determine additionally – though mostly only approximately – the flight time of each ion in a field–free region and thus the true ion velocity. The ion's charge $q$ then is found from a comparison of this flight time and the one in the TOF mass analyzer.

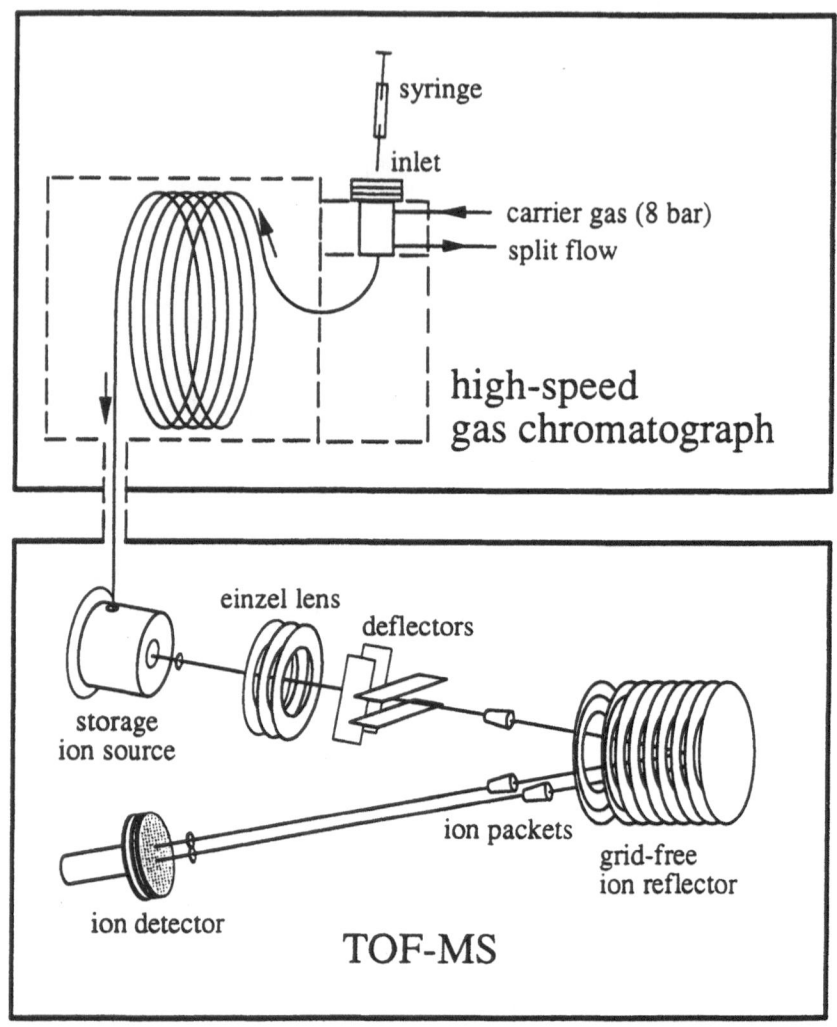

Figure 4. A time–of–flight mass analyzer is shown including a pulsed gas ion source and a particle detector as well as a grid–free ion mirror and the necessary adjustment devices, i.e. a beam deflector and an adjustment lens. Also shown is a gas chromatograph the effluent of which is to be mass analyzed by the TOF system [39]. Note, that the shown ion bunches are short at both the front end and the back end as well as close to their apices in the middle of the repeller field, but that they are long in between these positions.

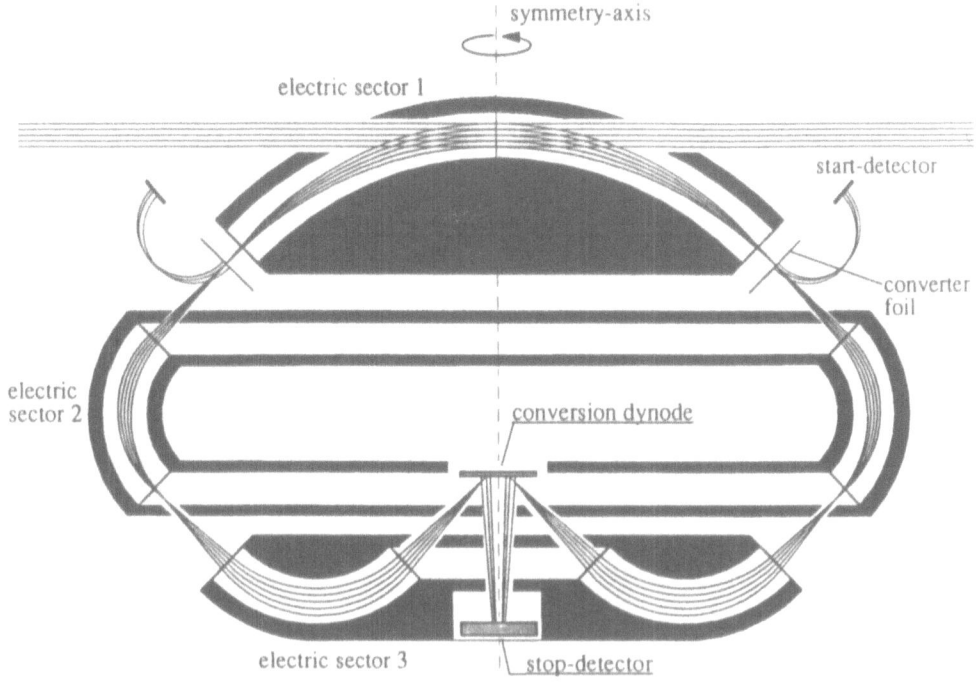

Figure 5. An energy–isochronous time–of–flight mass spectrometer for upper atmosphere or solar wind investigations. The ions under consideration here pass through a thin foil at the surface of which they create secondary electrons that serve as start signal and then flie energy–isochronously to a stop detector. The energy isochronicity here is achieved by electrostatic sector fields. Note that in this special system [18,53] poloidal electrostatic sector fields are used that are rotationally symmetric to the shown axis of rotation. Thus all ions are mass analyzed at the end that are initially perpendicular to the shown axis of rotation, where the azimuthal arrival position characterizes the initial azimuthal angle of incidence.

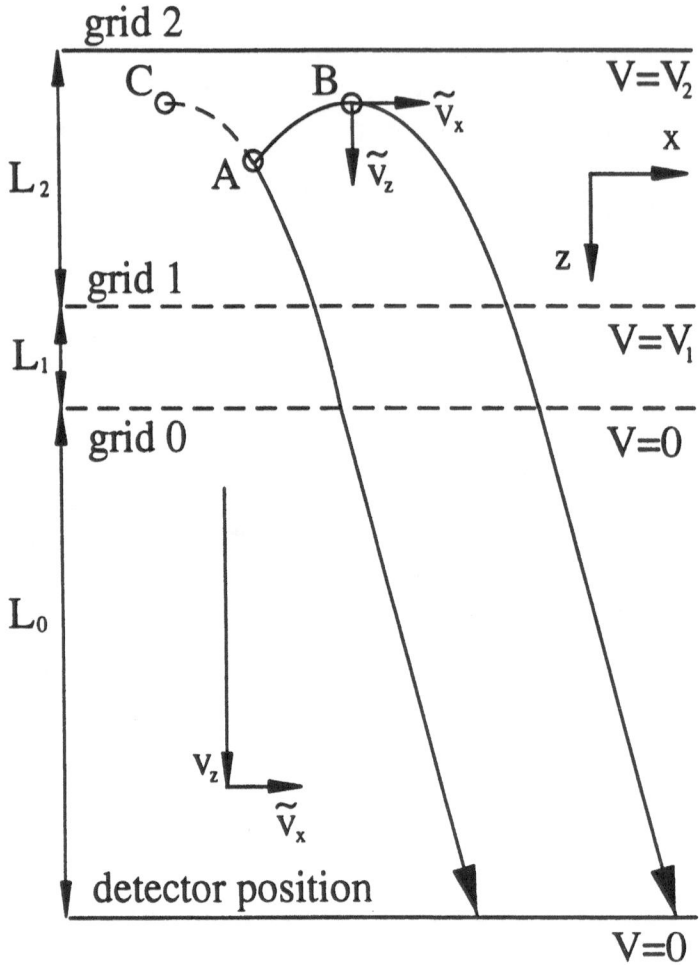

Figure 6. Sketch of an ion buncher in which the x-component of the ion velocity $\tilde{v}_x$ stays undisturbed while the z-component $\tilde{v}_z$ is changed to $v_z$ by a pulsed field $(V_2 - V_1)/L_2$. The magnitude of this field must be chosen such that the ion flight times to the ion detector – a distance $L_0$ downstream – become independent of the initial z-position of the ions under investigation. If an ion, that starts from point A, is not at rest initially but moves towards point B, it arrives at the final ion detector delayed. This delay time is equal to the time the ion needed to reach point B where the z–component of the ion's velocity vanishes. If an ion that starts from point A seems to come from point C, it arrives early at the final ion detector with the difference in arrival times being the time the ion would have needed to move from point C to point A.

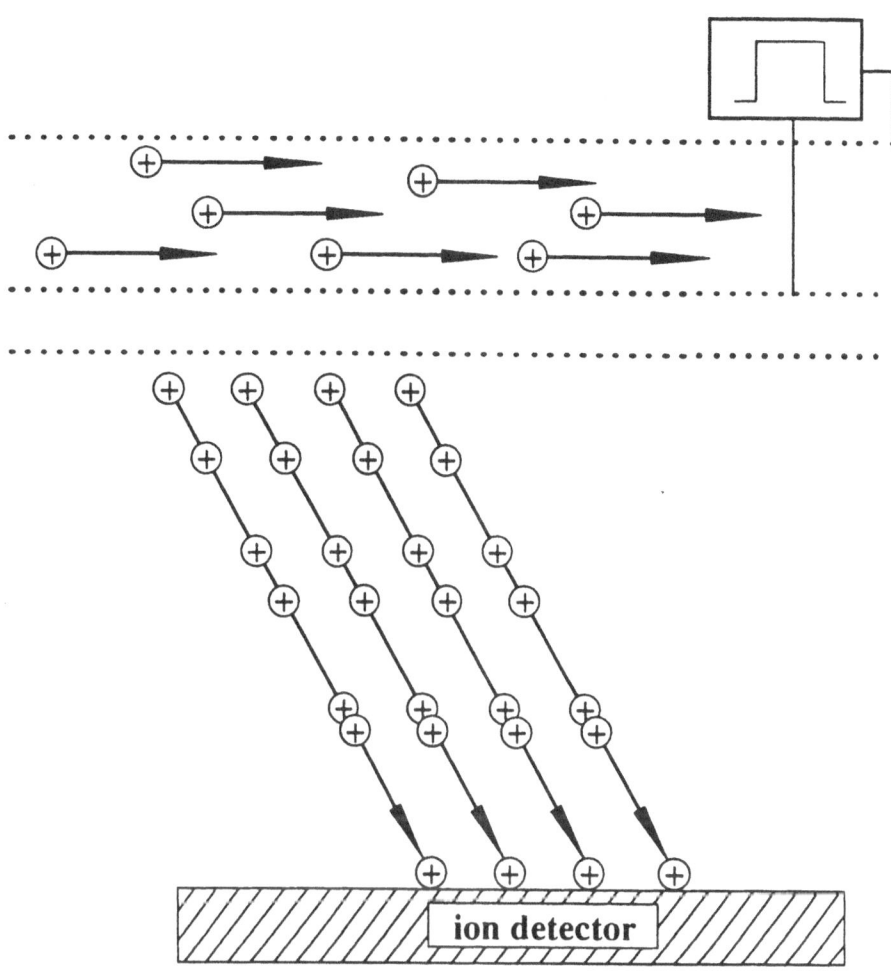

Figure 7. "Orthogonal" time–of–flight spectrometer in which the ion beam initially moves perpendicularly to the electric field of the ion buncher. Note that the ions initially should move as parallely as possible at rather low velocities. Note further that the final ion trajectories are not perpendicular to the incoming ion beam even if the $z$-component $\bar{v}_z$ of the ion velocity has been enlarged considerably to $v_z$ by the electrostatic field between three grids indicated. Note finally that the ions are bunched at the position of the ion detector since the ions that initially are further away from this detector receive a little more energy by the pulsed potential on the grids.

136

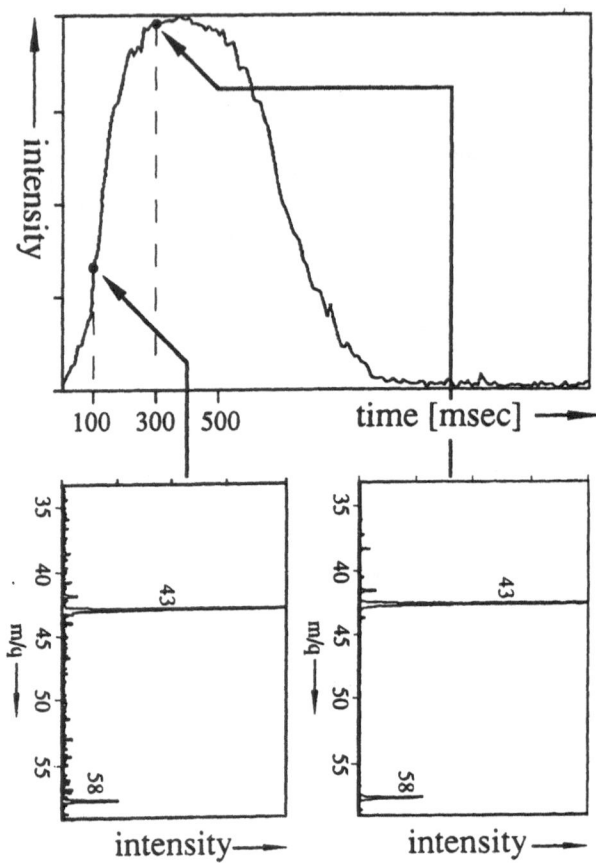

Figure 8. TOF–Mass analysis of the GC–peak of acetone [36]. Shown is the GC–peak – that has a base width of about 1 second. Note also that the mass spectra obtained at 100 msec and at 300 msec are almost identical. The only difference is that in the mass spectrum at 300 msec the signal-to-noise ratio is considerably better than in the mass spectrum at 100 msec. Summing all TOF–mass spectra over the duration of the GC–peak furthermore reveals a very sensitive recording of the overall mass spectrum with a $\approx \sqrt{1000}$–times improved signal-to–noise ratio. This situation is drastically different from the single mass spectrum, one would have obtained by a slow scanning quadrupole mass spectrometer in which the intensity of the mass spectrum is folded with the GC–peak height.

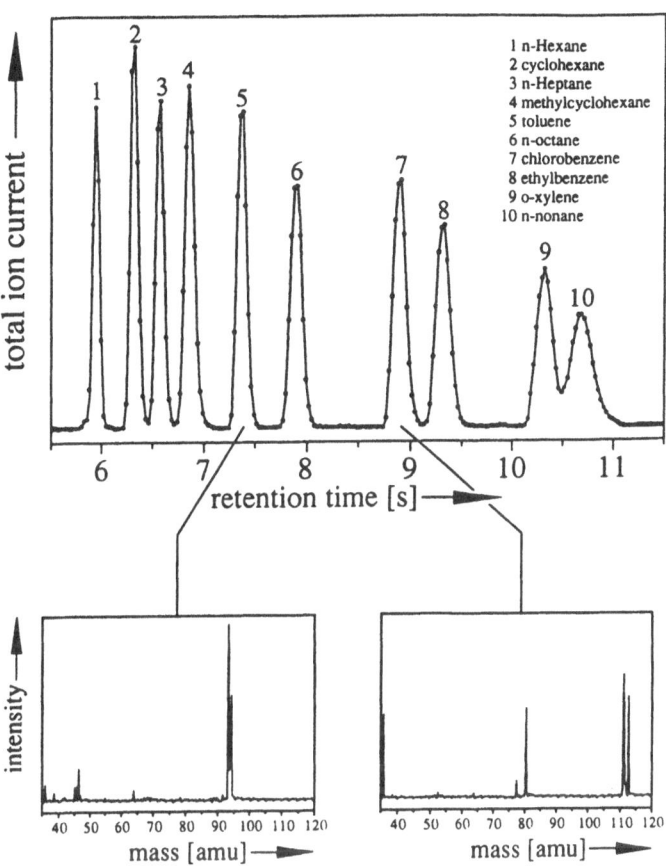

Figure 9. Fast GC–recording of a mixture of 10 molecules for which the intensity of all ions between 30 and 150 were added [39]. Note that in order to have about 6 intensity points per GC–peak the overall time for a GC–chromatogram of about 5 seconds required to sample 35 mass spectra per second. Each of these mass spectra consisted of the sum of a large number of TOF mass spectra, each of which was summed over about 20 mass spectra each of which was recorded within about $100\mu sec$, after the ions of interest had been collected in the storage ion source. For each mass spectrum then the signal strengths between the masses of 30u and 150u were added to obtain the shown GC–chromatogram. Parallely also the mass spectra for GC–peaks were summed. For two of the GC–peaks the resulting mass spectra are shown.

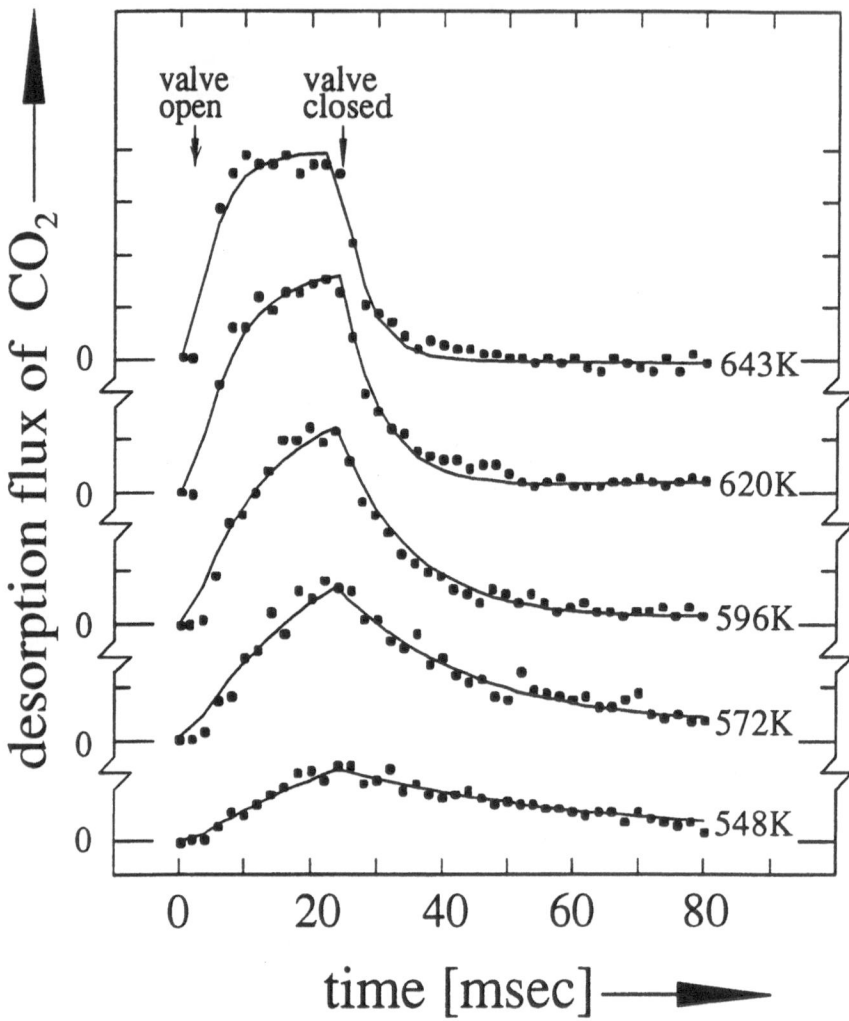

Figure 10. The time dependence of the intensity of $CO_2$ molecules is recorded by a TOF mass analyzer after a puff of acetic acid molecules was applied to a palladium surface that was heated to different temperatures. The number of molecules of acetic acid here could be kept so small that at most 1/1000 of a monolayer of such molecules existed at the palladium surface during the investigation.

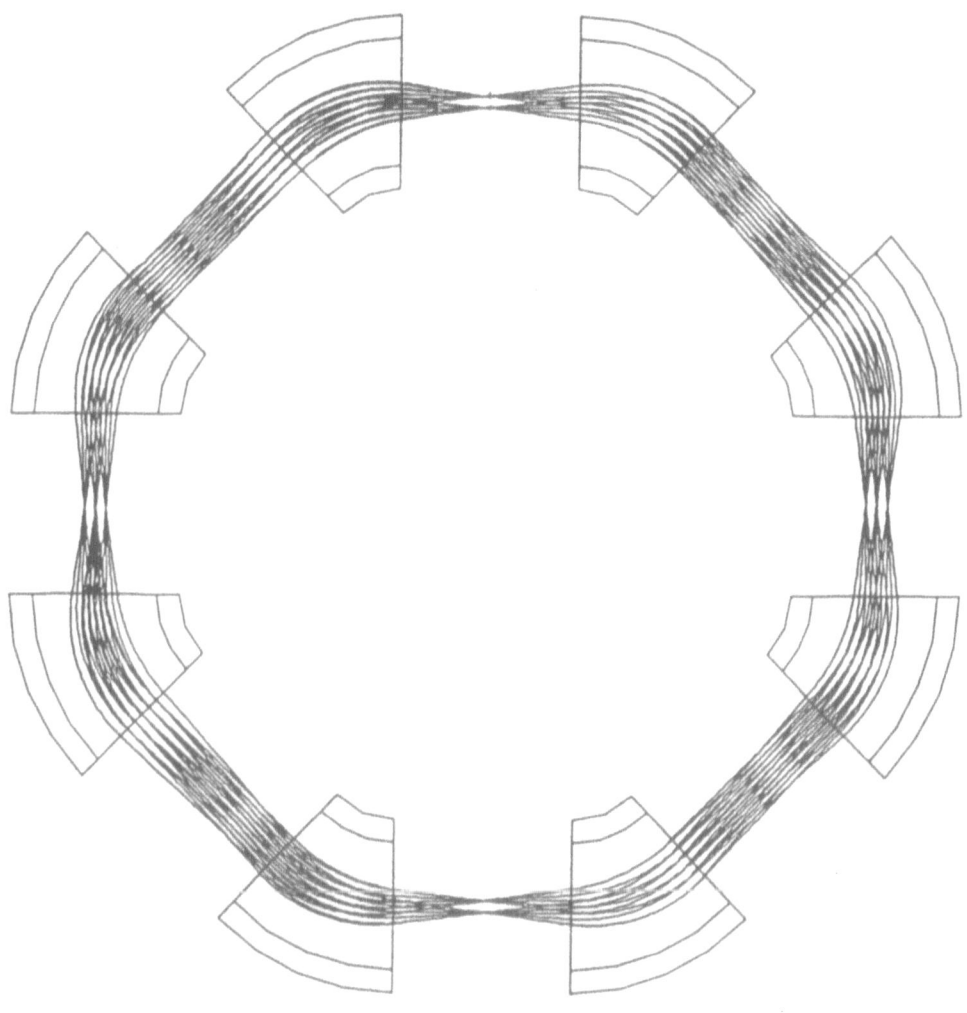

Figure 11. An energy–isochronous storage ring for ions is shown that consists of 8 electrostatic sector fields of 45° deflection angle each.

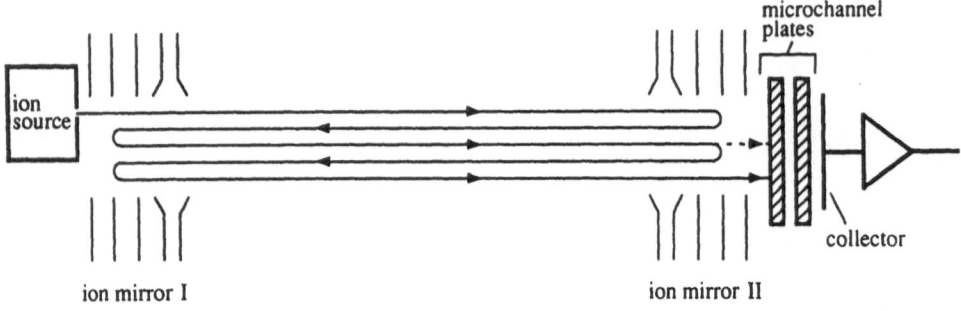

Figure 12. A TOF mass analyzer is shown in which the ions are passed back and forth between two ion mirrors repeatedly.

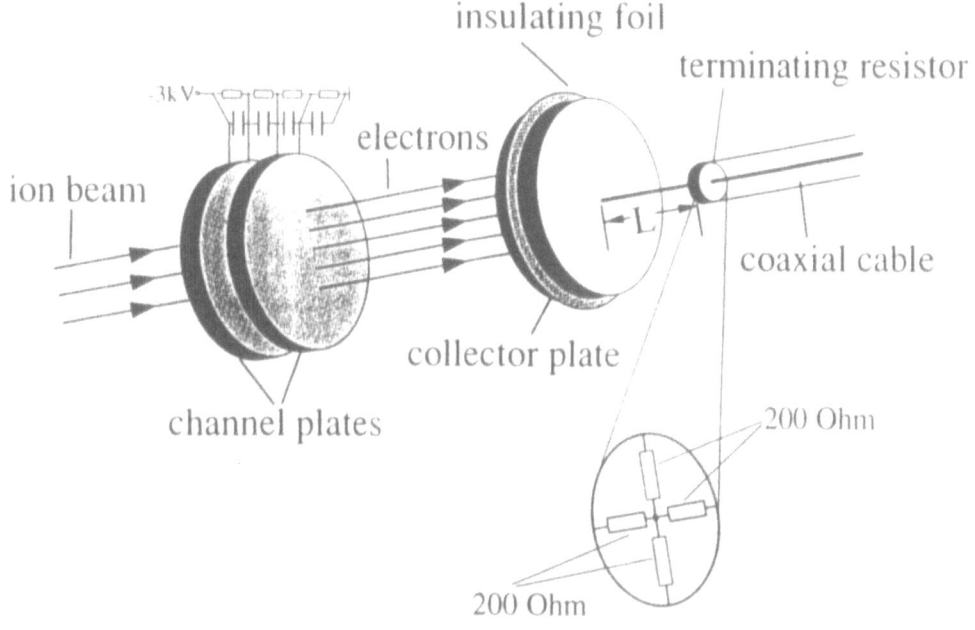

Figure 13. An ion-registration system is shown consisting of a double channel plate amplifier that usually has a gain of about $10^6$ or $10^7$. Note the capacitor formed by a plastic foil between two relatively large conductive plates, that couples the fast signal to a coaxial cable. Note also the short distance "$L$" between the final collector plate and the $50\Omega$ terminator of the coaxial cable as well as the shown capacitors.

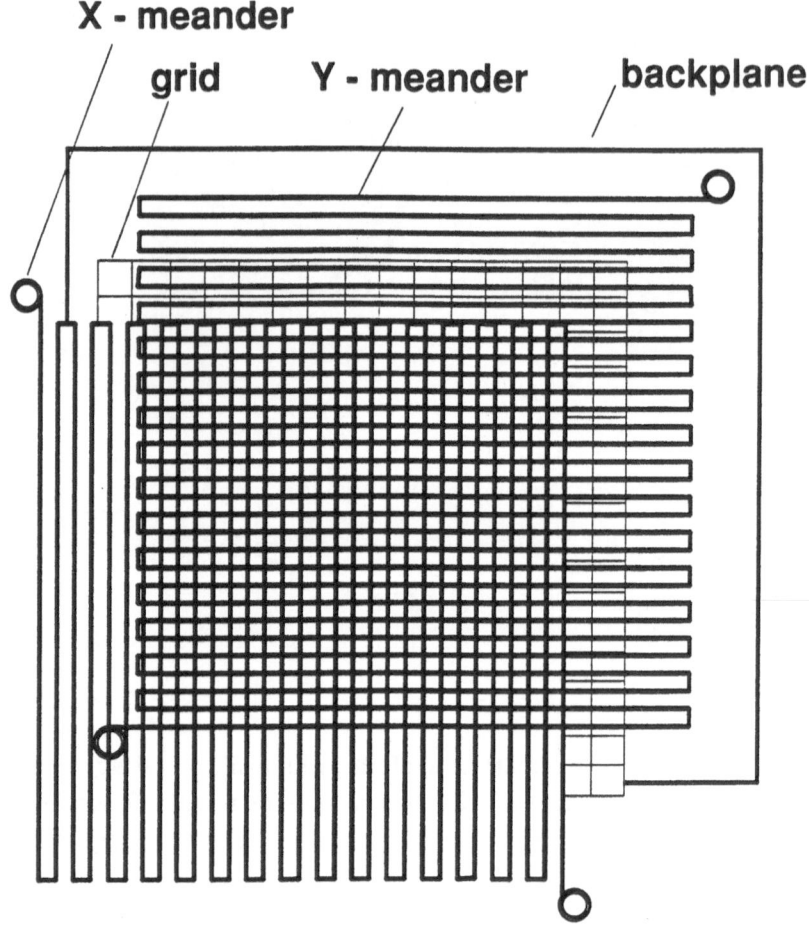

**X - meander**

**grid**    **Y - meander**    **backplane**

Figure 14. A position sensitive fast timing detector. The time defining pulse here is taken from the center grid, while the x- and y-positions are determined from the observed delays of the signals on the meander–like grids. Over a distance of $40mm$ this detector has resolved $\pm 1mm$ both in x- and y-position [52].

3. its efficient use of all formed ions and the resulting high sensitivity.

Especially the application of the first property has come of age. The development of the PDMS [19] and the MALDI [27,28,29] technique now allows routinely to determine molecules of weights of $10^4$ and $10^5$ mass units. This technology has so far been established in many installations worldwide and only waits for an improved ion–detection technology that increases the so far limited mass resolving power for very heavy molecular ions.

The fast scanning capabilities of TOF systems have only recently been put to use. Combined with the high detection sensitivity of TOF systems this fast scanning capability should especially improve the range of the mass spectrometric investigations of surface reactants [41]. Especially in view of the analysis of the effluents of fast gas chromatrographs [39] or of new capillary zone electrophoresis systems, time–of–flight mass analyzers may become indispensible tools for the chemical analysis.

# 7 Acknowledgement

For fruitful discussions I am indebted to A. Kraft, R. Becker and R.Scherer. For financial support I am thankful to the "Bundesminister für Forschung und Technologie" and to the "Deutsche Forschungsgemeinschaft".

# 8 References

1. F.W. Aston, Philos. Mag. **38**(1919)709

2. J. Mattauch, R. Herzog Z.Phys. **89**(1934)786

3. H. Wollnik, "Optics of Charged Particles", 1987 Acad. Press, Orlando, pp 265,28,220

4. E. O. Lawrence and M. S. Livingston, Phys. Rev. **37**(1931)1707

144

5. A. G. Marshall and L. Schweikhard, Int. J. Mass Spectr. and Ion Proc. **118/119**(1992)37

6. W. C. Wiley and I. H. Mclaren, Rev. Sci. Instr. **26**(1955)150

7. S. G. Alikanov, Soviet. Phys. JETP **4**(1957)452

8. W. P. Poschenrieder, Int. J. Mass Spectr. and Ion Phys. **9**(1972)35

9. H. Wollnik, Nucl. Instr. and Meth. **186**(1981)441

10. H. Wollnik and T. Matsuo, Int. J. Mass Spectr. and Ion Phys. **37**(1981)209

11. J. Wouters, D. J. Vieira, H. Wollnik, H. A. Enge, S. Kowalski, K. L. Brown Nucl. Instr. and Meth. **240**(1985)77

12. H. Wollnik, J. M. Wouters, D. J. Vieira, Nucl. Instr. and Meth. **A258**(1987)331

13. H. Wollnik, Nucl. Instr. and Meth. **A258**(1987)289

14. B. A. Mamyrin, V. I. Karataev, D. V. Shmikk, V. A. Zagulin Soviet Phys. JETP **37**(1973)45

15. W. Gohl, R. Kutscher, H. J. Laue, H. Wollnik, Int. J. Mass Spectr. and Ion Physics **48**(1983)411

16. T. Bergmann, T. P. Martin, H. Schaber, Rev.Sci. Instr. **60**(1989)347,792

17. H. Wollnik, U. Grüener and G. Li, Ann. Phys. **48**(1991)215

18. D. Young, J.A. Marschall Nucl. Instr. and Meth. **A298**(1991)227

19. D. F. Torgerson, R. P. Skowronski. R. D. Macfarlane, Bioch., Biophys. Res. Comm. **60**(1974)616

20. A. Benninghoven, D. Jaspers, W. Sichtermann, Appl. Phys. **11**(1976)35

21. R. Kaufmann, F. Hillenkamp, R. Wechsung, Med. Progr. Technol. **6**(1979)109

22. A. Duhr, R. Guckert, H. Wollnik, ASMS Conf. Proc. (1993) 929A

23. H. Wollnik, R. Grix, G. Li, R. Kutscher, P. Feigl
Proc. Japan–China Symp. on Mass Spectr., eds. H. Matsuda and
X. Liang, (1987) Bando Press, Osaka, p. 181

24. R. Grix, R. Kutscher, G. Li, U. Grüner, H. Wollnik, Rapid Comm. in
Mass Spectr. **2**(1988)83

25. R. J. Cotter, Anal. Chem. **64**(1992)1027A

26. U. Boesl, J. Grotemeyer, K. Müller-Dethlefs, H. J. Neusser, H. L. Sel-
zle, Int. J. Mass Spectr. and Ion Proc. **118**(1992)191

27. K. Tanaka, Y. Ido, S. Akita, Y. Yoshida, T. Yoshida,
Proc. Japan–China Symp. on Mass Spectr., eds. H. Matsuda and
X. Liang, (1987) Bando Press, Osaka, p. 185

28. A. Overberg, A. Hassenbürger, F. Hillenkamp, in "Mass Spectr. in
Bio. Sci: A Tutorial" ed., Michael L. Gross, (1992) Kluwer Acad.
Publishers, Dordrecht, p. 181

29. F. Hillenkamp and M. Karas, in: "Methods in Enzymology" ed., J.A.
McCloskey, (1990) Acad. Press, New York, pp. 280

30. S. M. Michael, D. M. Lubman, Rev. Sci. Instr. **63**(1992)4277

31. J. Tammoscheit, thesis 1992, Uni Giessen

32. G. J. O'Halloran, R. A. Fluegge, J. F. Betts, W. L. Everett, (1964)
ASD TDR report 62–644, Bendix Co.

33. J. H. J. Dawson and M. Guilhaus, Rapid Comm. in Mass Spectr.,
**3**(1989)155

34. A. F. Dodonov, I. V. Chernushevitch, V. V. Laiko
12th Int. Mass. Spectr. Conf., (1991) 159

35. P. A. Leclercq and L. A. Cramers, J. High Resol. Chromat **11**(1988)845

36. R. Grix, R. E. Tecklenburg, J. F. Holland, 2. Int. Symp. Applied Mass
Spectr. in Health Sci., Barcelona, (1990)259

146

37. W. Paul and H. Steinwedel, Z. Naturforschung **A8**(1953)448

38. P. H. Dawson, "Quadrupole Mass Spectrometry and its Applications", 1976 Elsevier, Amsterdam

39. H. Wollnik, R .Becker, H. Götz, A. Kraft, H. Jung, C.-C. Chen, P. G. Ysacker, H.-G. Janssen, H. M. .J. Snijders, P. A. Leclerq, L. A. Cramers, Int. J. Mass Spectr. and Ion Proc., submitted

40. J. F. Holland,B. Newcombe,R. E. Tecklenburg, M. Devenport,// J. Allison, J. T. Watson, C. G. Enke, Rev. Sci. Instr. **62**(1991)69

41. H. J. Kelleter, G. Schüll, C. D. Kohl, U. Grüner, A. Kraft, H. Wollnik, Surface and Interface Analysis, **19**(1991)581

42. H. Wollnik, Nucl. Instr. and Meth. **B26**(1987)267

43. M. A. May, A. G. Marshall, H. Wollnik, Anal. Chem.,**64**(1992) 1601

44. D. Liesen, GSI-Rep. 88-42 (1988)

45. C. Su, Int. J. Mass Spectr. and Ion Proc. **88**(1989)21

46. H. Wollnik and M. Przewloka, Int. J. Mass. Spectr. and Ion Proc. **96**(1990)267

47. N. R. Daly, Rev. Sci. Instr. **31**(1960)264

48. H. C. Seifert, D. J. Vieira, H. Wollnik, J. M. Wouters Nuc. Instr. and Meth. **A292**(1990)533

49. G. Beck, Rev. Sci. Instr. **47**(1976)849

50. H. Wollnik, Mass. Spectr. Reviews, in print

51. A. Esper, Thesis 1991, TH Aachen

52. C. Klein, J. Trötscher, H. Wollnik Nucl. Instr. and Meth., **A335**(1993)146

53. M. Yavor, B. Hartmann and H. Wollnik, Nucl. Instr. and Meth., in print

# THE DETECTION OF HIGH MASS-TO-CHARGE BIOLOGICAL IONS BY FOURIER TRANSFORM MASS SPECTROMETRY: ISSUES AND ROUTES FOR INSTRUMENT IMPROVEMENT

CHRISTOPHER L. HOLLIMAN, DON L. REMPEL,
and MICHAEL L. GROSS
*Midwest Center for Mass Spectrometry, Department of Chemistry*
*University of Nebraska-Lincoln*
*Lincoln, Nebraska, 68588-0362*
*USA.*

ABSTRACT. Although Fourier transform mass spectrometry (FTMS) has demonstrated the ability to trap biomolecular ions that have mass-to-charge ratios on the order of 150000, FTMS is seldom employed as a tool for the characterization of biological molecules. This is because the superior detection performance that is associated with FTMS rapidly degrades for ions with mass-to-charge ratios greater than 5000. The purpose of this chapter is to introduce the reader to the issues that stand in the way of the development of routine high mass-to-charge FTMS and to the recent advances and strategies that are extending the impressive capabilities of FTMS to singly-charged, biological molecules.

## 1. Introduction.

Literature reports often tout the superior resolution, sensitivity and mass-accuracy capabilities of Fourier transform mass spectrometry (FTMS). These claims are indeed true for ions of *m/z* less than a few thousand. The organic mass spectrometry literature is replete with examples of the utility and importance of FTMS for the elucidation of ion-molecule reaction pathways and kinetics. On the other hand, the biological mass spectrometry literature demonstrates that FTMS is seldom employed for biological applications. The absence of FTMS applications in the biological literature has been, in part, due to the inability to realize the potential of the technique for ions of high mass-to-charge ratio. We expect, however, that reports of the application of FTMS coupled with electrospray ionization will multiply rapidly owing to the high resolving power that can be achieved for multiply charged ions on the order of *m/z* 1000 as demonstrated by resolving powers of 103000 for the 13+ state of equine myoglobin (*m/z*

*R. M. Caprioli et al. (eds.), Mass Spectrometry in Biomolecular Sciences, 147–175.*
© 1996 *Kluwer Academic Publishers.*

1300 MW 16951) [1] and 500000 for the +21 state of carbonic anhydrase (*m/z* 1380, MW 29025) [2].

The purpose of this chapter is to introduce the reader to the issues that stand in the way of the development of routine high mass-to-charge FTMS and to the recent advances and strategies that are extending the impressive capabilities realized for low-mass-to-charge (< 2000) organic molecules to high-mass-to-charge biological molecules. Our intentions do not include, however, to provide a review of electrospray (see the chapter by Caprioli) or other biological applications of FTMS.

## 2. FTMS Principles.

The principles of the FTMS experiment and the mechanism of ion trapping are briefly described to lay a foundation for strategies to develop high-mass-to-charge FTMS. Equations are presented, but little attention is paid to their derivation so that those parameters that have an effect on high-mass-to-charge performance can be highlighted. For a more rigorous description of the FTMS experiment and a more detailed treatment of theory, the reader is referred to several comprehensive reviews [34567] and to the monograph by Marshall and Verdun [8].

### 2.1. THE FTMS EXPERIMENT.

Fourier transform mass spectrometry is a tandem-in-time technique, meaning that all manipulations of the ions take place within the confines of a single space (an ion trap or dual ion trap), but in temporally separate events. The ion trap, or cell, is usually of cubic geometry (Figure 1) and is placed in the homogeneous region of a static magnetic field; ions are formed directly or injected from an external source. An ion in a magnetic field will undergo cyclotron motion in a plane perpendicular to the field at a frequency characteristic of its mass-to-charge ratio. Confinement parallel to the magnetic-field lines is achieved by a static electric field generated by applying a small voltage ($\sim$ 1 V) onto two "trapping plates" that are orthogonal to the magnetic field. The interactions of the ions with both the magnetic and electric fields permit the ions to be trapped, manipulated, and ultimately detected.

The basic ion manipulation that is essential in every FTMS sequence of events is to increase the radius of the cyclotron orbit in order to eject unwanted ions, to submit ions of interest to collisional activation, or to detect ions. This is achieved by the application of a radio-frequency (rf) voltage onto two "transmitter plates" located opposite each other in a cubic trap. When the radio frequency is resonant with the cyclotron frequency of the ion, the ion will absorb energy from the dipolar rf-field. This results in an increase of its translational energy and orbital radius of the ion. All ions within the cell can be promoted to higher cyclotron radii by excitation with an rf-chirp [9], or an ion of interest can be isolated from a large population of other ions by ejecting the unwanted ions with frequency sweeps or other computer-generated waveforms [10].

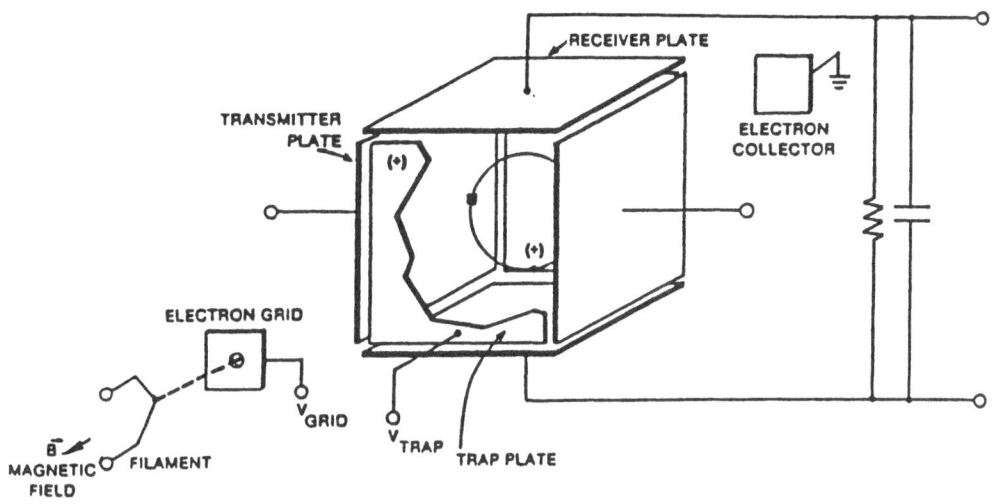

Figure 1. The components of the FTMS cell with electron impact ionization. The magnetic field lines go into the plane of the page. Reprinted with permission from reference [11].

Fourier transform mass spectrometry is unique in that it is the *only* mass spectrometric technique that non-destructively detects ions. Rather than directing the ion beam onto an electron multiplier, Faraday cup, or conversion dynode, ions in FTMS are detected by the differential charge they induce onto two opposed "receiver plates". These plates make up the remaining sides of the cubic cell. After ions are formed or introduced to the center of the cell, ions of a given $m/z$ will have the same cyclotron frequency but random phases. As a result there is no net signal on the receiver plates because, for any given ion inducing a current on the top plate, there is another inducing an equal current on the bottom plate. A differential charge is created by displacing the ions from the center of the cell as a phase-coherent ion "packet" by applying an excitation to the transmitter plates so that the radius of the cyclotron trajectory is increased to about 50% of the limit of the cell. As the phase-coherent packet passes the receiver plates, it induces an image current whose frequency is the cyclotron frequency and whose amplitude is proportional to the number of ions in the packet. The image current is the superposition of the cyclotron frequencies of all the ions excited for detection in the cell, and it is collected as a transient current, converted to a voltage, and digitized by a computer. The transient is deconvoluted by fast-Fourier transform into its frequency components which reveal mass and ion-abundance information.

The high performance detection capabilities of FTMS stem from this non-destructive image-current detection scheme that was introduced by Comisarow and Marshall [12] in 1974. Detection performance increases with the time that the ion motion remains coherent and unparalleled performance can be realized [13] when processes that result in the loss of coherent ion motion are moderated. For example, a resolution of greater than 200,000,000 ($m/\Delta m$ FWHM) for the detection of the argon ion of $m/z$ 40 was demonstrated by Wanczek and co-workers [14] when the scattering of ions out of the coherent "packet" by ion-molecule collisions was reduced by operating at the unusually low pressure of $5 \times 10^{-12}$ mbarr . FTMS inherently has the potential for superior precision for mass measurement because the frequencies of the image current can be measured to better than nine significant figures, corresponding to a mass measurement accuracy on the order of 1 part per billion (ppb) provided a suitable mass calibration law can be developed.

Although high-performance detection of low-mass-to-charge molecules is relatively routine in FTMS, the detection of high-mass-to-charge molecules has been relatively poor. To understand the mass-dependent detection performance of FTMS, it is useful to examine the forces by which ions are trapped in an FT mass-spectrometer cell.

## 2.2. ION TRAPPING WITH STATIC FIELDS.

2.2.1. *The Magnetic Field.* Ions in a magnetic field will experience an inward Lorentz force perpendicular to the velocity of the ion and the magnetic field (Figure 2). For example if the field is in the z-direction (this is the usual convention), the Lorentz force causes the ion to cyclotron in the x-y plane. Ions are trapped in the x-y plane (the radial direction) and do not collapse to the z-axis because the inward force from the magnetic field is balanced by an outward centrifugal force. Additional understanding can be derived by a straightforward balancing of the equations (see equation. 1) that describe these forces.

Inward Force = Outward Force

$$qv_{xy}B = mv_{xy}^2/r \qquad (1)$$

Equation (1), can be rearranged to give the relationship between the mass-to-charge ratio and cyclotron frequency of the ion (equation 2)

$$v_{xy}/r = qB/m = \omega_c \qquad (2)$$

where $q$ is the charge on the ion in Coulombs, $v_{xy}$ is the velocity of the ion in the x-y plane in meters per second, $B$ is the magnetic field strength in tesla, $m$ is the mass of the ion in kilograms, $r$ is the radius of the cyclotron orbit in meters, and $\omega_c$ is the cyclotron frequency in Hertz.

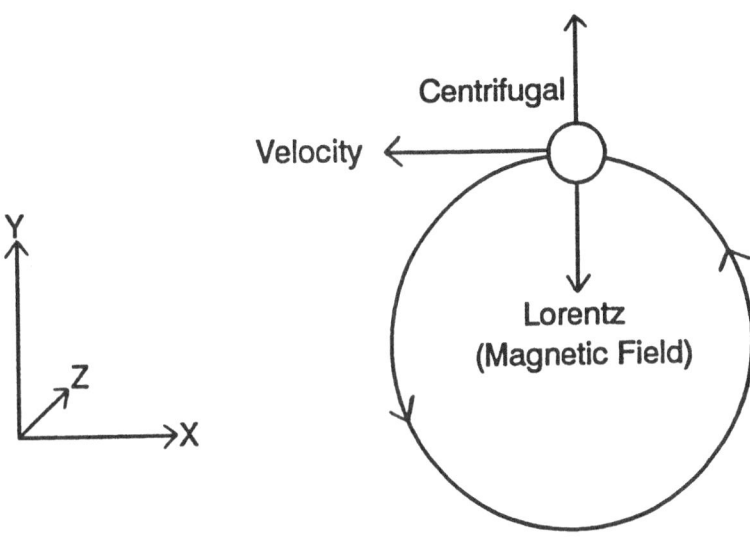

Figure 2. The forces that act on an ion in a magnetic field. The magnetic field lines go into the plane of the page.

2.2.2. *The Electric Field.* The magnetic field constrains the ions in the x-y plane, but the ions are free to drift along the z-axis, parallel to the magnetic field. Escape in the axial z-direction is prevented by an electric field that is generated by applying a small voltage to the two "trapping plates." Because the excitation and receiver plates are at ground, an electric trapping well is formed along the z-axis.

Owing to Gauss' law of electrostatics, the axially inward electric field must also have a radially outward component, or in other words, the electric field will "flow" from the charged trapping plates to the grounded excitation and receiver plates. This outward component increases the complexity of the FTMS experiment considerably, and we propose that the radial electric field in the regions of the cell where the ions are detected is the primary factor that has limited high-mass performance.

The radial component of the electric field is an additional outward force on the ion that is opposite to the force of the magnetic field (Figure 3), and, therefore, reduces the effective magnetic field strength. If one assumes a perfectly quadrupolar electric-trapping field, the magnitudes of the various trapping forces on the ion can be defined, and the maximum mass-to-charge ratio that an ion can have and still maintain a stable trajectory in a cell can be calculated. This limit is called the critical mass of the cell, $m_c$ [15] and is given by equation (3),

$$m_c = qB^2a^2/8\alpha V_T \qquad (3)$$

where $a$ is the length of the cell, $\alpha$ is a factor based on the cell geometry (1.3869 for the cubic cell), and $V_T$ is the trapping voltage. A system with a three-tesla magnet and a 2.54-cm cubic cell, operated with 1 V on the trapping plates would have a critical mass-to-charge ratio of 50470. The actual usable mass range of a cell may be only 10% of its critical mass because the critical mass calculation is based on the assumption that the electric field is perfectly quadrupolar, which is difficult to realize. Nevertheless, $m_c$ is useful for developing theory and for directing strategies for the improvement of performance.

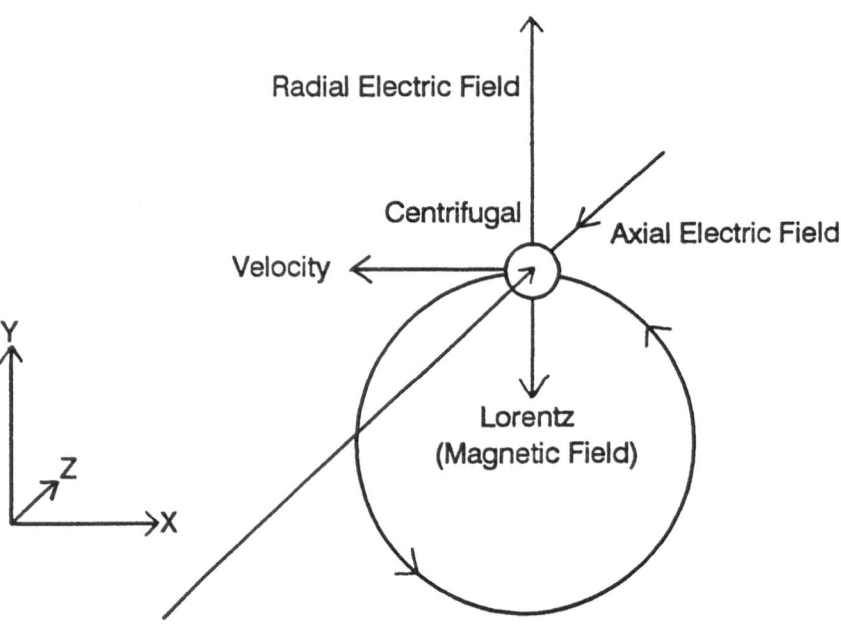

Figure 3. The forces that act on an ion in the combined static magnetic and electric fields of the FTMS cell.

The electric field has a strong effect on the ion's motion (Figure 4) and, hence, on its detectability. The additional outward force from the electric field decreases the cyclotron frequency to give a reduced ion cyclotron frequency $\omega_+$ (equation 5), and causes two new ion motions. The first is the magnetron motion, $\omega_-$ (equation 6), which is a slow precession of the ion cloud in the xy-plane about the equipotential lines of the trapping field. The second is the trapping motion, $\omega_z$ (equation 8), which is a harmonic oscillation of the ions in the z-direction between the two trapping plates. The frequencies of the reduced cyclotron and magnetron motions are found by assuming a quadrupolar trapping potential and using the force-balance procedure to give equation 4,

$$\text{Inward Force} = \text{Outward Force}$$

$$qv_{xy}B = mv^2/r + qE_{radial} \tag{4}$$

which leads to the equations that describe the frequencies of the reduced cyclotron and magnetron motions, equation (5) and equation (6) respectively,

$$\omega_+ = qB + \sqrt{q^2B^2 - 4mqE_0}\,/2m \tag{5}$$

$$\omega_- = qB - \sqrt{q^2B^2 - 4mqE_0}\,/2m \tag{6}$$

where the value of the electric field strength is given by equation (7).

$$E_0 = 2\alpha V_T/a^2 \tag{7}$$

The frequency of the third motion, the trapping motion, is given by equation (8).

$$\omega_z = 2/a\sqrt{q\alpha V_T/m} \tag{8}$$

   In the analysis of ion motion, it is often advantageous to separate the ion trajectory into its axial and radial components. The ion is displaced radially from the center of the cell owing to the superposition of the cyclotron and magnetron frequencies which is periodic at a combination frequency (radial frequency). The radial frequency can be considered a fourth motion of the ion and is described by equation (9).

$$\omega_{radial} = \omega_+ - \omega_- \tag{9}$$

This fourth motion is useful for explaining ion-motion features which also occur in traps that are rotationally symmetric about the z-axis.

2.2.3. *Ion Motion and Interactions with the Electric Field.* To maximize the coherency lifetime of, and hence the ability to detect, a given mass-to-charge ratio ensemble of ions, it is desirable that all the ions of the ensemble have the same frequencies of motion during detection. In a perfectly quadrupolar trapping field, the magnitude of the radial component of the electric force on the ion varies linearly with the displacement of the ion from the center of the cell. As a result, the frequencies of motion are constant regardless of the position of the ion in the electric field. The trapping potentials in ion traps of simple geometry are thought to approximate a quadrupolar potential at the center of the cell and have proven to be satisfactory for the trapping and the manipulation of ions. The trapping

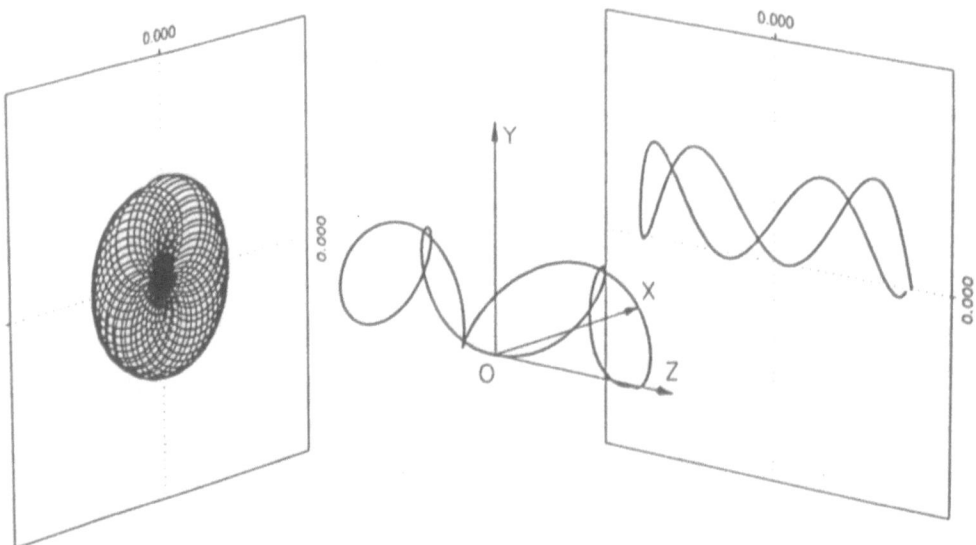

Figure 4. The trajectory of an ion over one period of the z-motion is shown in the volume of the cell and on the y-z plane. One period of the magnetron motion is shown on the x-y plane. The magnitude of the magnetron motion relative to the cyclotron motion has been increased to show the magnetron contribution to the trajectory of the ion.

potential, however, increasingly deviates from the quadrupolar ideal, or becomes anharmonic away from the center of the cell where the ions are detected. The force of the electric field on the ion in the detection regions of the cell do not vary linearly with orbit radius, and as a result, there is a position-dependent shift in the frequencies of the ion motions. For the ions of a given mass-to-charge to have common frequencies, they must be excited to *exactly* the same orbit for detection, otherwise, as is the actual case, the excited ensemble will consist of ions with a distribution of frequencies. The slight differences in frequencies at the detection orbit cause broadening of the detected peak as well as increase the rate at which the ions of the ensemble de-phase during detection [16]. Wang and Marshall [17] have used ion trajectory calculations to show that the relative distances between ions are maintained during and after the excitation of the cloud to higher orbits. Therefore, ion clouds whose pre-excitation states are radially large are expected to sample more of the non-linear radial electric field during detection, de-phase more quickly yielding shorter transient lifetimes, and result in poorer detection performance than radially small clouds do.

Clouds of high mass-to-charge ions have higher initial orbits and are formed by processes imparting more translational energy than are the low-mass-to-charge ions formed by electron ionization for which FTMS has demonstrated high-performance detection. Efforts to remove the excess energy by collisions with neutrals are complicated by the unique behavior of the magnetron motion. As an ion loses translational energy, the magnitudes of the cyclotron and trapping motions are damped and decrease, but the magnitude of the magnetron motion *increases,* resulting in an overall increase in the radial extent of the ion cloud. Mechanisms that remove energy from the ion cloud, such as collisions with neutrals or resistive-wall destabilization through the detection circuitry [18], result in the expansion of the ion-cloud radius, and the associated degradation of detection performance.

## 2.3. SPACE-CHARGE EFFECTS.

So far it has been assumed that all forces acting on an ion are the results of its interactions with the magnetic and electric fields. This is true only when one ion occupies the cell. The introduction of a second results in a mutual coulombic repulsion between the two ions. For large populations of ions, the result of these repulsions are referred to as space-charge effects. Space-charge effects result in an overall radial outward force on the ion cloud, giving rise to an ion-number-dependent shift in the motion frequencies [19]. For the detection of low mass-to-charge ions, the primary effect of space charge on detection is a reduction of mass accuracy.

Space-charge effects appear to play a larger role in the detectability of high mass-to-charge ions. Early reports of the high-mass resolving power of FTMS were made almost exclusively by using cesium iodide or an easily ionized polymer such as polyethylene glycol. These compounds produce high-mass cluster ions that are consistently detected at higher resolving power than are biomolecule ions with comparable mass-to-charge ratios [20]. Rempel et al. [21] showed by using the benzene molecular ion under scaling conditions (see next section) that the peak shape of high $m/z$ ions improves with ion number. A similar improvement in peak shape was achieved by keeping the ion number constant and shimming the electric field in the detection regions of the cell by using modified trapping plates. On the basis of the parallel behavior, they proposed that the integration of space charge with the electric trapping field can improve the detectability of ions by compensating the non-linearities of the electric fields that are sampled by the ions during detection.

2.4. A SCALING TECHNIQUE FOR MODELING HIGH-MASS BEHAVIOR.

The study of electric field effects (coulombic and trapping potential) on the detection of high mass-to-charge ions has proven to be difficult because current high-mass ionization techniques do not generate reproducible populations of ions and those high-mass ions that are produced are often reactive or metastable. Furthermore, a systematic study directed at identifying mechanisms for poor performance requires a set of high-mass models so that the evolution of electric field effects with increasing mass-to-charge can be observed. Scaling can be explained in the following way: the electric field is static, and its force on the ion is nearly mass-independent. Thus, the effects of the field on the trajectory of a low-mass ion will be the same as those on a high-mass ion, provided that the ions have the same charge and take similar paths through the electric field. The mass-dependent forces that determine the trajectory of the ion through the electric field are the result of the magnetic field induction, $B$, so that for a given set of electric-field conditions, the equations of motion of all ions are equivalent if the time parameters and the magnetic-field strength are scaled. Consequently, the problems of detecting high-mass ions in a high-magnetic field can be investigated with low-mass ions in a magnetic field that has been scaled so that the trajectory through the electric field by the low mass-to-charge species is the same as that of the high mass-to-charge species. Low-mass model ions such as those of benzene are stable, inert, and are produced reproducibly by electron ionization so that the only variable in the experiment is the regions of the electric field that are sampled.

The scaling of the magnetic field to model the detection performance of a high mass compound in a high strength magnetic field (denoted as the target) with the benzene molecular ion (denoted as the model) and a 1.2-tesla magnetic field based FT-mass spectrometer [21] is given by the relationship in equation (10).

$$B_{model} = \sqrt{mass_{model}(78)/mass_{target}}\, B_{target} \qquad (10)$$

In the first paper on scaling [21], the detection performance of the FT-mass spectrometer was shown to degrade as the instrument parameters were scaled to mimic ions of increasing mass-to-charge ratios.

## 3. Improving Detection Performance for High Mass-To-Charge Ions.

3.1. BACKGROUND PRESSURE.

One route to extend the lifetime of coherent ion motion and thereby improve detection performance is to reduce the background pressure of neutrals during detection. Some relief is offered by coupling FTMS to high-mass ionization techniques that do not co-produce large neutral populations. Examples of such are Californium-252 plasma desorption [22] and laser ablation [23]. These techniques are compatible with the temporal and low-pressure nature of FTMS and provide some of the earliest examples of the detection of biological molecules by FTMS. The maintenance of low pressures, however, becomes more difficult when FTMS is coupled to techniques such as FAB and electrospray, both of which produce large populations of neutrals and ions. The interfacing these ionization techniques to FTMS has lead to several strategies for the removal of neutrals from the FTMS cell.

One straightforward strategy for the removal of neutrals is to separate physically the high-pressure ionization source from the analyzer by one or more stages of differential pumping. This strategy is manifest in its simplest form by the dual-cell design of Ghaderi and Littlejohn [24]. The dual cell consists of two cubic cells that share a common trap plate that has a small (2-mm diameter) conductance limit and divides the vacuum housing into two sections. The sections are differentially pumped to maintain a pressure ratio of approximately 1000. Ionization occurs on one side of the cell under relatively high pressure ($1 \times 10^{-5}$ torr) conditions, and ions are transferred through the conductance limit to the low pressure side for detection. The main advantage of a dual-cell design is that it is not necessary to transfer ions through the fringing fields of the magnet because ionization and detection both take place within adjoining regions at the same magnetic induction.

McIver and co-workers [25] showed that the source and ion cell could be separated by several stages of differential pumping if ionization takes place outside the magnetic field and ions are injected into the cell by using rf-quadrupoles. Ions guided by the quadrupoles overcome magnetic bottle effects, and this design has the additional advantage that the quadrupoles can be used as mass filters to prevent co-desorbed low-mass ions and neutrals from reaching the analyzer cell. Several high mass CsI clusters have been detected by using this strategy, including the $Cs(CsI)^{+}_{122}$ cluster of $m/z$ 31832 [26] which for a period of four years was the highest $m/z$ ion to be detected by FTMS. Other successful strategies to guide externally created ions into the cell include electrostatic lenses [27] and electrostatic ion guides [28]. Although these guide methods add mechanical complexity to the FTMS instrument, these schemes, as well as that associated with the dual cell, have satisfactorily addressed the issue of the elevated background pressures associated with employing certain high-mass ionization sources.

## 3.2. THE INWARD FORCE ON THE ION

### 3.2.1. *The Magnetic Field B.*    Another strategy to improve the detection of high-mass-to-charge ions and to extend the $m/z$ range of the instrument is to increase the inward force on the ion. This reduces the initial cyclotron orbit radius of high $m/z$ ions and the pre-excitation radius of the ion cloud. Recall that a radially small pre-excitation distribution improves detection because the excited cloud will sample a smaller portion of the non-linear trapping fields during detection, thus increasing the lifetime of coherent ion motion. Equation 1 suggests two routes to increase the inward force on the ion. The most straightforward, although certainly not the least expensive, is to increase the strength of the magnetic field. One design consideration is to maintain a magnet bore sufficiently large to house the cell because a reduction in the cell size, $a$, would negate, in part, the advantages

of using larger magnetic-field strengths. Clearly, this route is ultimately limited by magnet technology. It is expected that advances in high temperature superconducting materials will bring field strengths up and prices down. Today several instruments are based on superconducting magnets with three-tesla fields and 15.24-cm bores. Instruments for high mass applications are usually designed around six or seven-tesla magnets, and one instrument company recently announced the availability of a 9.4-tesla magnet. Considering critical mass limitations, we note that increasing $B$ from 3 to 9.4 T increases $m_c/z$ by a factor of ten to 495,562. Currently, the ultimate extension of this strategy is to conduct FTMS experiments with the 14-tesla, 105- mm bore and a 20-tesla, 50-mm bore magnets that are under development at the National High Magnetic Field Laboratory at Florida State University.

*3.2.2. Charge on the Ion q.* The second route that increases the inward force on the ion is increasing the charge on the ion $q$. Formerly, this was not a practical strategy for mass range improvement owing to a lack of ionization methods that produce multiply charged ions. This strategy has become more viable with the introduction by Fenn and co-workers [29] of electrospray as a mass spectrometry ionization source. McLafferty, Hunt, and co-workers [30] demonstrated that electrospray-FTMS is feasible when they coupled the two via quadrupole ion injection. Further refinement of the technique by McLafferty and co-workers [2] and Hofstadler and Laude [31] have shown that once the pressure differential is resolved, the combination is very powerful. The Cornell group [32] has generated impressive FTMS data for biomolecules; one example is the trapping of the bovine albumin $[M + 37H]^{+37}$ ion (MW = 66,267) at $m/z \sim 1800$. The high-performance characteristics of FTMS are a significant advantage for the analysis of spectra produced by collisional activation [33] or for the study of mixtures of electrospray-generated ions because the assignments of charge state (and then mass) can be achieved directly if the isotopic cluster peaks are resolved [30].

3.3. OPTIMIZATION OF THE ELECTRIC TRAPPING FIELD .

Although minimizing background pressure and maximizing charge and/or the magnetic field do extend the working range of the instrument, they do not address the mechanisms that result in poor mass performance for high mass-to-charge ions (>5000). Rather, the problems are pushed higher up the $m/z$ scale. The electric trapping field, however, is the factor that has introduced complexity to the FTMS experiment. To fundamentally improve the detection performance of FTMS will require the *development and execution* of strategies to optimize the electric field for the detection of high-mass-to-charge ions.

Optimization of the trapping fields for detection has proven to be difficult in part due to the complexity of ion motion through the fields, and the difficulty in modeling accurately the fields generated from trapping electrodes of finite dimensions. As such, strategies for optimizing the electric field are usually derived from fundamental studies involving empirical observation of detection performance coupled with numerical calculations. Over the course of many investigations in a number of laboratories, two ideals for the electric field have emerged. Both ideals seek to eliminate the ion-motion frequency dependence on position. The first, as described previously, is the quadrupolar field, which has as a feature that the radial component of the field varies linearly with radius. The second is the

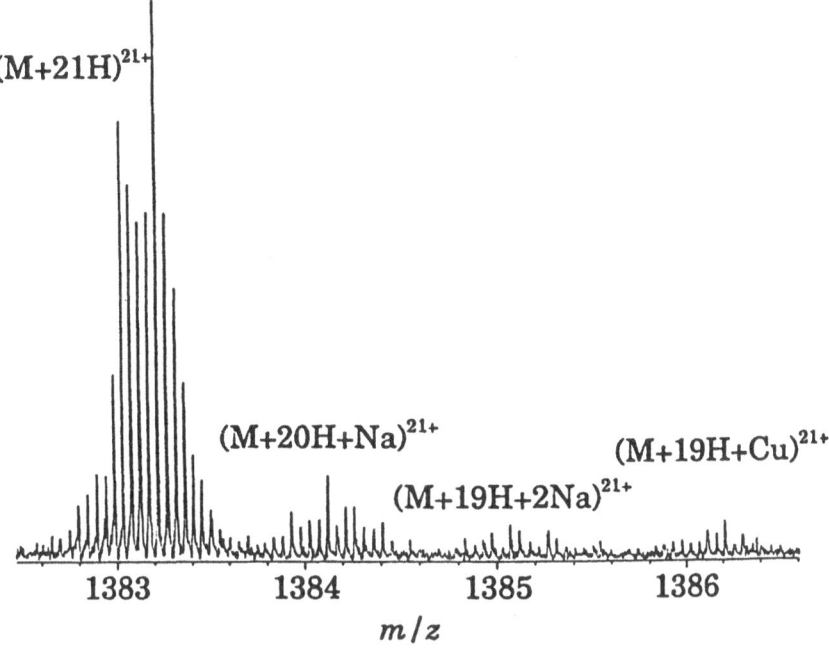

Figure 5. Partial electrospray-FTMS mass spectrum of carbonic anhydrase (MW 29025) showing a mass resolution of 500000. Reprinted with permission from [2].

particle-in-a-box, ideal which seeks to eliminate the radial-electric field in the cell altogether, or at least to confine it to the boundaries of the cell.

3.3.1. *The Quadrupolar Ideal: Linearized Radial Fields.* Several advantages are gained as the dimensions of the cell are extended. The quadrupolar region of the electric field becomes larger, the ion capacity before the onset of space charge is enlarged, and the degree of ion-cloud phase coherency that can be achieved is increased [34]. Unfortunately, the strategy to increase the cell size quickly finds its limits because the increase in the magnet bore necessary to accommodate a larger cell results in a reduction of the magnetic field strength across the bore. Generally, FTMS cells do not exceed the cross section of a 5.08 cm cube. Therefore, the pursuit of a quadrupolar potential has turned to modifying the trapping plates and adopting new cell geometries.

Although a truly quadrupolar potential in a cubic cell requires infinite electrodes, a quadrupolar field can be generated over most of the cell volume by a trap consisting of an hyperbolic ring and two endcap electrodes whose shapes follow the rotation of an hyperbola. Such a geometry is used for the dynamic trapping of ions in quadrupole ion storage traps (QUISTORs) [35] and was first utilized in FTMS by Rempel et al.[36]. In the hyperbolic FTMS cell, the endcaps are employed as the trapping and excitation electrodes, and the unbroken ring for detection. Because no separate excitation and receiver plates exist with this geometry, the development of a new mode of ion excitation and detection was required; this mode is parametric excitation. Evaluation showed reduced sensitivity to space charge, yielding a simpler mass calibration law and the expected independence of ion cyclotron frequency with position. Conventional FTMS excitation and detection can be employed with the hyperbolic geometry if the ring electrode is split into four quadrants, creating excitation and receiver electrodes. A comparison of detection performance between this and a cubic cell was made by Marshall and co-workers [37] who demonstrated several advantages that have been predicted for an FTMS cell with a linearized radial electric field.

Hyperbolic cells have not seen wide use in FTMS because they have a propensity to eject ions from the cell in the z-direction during excitation and because the strict requirements of the hyperbolic geometry cause the cell to be not as versatile as the simpler geometries, especially for using novel events or coupling interfaces to new ionization techniques. Furthermore, the curved electrodes do not efficiently use the bore volume of superconducting magnets; that is a given magnet bore can accommodate a larger volume cubic than hyperbolic cell.

3.3.2. *Modified Cells of Simple Geometries.* Simple cell geometries are adequate for the trapping and manipulation of the ions, but detection is compromised because it takes place in regions where the radial fields are non-ideal. Another strategy to improve detection has been to expand the quadrupolar region of the trapping potential from the center of the cell. This has been approached by using auxiliary trapping electrodes that compensate for the non-linearity of the field [38] or "by shimming" the electric field for frequency homogeneity much like the magnetic field in NMR is shimmed for homogeneity. The implementation of this strategy with cylindrical [39] FTMS cells involves splitting the trapping plates into concentric rings over which the applied voltages may be independently controlled. Inoue et al. [40] reported that peaks are 10% narrower, and peak heights increase by 65% when the trapping fields of the cylindrical geometry are compensated by using a three-electrode scheme.

Marshall et al. [41] split the excitation and receiver plates of a cubic cell into five sections coupled through a resistive/capacitive network (Figure 6). The value of the resistor associated with each segment was chosen on the basis of field calculations to generate as closely as possible a quadrupolar potential in the cell. Although a quadrupolar potential could not be achieved near the edges of the trap, it was demonstrated that there is an increase in detection performance and the elimination of the peak-splitting that is associated with non-quadrupolar trapping potentials.

Rempel et al. [42] have proposed that it is unnecessary to linearize the trapping field throughout the cell since the electric fields that the ions sample during detection are expected to contribute most to the loss of coherent motion. This is particularly true for clouds in which the ions are thermalized, The authors have compensated the cubic cell with the goal of linearizing the trapping field only over the subsets of radii and axial

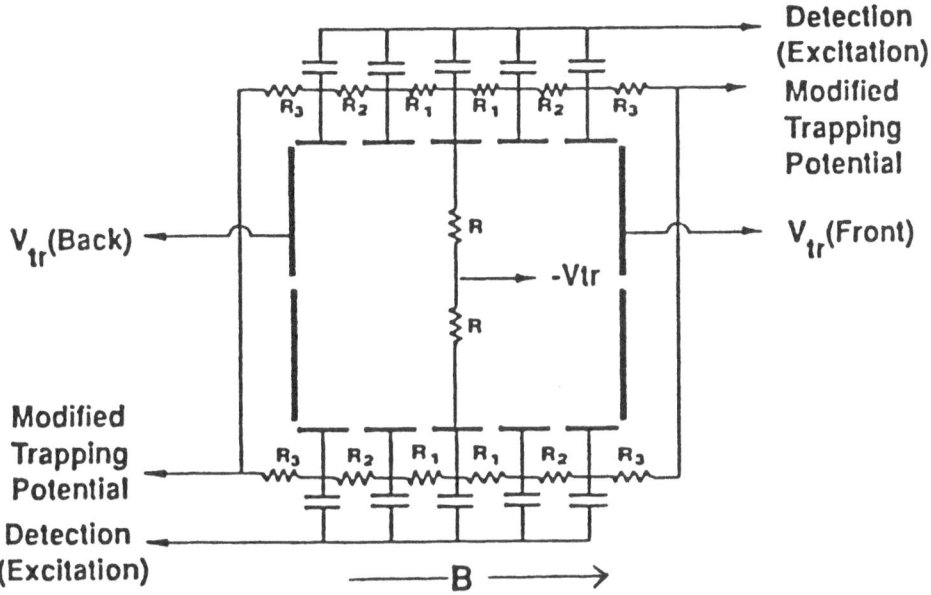

Figure 6.  The linearized cell of Marshall and co-workers.  Reprinted with permission from the author. [41].

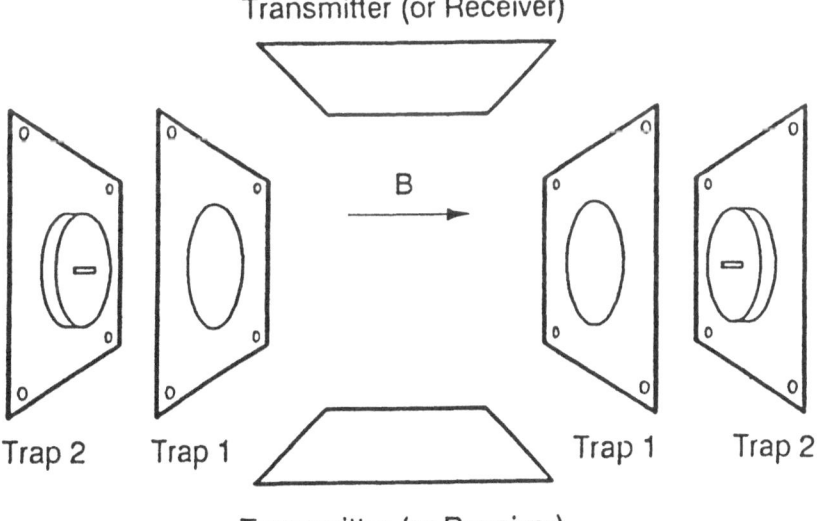

Figure 7.  The compensated cubic cell of Rempel, Gross, and co-workers.  Reprinted with permission from reference [21].

amplitudes that are sampled by the ions during detection. The less rigorous requirements for the shimming of the field permit a simpler cell design (Figure 7) while demonstrating detection improvement comparable to the more complicated schemes; the peak splitting and broadening associated with non-linear radial fields are eliminated, and there is an increase in the signal-to-noise ratio by a factor of two [43].

3.3.3. *The particle-in-a-box ideal: Elimination of the radial component.* Although the electric field cannot be eliminated from FTMS, its strength at the center of the cell can be reduced by strategies that prevent the penetration of the electric field into the cell. In this way the radially outward force on the ion is reduced, increasing the critical mass of the cell and reducing the dependence of the motion frequencies on orbit radius. Hunter et al. [44] used numerical calculations to show that the potential of 0.33 V at the center of a cubic cell at 1 volt trapping potential is reduced to 77 $\mu$V by elongating the trap in the z-direction by a factor of 4.75 greater than the radial dimension. Grosshans and Marshall [45] calculated that a cell with an aspect ratio of 6:1 reduces $\alpha$ by a factor of 4, and in a three-tesla magnetic field would have a critical *m/z* of approximately 100000.

Reduction of the electric field by elongating the cell in the z-direction finds its limit in the particle-in-a-box potential where the electric field within the cell is reduced to zero, thus removing the position dependence of the motion frequencies. Ions experience a restoring force in the axial direction only at the boundary of the cell. Although it is extremely difficult to form the sharp field gradients at the edges of the cell, Wang and Marshall [46] reported the particle-in-a-box potential is approximated in a cubic cell by placing grounded screens in front of the trapping plates. Wilkins and co-workers [47] evaluated the use of the screened cell with ions produced by matrix assisted laser desorption ionization (MALDI) and compared its performance with that of the cubic cell. They reported that the screens do indeed reduce the position-dependent shift in the cyclotron frequency (by a factor of ~25) as the ions are excited to higher orbits for detection, but no significant improvement of mass resolution was observed when compared to the cubic cell without screens.

Ion behavior in the region next to the screens is extremely difficult to predict because of the complexity in modeling accurately the electric fields in the vicinity of the screens, but some insight can be gained by examining the ion behavior by using a particle-in-a-box model.

3.3.4. *Parametric Resonances in cells with non-quadrupolar potentials.* Recall that one of the fundamental motions of the ion is a harmonic oscillation in the z-direction ($\omega_z$, see equation 8). In a quadrupolar field, the radial force on the ion is independent of position so that as the ion transverses the z-direction, the radial force on the ion is constant. In a field that deviates from the ideal, however, the radial force will increase as the ion approaches a trapping plate. As a result, the radial force on the ion is modulated as a result of the z-motion at a frequency equal to $2\omega_z$. Recall also that the radial distance of the ion from the center of the cell is modulated owing to the superposition of the cyclotron and magnetron motions at a frequency $\omega_{radial}$. Because both frequencies are mass-dependent, there exist mass ranges over which these frequencies become equal, or are resonant, thus opening up a channel for energy transfer between the motions. We have called these parametric resonances, and ion-trajectory calculations have shown that energy is transferred from the magnetron mode to the cyclotron mode. This transfer of energy results in an increase in the magnitude of both motions and leads to the radial expansion of the ion cloud [48].

In the same study, experiments in which the behavior of high-mass ions were modeled via scaling showed there is a loss of sensitivity over the *same* mass ranges that the resonance condition is predicted to occur. The mass ranges over which parametric resonances occur are expected to increase as the cell deviates from the quadrupolar ideal. Furthermore, ion trajectory calculations predict that because the axial restoring force in the particle-in-a-box potential approximates is confined to the cell boundaries, even sub-thermal ions must sample the strong radial fields at the trapping plates. This results in the parametric radial expansion of the ions in the cell [49] and would be especially pronounced for ions with large axial energies, such as those formed by matrix-assisted laser desorption [50]. Therefore, we conclude that the particle-in-a-box trapping potential is not the proper ideal for cells designed to trap high mass-to-charge ratio ions.

## 4. Ancillary Techniques that Increase Performance for High Mass-to-Charge Ions.

### 4.1. ION COOLING STRATEGIES.

Beavis and Chait [50] showed the axial velocity of peptide ions desorbed by MALDI is approximately 760 m/s and independent of mass. Thus, the average kinetic energy for ions produced by MALDI increases with mass, and deep trapping wells are required to collect a significant population of the desorbed ions. The high-trapping voltages necessary to form these wells, however, also increase the strength of the radial component and, make more likely the degradation mechanisms associated with the electric field. Clearly if the ions formed by MALDI are to be collected efficiently and detected with good performance, the FTMS experiment must be modified to accommodate the translational energies imparted to ions by energetic ionization sources.

4.1.1. *Passive Cooling Techniques.* Soulki and Russell [51] showed, as predicted from the previously cited Chait studies on desorbed ion velocities, that the optimum electric potential for trapping ions produced by MALDI increases with the mass of the analyte. Efforts to reduce the energy of the ions by using collisions with pulsed neutrals at elevated pressures were relatively unsuccessful owing to the expansion of the magnetron-mode as the ions lost energy. To form a localized region of high pressure at the MALDI probe tip, a "waiting room" was constructed by capping the probe with a hollow brass cylinder. At the end of the cylinder is a small orifice to allow the laser beam to enter and ions to exit. Following laser ablation, the small volume of the waiting room becomes a high-pressure region where analyte molecules undergo energy-reducing collisions with matrix and analyte neutrals before being pulsed into the analyzer cell. Using this strategy, the authors reported that they could reduce trapping voltages and still detect the proton-bound bovine insulin dimer of $m/z$ 11469 with a resolution of approximately 20.

Cooling the ions by collisions with pulsed neutrals has been more successful when used in conjunction with external ionization and an rf-only quadrupole ion guide. McIver et al. [52] pressurized the FTMS cell with a 2-ms pulse of argon before ions were desorbed by MALDI and guided to the cell by the quadrupoles. As ions were introduced into the cell, the front trapping plate was grounded while the rear plate was held at 10 V. After most of the ions had entered the cell, the front plate was raised to 10 V and the ions were allowed to relax collisionally for 10-15s. Following cooling, both trap plates were ramped down to 0.7 V for the detection of the ions. The authors demonstrated that biomolecule ions are stable on the time scale of seconds, and were able to detect the [M + Na]$^+$ gramicidin S ion of $m/z$ 1163.7 with a mass resolution of 1100000 and the [M + H]$^+$ bovine insulin ion isotope cluster centered at $m/z$ 5735 with a resolution of 90000.

Wilkins and co-workers [53] demonstrated that the addition of sugars to the MALDI matrix results in an increase in the detection performance of high-mass molecules. Specifically, they detected the horse cytochrome $c$ molecular ion of $m/z$ 12349 with a half-height resolution ($m/\Delta m$) of 12000, although the base-line resolution was poor (Figure 8). The authors hypothesized that the decomposition of the sugars during ablation results in a locally high pressure of $CO_2$ and $H_2O$ neutrals and that ions are cooled by collisions as they leave the probe tip.

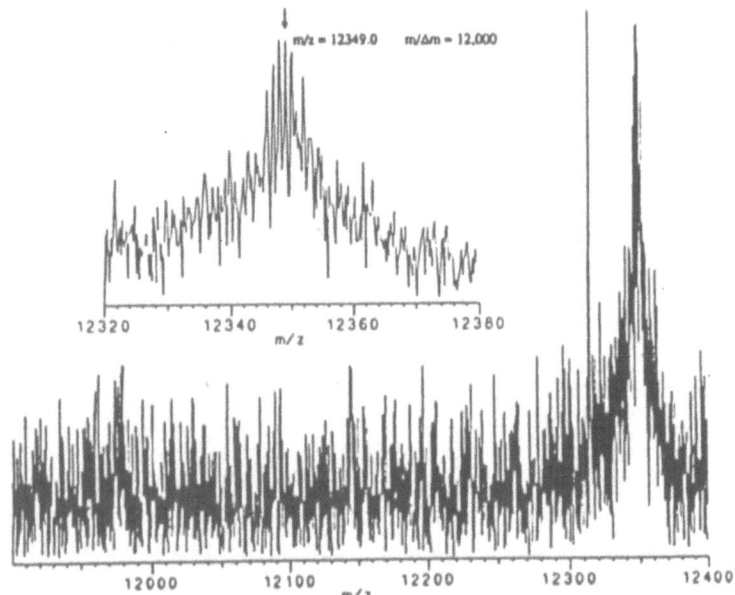

Figure 8. MALDI-FTMS of horse cytochrome $c$ using a sucrose co-matrix. Reprinted with permission from reference [53].

4.1.2. *Active Cooling Techniques.* In the static trapping fields of the FTMS cell, the beneficial effects of ion collisions with neutrals are often offset by the resulting radial expansion of the ion cloud and the associated decrease in detection performance. Two recently developed techniques circumvent the metastable behavior of the magnetron mode by the addition of auxiliary fields to the normal FTMS trapping fields during collisional cooling. Both have demonstrated the ability *to focus* ions to the center of the cell as energy is removed by collisions with neutrals and should significantly increase the utility of FTMS by enabling the operation at moderate to high pressures ($10^{-5}$ to $10^{-2}$ torr).

4.1.3. *Quadrupolar Axialization.* Savard et al. [54] have shown that the diameter of the ion cloud can be reduced by collisions with neutrals during quadrupolar excitation at the combination frequency that is the sum of the cyclotron and magnetron frequencies ( $\omega_+ + \omega_-$ ) of the ion. Schweikhard, Guan and Marshall [55] have incorporated this into the FTMS sequence as a means for the axialization of ions in the FTMS cell. During excitation in this manner the motion of the ion in the x-y plane is periodically converted between pure cyclotron and pure magnetron motion. Because the rate of conversion between the motions is faster than the rate that the magnetron mode increases, collisions with a buffer gas remove energy from the ions primarily via the cyclotron mode and, as a result, cause the ion cloud to collapse to the center of the cell. In their original demonstration, benzene ions were trapped for more than 20 seconds at $1.5 \times 10^{-5}$ torr and could still be transferred through a 2-mm conductance limit into an adjoining cell. In another paper incorporating a dual cell [56], the mass resolving power for detecting the laser-formed Rhodamine 6G $[M + H]^+$ ion of $m/z$ 443 was improved from 2400 when detected at $8 \times 10^{-8}$ torr in the source side of the cell, to 144000. The improvement in detection performance was acheived in the analyzer cell at $0.2 \times 10^{-8}$ torr following cooling and focusing with a 5-s axialization event in the presence of $7 \times 10^{-7}$ torr argon (Figure 9). The ions could not be transferred to the more favorable detection conditions of the analyzer cell without the focussing event.

4.1.4. *RF-Only Mode Event.* The inability to trap ions at pressures greater than $10^{-5}$ torr is not a characteristic of all ion trapping techniques. Quadrupole ion storage devices or QUISTORs confine ions by their interaction with oscillating or dynamic electric fields (Figure 10). Unlike the ions trapped by the static trapping fields employed in FTMS, ions trapped by the dynamic trapping fields are *focussed* to the center of the trap as they lose kinetic energy by collisions with neutrals. The number of collisions that the ions undergo is increased by the introduction of a bath gas, such as helium, so that a QUISTOR cell is operated at a total pressure in the range of $10^{-3}$ to $10^{-2}$ torr, considerably higher than an FTMS cell even operating in the axialization mode.

Rempel and Gross [57] recently introduced an RF-only-mode event (QUISTOR event) to the FTMS sequence (Figure 11). This event includes the transition from the static trapping mechanism of FTMS to a QUISTOR-like dynamic trapping mechanism with efficient retention of the ion population. The ability of the RF-only-mode event to remove excess internal energy from ions was demonstrated in two reports where adducts that

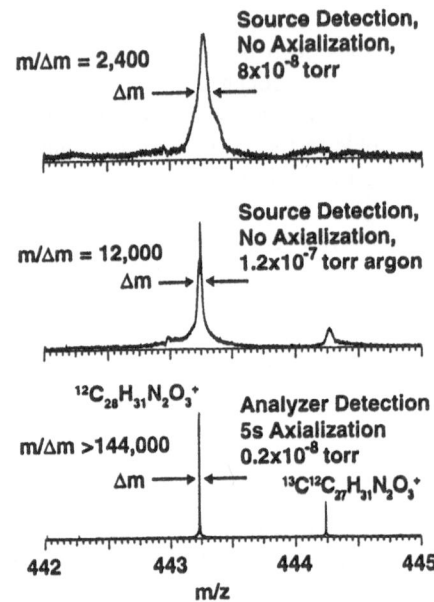

Figure 9. The improvement in resolution obtained by quadrupolar axialization of the ion cloud before detection. Reprinted with permission from reference [56].

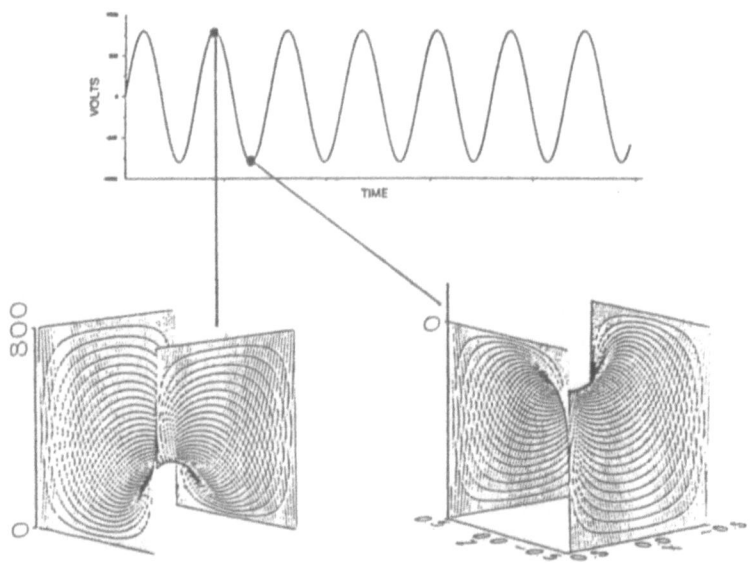

Figure 10. The oscillating electric field of the QUISTOR traps and focuses ions to the center of the cell by alternating the trapping well between the axial and radial directions.

decompose before detection under normal FTMS operating conditions were stabilized and detected in the FT-mass spectrometer following the RF-only-mode event [58].

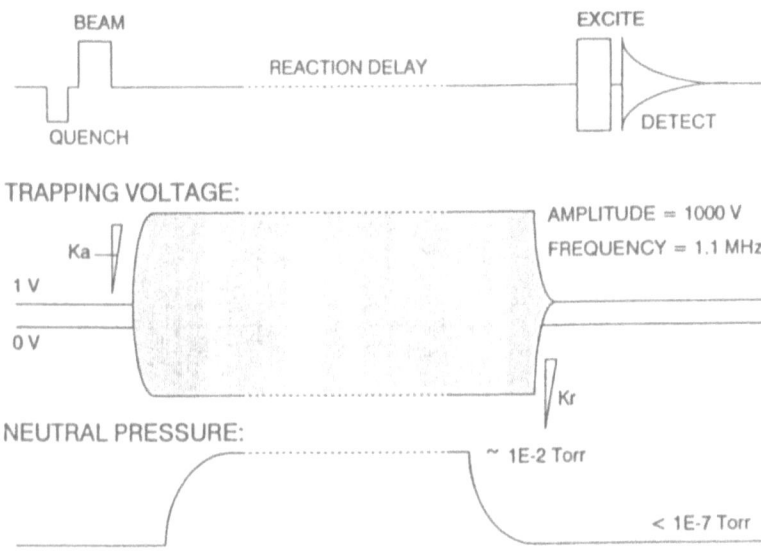

Figure 11. An FTMS sequence of events incorporating the RF-only-mode event.

## 4.2. ION CLOUD FOCUSING.

4.2.1. *Quadrupolar Axialization.* Techniques that focus ions to the center of the cell present the opportunity to manipulate the ion cloud so that its shape is optimum for excitation and detection. Amster and co-workers [59] showed that coherent motion of the ion cloud formed by laser desorption of Rhodamine 6G (*m/z* 443) can be detected for more than 64 seconds if it is focussed to a small radius by quadrupolar axialization before it is excited to the detection orbit. Guan and Marshall [60] demonstrated that a range of ions that have radially expanded beyond the 2-mm conductance limit of a dual cell can be brought back to the center of the cell by broadband sweep of the quadrupolar axialization frequency.

4.2.2. *RF-Only Mode Event.* The complimentary nature of decreasing the pre-excitation radius of the ion cloud and compensating the non-linearites of the trapping field at the detection radius was demonstrated by Gross and co-workers [61]. Using the benzene molecular ion, they overtly radially expanded the ion cloud so that it could not be detected. By use of a new ion cloud sectioning method, they were able to establish that the cloud is

168

indeed expanded to a diameter that is approximately 50% of the cell diameter. Use of either cell compensation or the RF-only-mode event separately offers small improvement for the detection of the ions, but when used together, signal-to-noise ratio of ≅18 and a mass resolution of 32500 was achieved for the formerly diffuse cloud of benzene ions. A similar demonstration with the molecular of ion of $C_{60}$ buckminsterfullerene showed a signal-to-noise ratio of 80 and resolution of 30000 (Figure 12). Applying the scaling relationships to these results predicts that this strategy should yield resolutions on the order of 30000 for the same size population of *m/z* 24500 ions.

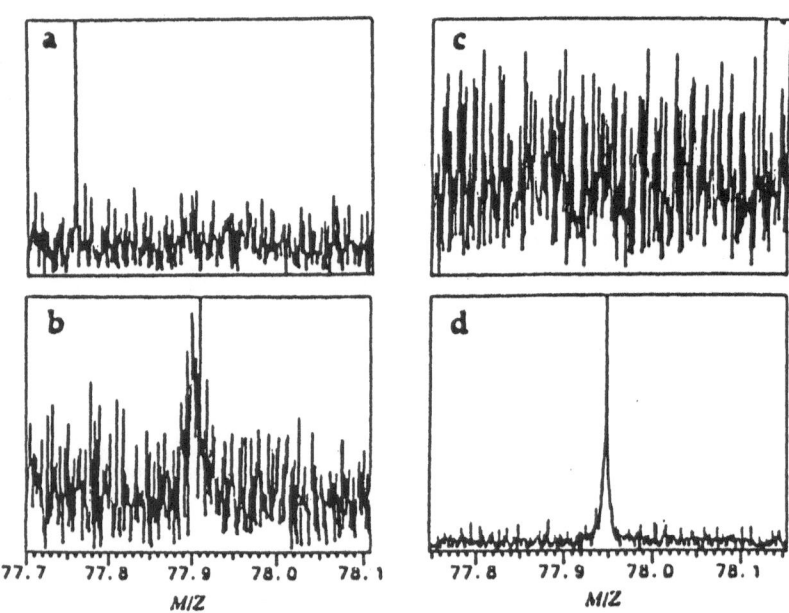

Figure 12. The complementary nature of ion-cloud focussing and electric-field compensation for detecting an overtly expanded cloud of benzene molecular ions. a) no compensation or RF-event. b) only and RF-event. c) only compensation. d) both compensation and RF-event. Reprinted with permission from reference [61].

Techniques that create a reproducible cloud shape should permit the optimization or "tuning" of the trapping fields with easily produced reference ions before attempting analysis of trace levels of precious samples, thus standardizing the operation of FTMS for high mass-to charge ions.

4.2.3. *Cell Internal Ion Guide.* The improvement in the detectability of high mass-to-charge ions that have been focussed to the center of the cell prior to excitation for detection was dramatically demonstrated by Russell and co-workers [62]. In their focussing scheme, a wire ion guide was suspended between the trapping plates and traverses across the center of the cell. The wire is held at an potential opposite in polarity to the trapping plates, creating a deep well that not only traps ions, but focuses then around the wire at the center of the cell. By using this scheme, the transferrin dimer of $m/z$ greater than 157000 was trapped and detected with a signal-to-noise ratio on the order of 15.

## 4.3. ION REMEASUREMENT.

The ability to focus an ion cloud, independent of its position, to the center of the cell coupled with the non-destructive image current detection opens up the possibility of redetecting an ion population after it has lost phase coherency. Williams et al. [63] demonstrated that $m/z$ 1904 ions [M + Na]$^+$ of gramicidin D could be remeasured 31 times with an overall efficiency of 98% if long delays (approximately 120 seconds for the $m/z$-1164 ion at $5 \times 10^{-9}$ torr) were inserted into the FTMS sequence following the detection event. They proposed that, unlike low-mass ions, high mass ions are not significantly scattered when they collide with low-mass neutrals. As a result, the ions lose energy through the cyclotron mode with little deviation of their trajectory and spiral to the center of the cell where they can be excited again for remeasurement. Laude and co-workers [64] showed that rate that ions relax to the center of the cell increases with mass-to-charge ratio. In experiments with electrosprayed bovine serum albumin dimer (MW = 132532), the pressure-optimized-relaxation delay is reduced to two seconds, and remeasurement efficiencies of nearly 100% were obtained when a narrow range of excitation voltages were used. The dimer was remeasured 250 times in an 8-minute experiment, and a 16-fold improvement in signal-to-noise versus that of a single measurement was observed.

Clearly quadrupolar axialization and the RF-only-mode event are poised to make an impact on the efficiency of ion remeasurement, especially for small molecules whose relaxation rates are very slow. Amster and co-workers [60] used quadrupolar axialization to remeasure the laser desorbed Rhodamine 6G molecular ion ($m/z$ = 443) 200 times with a remeasurement efficiency greater than 99.5%. In a demonstration of interactive parameter adjustment, the remeasurement of the same population of gramicidin S molecular ions ($m/z$ 1142) was made in a series of experiments that showed that the mass spectral resolution can be systematically increased by a factor of 500.

## 5. Summary.

There is potential for the application of FTMS to the characterization of biological molecules. The coupling of FTMS to electrospray ionization is a powerful combination. The direct identification of the charge state of sprayed molecules by high resolution and the ability to couple to high efficiency separation techniques should prove an important strategy for the trace level analysis of proteins.

Ion cloud focusing and optimized electric fields are expected to increase the usable mass-to-charge range of FTMS. That will bring the high-performance detection that is routinely achieved with low-mass-to-charge ions to the high mass-to-charge ions that are formed by MALDI. The ability to include high pressure events in the FTMS sequence should generate the development of new coupling strategies to high-pressure and energetic ionization techniques perhaps moving towards simplification of the instrument and away from complex differential pumping schemes. Finally the ability to focus ions back to the center of the cell enables the exploitation of the non-destructive detection scheme of FTMS and should lead to experimental sequences where a single population of peptide ions undergoes several characterizing manipulations before they are lost.

## 6. Acknowledgements.

We acknowledge the National Science Foundation (CHE-9017250) and the National Institutes of Health (2P41RR00954) for support during the preparation of this manuscript. C.L.H. thanks the organizers and the National Science Foundation for support and travel funds to attend the NATO-ASI on Biological Mass Spectrometry and the Eastman Chemical Company for the sponsorship of an American Chemical Society Analytical Division Fellowship. An extended version of this manuscript has been submitted to *Mass Spectrometry Reviews*.

## 7. References.

1  B. E. Winger, S. A. Hofstadler, J. E. Bruce, H. R. Udseth, and R. D. Smith, "High-Resolution Accurate Mass Measurements of Biomolecules Using a New Electrospray Ionization Ion Cyclotron Resonance Mass Spectrometer", *J. Am. Soc. Mass Spectrom.*, **4**, 566-577, (1993).

2  S. C. Beu, M. W. Senko, J. P. Quinn, F. M. Wampler, and F. W. McLafferty, "Fourier-Transform Electrospray Instrumentation for Tandem High-Resolution Mass Spectrometry of Large Biomolecules" *J. Am. Soc. Mass Spectrom.* **4**, 557-565, (1993).

3  M. V. Buchanan, and R. L. Hettich, "Fourier Transform Mass Spectrometry of High-Mass Biomolecules", *Anal. Chem,* **65**, 245A-259A, (1993).

4  C. B. Jacoby, C. L. Holliman, and M. L. Gross, in Mass Spectrometry in the Biological Sciences: a Tutorial M. L. Gross ed., NATO ASI Series, Kluwer Academic Publishers. Dordrecht, 1990.

5  C. Köster, M. S. Kahr, J. A. Castoro, and C. L. Wilkins, "Fourier Transform Mass Spectrometry", *Mass Spectrom. Rev.*, 495-512, (1992).

6  A. G. Marshall, and L. Schweikhard, "Fourier Transform Ion Cyclotron Resonance Mass Spectrometry: Technique Developments", *Int. J. Mass Spectrom. Ion Processes*, **118/119**, 37-70, (1992).

7   A. G. Marshall, P. B. Grosshans, "Fourier Transform Ion Cyclotron Resonance Mass Spectrometry: The Teenage Years", *Anal. Chem.* **63**, 215A-229A, (1991).

8   A. G. Marshall, and F. R. Verdun, <u>Fourier Transforms in NMR, Optical and Mass Spectrometry</u>, Elsevier Scientific Publishers, Amsterdam, 1990.

9   M. B. Comisarow, and A. G. Marshall, "Frequency-Sweep Fourier Transform Ion Cyclotron Resonance Spectroscopy", *Chem. Phys. Lett.* **26**, 489-490, (1974).

10  A. G. Marshall, T-C. L. Wang, and T. L. Ricca, "Tailored Excitation for Fourier Transform Ion Cyclotron Resonance Mass Spectrometry", *J. Am. Chem. Soc.* **107**, 7893-7897, (1985).

11  Nicolet (now Extrel) FTMS-2000 Operation Manual, Madison Wisconsin, 1985.

12  M. B. Comisarow, and A. G. Marshall, "Fourier Transform Ion Cyclotron Resonance Spectroscopy", *Chem. Phys. Letters*, **25**, 282-283, (1974).

13  P. A. Limbach, P. B. Grosshans, and A. G. Marshall, "Experimental Determination of the Number of Trapped Ions, Detection Limit, and Dynamic Range in Fourier Transform Ion Cyclotron Resonance Mass Spectrometry", *Anal. Chem*, **65**, 135-140, (1993).

14  M. Bamberg, M. Allemann, and K. P. Wanczek, "Proceedings of the 35th. ASMS Conference on Mass Spectrometry and Allied Topics", Denver, Colorado, p.1116, (1987).

15  E. B. Ledford, D. L. Rempel, and M. L. Gross, "Space-Charge Effects in Fourier Transform Mass Spectrometry: Mass Calibration", *Anal. Chem.*, **56**, 2744-2748, (1984).

16  M. B. Comisarow, in <u>Lecture Notes in Chemistry: Ion Cyclotron Resonance Spectrometry II</u> H. Hartmann and K.-P. Wanczek, Springer-Verlag Publishers, Berlin, 1982.

17  M. Wang, and A. G. Marshall, "Laboratory-Frame and Rotating-Frame Ion Trajectories in Ion Cyclotron Resonance Mass Spectrometry" *Int. J. Mass Spectrom. Ion Processes.*, **100**, 323-346, (1990).

18  S. C. Beu, D. A. Laude, "Radial Transport Due to Resistive-Wall Destabilization in Fourier Transform Mass Spectrometry", *Int. J. Mass Spectrom. Ion Processes,*, **108**, 255-268, (1991).

19  E. B. Ledford, S. Ghaderi, R. L. White, R. B. Spencer, P. S. Kulkarni, C. L. Wilkins, and M. L. Gross, "Exact Mass Measurement by Fourier Transform Mass Spectrometry", *Anal. Chem.*, **52**, 463-468 (1980).

20  D. F. Hunt, J. Shabanowitz, J. R. Yates, N-Z. Zhu, D. H. Russell, and M. E. Castro, "Tandem Quadrupole Fourier-transform Mass Spectrometry of Oligopeptides and Small Proteins", *Proc. Natl. Acad. Sci. USA*, **84**, 620-623, (1987).

21  D. L. Rempel, R. P. Grese, and M. L. Gross, "A Scaling Technique for Studying the Dynamics of High Mass Ions in Fourier Transform Mass Spectrometry: A Preliminary Report", *Int. J. Mass Spectrom. Ion Processes*, **100**, 381-395, (1990).

22  J. C. Tabot, J Rapin, M Poreti, and T Gäumann *Chimia*, **40**, 169-171, (1986).

23  D. A. McCrery, E. B. Ledford, and M. L. Gross, "Laser Desorption Fourier Transform Mass Spectrometry", *Anal. Chem.*, **54**, 1435-1437, (1982).

24  S. Ghaderi, and D. P. Littlejohn, "Proceedings of the 33rd ASMS Conference on Mass Spectrometry and Allied Topics" San Diego California, p. 727 (1985).

25  R. T. McIver, R. L. Hunter, and W. D. Bowers, "Coupling a Quadrupole Mass Spectrometer and a Fourier Transform Mass Spectrometer", *Int. J. Mass Spectrom. Ion Processes*, **64**, 67-77, (1985).

26  C. B. Lebrilla, D. T-S. Wang, R. L. Hunter, and R. T. McIver, "Detection of Mass 31830 Ions with an External Source Fourier Transform Mass Spectrometer", *Anal. Chem.*, **62**, 879-880, (1990).

27  P. Kofel, M. Allemann, Hp. Kellerhals, and K. P. Wanczek, "External Generation of Ions in ICR Spectrometry", *Int. J. Mass Spectrom. Ion Processes*, **65**, 97-103, (1985).

28  P. A. Limbach, A. G. Marshall, and M. Wang, "An Electrostatic Ion Guide for Efficient Transmission of Low Energy Externally formed Ions into a Fourier Transform Ion Cyclotron Resonance Mass Spectrometer" *Int. J. Mass Spectrom. Ion Processes*, **125** 135-144, (1993).

29  J. B. Fenn, M. Mann, C. K. Meng, S. F. Wong and C. M. Whitehouse, "Electrospray Ionization for Mass Spectrometry of Large Biomolecules", *Science*, **246**, 64-70, (1989).

30  K. D. Henry, E. R. Williams, B. H. Wang, F. W. McLafferty, J. Shabanowitz, and D. F. Hunt, "Fourier-Transform Mass Spectrometry of Large Molecules by Electrospray Ionization", *Proc. Natl. Acad, Sci, USA*, **86**, 9075-9078, (1989).

31  S. A. Hofstadler, and D. A. Laude, "Trapping and Detection of Ions Generated in a High Magnetic Field Electrospray Ionization Fourier Transform Ion Cyclotron Resonance Mass Spectrometer", *J. Am. Soc Mass Spectrom.*, **3**, 615-623, (1992).

32  K. D. Henry, and F. W. McLafferty, "Electrospray Ionization with Fourier-Transform Mass Spectrometry. Charge State Assignment from Resolved Isotopic Peaks" *Org. Mass Spectrom.*, **25**, 490-492, (1990).

33  M. W. Senko, S. C. Beu, F. W. McLafferty, "High-Resolution Tandem Mass Spectrometry of Carbonic Anhydrase", *Anal. Chem.*, **66**, 415-417, (1994).

34  C. D. Hanson, M. E. Castro, and D. H. Russell, "Phase Synchronization of an Ion Ensemble by Frequency Sweep Excitation in Fourier Transform Ion Cyclotron Resonance", *Anal. Chem.*, **61**, 2130-2136, (1989).

35  R. E. March, R. J. Hughes, <u>Quadrupole Storage Mass Spectrometry</u>, John Wiley & Sons, New York, 1989.

36 D. L. Rempel, E. B. Ledford, S. K. Huang, and M. L. Gross, "Parametric Mode Operation of a Hyperbolic Penning Trap for Fourier Transform Mass Spectrometry", **59**, 2527-2532, (1987).

37 W. W. Yin, M. Wang, A. G. Marshall, and E. B. Ledford, "Experimental Evaluation of a Hyperbolic Ion Trap for Fourier Transform Ion Cyclotron Resonance Mass Spectrometry", *J. Am. Soc. Mass. Spectrom.*, **3**, 188-197, (1992).

38 G. Gabrielse and F. C. Mackintosh, "Cylindrical Penning Traps with Orthogonalized Anharmicity Compensation". *Int. J. Mass Spectrom. Ion Processes*, **57**, 1-17, (1984).

39 Y. Naito, and M. Inoue, "Improvement of the Electric Field in FT-ICR Trapped Ion Cell", *Proceedings of the 36th ASMS Conference on Mass Spectrometry and Allied Topics, San Francisco, California*, 608-609, (1987).

40 Y. Naito, M. Fujiwara, and M. Inoue, "Improvement of the Electric Field in the Cylindrical Trapped-Ion Cell", *Int. J. Mass Spectrom. Ion Processes*, **120**, 179-192, (1992).

41 H. S. Kim, R. Chen, and A. G. Marshall, "Quadrupolarized ICR Cubic Ion Trap", *Proceedings of the 41st ASMS Conference on Mass Spectrometry and Allied Topics San Francisco, California*, 448a-448b, (1993).

42 D. L. Rempel and M. L. Gross, "An Improved Calculation of Peak Shape in FTMS", *Proceedings of the 41st ASMS Conference on Mass Spectrometry and Allied Topics, San Francisco, California*, 459a-450b, (1993).

43 D. L. Rempel, "Improved Peak Shapes form a Modified FTMS Cubic Cell Incorporating Segmented Trap Plates", *Proceedings of the 35th ASMS Conference on Mass Spectrometry and Allied Topics, Denver, Colorado*, 1124-1125, (1987).

44 R. L. Hunter, M. G. Sherman, R. T. McIver, "An Elongated Trapped-Ion Cell for Ion Cyclotron Resonance Mass Spectrometry with a Superconducting Magnet", *Int. J. Mass Spectrom. Ion Processes*, **50**, 259-274, (1983).

45 P. B. Grosshans, and A. G. Marshall, "Theory of Ion Cyclotron Resonance Mass Spectrometry: Resonant Excitation and Radial Ejection in Orthorhombic and Cylindrical Ion Traps", *Int. J. Mass Spectrom Ion Processes*, **100**, 347-379, (1990).

46 M. Wang, A. G. Marshall, "A "Screened Electrostatic Ion Trap for Enhanced Mass Resolution, Mass Accuracy, Reproducibility, and Upper Mass Limit in Fourier Transform Ion Cyclotron Resonance Mass Spectrometry", *Anal. Chem.*, **61**, 1288-1293, (1989).

47 J. A. Castoro, C. Köster, and C. L. Wilkins, "Investigation of a "Screened" Electrostatic Ion Trap for Analysis of High Mass Molecules by Fourier Transform Mass Spectrometry", *Anal. Chem.*, **65**, 784-788, (1993).

48 C. L. Holliman, D. L. Rempel, and M. L. Gross, "A Mechanism for Poor High Mass Performance in Fourier Transform Mass Spectrometry", *J. Am. Soc Mass Spectrom,* **3**, 460-463, (1992).

49 C. L. Holliman, D. L. Rempel, and M.L. Gross, "Ion Cloud Radial Expansion Due to Resonant Coupling for Anharmonic Trapping Wells", *Proceedings of the 40th ASMS Conference on Mass Spectrometry and Allied Topics, Washington D. C.*, 736-737, (1992).

50 R. C. Beavis, and B. T. Chait, "Velocity Distributions of Intact High Mass Polypeptide Molecules Ions Produced by Matrix Assisted Laser Desorption", *Chem. Phys. Letters*, **181**, 479-484, (1991).

51 T. Solouki, and D. H. Russell, "Laser Desorption Studies of High Mass-Biomolecules in Fourier Transform Ion Cyclotron Resonance Mass Spectrometry", *Proc. Natl. Acad. Sci. USA.*, **89**, 5701-5704, (1992).

52 R. T. McIver, Y. Li, and R. L. Hunter, "High Resolution Laser Desorption Mass Spectrometry of Peptides and Small Proteins", *Proc. Natl. Acad. Sci. USA*, **91**, 4801-4805, (1994).

53 C. Köster, J. A. Castoro, and C. L. Wilkins, "High-Resolution Matrix-Assisted Laser Desorption/Ionization of Biomolecules by Fourier Transform Mass Spectrometry", *J. Am. Chem. Soc*, **114**, 7572-7574, (1992).

54 G. Savard, St. Becker, G. Bollen, H.-J. Kluge, R. B. Moore, Th. Otto, L. Schweikhard, H.; Stolenberg, U. Wiess, "A New Cooling Technique for Heavy Ions in a Penning Trap", *Physics Letters A*, **158**, 247-252, (1991).

55 L. Schweikhard, S. Guan, and A. G. Marshall, "Quadrupolar Excitation and Collisional Cooling for Axialization and High Pressure Trapping of Ions in Fourier Transform Ion Cyclotron resonance Mass Spectrometry", *Int. J. Mass Spectrom. Ion Processes,* **120**, 71-84, (1992).

56 S. Guan, M. C. Wahl, T. D. Wood, and A. G. Marshall, "Enhanced Mass Resolving Power, Sensitivity, and Selectivity in Laser Desorption Fourier Transform Ion Cyclotron Resonance Mass Spectrometry by Ion Axialization and Cooling", *Anal. Chem.,* **65**, 1753-1757, (1993).

57 D. L. Rempel, and M. L. Gross, "High Pressure Trapping in Fourier Transform Mass Spectrometry: A Radiofrequency-Only-Mode Event", *J. Am. Soc. Mass Spectrom.,* **3**, 590-594, (1992).

58 S. J. Yu, C. L. Holliman, D. L. Rempel, and M. L. Gross, "The β-Distonic Ion from the Reaction of Pyridine Radical Cation and Ethene: A Demonstration of High-Pressure Trapping in Fourier Transform Mass Spectrometry", *J. Am. Chem. Soc.* **115**, 9676-9682, (1993).

59 J. P. Speir, G. S. Gorman, C. C. Pitsenberger, C. A. Turner, P. P. Wang, and I. J. Amster, "Remeasurement of Ions Using Quadrupolar Excitation Fourier Transform Ion Cyclotron Resonance Spectrometry", *Anal. Chem.* **65**, 1746-1752, (1993).

60 S. Guan, and A. G. Marshall, "Bloch Equations Applied to Ion Cyclotron Resonance Spectroscopy: Broadband Interconversion Between Magnetron and Cyclotron Motion for Ion Axialization", *J. Chem. Phys,* **98**, 4486-4493, (1993).

61 C. B. Jacoby, C. L. Holliman, D. L. Rempel, and M. L. Gross, "Ion Cloud Manipulation Using the Radiofrequency-Only-Mode Event as an Improvement for High Mass Detection in Fourier transform Mass Spectrometry", *J. Am. Soc. Mass Spectrom.*, **4**, 186-189, (1993).

62 T. Solouki, K. J. Gillig, and D. H. Russell, "Detection of High Mass Biomolecules in Fourier Transform Ion Cyclotron Resonance Mass Spectrometry: Theoretical and Experimental Investigations" *Anal. Chem.*, **66**, 1583-1587 (1994).

63 E. R. Williams, K. D. Henry, and F. W. McLafferty, "Multiple Remeasurement of Ions in Fourier-Transform Mass Spectrometry", *J. Am. Chem. Soc.,* **112**, 6157-6161, (1990).

64 Z. Guan, S. A. Hofstadler, and D. A. Laude, "Remeasurement of Electrosprayed Proteins in the Trapped Ion Cell of a Fourier Transform Ion Cyclotron Resonance Mass Spectrometer", *Anal. Chem.*, **65**, 1588-1593, (1993).

# QUADRUPOLE MASS FILTERS, QUADRUPOLE ION TRAPS AND FOURIER TRANSFORM ION CYCLOTRON RESONANCE SPECTROMETERS

LAURA OLIMPIERI and PIETRO TRALDI
*CNR, Research Area,*
*Corso Stati Uniti 4,*
*35100 PADOVA (Italy)*

ABSTRACT. Quadrupole mass spectrometers are instruments equipped with the simplest available mass analyser. The development of quadrupole mass filters and ion trap was mainly due to the Pauli's group. Quadrupoles are widely employed in routine mass spectrometric analysis as well as in sophisticated applications. The principle of operation of mass filters and ion traps will be discussed. A short description of the ICR cell will be also presented

## 1. Introduction

The first applications of magnetic sector mass spectrometry to the organic field, starting in 1940s and growing in the 1950s, produced the demand for new instruments with new performances and, as an effect, interest on dynamic analysers showed a clear increase. Such analysers, in which the ion separation is based on the time dependence of one of the system parameters, energy-balance spectrometers (e.g. omegatron and ICR), path-stability spectrometers (quadrupoles)

Quadrupole Mass Spectrometers are, at first sight, the simplest mass analyser available. They are built up of few mechanical parts (four rods in the case of mass filters, three hyperbolic-shaped electrodes for ion traps), are of small dimension (thus requiring small pumping systems), are easy to be operated via suitable software and exhibit performances, especially in terms of sensitivity, which well explain the wide diffusion and marketing of such devices. However, as it will seen later, their behaviour is not so easy to be described in a simple way. the development of quadrupole mass filter and ion trap was mainly due to the Paul's group [1,2] in the second half of 1950s: their work led to construction of a 5.82 m long mass filter with 10.000 resolution for very high resolution studies [3]. At the beginning, quadrupole mass filters were mainly commercialised and employed as gas analysers, not sufficient to measure a total pressure but essential to identify its components. Further developments saw the quadrupole flying out for the Earth atmosphere, for ionospheric measurements, for the analyses of space-cabin gases, astronaut's breath and planetary atmospheres.

*R. M. Caprioli et al. (eds.), Mass Spectrometry in Biomolecular Sciences, 177–200.*
© *1996 Kluwer Academic Publishers.*

178

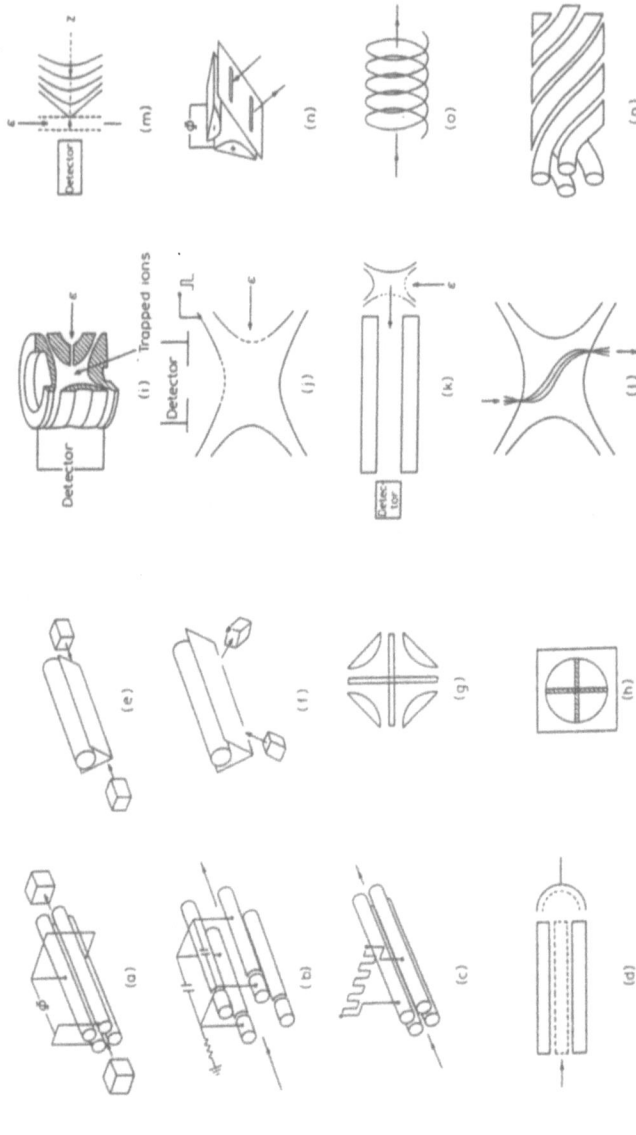

Fig. 1. Some quadrupole field device. (a) The mass filter ; (b) The mass filter with a delayed d.c. ramp ; (c) The mass filter driven rectangularly ; (d) The mass filter with A high-resolution attachment ; (e) The monopole ; (f) The focussing monopole ; (g) the quadrupole monopole, end view ; (h) A four-fold monopole, end view ; (i) The three-dimensional quadrupole ion trap with resonant ion detection ; (j) THe ion trap with external ion detection ; (k) The ion trap as a mass filter source ; (l) A static three-dimensional quadrupole field used as an energy analyser ; (m) A time-of-flight spectrometer ; (n) A focussing "dipole" ; (o) A solenoid mass spectrometer ; (p) A ststic twisted quadrupole view transport system.

A review in 1968 well describes the continue, but slow, evolution of quadrupole field devices [4], and the reached know-how inspired a number of possible quadrupole structures (see figure 1), most of which remained only theoretical ideas.

For such reasons we will focalize our attention on the two devices more common and available on the market, i.e. the mass filter (a of figure 1) and the ion trap (J of figure 1). For an in-deep study of their behaviour we suggest the extensive book of Dawson [5], which represents a particularly valid description of the state-of-out of the field.

The history of FT ICR starts from more far : it was in 1931 that Lawrence and Livingston conceived the cyclotron principle, demonstrating that a charged particle constrained in a magnetic field B covers a circular trajectory and exhibits an angular frequency (called cyclotron frequency) dependent from B and from its m/z ratio. When the particle is irradiated by an rf field with a frequency identical to the cyclotron frequency of the particle, it will trace a spiral path and it can be detected.

The practical use of ICR in the mass spectrometric field is mainly due to the developments carried out in the early 1970s by Mc Iver [6,7] and Comisarow and Marshall [8], introducing pulsed techniques, new analyser cells and the Fourier transform method.

### 2.Few considerations about quadrupole field

The linear dependence on the coordinate position describes a quadrupole field. In a Cartesian, three dimensional space it may be written as

$$\underline{E} = E_0 ( \lambda x + \sigma y + \gamma z ) \tag{1}$$

where $\quad \lambda, \sigma$ and $\gamma$ are weighting constants

$E_0$ is a factor independent from the position but which can be time-dependent

When an ion is subjeted to $\underline{E}$, the force which it experiments, $e\underline{E}$, increase with respect to its displacement from the origin.

$\underline{E}$ is subjected to the restraints deriving from the Laplace's equation

$$\nabla\underline{E}=0$$

and consequently it must be

$$\lambda + \sigma + \gamma = 0 \tag{2}$$

At this point one could choose an infinite set of $\lambda$, $\sigma$ and $\gamma$ values satisfying eq. (2), but let us to consider a first simple choice, placing

$$\lambda = -\sigma \quad , \quad \gamma = 0$$

By this values, eq. (1) becomes

$$\underline{E} = E_0 \lambda ( x - y ) \tag{3}$$

To generate $\underline{E}$, we must necessarily apply a potential $\Phi$ of suitable form. Since

$$E_x = - \partial \Phi / \partial x \quad , \quad E_y = - \partial \Phi / \partial y$$

by integration of eq. (3) it can be obtained

$$\Phi = - 1/2 [ E_0 \lambda ( x^2 - y^2 ) ] \tag{4}$$

Eq. (4) describes a set of equipotential lines in the x, y plane with a four-fold symmetry about the z axis; they can be easily generated by a set of four hyperbolic rods with adjacent electrodes oppositively charged (see fig. 2).

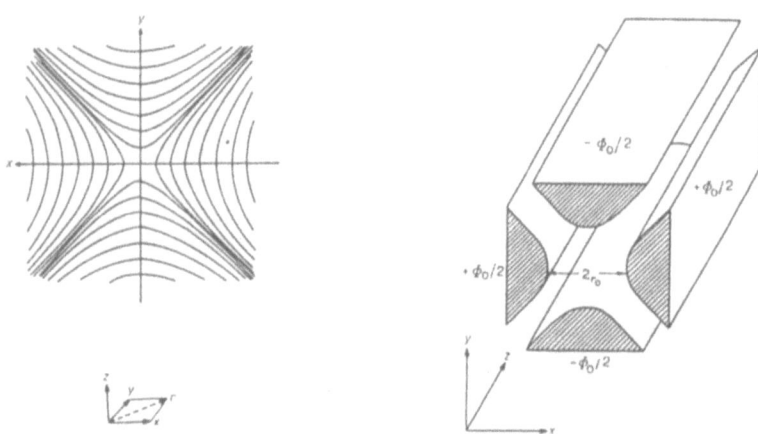

Fig. 2. (a) Equipotential lines for a quadrupole field where $\Phi = -1/2 \, [ \, E_0 \, \lambda \, ( \, x^2 - y^2 \, ) \, ]$.

(b) The electrode structure required to generate the potential shown in (a).

If we consider $2r_0$ and $\Phi_0$ as the minimun distance and the potential between two opposite electrodes respectively, the eq. (4) becomes

$$\Phi = \Phi_0 \, ( \, x^2 - y^2 \, ) \, / \, 2 \, r_0^2 \qquad\qquad (5)$$

Now we go back to eq. (2) and we make a second simple choice of $\sigma$, $\lambda$ and $\gamma$ values satisfying it. Consider now to place

$$\lambda = \sigma \qquad , \qquad\qquad \gamma = -2 \, \sigma \qquad\qquad (6)$$

The potential $\Phi$ thus becomes

$$\Phi = -1/2 \, [ \, E_0 \, \lambda \, ( \, x^2 + y^2 - 2 \, z^2 \, ) \, ]$$

and, passing in cilindrical coordinates (i.e. placing $x^2 + y^2 = r^2$)

$$\Phi = -1/2 \, [ \, E_0 \, \lambda (r^2 - 2 \, z^2 \, ) \, ] \qquad\qquad (7)$$

This equation describes equipotential lines as those described in fig. 3 for the x,y and r,z plane. It can be obtained by an hyperboloid of one sheet forming a ring electrode and an hyperboloid of two sheet forming two end-cap electrodes.

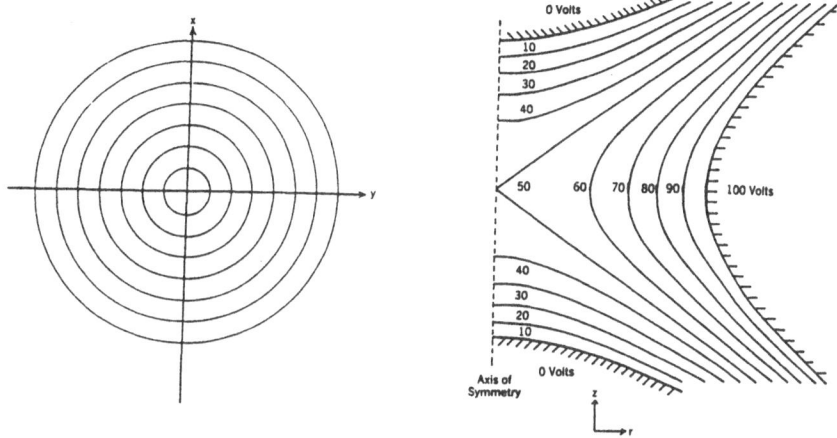

Fig. 3. (a) Equipotential contours of the tri-dimensional quadrupole field in the x y plane.

(b) Equipotential contours of the tri-dimensional quadrupole field in the r z plane.

Again, if $\Phi_0$ is the potential applied between ring electrode and end-cap electrodes, eq. (7) becomes

$$\Phi = \Phi_0 (r^2 - 2 z^2) / 2 r_0^2 \tag{8}$$

At this point it is necessary to define the form of the applied potential $\Phi_0$ : it is usually choosen as

$$\Phi_0 = U - V \cos \omega t$$

i.e. due to a direct voltage of amplitude U and a sinusoidal voltage of amplitude V and angular frequency $\omega$.

### 3. The equation of motion

Once defined $\Phi_0$, we are now able to write the equations of motion.
For the mass filter they can be expressed as

$$\ddot{x} + (e / m \, r_0^2) (U - V \cos \omega t) x = 0 \tag{9}$$

$$\ddot{y} - (e / m \, r_0^2) (U - V \cos \omega t) y = 0 \tag{10}$$

if we define three new quantities as

$$a_u = a_x = -a_y = 4 e U / m \, \omega^2 r_0^2 \tag{11}$$

$$q_u = q_x = -q_y = 2 e V / m \, \omega^2 r_0^2 \tag{12}$$

$$\xi = \omega t / 2$$

both equations (9) and (10) have the form

$$d^2 u / d \xi^2 + (a_u - 2 q_u \cos 2 \xi) u = 0 \tag{13}$$

where u represents either x or y.

For the <u>ion trap</u> the equations of motion are

$$\ddot{z} - (2 e / m r_0^2)(U - V \cos \omega t) z = 0$$

$$\ddot{r} + (e / m r_0^2)(U - V \cos \omega t) r = 0$$

and defining, analogously with what done for the mass filter,

$$a_z = -2 a_r = -8 e U / m r_0^2 \omega^2 = -4 e U / m z_0^2 \omega^2 \qquad (14)$$

$$q_z = -2 q_r = -4 e V / m r_0^2 \omega^2 = -2 e V / m z_0^2 \omega^2 \qquad (15)$$

$$\xi = \omega t / 2$$

the equation of motion for the ion trap became

$$d^2 u / d \xi^2 + (a_u - 2 q_u \cos 2 \xi) u = 0 \qquad (16)$$

where u represent either r or z.

Equation (16) is of the same form as the (13) one. Hence quadrupole mass filter and quadrupole ion trap have in common that the singol equation (16) describes the motion of the ions in both coordinate directions. The motion in the two directions is independent, except for the constraints

$$a_z = -2 a_r \quad , \quad q_z = -2 q_r \qquad \text{for the ion trap}$$

$$a_x = -a_y \quad , \quad q_x = -q_y \qquad \text{for the mass filter}$$

which derive directly from the Laplace equation.

What we like to emphasize is the dependence of the $a_u$ quantities from the direct potential U and the dependence of the $q_u$ ones from the rf potential V. Such considerations will be precious later.

Equation (16) is the canonical form of the Mathieau equation. In the general case the complete solution $u(\xi)$ is composed of two linearly independent solutions, $u_1(\xi)$ and $u_2(\xi)$

$$u = \Gamma u_1(\xi) + \Gamma' u_2(\xi)$$

where $\Gamma$ and $\Gamma'$ are constant of integration depending upon the initial conditions ($u_0$, $\dot{u}_0$, $\xi_0$). Furthermore a corollary of the Floquet's theorem states there will be a solution of eq. (16) of the form

$$u(\xi) = e^{\mu n} \Psi(\xi) \qquad \text{where } \mu = \text{const}$$

From the Fourier's theorem (a periodic function may be expressed as an infinite sum of exponential terms) we can write

$$\Psi(\xi) = \sum C_{2n} \exp(2 n i \xi)$$

$$\Psi(-\xi) = \sum C_{2n} \exp(-2 n i \xi)$$

Some other conditions lead to the general solution

$$u(\xi) = \Gamma e^{\mu \xi} \sum C_{2n} \exp(2 n i \xi) + \Gamma' e^{-\mu \xi} \sum C_{2n} \exp(-2 n i \xi) \qquad (17)$$

where the $C_{2n}$ coefficents are factors describing the amplitude of ion motion, which depend from $a_u$ and $q_u$

$\mu$ is called <u>characteristic exponent</u>. It can be real immaginary or complex and it may be generally expressed as $\mu = \alpha + i \beta$

Two possible solutions can be present :

i) <u>stable</u>    when u remains finite as $\xi$ increases ;

ii) <u>instable</u>  when u increases without limit as $\xi$ increases.

It must be emphasized that, looking at eq. (17), only solutions where $\alpha = 0$ are stable. The four possibilities have been described by Dawson [5], and they can be summarized as follows :

1.  $\mu$ is real and not zero. One of the term $e^{\mu\xi}$ or $e^{-\mu\xi}$ will increase without limit, and the solution is not stable.

2.  $\mu$ is complex and the solution are not stable.

3.  $\mu = i\,m$ , where m is an integer. The solution are periodic but unstable. These solution are called Mathieu functions of integral order and form the boundaries between stable and unstable region on the stability diagram. The boundaries are referred to a *characteristic curves* or *characteristic values*.

4.  $\mu = i\,\beta$ , which is imaginary, and $\beta$ is not a whole number. These solution are periodic and stable.

Considering $\alpha = 0$,  exp $i\beta$ = cos $\beta$+ i sen $\beta$  and by placing

$$A = (\Gamma + \Gamma')\ ,\quad B = i(\Gamma - \Gamma')\ ,\quad \text{eq. (17) becomes}$$

$$u(\xi) = A\ \Sigma\ C_{2n}\ \cos(2n+\beta)\ \xi + B\ \Sigma\ C_{2n}\ \text{sen}(2n+\beta)\ \xi$$

This equation is an expression for the frequency spectrum of stable trajectories of the ions in which the $C_{2n}$ coefficients represent the amplitudes of oscillations and the $(2n+\beta)\xi$ terms the respective frequencies of the infinite number of components. The quantity $\beta$ and $C_{2n}$ may be calculated from a and q using a series of recurrence relation in the form of continued fractions.

$$\omega_{u,n}\, t = (2n + \beta_u)\xi\ ,\quad \text{i.e.}$$

$$\omega_{u,n} = (n + \beta_u/2)\,\omega$$

The strongest component occours when n = 0 , so that the ion possesses the fundamental fraquency $\omega_{u,n} = \beta_u/2$ .

Hence, there will be low frequency macroscopic "secular" components, superimposed upon which is a high frequency ripple ("micromotion").

The stable solutions can be plotted using $a_u$ and $q_u$ as coordinate axes.

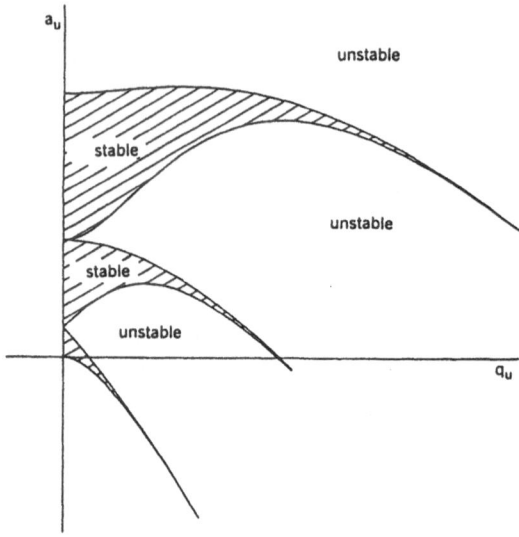

Fig. 4. Graphical representation of stable solution of the Mathieu equation plotted in ( a , q ) space.

Such graphical representations are called stability diagrams. One example is reported in fig. 4. Regions in which the values of $a_u$ and $q_u$ represent stable solutions are shaded, while those unshaded represent unstable solutions.

The ion trajectory will be stable if the related value of mass, charge, frequency and voltage lead to a and q values lying into the stability region. However it must be emphasized that a stability diagram as that shown in fig.4 describes the stability in one dimension only. In the case of both mass filter and ion trap, for which the stability must be necessarily present in two dimension, two separate diagrams must be constructed and overlapped. The overlapping region will ensure simultaneous stability in both the dimensions.

As it will be seen later, the stability diagram will be the key for an easy description of the operative conditions of both mass filter and ion trap.

### 4.Quadrupole mass filter

The condition for which an ion can pass through the four rod of a quadrupole mass filter is that its trajectory in both x and y directions are stable. If we overlap the two stability diagrams corresponding to ( $a_x$ , $q_x$ ) and ( $a_y$ , $q_y$ ) reported in fig. 5, we can obtain the Mathieu stability diagram in two dimensions ( x and y ) (see fig. 6). Only for a , q values lying into the region of simultaneous overlap ( A , B , C and D of fig. 6 ) stable trajectories are present.

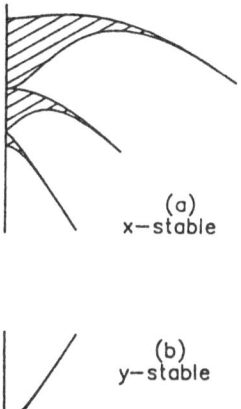

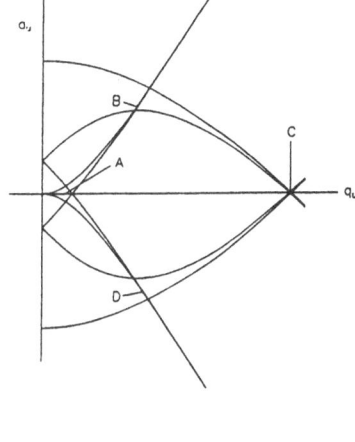

Fig. 5. Several Mathieu stability regions of the two-dimensional quadrupole field : (a) diagrams in the x direction ; (b) diagrams in the y direction.

Fig. 6. The Mathieu stability diagram in two dimensions (x and y). Regions of simultaneous overlap are labeled A, B and C.

Region A is that usually employed and in fig. 7 it is reported, bounded by the characteristic curves for which $\beta_x = 0.1$ and $\beta_y = 0.1$. Within these stable region $0 < \beta_x < 1$ and $0 < \beta_y < 1$ and along the iso-$\beta$ lines ions will posses similar frequency components but can have different trajectory.

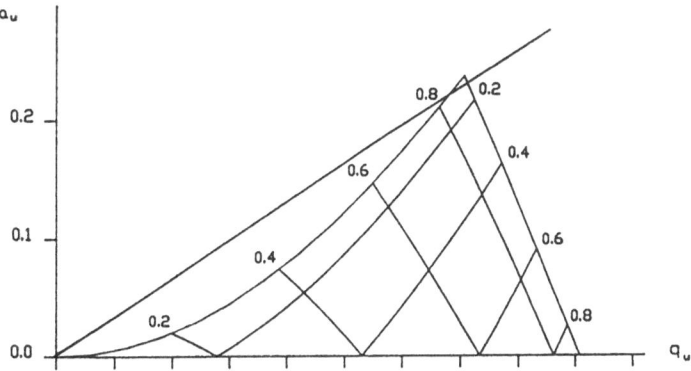

Fig. 7. A portion of the stability region nearest the origin of the quadrupole mass filter.

In the same diagram is also represented what is usually called the scan (or operating) line. It representes the values of a and q to which correspond various combinations of U , V (at a fixed ratio) and mass.

If U and V are mantained at a fixed ratio, the operating line has a fixed slope of a / q. In fact

$$a_x / q_x = a_y / q_y = ( 8 e U / m r_0^2 \omega^2 ) ( m r_0^2 \omega^2 / 4 e V ) = 2 U / V$$

As the values of U and V are changed, so too are the masses which are stable in the device.

As suggested by March and Hughes [9] , the graphical descriptions of mass scanning can be more easily understood by considering the stability diagrams in a ( U , V ) space instead of the ( a , q ) space; the former are practically identical in shape to the latter, as shown in fig. 8. Three ions ( $m_3 > m_2 > m_1$ ) are considered each one exhibiting its own stability diagram. The higher mass ions are transmitted at higher values of U and V, but it must be noted that at the intersection of the scan line with the stability diagram the three ions posses identical a and q values.

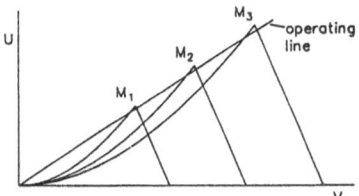

Fig. 8. Stability diagrams transformed to ( U , V ) space.

In summary trasmission of an ion occurs when U and V have such values that the ion has particularly values of a and q lying within the stable region.

To further clarify this point we think of interest to report the Dawson's arguments :

" Paul et al., however, pointed out that it is possible to use the tip of the first stability region, as shown in fig. 7 , to obtain mass resolution. For fixed value of $r_0$ , $\omega$ , U , and V , all ions of the same m/e have the same operating point ( a , q ) in the stability diagram. Since a / q is equal to 2 U / V and does not depend on m / e , the operating points for all ions lie on the same line of constant a / q , passing through the origin of the stability diagram. This is called the mass scan line, mass sampling line, or operating line. Now, when a ≠ 0 , only those ions with operating points lying betwen the intersections of the mass scan with $\beta_y = 0$ and $\beta_x = 1$ will have stable trajectories in both x and y directions and only those ions will pass through the filter. By increasing the U / V ratio, the mass scan line approaches closer to the tip of the stable region and only a narrow range of m / e values will be associated with stable trajectories. Ions of lower mass will be unstable in the x directions and ions of higer mass unstable in the y direction The ions with unstable trajectories will strike the hyperbolic electrodes or exit *laterally from the field. The mass number corresponding to the stable region can be changed (that is, the mass spectrum can be scanned) by varying the magnitudes of U and V but mantaining their ratio constant in order to mantain a constant mass resolution. "*

Just from the pictorical point of view, we can now consider the kind of trajectories of some ions in a quadrupole mass filter.

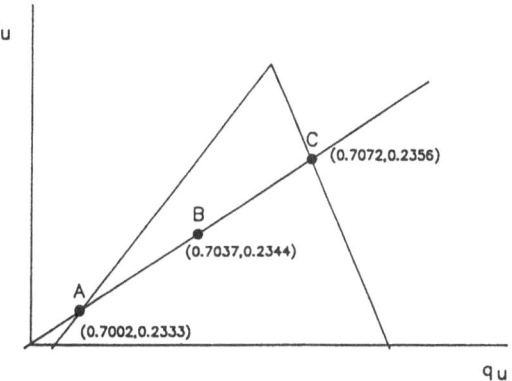

Fig. 9. The apex of the a , q stability region with a mass scan line corresponding to a
resolution of 100.

In fig. 9 we report the apex of the a , q stability diagram with a mass scan line corresponding to a resolution of 100 , while in fig. 10 the trajectories of the m/z 100 ion calculated at points A , B and C of fig. 9 are shown.

188

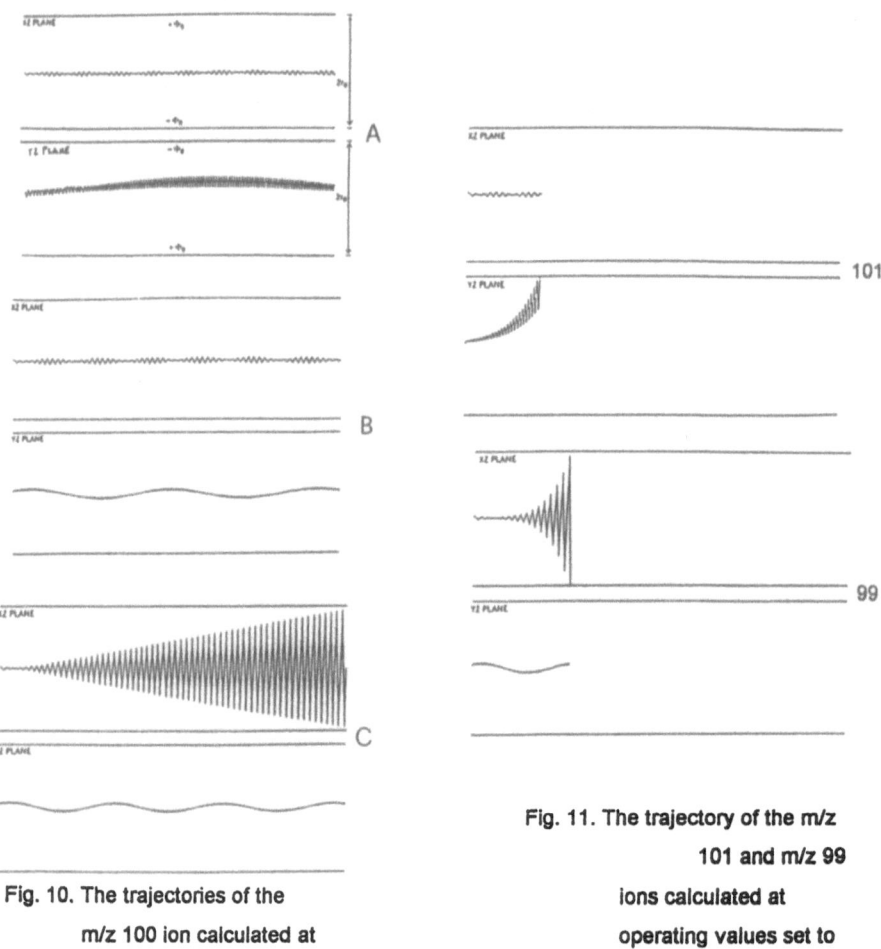

Fig. 10. The trajectories of the
m/z 100 ion calculated at
points A , B and C in fig. 9.

Fig. 11. The trajectory of the m/z
101 and m/z 99
ions calculated at
operating values set to
transmit only m/z 100.

The trajectories of the m/z 101 and m/z 99 ions calculated at the operating values set to transmit only the m/z 100 ions are reported in fig. 11. As it can be seen, the ions motions are unstable and they go to cross the rods.

### 5.Ion trap

In the case of ion trap $a_z = -2 a_r$ and $q_z = -2 q_r$ : consequently the stability parameters for the z and r directions differ by a factor of -2. The two stability diagrams, corresponding to stability in z and r directions, will be not simmetrical, as in the case of mass filter. They are depicted in fig. 12 and, again, ion stability corresponds to a , q values within the overlapping region of the two stability diagrams; in other words ions can be stored in the ion trap only if they are stable in both r and z directions simultaneously.

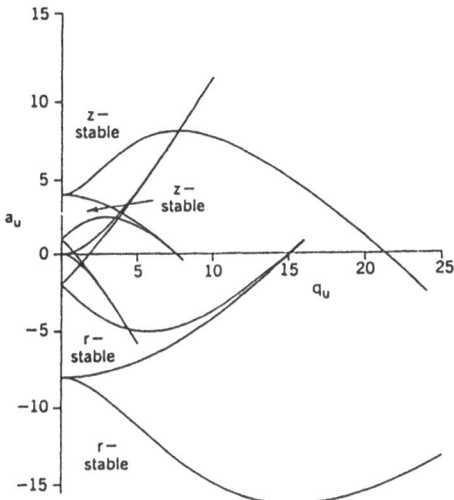

Fig.12. Mathieu stability diagram for the thtee-dimensional values set to a resolution of 100.

As in the case of mass filter, the operative region usually employed is that close to the origin, and shown in fig. 13. Also in this case mass scanning can be achieved by scanning the applied potential, mantaining constant the U / V ratio, so to bring successive masses into the apex of the stability region, but, as it will be seen below, some others scan functions are preferred.

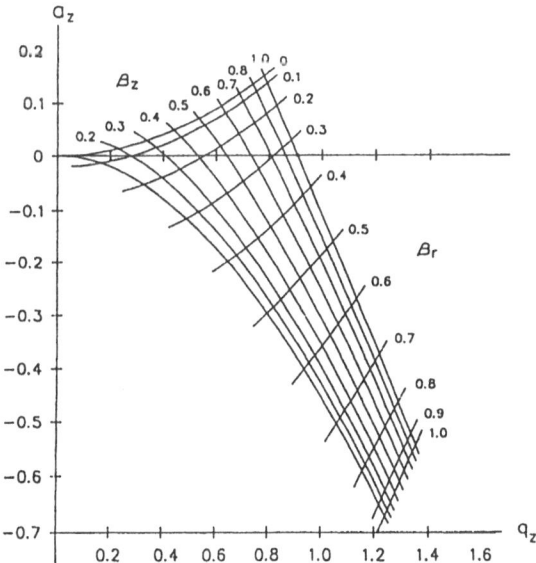

Fig. 13. Stability region near the origin for the three-dimensional ion trap showing the

iso-β lines.

Three different operating modes have been proposed for ion trap, i.e. :

<u>Mode I</u> . The potential $\Phi_0$ is applied to the ring electrode while the potential $-\Phi_0$ is applied to the end-cap electrode [10,11].

<u>Mode II</u> . The end-caps are earthed and $\Phi_0$ is applied to the ring electrode only [12].

<u>Mode III</u> . The DC component, $-U$ , is applied to both end-cap electrodes while RF component ( $V \cos \omega t$ ) is applied to the ring electrode [13].

As it will be seen later, mode II is that usually employed in the commercially available devices.

Ion trap has received in the last decade particular attention from the scientific community due to the features that year after year have been evidenced. Many exhaustive reviews [14,15] and a book [9] are available: what we would like to refert here are just few examples on the potentialites of such device which make it, in our opinion, unique.

In table 1, from ref. 14, the development of ion trap is reported; in function of its age it has be subdivided in three main periods, each dedicated to the development of specific operative behaviours. They correspond to "mass selective detection", in which the frequency components of the motion are used to characterize the $m/z$ value of the ion of interest, "mass selective storage" where only ions with a preselected $m/z$ value are effectively trapped (i.e. held in stable trajectoires) and "mass selective ejection" which derives by sequentially moving the ( $au$ , $qu$ ) coordinates for a series of ions across the stability boundary.

Table I.   The ages of the ion trap mass spectrometer.

| | | |
|---|---|---|
| 1953 | First disclosure; Paul and Steinwedel | |
| 1959 | Storage of microparticles; Wuerker, Shelton and Langmuir | Mass-Selective |
| 1959 | Use as mass spectrometer; Fischer | Detection |
| 1962 | Storage of ions for rf spectroscopy; Dehmelt and Major | |
| 1968 to 1975 | Ejection of ions into an external detector, use as a mass spectrometer, trajectory computations; Dawson, Whetten, Lambert | |
| 1972 to 1980 | Combination of QUISTOR with quadrupole mass filter for analysis of ejected ions, characterization of the trap, CI, ion-molecule reaction kinetics, negative ions, weak mass spectral peak enhancement; Todd et al. | Mass-Selective Storage |
| 1976 to 1982 | Collisional focusing of ions, selective ion reactor, resonant ejection of ions, use as a GC detector, multiphoton (IR) dissociation of ions; March et al. | |
| 1984 | Disclosure of the ion trap detector (ITD®); Stafford, Kelley, Syka, Reynolds, and Todd | |
| 1985 | CI, ion trap mass spectrometer (ITMS®) for MS/MS work; Kelley, Stafford, Syka, Reynolds, Louris, and Todd | |
| 1987 | Automatic gain control, negative ions, photodissociation of ions, injection of ions, mass-range extension, Fourier transform; Stafford, Kelley, Glish, McLuckey, Syka, Fies, Weber-Grabau, Cooks, Louris, Todd, | Mass-Selective Ejection |
| 1988 | Axial modulation of ions on ejection; Weber-Grabau | |
| 1989 | Ions up to $m/z$ 45,000 trapped and analyzed; Cooks et al. | |
| 1990 | Successful injection of multiply charged ions from electrospray source; Glish and McLuckey | |

As it can be seen the most of the performances of interest for a mass spectrometrist (higher mass, high resolution, MS-MS facilities, different ionization methods, etc.) have been gained in the last operative mode and for such reason we will focalize our attention on it.

Briefly, mass-selective ejection consists in the controlled destabilization along the z axis of gaseous ions stored in the trap in function of their mass-to-charge-ratio.

For this aim consider the trap operating with the two end-caps grounded and the RF potential ( V cos ωt ) only applied to the ring electrode.

Reminding that

$$a_z = -8 U / m r_0^2 \omega^2 \qquad q_z = -4 V / m r_0^2 \omega^2$$

it will follows that all the ions will lie on the $q_z$ axis of the stability diagram (in fact, being U = 0 , it follows $a_z$ = 0 ) and, being $q_z$ inversely proportional to m , the ions of higher mass will lie closer to the origin; the $q_z$ values of different ions will increase decreasing the ion mass value. This aspect is depicted in fig. 14.

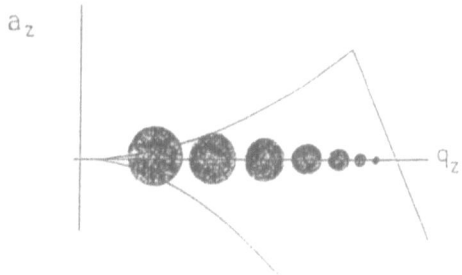

Fig.14. Schematic representation of working points in the stability diagram for several

species of ions stored concurrently. The arrangement of working points with

respect to mass-to-charge ratio is depicted by figures which differ in size.

If now the V value is scanned, the $q_z$ value related to an ion of mass m will increase until the ion reaches the boundary of the stability diagram corresponding to $\beta_z$ = 1 . At this point the ion trajectory is destabilized axially and the axial excursion from the center of the trap increases until they exceed the dimension $z_0$ and strike the end-cap electrodes (see fig. 15). If one of the electrode is perforated, the ion can leave the trap, and be detected by an electron multiplier. By a linear scan of V , all the ions confined into the trap can be succeccively ejected and detected. This is the phenomenon on which the commercially available ITD® , ITMS® and Saturno® instruments are based.

192

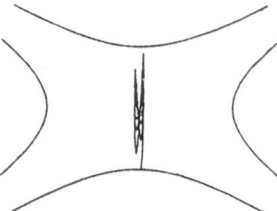

Fig. 15. The trajectory of an ion, focussed initially near the centre of the trap, and exited
axially.

A more detailed description of the operative steps of the ion trap can be easily obtained by fig. 16, in which both the schematic diagram and timing sequence for ITD® are reported. An electron pulse is injected into the ion trap, in which neutral species are present; the ions so generated remain trapped into the trap until the RF voltage is scanned. The two phenomena of ion production and ion analysis, occurring in usual mass spectrometers into different phisical space, in ion trap occur in the same phisical space, i.e. within the ion trap, and hence they must be time-separated.

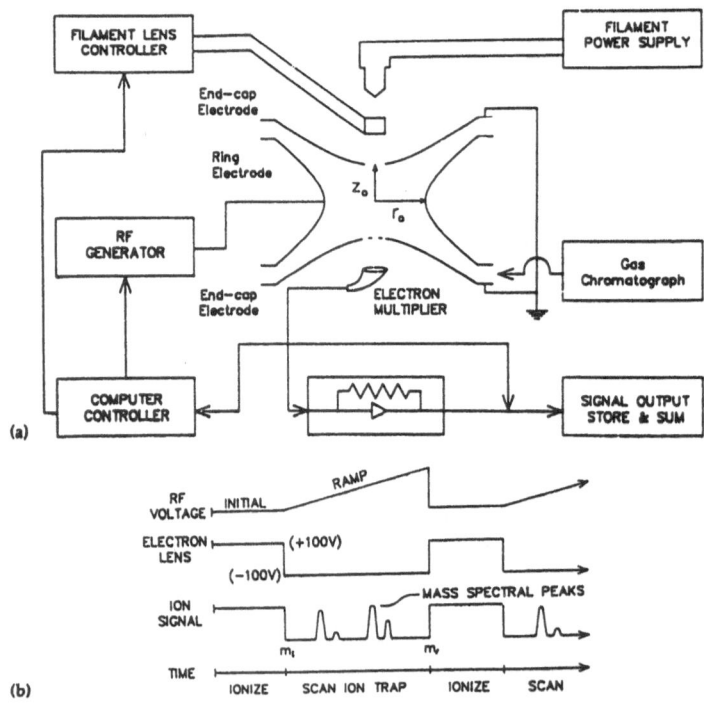

Fig. 16. Schematic diagram (a) and timing sequence (b) for the ion trap detector
(ITD™).         reprinted with permission of reference [14]

### 5.1 Ion isolation

Effective ion isolation can be performed by different procedures, all based on making unstable the trajectories of all the ions but that of interest.

One of these ways[16] is to move the ion to be isolated along the $q_z$ axis (by means of the apt V value) at the B point of fig. 17, just at the base of the apex of the stability diagram. After this, the use of a suitable U (DC voltage) value make possible to move the ion just on the apex (point C of fig. 14). In this conditions all ions, but that of interest, lie out of the stability diagram and are necessarily ejected from the trap. The only ion which will remain trapped will be that of interest.

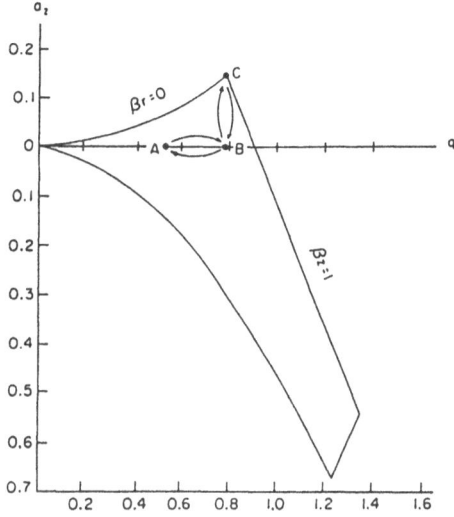

Fig. 17. Stability diagram for the quadrupole ion trap showing the change in location of the working points for an ion undergoing isolation.

The above described isolation method is not always so effective, being not so able to isolate an ionic species $M^+$ only : usually $M^+$ is accompagnished by $[M+1]^+$ and $[M-1]^+$ ions. This undesired behaviour is due to the real shape of the stability diagram for the commercial version of ion trap [17] which show some geometrical difference with the Paul one. The related stability diagram shows a rounded-shaped apex, thus giving account for the low efficiency in ion isolation. For such reason other ion isolation methods have been proposed, among which the two-step isolation method, consisting in selective ejection of ions at higher mass of M followed by ejection of the lower masses, results particularly effective[18].

The ion thus isolated and trapped can be the object of many further experiments : first of all CID experiments, usually performed by resonant excitation [19] or by border effect [20] , but also surface induced dissociation by fast DC pulses [21] and photodissociation by laser irradiation [22]. Due to the high efficiency of the collisional process, $MS^n$ experiments can be easily performed.

### 5.2 Mass range extention

The commercially available ion traps have a common mass range from 0 to 650 Da, but recently many papers have appeared on the possibility to easily extend it. From the already discussed equation

$$q_z = -4 \, e \, V \, / \, m \, r_0^2 \, \omega^2, \quad \text{evidencing m/e}$$

$$m/e = -4V \, / \, q_z \, r_0^2 \, \omega^2$$

it follows that the mass range can be extended in four possible ways :

i)  by ejection at a $q_z$ value less than 0.908, which is the highest $q_z$ limits of the stability diagram ;

ii)  by increasing V (the RF voltage) ;

iii)  by decreasing $r_0$ ;

iv)  by decreasing the main RF frequency $\omega$.

The best results have been achieved by the first approach, by using the axial modulation frequency employed for the tickle experiments. A supplementary RF voltage of 6 - 10 V amplitude and of fixed frequency is applied to the end-cap electrodes : when the RF drive potential is scanned the fundamental axial secular frequencies come in resonance with the supplementary RF voltage : the ions are rapidly excited and ejected from the trap.

As an example we report in fig. 18 the spectrum of $Cs_{n+1} \, I_n^+$ cluster ions from FIB of CsI obtained when they are resonantly ejected at $q_z = 0.027$ using a supplementary frequency of 11 KHz : the upper limit of the detected ions correspond to n = 80 , i.e. 20918 Da ! [23].

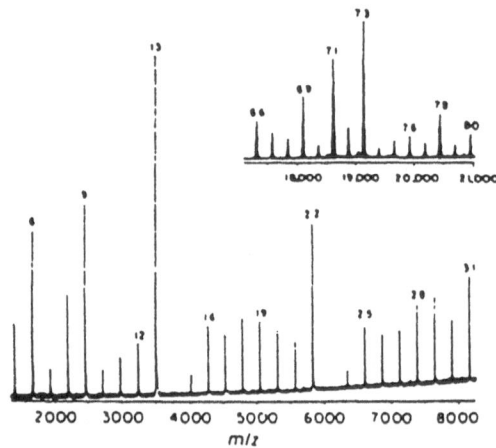

Fig. 18. Mass spectrum of Cs I using an axial modulation frequency of 28 KHz

( $q_z$ = 0.071, $\beta_z$ = 0.050 ). The insert show data recorded at 11 KHz

( $q_z$ = 0.071, $\beta_z$ = 0.020 ). Cluster sizes of $Cs^+$ $(Cs \, I)_n$ are indicated by values

of n.

### 5.3. Resolution

In the last two years the mass resolution achievable by ion trap has been enormeously increased, just by reducing the RF voltage scan rate. Thus Schwartz et al [24] reduced the RF voltage scan rate from its usual value (5555 Da/s) by a factor of 20 : in this condition the time required to scan one mass unit was increased from 190 μs/Da to about 3.8 μs/Da and the resolution jumped up to 33000 (see fig. 19). Cooks et al [25] , by reducing the scan rate by a factor of 333 obtained the exceptional result reported in fig. 20 : for the Cs I cluster ion at m/z 3510 , a resolution of 1,130,000 was achieved.

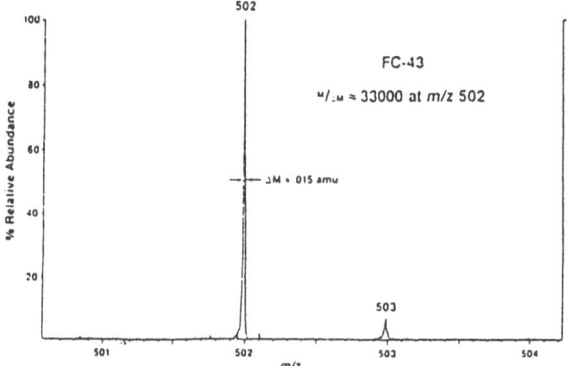

Fig.19. The mass to-charge-ratio 502 region of FC-43 obtained with a scan rate of 27.8 Da s-1.

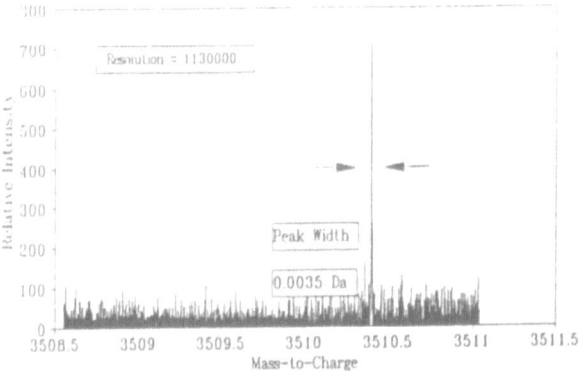

Fig.20. The Cs I cluster ion at m/z 3510 showing mass resolution in excess of $10^6$.

In conclusion the above reported results, together with the easy coupling with external ion sources ( FIB , MALDI , API , Electrospray ) [15] make ion trap one of the most interesting devices nowadays available.

It should be hopeful that these new facilities, discovered and described in literature, in particular high mass range and high resolution, would be soon commercially available.

### 6. ICR and FTICR

Ion cyclotron resonance consists in the irradiation (by means of a suitable RF field) of an ion subjected to cyclotron motion. In ICR spectrometry such resonance phenomenon is applied for the mass analysis of ionic species.

If we consider an ion of mass $m$, charge $z$, velocity $\underline{v}$, subjected to the action of a magnetic field $\underline{B}$, the force acting on it is given by

$$\underline{F} = m\,\underline{a} = z\,\underline{v} \times \underline{B} \qquad (18)$$

The direction of $\underline{F}$ is perpendicular to both $\underline{B}$ and $\underline{v}$ and for enough high $\underline{B}$ values the ion trajectory will be a circumference in a plane perpendicular to $\underline{B}$. For such a motion $|\underline{a}| = v^2 / r$ and hence eq. (18) can be written as

$$m\,a = m\,v^2 / r = z\,v\,B$$

and, considering that the angular frequency is defined as $\omega_c = v / r$, we obtain

$$m\,\omega_c = z\,B, \qquad \text{from which}$$

$$\omega_c = z\,B / m \qquad (19)$$

Equation (19) is the base equation of the cyclotron motion and it shows that an ion placed in a magnetic field is subjected to a circular motion whose angular frequency depends from $m/z$ ratio and from the magnetic field strenght $\underline{B}$ only.

Usually, instead of $\omega_c$, it is used the cyclotronic frequency $\nu_c$ ( $\omega_c = 2\,\pi\,\nu_c$), given by

$$\nu_c = z\,B / 2\,\pi\,m \qquad (20)$$

If we now put in evidence in eq. (20) the $m/z$ ratio, it is obtained

$$m/z = B / 2\,\pi\,\nu_c$$

which shows explicitly the analyzing capabilities of an ICR spectrometer : to every $m/z$ value a well characteristic $\nu_c$ is associated.

The salient advantages to this type of instrumentation, now available in commercial systems, have been described by White and Wood [26] :

1)  Ion generation and mass analysis (can) occur within the same region.
2)  All ions are detected simultaneously.
3)  Negative or positive ion spectra can be obtained with equal ease and rapid switching between the two.
4)  Ion formation and detection are separated in time.
5)  Complex ion optics and high voltages are not needed.
6)  Ion-molecule reactions can be studied at pressures as low as $10^{-9}$ torr - far below pressures usually required for chemical ionization.
7)  A resolution exceeding 1,000,000 is possible.
8)  Samples may be run without an internal mass standard, once the instrument is calibrated.
9)  Sensitivity increases with resolution, in contrast to conventional instruments.

A schematic view of the cubic trapped-ion cell, employed in FTCIR spectrometry, is reported in fig. 21. As in the case of the ion trap, ionization, mass analysis and ion detection occur in the same phisical space and hence they must be time-separated.

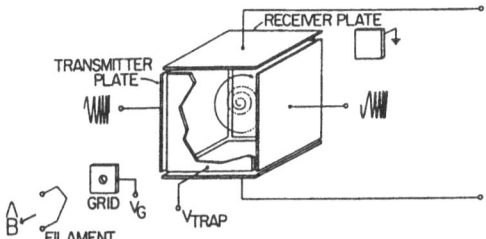

Fig. 21. Schematic of the cubic ion cyclotron risonance cell. Ions are formed by an electron beam pulse, and the coherent motion of ions undergoing resonance is detected by image currents induced in the receivers plates. (Nicolet Analytical Instruments).

To obtain an EI mass spectrum, a short pulse of electrons is admitted into the cell, thus ionizing the neutral molecules present within it. The ions so formed undergo a circular motion, as above described, and possible ion motion parallel to the magnetic flux lines is inhibited by a small electrostatic potential applied to the trapping plates, wich also determine the polarity of the trapped ions. To detect the trapped ions, the cyclotron motions of the ions are excited to large orbits, by applying a swept frequency electric field to the transmitted plates of the cell. This sweep is very fast (~1µs). The cyclotron motions persists after the excitation, and the coherent motion of the ions induced an analogous signal at the resonant frequency in the receiver plates. This sequence is shown in fig. 22, where the treatment of the analog signal coming from the receiver is also reported, consisting in the early steps of amplification, digitalization and accumulation in a computer. The summed digital data are subjected to Fourier transform to produce a frequency domain spectrum which, through frequency/mass conversion, leads to the corresponding mass spectrum.

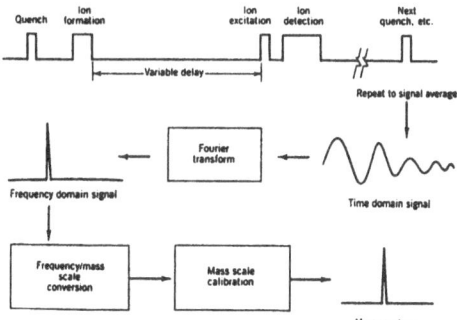

Fig. 22. FTMS sequence of ion formaton, ion detection, and conversion of a time domain signal to a mass spectrum.

The resolution of such instruments can be expressed either in terms of the time constant $\tau$ of the signal decay

$$R \approx 1/2\,\omega\,\tau = \pi\,f_0\,t$$

or in terms of $B$, $\omega$ and m/z

$$R \approx m/\Delta m = \omega_c/\Delta\omega_c = z\,B/m\,\Delta\omega_c$$

Mass resolution in FT ICR is linearly related to the duration of signal aquisition. Using a 4 - Mhz analog-to-digital converter and a 7 T magnet a mass resolution of 10.000 at m/z 500 was obtained digitizing the transmitted signal for about 60 ms. This limitation can be overcome over a limited mass range by narrow-band data aquisition. With the aquisition of only a narrow mass range (about 10 Da), mass resolution greater of 1,500,000 has been obtained, as shown in fig. 23.

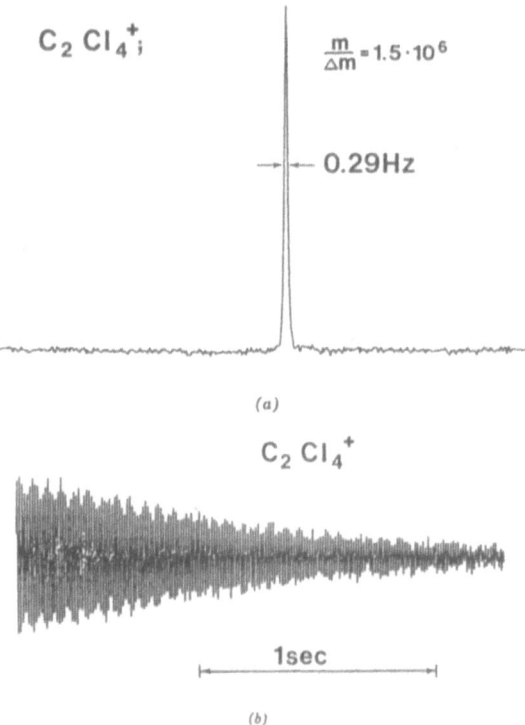

$C_2\,Cl_4^+{}_i$ $\qquad \dfrac{m}{\Delta m} = 1.5\cdot 10^6$

$\dashrightarrow\!\!\longleftarrow$ 0.29Hz

*(a)*

$C_2\,Cl_4^+$

1sec

*(b)*

Fig. 23. (a) Signal for m/z 166, from the spectrum of $C_2Cl_4^+$, taken at a resolution power of 1.5 x $10^6$. (b) First part of the transient $C_2Cl_4^+$ signal shown in a.

However the single cell above described suffers severe limitations, being vulnerable to high sample or backgroun pressure conditions, leading to a degradation of both resolution and sensitivity. For such reasons the development of other instumental configurations, as the differentially pumped dual cell from Nicolet (fig. 24) or the FTMS with external ion source developped by Bruker (fig. 25) led to systems of higher performances.

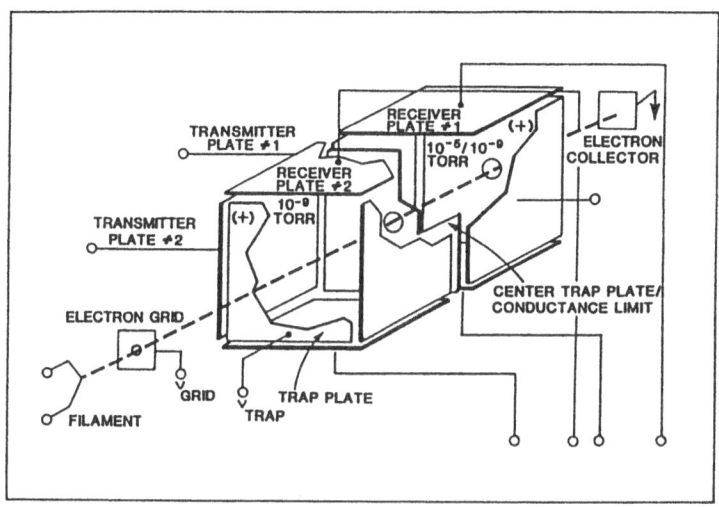

Fig. 24. Schematic diagram of a differentially pumped FTMS dual cell. (Nicolet Analytical Instruments).

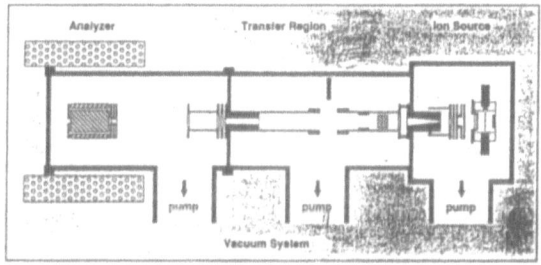

Fig. 25. FTMS with extended ion source developped by Bruker.

200

## References

1. W, Paul and H. Steinwedel, Ger. Pat. 944,900 (1956) ; U.S. Pat. 2, 939, 952 (1960).
2. W. Paul, H.P. Reinhard and U. von Zahn, *Z. Phys*, **152**, 153 (1958).
3. U. von Zahn, *Z. Phys.*, **168**, 129, (1962);
   U. von Zahn, S. Ghebauer and W. Paul, 10th Annu. conf. Mass Spectrom., New Orleans, 1962 (not generally available).
4. P.H. Dawson and N.R. Whetten, *Advan. Electron. Phys.*, **27**, 59, (1969).
5. Quadrupole Mass Spectrometry and Its Applications, P.H. Dawson (Ed.), Elsevier, Amsterdam (1976).
6. R.T. Jr. McIver, *Rev. Sci. Instrum.*, **41**. 555, (1970).
7. R.T. Jr. McIver, Ph. D. Thesis, Stanford University (1971).
8. M.B. Comisarow and A.C. Marshall, *Chem. Phys. Lett.*, **25**, 282, (1974).
9. R.E. March and R.J. Hughes, Quadrupole Storage Mass Spectrometry, John Wiley & Sons, New Jork (1989).
10. W. Paul, O. Osberghaus, E. Fischer, Forschungsberichte des Wirtschaft und Verkehrministeriums Nordrhein Westfalen, No 415, Westdeutscher Verlag, Koln and Opladen, 1958.
11. E. Phischer, *Z. Phys.*, **156**(1), 1, (1959).
12. P.H. Dawson, and N.R. Whetten, *J. Vac. Sci. Technol.*, **5**, 11, (1968).
13. J.E. Fulford and R.E. March, *Int. J. Mass Spectrom. Ion Phys.*, **26**, 155, (1978).
14. J.F.J. Todd, *Mass Spectrom. Rev.*, **10**, 3, (1991).
15. R.E. March, *Int. J. Mass Spectrom. Ion Processes*, **118/119**, 71, (1992).
16. J.N. Louris, J.S. Brodbelt-Lusting, R.G. Cooks, G.L. Glish,G.J. Van Berkel,S.A. McLuckey, *Int. J. mass Spectrom. Ion Proc.*, **96**, 117, (1990).
17. J.V. Johnson, R.E. Peddes, R.A. Yost, *Rapid Commun. Mass. Spectrom.*, **6**, 322, (1992).
18. C.E. Ardanaz, P. Traldi, U. Vettori, J. Kovka and F. Guidugli, *Rapid Commun. Mass Spectrom.*, **5**, 5, (1991).
19. J.E. Fulford, D.N. Hoa, R.J. March, R.F. Bonner and G.J. Wong, *J. Vac. Sci. Technol.*, **17**, 829, (1980).
20. O. Curcuruto, S. Fontana, P. Traldi and E. Celon, *Rapid Commun. Mass Spectrom.*, **6**, 322, (1992).
21. S.A. Lammert, R.G. Cooks, *J. Am. Soc. Mass Spectrom.*, **2**, 487, (1991).
22. C. Lifshitz, *Int. J. Mass Spectrom. Ion Processes*, **106**, 159, (1991).

# ENERGY SHIFTS IN COLLISIONAL ACTIVATION

HELEN J. COOPER AND PETER J. DERRICK

*Institute of Mass Spectrometry and Department of Chemistry*

*University of Warwick*

*Coventry CV4 7AI*

*United Kingdom*

ABSTRACT. A chronological account of the work on energy shifts accompanying collisional activation is given. The direct momentum transfer mechanisms for collisional activation, known as impulsive collision transfer theory and the quasi-diatomic approximation, are investigated. Current studies on target capture and angle-resolved translational energy spectra are discussed.

## 1. Introduction

Collisional activation[1] is the basis of tandem mass spectrometry in which ions are selected by a mass spectrometer and collided with an inert gas. A fraction of the initial translational energy of the ion is lost and is imparted either to the gas atom as translational energy or to the ion as internal energy. This internal energy uptake results in fragmentation and hence structural information. A second mass spectrometer scans the fragments providing a collisionally-activated decomposition (CAD) spectrum of the ion. Such spectra are also known as collision-induced decomposition (CID) spectra.

Collisional activation and subsequent fragmentation are usually considered to be two separate processes in accordance with Lindemann theory. The fragmentation is therefore a unimolecular process. This model is known as the 'two step model'. If ion $AB^+$ collides with a gas G, the mechanism is as follows:

$$AB^+ + G \quad \text{---------->} \quad (AB^+)^*$$
$$(AB^+)^* \quad \text{---------->} \quad A^+ + B$$

The asterisks indicate excitation without implication to the nature of this excitation.

201

*R. M. Caprioli et al. (eds.), Mass Spectrometry in Biomolecular Sciences, 201–259.*
© 1996 *Kluwer Academic Publishers.*

The usefulness of tandem mass spectrometry in the determination of peptide and biomolecular structures is limited as yet. An unknown peptide of RMM approximately 500 Da could be fully sequenced using tandem mass spectrometry, whereas for a peptide of around 5000 Da tandem mass spectrometry would typically provide no more information than could be obtained from a standard double-focusing mass spectrometer. The usual explanation is that this is the result of too little internal energy being deposited into the parent ion on collision. The nature and mechanism of the internal energy uptake are therefore of particular interest. The view held for a long time was that the collision resulted in electronic excitation of the ion. Cooks et al.[2] state that "the excitation step involves energy transfer mainly to the electrons of the ion; there is very little momentum transfer". This electronically excited ion was considered to undergo radiationless transitions leaving a vibrationally excited ion which subsequently fragmented.

The data available during this period concerned diatomic ions and small organic ions such as benzene. For these cases, the above explanation is often adequate. Techniques for producing large molecule-ions have since been developed. Experimental data[3-8] relating to these large ions is less consistent with the electronic mechanism and more in line with a direct momentum transfer mechanism.[9-11]

In this review, a chronological discussion of the work concerning large ions in keV collisions is given along with an investigation of the suggested direct momentum transfer mechanisms for large ions, in particular impulsive collision transfer (ICT) theory.[10,11] Coverage of the literature is not exhaustive. The approach adopted is pedagogic and draws heavily on results from our own laboratory.

## 2. Energy Shifts in Collision-Induced Decomposition Spectra

It has already been stated that, following a collision, an ion will lose a fraction of its translational energy. If an electric sector is being used to analyse the fragment ions produced by decomposition of the parent, this energy loss will affect the positions of the peaks in the spectrum. This peak shift has been noticed for peptides[3,12] cationised saccharides[13] and synthetic organic polymers.[14]

Consider a singly charged parent ion, mass $m_1^+$, which decomposes to a singly charged fragment ion, mass $m_2^+$. The translational energy, $E_0$, of the ion $m_1^+$ is equal to $eV_{acc}$ ($V_{acc}$ is the accelerating potential). If there were no translational energy loss on collision, the fragment ion $m_2^+$ would pass through the electric sector at a potential $V_1 = \left(\dfrac{m_2}{m_1}\right) V_0$ where $V_0$ is the potential at which the parent ion would be transmitted in the absence of collision gas.

If however the parent ion does suffer translational energy loss, $\Delta E$, in the collision then the fragment will require a lower sector potential $V_2$ for transmission. This peak shift can be related to the translational energy loss $\Delta E$ as follows:

$$\Delta E = e\left(\frac{V_{acc}}{V_0}\right)\left(\frac{m_1}{m_2}\right)(V_1 - V_2) \qquad (1)$$

This expression can be written more simply in terms of the energy of the fragment ion in the absence of collision gas, $E_1$, and $E_2$ the measured energy of the fragment.

$$\Delta E = \left(\frac{m_1}{m_2}\right)(E_1 - E_2) \qquad (2)$$

By measuring the peak shift for a fragment ion, it is possible to calculate the translational energy loss suffered by the parent ion that decomposed to give *that* fragment. This use of an electric sector to measure energy shifts is known as MIKES i.e. mass-analysed ion kinetic energy spectroscopy.

Peaks in MIKE spectra are broader than peaks in standard double-focusing mass spectra. This is a result of translational energy release i.e. on decomposition some of the internal energy of the parent is converted to translational energy of the fragment. This energy release however *does not* affect the measurement of the overall energy shift. Some fragments will be given translational energy in the forward direction and some in the backward direction. Since the direction of this translational energy release is random, the effect on the fragment peak is one of broadening rather than further shifting of its position.

### 3. Optimum Pressure for Collision-Induced Decomposition

Except where otherwise stated, the work from our laboratory described here was performed at the 'optimum pressure'.[15] The collision gas pressure has been set so that the probability of a single collision is highest.

This probability is found using Poisson statistics.[4]

$$P(x) = \frac{\lambda^x e^{-\lambda}}{x!} \qquad (3)$$

$P(x)$ is the probability of x collisions occurring and $\lambda$ is the average number of collision and is proportional to pressure. So the probabilities of 1, 2, 3 .... collisions occurring are

$$e^{-\lambda}, \lambda e^{-\lambda}, \tfrac{1}{2}\lambda^2 e^{-\lambda}, \tfrac{1}{6}\lambda^3 e^{-\lambda}....$$

The probability of a single collision reaches a maximum when $\lambda = 1$. At this pressure 1/e (37%) ions do not collide, 1/e (37%) undergo one collision, 1/2e (18%) undergo two collisions, 1/6e (6%) undergo three collisions etc.

In order to find the pressure at which $\lambda$ has a value of one, an assumption is made. This is that the probability of the incident ion undergoing zero collisions, $P(0)$, can be equated to the proportion of incident ions reaching the detector. (This proportion of incident ions is known as the transmission). The transmission is varied until a value for $\lambda$ of one is obtained as calculated from equation (3). The collision gas pressure which gives this transmission is the optimum pressure. The optimum pressure is always found to correspond to an attenuation of the original ion beam to between 30% and 40%.

## 4. Effect of the mass of the incident ion on translational energy losses

There have been a number of studies on the effect of the mass of the incident ion on the translational energy loss, $\Delta E$. One study[3] is concerned with incident ion masses between 400 Da and 1200 Da, and a second[4] with ion masses between 400 Da and 1600 Da.

Both of these studies were conducted on a large scale reverse geometry mass spectrometer.[16] The collision cell employed was sited at the common focal point between the magnetic and electric sector. The mean radius of the electric sector was 1000 mm and had an energy resolving power of around $10^3$. In both cases, peptide ions were produced by field desorption.

### 4.1 MASS RANGE 400-1200 Da

The initial study used seven peptide ions. These were

| (a) | m/z 448 | [M+H]$^+$ from H-Pro-Phe-Gly-Lys-OH |
|-----|---------|-------------------------------------|
| (b) | m/z 452 | fragment ion [H-pGlu-His-Trp-NH$_2$+H]$^{+}$ [17] from |
|     |         | H-pGlu-His-Trp-Ser-Tyr-Gly-Leu-Arg-Pro-Gly-NH$_2$ |
| (c) | m/z 574 | [M+H]$^+$ from H-Tyr-Gly-Gly-Phe-Met-OH |
| (d) | m/z 707 | [M+H]$^+$ from H-Lys-Phe-Ile-Gly-Leu-Met-NH$_2$ |
| (e) | m/z 754 | [M+H]$^+$ from H-Phe-Phe-Phe-Phe-Phe-OH |
| (f) | m/z 1061 | [M+H]$^+$ from H-Arg-Pro-Pro-Gly-Phe-Ser-Pro-Phe-Arg-OH |
| (g) | m/z 1183 | [M+H]$^+$ from H-pGlu-His-Trp-Ser-Tyr-Gly-Leu-Arg-Pro-Gly-NH$_2$ |

The collision-induced decomposition spectra of these ions, with helium at optimum pressure as collision gas and an incident ion energy of 8 keV, are shown in *figure 1*[3].

In order to calculate translational energy losses, it is necessary to assign masses to the fragment peaks in each spectrum. There are a set of ions which have been determined as being the result of collisional activation.[12,17-19] In calculating energy shifts only ions in this set were considered.

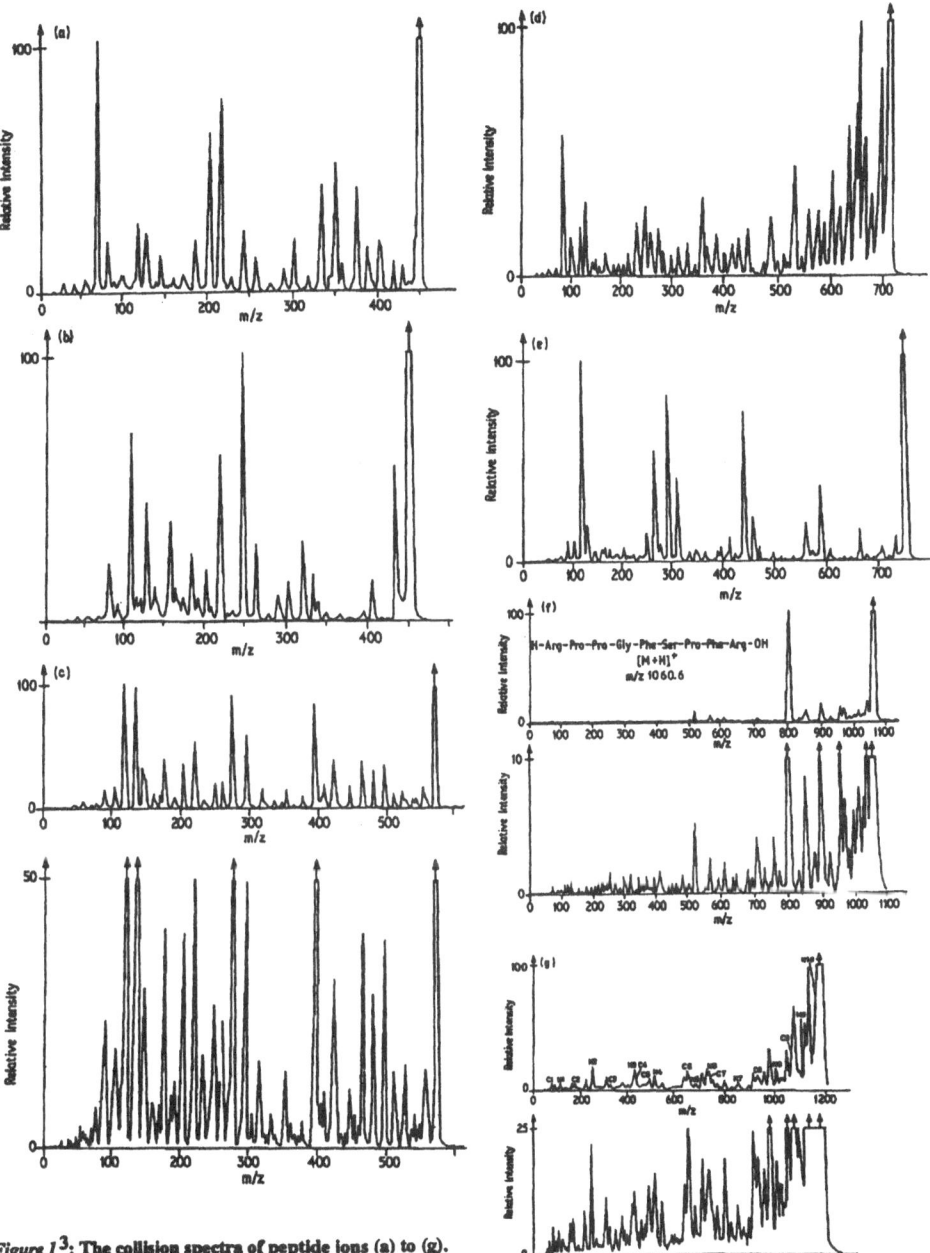

*Figure 1* [3]: The collision spectra of peptide ions (a) to (g).
The symbols N1, C1 etc in fig. 1(g) indicate N-sequence
and C-sequence ions.

Weak peaks were not utilised and nor were peaks that were difficult to centroid with accuracy. The six types of peaks considered were:

(i)     N-sequence ions, known as $b_n$ ions

(ii)    N-sequence ions minus CO, known as $a_n$ ions

(iii)   C-sequence ions, known as $y_n+2$ ions

(iv)    Side-chain or side-chain loss ions

(v)     Ions characteristic of certain amino acid residues

(vi)    Ions corresponding to the elimination of small neutral molecules

The mass assignments for the different spectra are given in *table 1*. Independent support for these mass assignments was originally given by Matsuo *et al.*[19] although much more evidence is available now.

| Assigned mass | Type of ion | Assigned mass | Type of ion |
|---|---|---|---|
| (a) m/z 448 | | | |
| 70 | N-sequence - CO | 245 | N-sequence |
| 120 | H-Phe - CO | 302 | N-sequence |
| 129 | H-Lys | 351 | C-sequence |
| 147 | C-sequence | 376 | Loss of Lys side-chain |
| 204 | C-sequence | 430 | Loss of $H_2O$ |
| 217 | N-sequence - CO | | |
| (b) m/z 452 | | | |
| 81 | His side-chain | 221 | N-sequence - CO |
| 110 | H-His - CO | 249 | N-sequence |
| 130 | Trp side-chain | 322 | Loss of Trp side-chain |
| 159 | H-Trp - CO | 435 | Loss of $NH_3$ |
| (c) m/z 574 | | | |
| 120 | H-Phe - CO | 397 | N-sequence - CO |
| 136 | N-sequence - CO | 425 | N-sequence |
| 221 | N-sequence | 467 | Loss of Tyr side-chain |
| 278 | N-sequence | 499 | Loss of Met side-chain |
| 297 | C-sequence | | |

| (d) m/z 707 | | | |
|---|---|---|---|
| 120 | H-Phe - CO | 531 | N-sequence - CO |
| 129 | N-sequence | 632 | Loss of Met side-chain |
| 248 | N-sequence - CO | 650 | Loss of Ile/Leu side-chain |
| 276 | N-sequence | 692 | Loss of $CH_3$ |
| 361 | N-sequence - CO | | |
| (e) m/z 754 | | | |
| 120 | H-Phe - CO | 460 | C-sequence |
| 267 | N-sequence - CO | 561 | N-sequence - CO |
| 295 | N-sequence | 589 | N-sequence |
| 313 | C-sequence | 663 | Loss of Phe side-chain |
| 442 | N-sequence | | |
| (f) m/z 1061 | | | |
| 527 | N-sequence - CO | 859 | N-sequence - CO |
| 614 | N-sequence - CO | 961 | Loss of Arg side-chain |
| 807 | C-sequence | 970 | Loss of Phe side-chain |
| (g) m/z 1183 | | | |
| 112 | N-sequence | 685 | N-sequence |
| 172 | C-sequence | 714 | N-sequence - CO |
| 221 | N-sequence - CO | 855 | N-sequence |
| 249 | N-sequence | 984 | N-sequence - CO |
| 328 | C-sequence | 1012 | N-sequence |
| 435 | N-sequence | 1152 | Loss of Ser side-chain |
| 522 | N-sequence | 1166 | Loss of $NH_3$ |

*Table 1*: Mass assignments for CID spectra in *figure1* [3].

For each CID spectrum, the energy shifts of the fragment peaks were measured. The overall translational energy loss of the parent is given as the average of the energy shifts for the fragments selected. It cannot be expected that different fragments should have the same energy losses. Different fragments form via different pathways requiring different energies. An ion formed via a low energy pathway presumably had less internal energy imparted to its parent than a fragment formed by a high energy pathway.

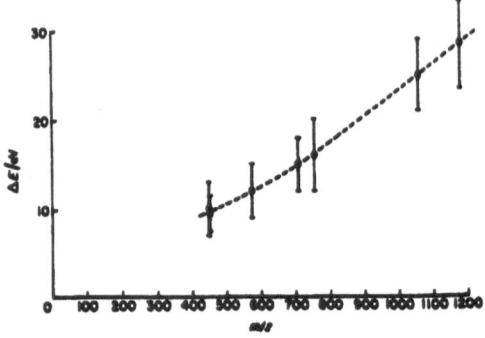

*Figure 2[3]*: **The translational energy loss** $\Delta E$ **suffered by singly charged peptide ins as a result of collision with helium against the mass-to-charge ratios of the peptide ions.**

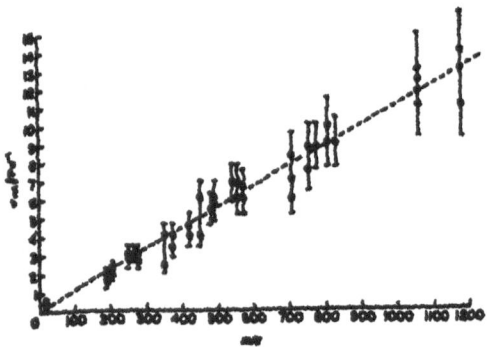

*Figure 3[3]*: **Relative cross-sections of peptide ions against mass-to-charge ratios of the ions. The points plotted refer to the six peptides used in drawing up** *figure 2[3]* **and other peptides.**

*Figure 2[3]* shows the relationship between the translational energy loss by the parent ion and its mass for this mass range. As the masses increase, translational energy losses reach values of up to 30 eV. The magnitude of these energy losses has caused much controversy and it is often suggested that multiple collisions have taken place particularly in the case of the more massive ions. In this experiment, there are three factors weighing against this idea. Firstly, the attenuation of the ion beam was the same in all the experiments which implies that every incident ion underwent the same average number of collisions.[20]

The relative collision cross-section versus incident ion mass is plotted in *figure 3[3]*. This cross section represents a probability of collision and is inversely proportional to collision gas pressure. This linear dependence is already stronger than expected[15] and the suggestion that the dependence is in fact stronger than this by saying heavier ions undergo a greater number of collisions than light ions is very difficult to justify.

The final argument against multiple collisions for heavier ions is that, for a particular collision-induced decomposition mass spectrum, the optimum pressure was the same for all the fragments. If the number of collisions was dependent on mass, then the more massive fragments would undergo more collisions before leaving the cell and the optimum pressure for these fragments would be lower than for the lighter fragments ions. This, however, is not the case.

Having decided that each incident ion probably undergoes one collision, we can then turn to the question of internal energy uptake. In order to calculate this Neumann et al.[3] used the first direct momentum transfer idea, known as the 'quasi-diatomic approximation' (QDA).[9] This considers the ion and the target gas as separate single entities and the collision to be of the 'hard-sphere' type. The gas is assumed to be stationary prior to the collision. This theory will be discussed in more detail later.

As mentioned in the introduction, the internal energy uptake, $Q$, can have a value lying between zero and a maximum that is slightly less than $\Delta E$, the translational energy loss. The remainder of $\Delta E$ is imparted to the target gas as translational energy. $Q$ has a maximum when there is zero deflection and falls as deflection increases. In calculating $Q$, the deflection angle was assumed to be zero. This assumption finds some support from Todd et al.[21] however it is not strictly accurate. The equations used to calculate $Q$ were:

$$Q = \left[ \left( m_i + m_g \right) \Delta E \Big/ m_g \right] - \left[ \left( 2 m_i E \Big/ m_g \right) \left( 1 - \left[ \left( E - \Delta E \right) \Big/ E \right]^{\frac{1}{2}} \cos \theta \right) \right] \qquad (4)$$

$$\Delta E = Q + \Delta E_g \qquad (5)$$

where $m_i$ = mass of ion; $m_g$ = mass of gas; $E$ = initial energy of ion; $\theta$ = scattering angle (set to zero in this case); $\Delta E$ = translational energy loss by the ion; $\Delta E_g$ = translational energy gain by the gas.

## 4.2 MASS RANGE UP TO 1600 Da

The centre-of-mass collision energy $E_{cm}$ represents the maximum amount of energy available for conversion to internal energy. If the gas is stationary prior to the collision (this assumption is standard when considering collisional activation), then $E_{cm}$ is given by:

$$E_{cm} = \frac{m_g}{m_i + m_g} E \qquad (6)$$

where $m_i, m_g$ and $E$ are as described above.

Clearly then if the mass of the ion increases and other factors are kept constant, $E_{cm}$ falls. It was already known that for the mass range 400-1200 Da, internal energy uptake, $Q$, rose approximately linearly. This makes the region above 1200 Da very interesting as it is in this region that $Q$ will begin to approach $E_{cm}$.

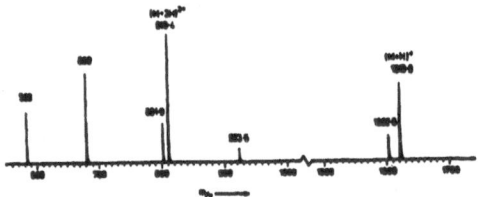

*Figure 4*[4]: **Field desorption mass spectrum of the peptide bombesin.**

In the second study[4], Neumann *et al.* used fragment ions from the field desorption of bombesin (H-pGlu-Gln-Arg-Leu-Gly-Asn-Gln-Trp-Ala-Val-Gly-His-Leu-Met-NH$_2$, RMM=1618.8 Da) as incident ions. The field desorption spectrum of bombesin is given in *figure 4*[4]. The incident ions employed were

(i)   m/z 1619.8     corresponding to [M+H]$^+$

(ii)  m/z 810.4      corresponding to [M+2H]$^{2+}$

(iii) m/z 680        N-terminal fragment ion (H-pGlu-Gln-Arg-Leu-Gly-Asn)$^+$

(iv)  m/z 583        N-terminal fragment ion ( H-pGlu-Gln-Arg-Leu-Gly-NH$_2$+H)$^+$

The collision-induced decomposition spectra at optimum pressure (target gas = helium, incident ion energy = 8000 eV) are shown in *figure 5*[4].

*Table 2* shows the translational energy loss, $\Delta E$, suffered by the incident ion and the internal energy uptake, $Q$, calculated from equation (4) with $\theta = 0$.

| m/z | Collision gas | Incident Ion Energy (keV) | $\Delta E$ (eV) | $Q$ (eV) |
|---|---|---|---|---|
| 583 | He | 8 | $12 \pm 3$ | $11.5 \pm 3$ |
| 680 | He | 8 | $14 \pm 3$ | $13 \pm 3$ |
| 810.4 | He | 8 | $25 \pm 15$ | $21 \pm 10$ |
| 1619.8 | He | 8 | $50 \pm 7$ | $18.5 \pm 2$ |
| 1619.8 | He | 20 | $<50 \pm 7$ | $18.7 \pm 3$ |
| 1619.8 | Ar | 8 | $19 \pm 4$ | $18.5 \pm 4$ |

*Table 2:* **Translational energy losses and internal energy uptakes for bombesin.**

Note: The large error margin in the translational energy loss for the [M+2H]⁺ ion is due to the smaller translational energy losses by its fragment ions (due to increased translational energy of the incident ion) and therefore smaller peak shifts.

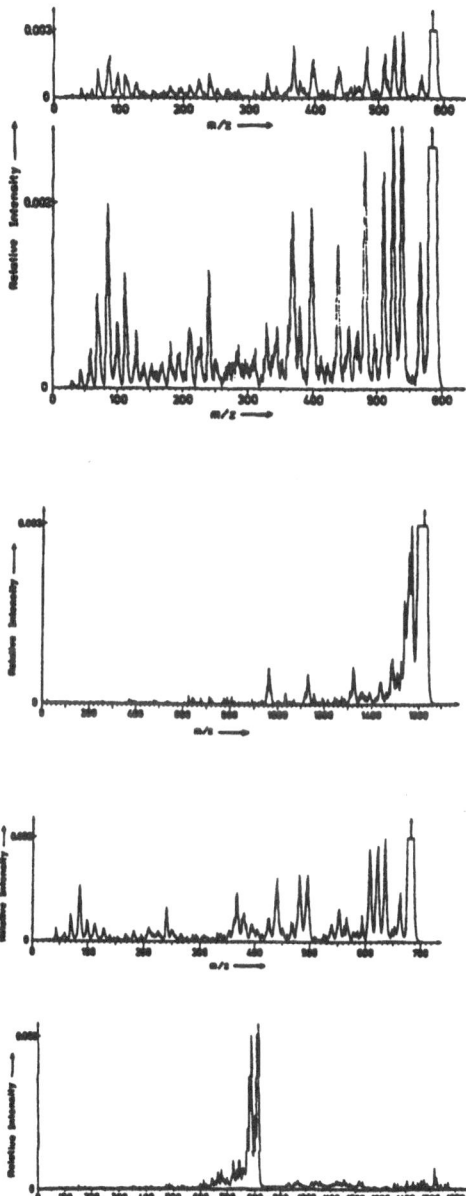

*Figure 5⁵*: CID spectra of the fragment ions (i) to (iv) produced during field desorption of bombesin. Collision gas = helium.

The effect of changing the collision gas on the CID spectrum is very interesting. The fragments present in the spectrum and their intensities relative to the parent ion were the same for both argon and helium but the *positions* of the fragment peaks varied markedly. The similarity of the spectra suggests that the internal energy uptake was roughly the same in both cases. On calculating $Q$ using QDA at $0°$ from the translational energy losses (50 ± 7 eV for helium; 19 ± 4 eV for argon), it is found to have the same value for both target gases (18.5 eV).

When using helium, a shoulder is found on the parent peak which has an intensity of between 4-5% of the parent ion intensity. This shoulder is not present in the absence of any collision gas nor in the presence of argon. See *figure 6*[4]. This shoulder is attributed to $[M+H]^+$ ions which have undergone collision, and so suffered translational energy loss, but which have not enough internal energy to fragment.

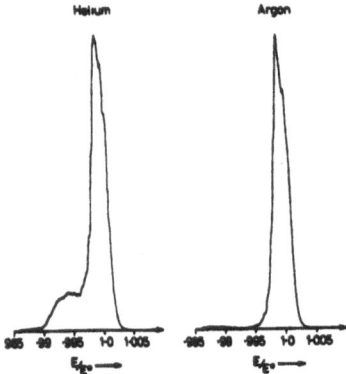

*Figure 6*[4]: **Parent peak from the CID spectrum of the $[M+H]^+$ ion of bombesin using helium and argon collision gases.**

This shoulder can be used to explain partly another characteristic of CID spectra. The attenuation of the beam at optimum pressure is 60-70% (see earlier) and yet the total yield of fragment ions is only 10%. See *figure 7*[4]. The remaining 50-60% of the ion current could either be lost due to excited parents not decomposing before they reach the analyser (as supported by the presence of the shoulder) or the incident ions undergoing charge transfer with the target gas.[2] Since the CID spectrum of $[M+2H]^{2+}$ ions contains singly-charged fragment peaks, this charge transfer idea is valid. Losses due to scattering do not need to be considered. For an ion of $m/z$ 1620 and a translational energy loss of 50 eV, all scattered ions should pass through the electric sector.

The relationship between internal energy uptake, $Q$, and incident ion mass is shown in *figure 8*[4]. Also shown on this figure are translational energy loss, $\Delta E$, and centre-of-mass collision energy, $E_{cm}$. Knowing $Q$ and $\Delta E$, it is possible to calculate the velocity of the helium atom using equation 5. This is shown in *figure 9*[4].

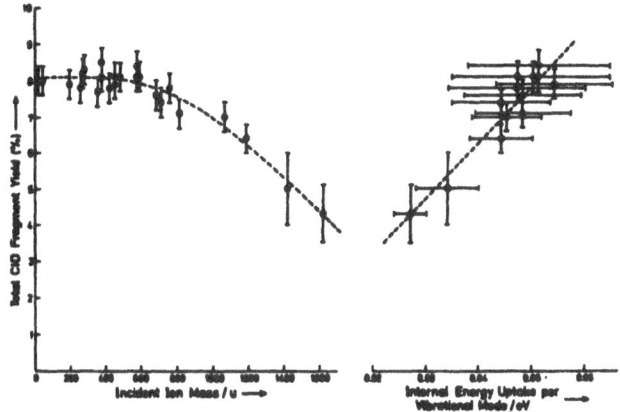

*Figure 7*[4]: Dependence of total CID fragment yield on incident ion mass.

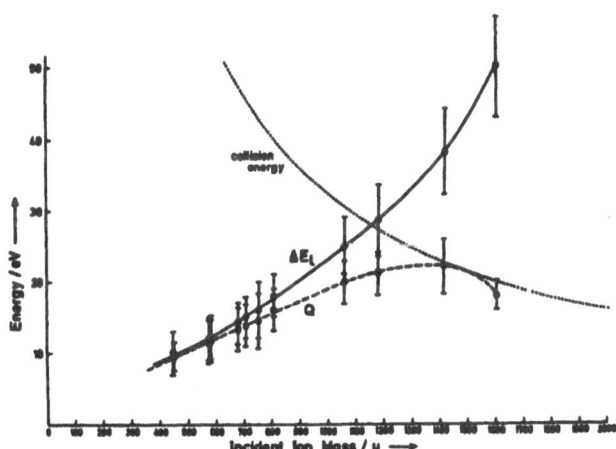

*Figure 8*[4]: Dependence of the translational energy loss suffered by the incident ion (solid line), the internal energy uptake ( broken line) and centre-of-mass collision energy (dotted line) on incident ion mass.( Collision gas = helium, incident ion energy = 8 keV).

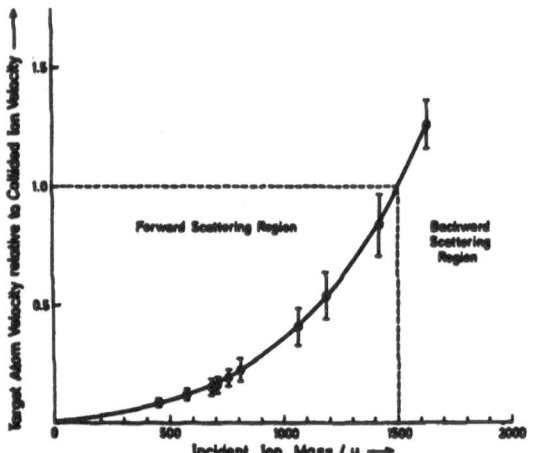

*Figure 9*[4]: Target atom velocity relative to collided ion velocity against incident ion mass.

For singly-charged incident ions of mass less than 1500 Da in collision with helium and with an initial energy of 8 keV, *figure 9*[4] shows that following collision these ions travel faster than the gas atom with which they collided. This is known as forward-scattering. It corresponds to the gas atom passing completely through the incident ion following a 'head-on' collision. To illustrate this scale drawings of the peptides bradykinin and luteinising hormone releasing hormone (LHRH) are shown in *figure 10*[4]. Clearly their physical cross-sections are much greater than the cross-sectional area of helium.

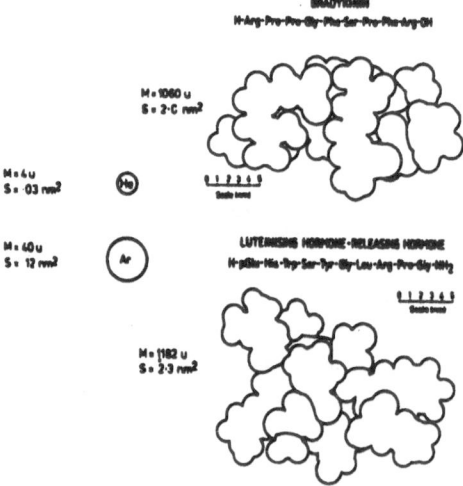

*Figure 10*[4]: Physical cross sections of the peptides bradykinin and luteinising hormone releasing hormone (LHRH) compared to helium and argon.

The reverse situation is known as back-scattering. In this case, the helium precedes the ion along the flight path of the instrument. The point at which the transition from forward- to back-scattering occurs is particularly interesting. At this point the ion and the gas atom have equal velocities and proceed together. This is known as target capture and will be discussed in greater detail later. *Figure 11*[22] depicts the ideas of forward- and back-scattering and target capture.

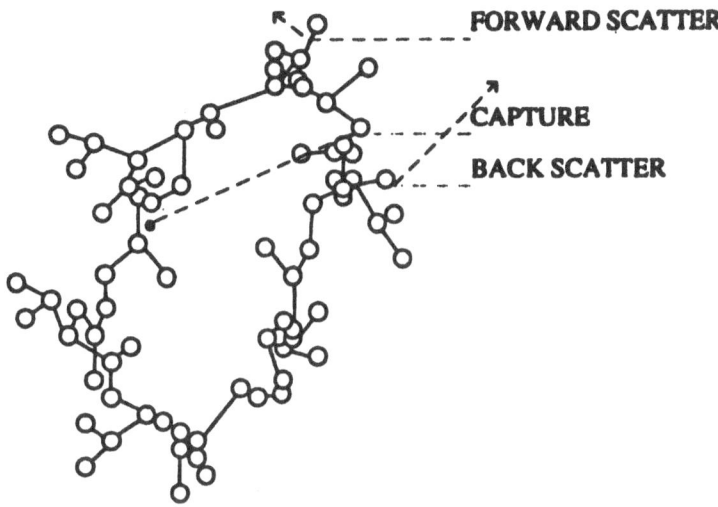

FORWARD SCATTER

CAPTURE

BACK SCATTER

**VALINOMYCIN**

*Figure 11*[22] : **Illustration of the idea of forward-scattering, back-scattering and target capture, using valinomycin.**

When target capture occurs all the centre-of-mass collision energy is converted to internal energy. At this point, then, the internal energy plot should touch the $E_{cm}$ plot. This can be seen in *figure 8*[4]. For the conditions here, i.e. collision gas helium, incident ion energy = 8 keV, target capture is predicted to occur for organic ions at incident ion masses of 1500 Da.

The centre-of-mass collision energy can be raised in two ways - either the target gas mass can be increased or the incident ion energy can be increased. This results in forward-scattering occurring at higher masses, and so a maximum in $Q$ is reached at higher masses. The implication of this work then is that for a particular incident ion, the target gas mass and the incident ion energy should be 'tuned' to get the maximum internal energy uptake possible and hence the most comprehensive spectrum.

### 5. Target gas excitation in collision-induced decomposition?

#### 5.1 ELECTRONIC EXCITATION OF THE TARGET GAS

The previous studies of the dynamics of CID of peptide ions[3,4] were extended by Bricker and Russell[30] to include those of the [M+H]$^+$ ion of chlorophyll-$\alpha$. Neumann et al.[3,4] reported large energy losses associated with CID and concluded that collisional activation occurs not by electronic excitation but rather by translational to vibrational (or rotational) excitation. The internal energy uptake, Q, was then calculated from the measured translational energy losses under the assumption that electronic excitation of the target atom was negligible.

Bricker and Russell[23] argued that the large energy losses were due either solely or partially to appreciable excitation of the target gas. In this case, the internal energy uptake by the ion would represent only a small fraction of the total translational energy loss.

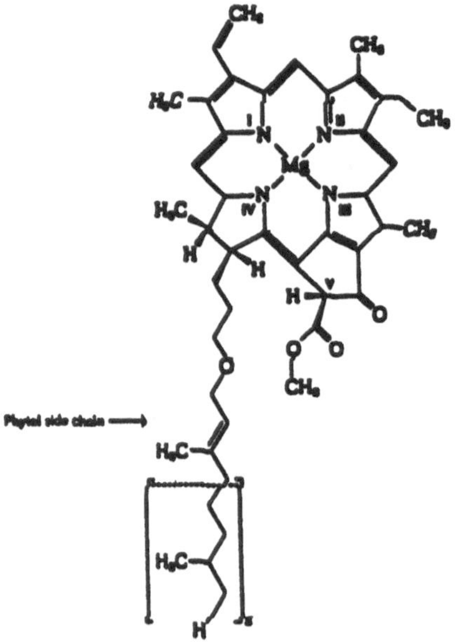

*Figure 12*[23]: Structure of chlorophyll-$\alpha$ showing the loss of the phytyl side-chain which gives rise to the ion at m/z 614.

The study of the chlorophyll-$\alpha$ [M+H]$^+$ ion was conducted on a triple analyser Kratos MS-50TA mass spectrometer. Fast-atom bombardment (FAB) ionisation was used. In the chlorophyll-$\alpha$ [M+H]$^+$ (m/z 893) spectrum a metastable ion is observed at m/z 614. This arises from the loss of the phytyl side chain. See *figure 12*[23]. This peak was employed for the measurement of energy lost by the parent according to the methods described by Neumann et al.[3,4] and equation (1). The effect on the CID spectrum of various target gases was investigated using He, Ne, Ar, Kr and N$_2$.

*Figure 13*[23] shows the translational energy losses measured versus the centre-of-mass collision energy. Also shown is the ionisation energy of the target gas. It appears that the measured energy loss increases as the ionisation energy increases. This idea of electronic excitation of the target gas, culminating in charge transfer, is contradictory in terms of an adiabatic model. However, Bricker and Russell argue that "if the collision complex is short-lived, charge distribution may not follow an adiabatic model"[23], i.e. if the interaction time is short then charge density (and energy) redistribution may not occur evenly. To illustrate this idea, *figure 14*[23] was plotted. This shows translational energy loss versus interaction time.

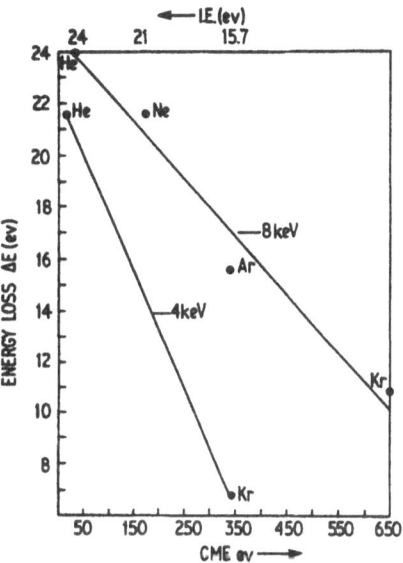

*Figure 13*[23]: **Translational energy losses by the chlorophyll-α [M+H]+ ion plotted against centre-of-mass collision energy. Also shown is the ionisation energy of the target gas.**

## 5.2 TARGET GAS MASS

The work on energy loss in collisional activation of chlorophyll-α was reinvestigated at a later date by Alexander, Thibault and Boyd[24]. Use of improved experimental techniques showed that *target mass* rather than target ionisation energy is the determining factor in energy shifts.

The short-comings of the original study are as follows. The fragment ion used (m/z 614) is the result of the loss of the phytyl side chain from chlorophyll-α. This ion forms readily from metastable dissociations and so each peak in the previously recorded MIKES spectra contained contributions from unimolecular dissociations as well as collision-induced decompositions. By floating the collision cell electrically, this ambiguity can be removed as the metastable and collisionally-activated components become easily distinguishable.

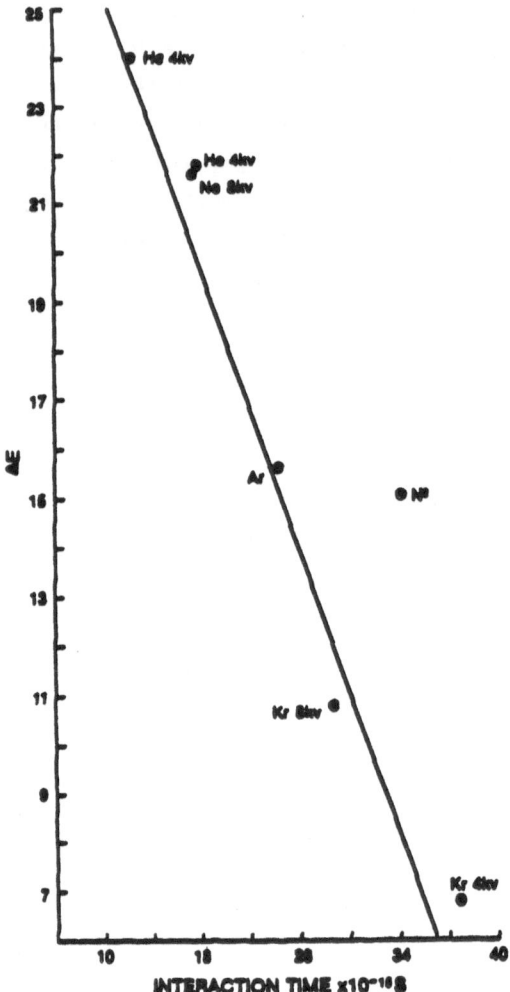

*Figure 14*[23]: Translational energy losses by the chlorophyll-α [M+H]$^+$ ion plotted against the interaction time of the ion and the neutral.

Secondly, the original spectra were the result of analogue recordings of single scans. Alexander *et al.* [24] used cumulative scans recorded digitally and thus improved the signal-to-noise ratio.

5.2.1 *Experiments using a floated collision cell.* Alexander *et al.* [24] used a VG Analytical ZAB-EQ instrument fitted with a short collision cell (2 cm) floated at -700 V. The target gases employed were Xe, Kr, $N_2$, Ar, Ne, He and $D_2$. The translational energy losses of the parent ion (m/z 893) as calculated from equation (1) are shown versus the target gas ionisation energy in

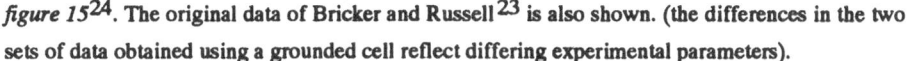

*figure 15*[24]. The original data of Bricker and Russell [23] is also shown. (the differences in the two sets of data obtained using a grounded cell reflect differing experimental parameters).

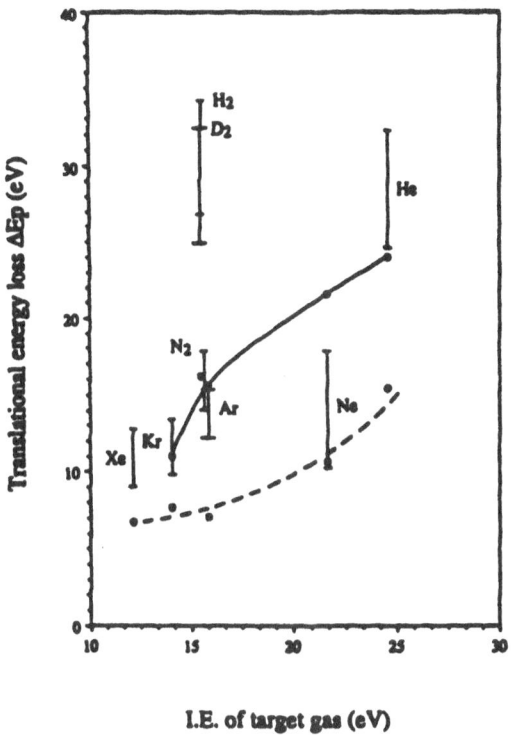

I.E. of target gas (eV)

*Figure 15*[24]: **Translational energy loss by the $M^{+\cdot}$ chlorophyll-$\alpha$ ion plotted against the ionisation energy of the target gas.**

What is particularly interesting here is the comparison of the He, $D_2$ and Ar results using the floated cell. The ionisation energies of He, $D_2$ and Ar are 24.6 eV, 15.4 eV and 15.9 eV respectively and yet $D_2$ behaves more like helium than argon. The conclusion drawn was that the target gas mass is the important factor in translational energy losses.

*5.2.2 Angular dependence.* The floated collision cell experiments were followed up with an investigation into the variation of translational energy loss with observation angle. This was done by collimating the ion-beam in the $z$-direction and using a movable $z$-aperture. The maximum observation angle for ions forming in the collision cell was 0.28° which corresponds to the $z$-aperture being moved by 5 mm.

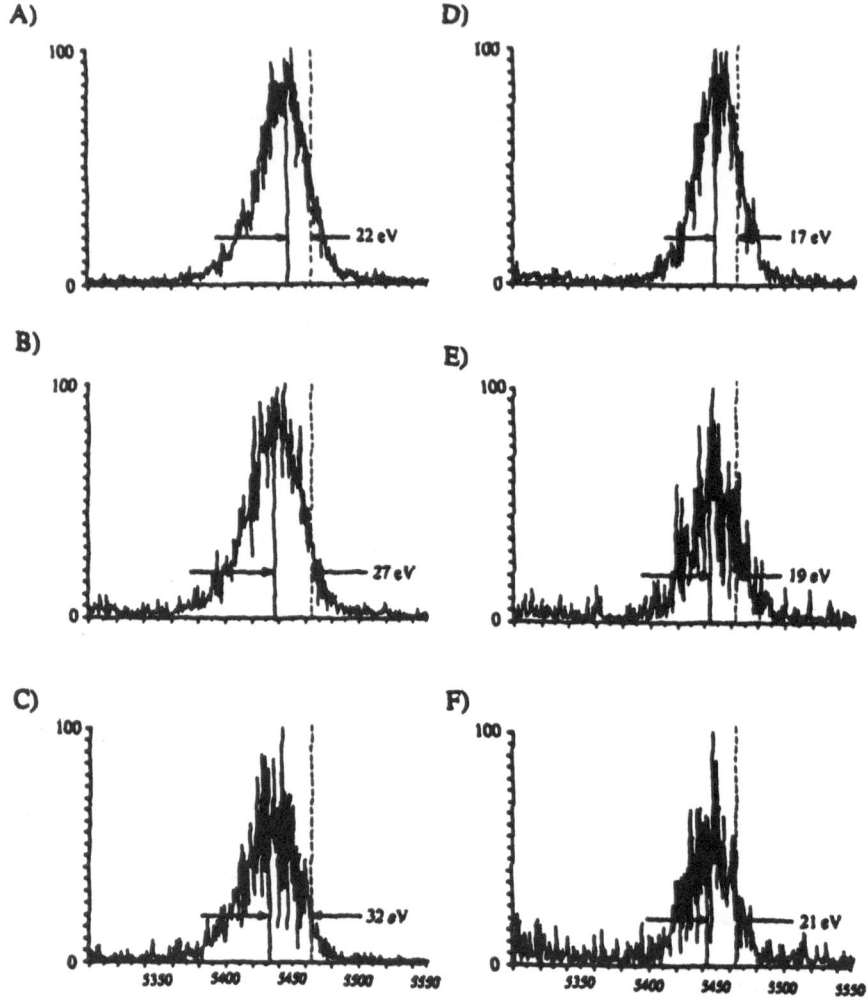

*Figure 16*[24]: MIKES spectra of the M[+·] chlorophyll-α ion for helium (A, B, C) and krypton (D, E, F). Z-displacements are 0 mm (A, E), 2.5 mm (B, D) and 4.0 mm (C, F).

*Figure 16*[24] shows the angle -resolved MIKES spectra of M[+·] chlorophyll-α ions with helium and krypton around the m/z 614 region. These results are shown graphically in *figure 17*[24]. Translational energy losses when helium is employed show an increase as the angle is increased. The translational energy losses when using krypton, however, appear less dependent on angle.

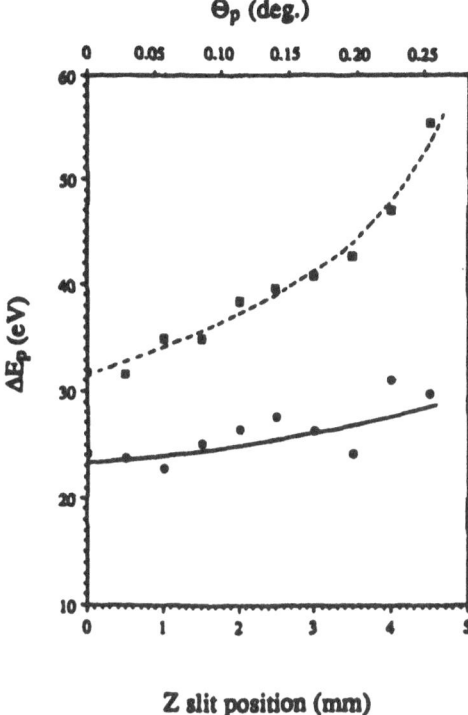

**Θ_p (deg.)**

**Z slit position (mm)**

*Figure 17[24]*: Translational energy losses for the $M^{+\cdot}$ ion of chlorophyll-α against observation angle. Circles and squares represent data obtained using helium and krypton, respectively, as collision gases.

In order to interpret these results, the theoretical results predicted by the 'quasi-diatomic model' (QDA)[9] were studied. As has been mentioned earlier, QDA considers the ion and the target gas to be separate single entities, i.e. structureless, which collide and exchange energy and momentum. Equation (4) shows the relationship between the internal energy uptake, the initial energy of the parent, the translational energy loss suffered by the parent and the scattering angle.

Two cases were considered theoretically: the chlorophyll-α $M^{+\cdot}$ ion with krypton and the $M^{+\cdot}$ ion with helium. The internal energy uptake, Q, was fixed arbitrarily at 10 eV. Translational energy losses then calculated for the krypton system fell in the range 10.1 eV (for scattering angle close to zero) and 11 eV (for scattering angles close to the instrumental maximum). In the helium case, the range of translational losses was 10.5 eV to 50 eV. These predictions are semiquantitatively in agreement with the experimental results depicted in *figure 17[24]*.

Another feature of the angle-resolved MIKES spectra noticed by Alexander *et al.*[24] was that the peaks were far less asymmetric than those in the ordinary CID-MIKES spectra. The broadness of MIKES peaks due to the release of internal energy was mentioned earlier. This release will result in symmetric broadening. Every dissociation will result in a single symmetric peak however the superposition of peaks corresponding to a range of translational energy losses explains the tailing of the overall peak towards the low energy end.

## 6. The relationship between translational energy loss and fragment ion mass.

Following the work of Neumann et al.[4] , Sheil et al.[5] varied the conditions for the peptide valine-gramicidin A (RMM 1881 Da) in order to obtain a 'good' CID spectrum. Its worth was judged on the amount of structural information it contained. During this study, measurements of translational energy losses were made and the trends shown by these proved to be very interesting.

It was stated earlier that in order to increase the internal energy uptake by an ion, it is necessary to increase the centre-of-mass collision energy (equation 6). This can be done by increasing either the target gas mass or the incident ion energy.

On changing the target gas from helium to argon, the centre-of-mass collision energy is increased by an order of magnitude. Experimentally, however, argon is not markedly superior to helium. This can be explained in terms of scattering losses[25,26] and also mass-matching (see section on ICT). Increasing the incident ion energy is then left as a method for increasing $Q$ , the internal energy uptake. Sheil et al. used an incident ion energy of 14.8 keV.

Other experimental conditions were as follows: The same reverse geometry double-focusing mass spectrometer used in Neumann's experiments was employed.[16] The ions were formed by field desorption. The collision gas was helium and the beam attenuation was 70% i.e. the pressure was slightly higher than the optimum pressure.

The CID spectrum of valine-gramicidin A [M+H]$^+$ is shown in *figure 18*[5]. Assignments of the peak using Roepstorff nomenclature[27] are given in *table 3*.

The $m/z$ 1882 peak represents two ions - the [M+H]$^+$ ion and the M$^{+\cdot}$ ion which has one $^{13}$C atom.[28] For valinomycin the CID spectra of M$^{+\cdot}$ and [M+H]$^+$ are very similar and this similarity is postulated for valine-gramicidin A. The uncertainties in the translational energy loss values reflect the possibility that a fragment may have been derived from M$^{+\cdot}$ containing $^{13}$C rather than the [M+H]$^+$ ion.

The translational energy losses , $\Delta E$, are plotted against fragment masses in *figure 19*[5]. On examining *figure 19*[5], there are two features to discuss. Firstly, the size of the energy losses and, secondly, their dependence on mass.

| Mass | $V_2$(V) | $V_1$(V) | $\Delta E$(eV) |
|---|---|---|---|
| 1882.1 | 494.90 | 495.00 | - |
| 1867.1 | 490.31 | 491.05 | $22 \pm 8$ |
| 1839.0 | 482.31 | 483.68 | $42 \pm 8$ |
| 1825.0 | 478.64 | 479.99 | $42 \pm 8$ |
| 1793.0 | 469.79 | 471.57 | $56 \pm 8$ |
| 1781.0 | 465.40 | 468.42 | $95 \pm 8$ |
| 1766.0 | 463.07 | 464.47 | $45 \pm 8$ |
| 1752.0 | 459.14 | 460.79 | $53 \pm 8$ |
| 1651.0 | 432.93 | 434.21 | $44 \pm 9$ |
| 1634.9 | 428.14 | 430.00 | $64 \pm 9$ |
| 1607.0 | 420.49 | 422.64 | $75 \pm 9$ |
| 1538.9 | 402.93 | 404.74 | $66 \pm 10$ |
| 1521.9 | 398.07 | 400.26 | $81 \pm 10$ |
| 1493.9 | 390.44 | 392.90 | $93 \pm 10$ |
| 1468.8 | 383.95 | 386.30 | $100 \pm 10$ |
| 1351.8 | 355.26 | 355.53 | $11 \pm 11$ |
| 1335.8 | 348.55 | 351.32 | $117 \pm 11$ |
| 1307.8 | 341.71 | 343.96 | $97 \pm 11$ |
| 1270.7 | 329.74 | 334.19 | $197 \pm 12$ |
| 1238.7 | 322.82 | 325.79 | $135 \pm 12$ |
| 1222.7 | 318.72 | 321.58 | $132 \pm 12$ |
| 1194.7 | 311.57 | 314.22 | $125 \pm 12$ |
| 1171.6 | 306.54 | 308.14 | $78 \pm 12$ |
| 1052.6 | 273.97 | 276.85 | $154 \pm 14$ |
| 1036.6 | 269.09 | 272.64 | $193 \pm 14$ |
| 1008.6 | 264.36 | 265.27 | $51 \pm 15$ |
| 985.5 | 254.62 | 259.20 | $262 \pm 15$ |
| 923.5 | 238.36 | 242.90 | $277 \pm 15$ |
| 895.6 | 231.75 | 235.54 | $238 \pm 16$ |
| 872.4 | 224.40 | 229.46 | $327 \pm 17$ |
| 753.4 | 194.70 | 198.17 | $259 \pm 19$ |
| 737.5 | 190.17 | 193.96 | $289 \pm 20$ |
| 709.5 | 184.96 | 186.60 | $130 \pm 21$ |
| 686.4 | 176.72 | 180.52 | $312 \pm 21$ |
| 654.4 | 169.09 | 172.04 | $254 \pm 22$ |

| | | | |
|---|---|---|---|
| 638.4 | 163.76 | 167.90 | 365 ± 23 |
| 610.4 | 158.50 | 160.54 | 188 ± 24 |
| 573.3 | 145.56 | 150.78 | 512 ± 26 |
| 555.4 | 142.14 | 146.06 | 397 ± 26 |
| 539.3 | 137.54 | 141.84 | 449 ± 27 |
| 511.3 | 129.36 | 134.48 | 564 ± 29 |
| 456.3 | 115.71 | 120.00 | 529 ± 32 |
| 440.3 | 110.59 | 115.79 | 665 ± 33 |
| 385.2 | 96.78 | 101.32 | 664 ± 38 |
| 366.2 | 92.62 | 97.10 | 683 ± 40 |
| 341.2 | 85.58 | 89.72 | 683 ± 43 |
| 272.2 | 66.22 | 71.58 | 1109 ± 54 |
| 256.1 | 62.97 | 67.36 | 965 ± 57 |
| 228.1 | 56.13 | 60.00 | 955 ± 64 |
| 185.1 | 44.83 | 48.68 | 1171±79 |

*Table 3:* **Assignments of fragments in CID spectrum of valine-gramicidin A. $V_2$ is the voltage of the centre of the peak at half-height. $V_1$ is the calculated position assuming $\Delta E=0$.**

For the low fragment ions, the translational energy losses appear to exceed 1000 eV. Values of 100's of eV are not necessarily in disagreement with energy uptake via direct vibrational excitation due to momentum transfer[17,23,29,30] but the fact remains that values of over 1000 eV are enormous. As in Neumann's work,[3,4] the collision is considered to contain several different interactions. The number of these interactions increases as the 'interaction region' increases. This 'interaction region' is the length of the ion and is dependent on mass. This hypothesis arose because even for elastic collisions when $Q = 0$ and $\Delta E$ is a maximum, it would not be possible to rationalise $\Delta E$s of this order of magnitude.

If this idea is applied to the low mass fragment ions, then to arrive at a translational energy loss of 1000 eV, nine interactions are required. This seems unlikely.

In fact, both the magnitude of the translational energy loss and the observation that this energy loss falls with increasing fragment ion mass can be justified with a single theory.

It was mentioned earlier that in these experiments, the transmission was lower than in previous experiments i.e. the collision gas pressure is higher. As a result the probability of multiple collisions is higher. A variation on the idea of multiple collisions is that of sequential collisions. This idea is presented below.

If an ion $m_i^+$ with energy $E_0$ collides and decomposes to $m_f'^+$ having lost energy $\Delta E_i$, and then $m_f'^+$ collides and decomposes to $m_f''^+$ having lost $\Delta E_f'$, the energy $E_2'$ of $m_f'^+$ is:

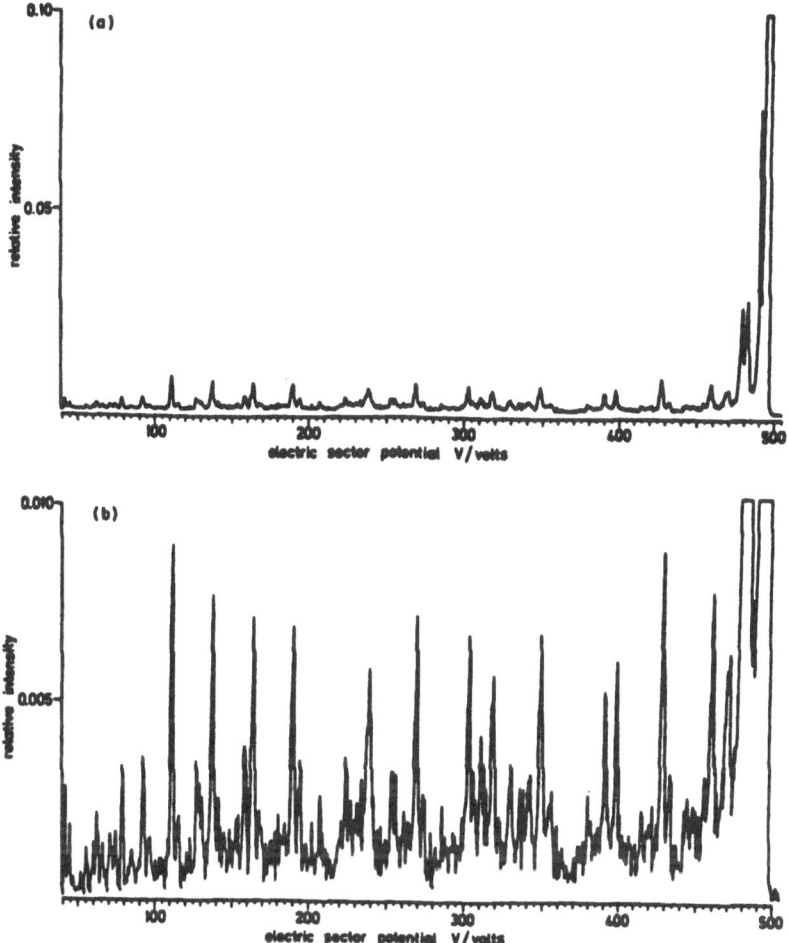

*Figure 18*[5]: MIKES spectrum of m/z 1882 formed by the field desorption of gramicidin. Collision gas = helium. Incident ion energy = 8 keV.

$$E_2' = \left(\frac{m_f'}{m_i}\right)\left(E_0 - \Delta E_i\right) \tag{7}$$

and the energy $E_2''$ of $m_f''^+$ is:

$$E_2'' = \left(\frac{m_f''}{m_f'}\right)\left(E_2' - \Delta E_f'\right) \tag{8}$$

226

Equations (7) and (8) can be combined to give:

$$E_2'' = \left(\frac{m_f''}{m_i}\right)E_0 - \left(\frac{m_f''}{m_i}\right)\Delta E_i - \left(\frac{m_f''}{m_f'}\right)\Delta E_f' \tag{9}$$

If the ion $m_f''^+$ had been formed directly from $m_i^+$, then its energy $E_1''$ would be:

$$E_1'' = \left(\frac{m_f''}{m_i}\right)E_0 \tag{10}$$

If the electric sector potential corresponding to energy $E_1''$ (direct decomposition) is $V_1''$ and the electric sector potential corresponding to $E_2''$ is $V_2''$, then the peak shift $(V_1'' - V_2'')$ is given by:

$$\left(V_1'' - V_2''\right) = \left(E_1'' - E_2''\right)\left(V_0/E_0\right) \tag{11}$$

where $V_0$ is the electric sector potential required for the incident ion; $E_0$ is the incident ion energy. Combining the equations (11), (10) and (9), the peak shift is given:

$$\left(V_1'' - V_2''\right) = \left(\frac{V_0}{E_0}\right)\left[\left(\frac{m_f''}{m_i}\right)\Delta E_i + \left(\frac{m_f''}{m_f'}\right)\Delta E_f'\right] \tag{12}$$

The translational energy loss, $\Delta E$, based on equation (3) earlier, is then:

$$\Delta E = \Delta E_i + \left(\frac{m_i}{m_f'}\right)\Delta E_f' \tag{13}$$

$\Delta E$ is therefore an *apparent* and amplified energy loss rather than being the energy loss following a single collision.

This sequential decomposition theory can be used to justify the large energy losses. If we consider an ion of mass 2000 Da decomposing (eventually) to ion $m_f''^+$, via ion $m_f'^+$ (200 Da), with energy losses $\Delta E_i = \Delta E_f' = 100$ eV, the apparent energy loss $\Delta E$ would equal 1100 eV.

The fact that the translational energy losses appear to be dependent on fragment mass can be explained as follows: the final peak shift will be dependent on the translational energy lost by the parent ion, which of course depends on the intermediate ion mass. The final peak shift is also dependent on the translational energy lost by the intermediate ion which in turn is dependent on the final ion mass.

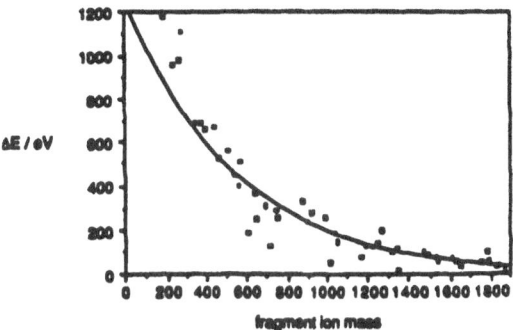

*Figure 19*[5]: Translational energy losses plotted against fragment ion mass.

## 7. Direct momentum transfer mechanisms

Having concluded that the mechanism of energy transfer in collisional activation of large ions was probably not one of electronic excitation[2] but one of direct momentum transfer, the model applied was the 'quasi-diatomic approximation'[9] (QDA). At this juncture, however, an alternative direct momentum transfer theory was put forward - impulsive collision transfer (ICT) theory.[10,11] From this point on in the study of collisional activation, it is this ICT theory which has been used when attempting to explain results.

### 7.1 QUASI-DIATOMIC APPROXIMATION

In this model, the ion is considered as a single entity colliding with a stationary inert gas atom. Conservation of energy and momentum give rise to the following relationship between internal energy uptake, $Q$, and the translational energy lost by the ion, $\Delta E$.

$$Q = \left[ \frac{(m_i + m_g)\Delta E}{m_g} \right] - \left[ \left( \frac{2m_i E}{m_g} \right) \left( 1 - \left[ \frac{E'}{E} \right]^{\frac{1}{2}} \cos \theta \right) \right]$$

where $m_i$ = mass of ion; $m_g$ = mass of gas; $E$ = initial energy of ion; $E'$ = translational energy of the ion after the collision; $\theta$ = scattering angle.

This will be recognised as equation (4) which Neumann[17] used in his work on incident ion mass and translational energy loss.

This quasi-diatomic approximation has recently been termed the 'limiting elastic model'.[30] This is somewhat misleading as the scattering is of course inelastic i.e. the internal energy uptake is not zero.

228

QDA is successful in that it predicts translational energy losses of the right order of magnitude (10s - 100s of eV). It cannot, however, distinguish between two incident ions of the same mass where one ion consists of a large number of small atoms an the other consists of a small number of large atoms nor does it describe accurately the target gas mass dependency of internal energy uptake.

## 7.2 IMPULSIVE COLLISION TRANSFER THEORY

ICT theory captures more experimental feature than the quasi-diatomic model and is a definite improvement, but it still does not describe multiatomic ion collisions completely.

### 7.2.1 *The Derivation Of Impulsive Collision Transfer Theory.* The key assumption of ICT theory is that, in a collision, energy and momentum transfer take place between the gas atom and a single atom within the ion. This atom has mass $m_a$, the mass of the ion is $m_i$ and the mass of the gas is $m_g$.

Initially the gas atom is stationary and the ion moves along the flight path ($x$-axis) with velocity $v$. On collision, the ion is deflected. Deflection is considered in the $y$- direction only (not the $z$-direction). This assumption is an attempt to simplify but not trivialise the derivation.

The situation before the collision is shown in *figure 20*.

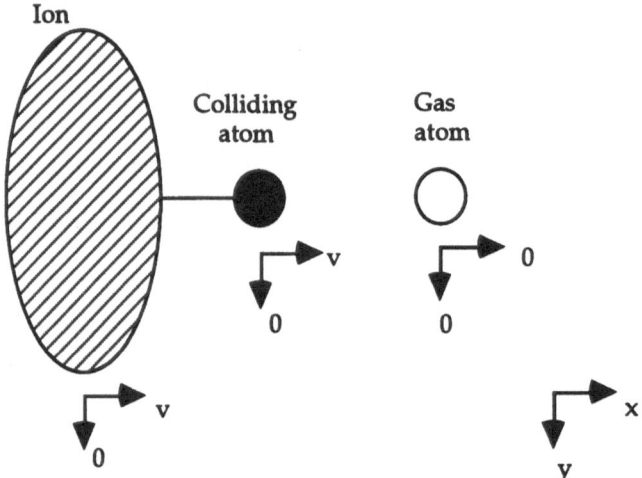

*Figure 20*: The situation before collision according to ICT theory.

Conservation of momentum and energy are used to obtain expressions for the velocities of the atoms involved in the collision, *immediately* after the collision. Since the collision is impulsive i.e. it takes place over an infinitesimal time period, the velocity of the rest of the ion is the same immediately after the collision as it was prior to the collision. *Figure 21* shows the situation after the collision. The translational energy, $E'(0^+)$, of the ion *immediately* after the collision is then:

$$E'(0^+) = \frac{1}{2}(m_i - m_a)v^2 + \frac{1}{2}m_a v_{a,x}'^2 + \frac{1}{2}m_a v_{a,y}'^2 \tag{14}$$

where $v_{a,x}'$ = velocity of atom involved in collision in x-direction; $v_{a,y}'$ = velocity of atom involved in collision in y-direction; $(0^+)$ is used to indicate the situation immediately after the collision.

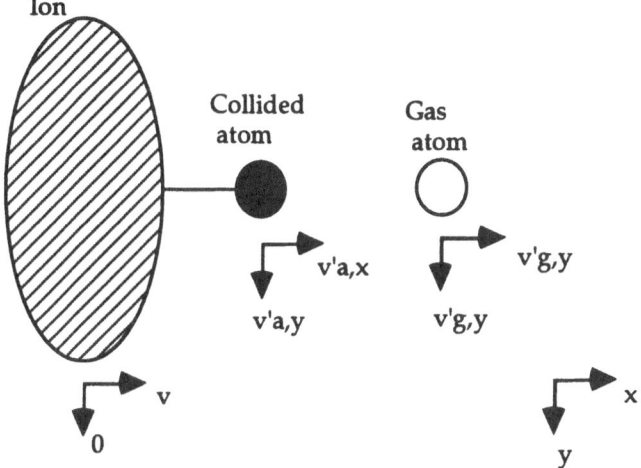

*Figure 21*: The situation *immediately* after collision according to ICT theory.

Clearly this situation cannot hold and energy redistribution occurs until all the atoms within the ion are travelling with the same velocity. This velocity is found by averaging the component velocities of the ion over the entire mass of the ion. In the *x*-direction, then, there is a mass $(m_i - m_a)$ travelling with velocity $v$ and a mass $m_a$ travelling with velocity $v_{a,x}'$.

$$v_x' = \frac{(m_i - m_a)v + m_a v_{a,x}'}{m_i} \tag{15}$$

The situation in the y-direction is simpler because prior to the collision, the ion was not moving in the y-direction. So, the velocity in the y-direction is given by:

$$v_y' = \frac{m_a}{m_i}v_{a,y}' \tag{16}$$

Using these velocities, it is possible to get an expression for the translational energy of the ion, $E'$, after energy redistribution has taken place.

$$E' = \tfrac{1}{2}m_i v_x'^2 + \tfrac{1}{2}m_i v_y'^2 \tag{17}$$

The internal energy uptake, $Q$, is the difference between the translational energy of the ion *immediately* after the collision and the translational energy of the ion after energy redistribution.

$$Q = E'(0^+) - E' \tag{18}$$

The translational energy lost by the ion, $\Delta E$, is the difference between the translational energy before the collision, $E(= \tfrac{1}{2}m_i v^2)$ and the translational energy after energy randomisation has taken place ($E'$).

$$\Delta E = E - E' \tag{19}$$

7.2.2 *Important Expressions Arising From ICT Theory.* The following expressions are the most important to arise from this derivation:

The internal energy uptake, $Q$, and the translational energy loss, $\Delta E$, can be related thus:

$$Q = \left(\frac{\mu}{2\varepsilon}\right)\Delta E \tag{20}$$

where $\mu = \dfrac{m_i(m_a + m_g)}{m_i(m_a + m_g) - m_a m_g}$ and $\varepsilon = \dfrac{(m_a + m_g)m_i}{2m_g(m_i - m_a)}$.

It is often the case when using this equation that an average value of $m_a$ is used i.e. a mean mass is found using the individual atoms within the ion. This will lead to errors. When using this equation, an average value for $(\mu/2\varepsilon)$ should be used. This is because $(Q/\Delta E)$, although being a function of $m_a$ is not *directly proportional* to $m_a$ whereas it is directly proportional to $(\mu/2\varepsilon)$.

The expression given for $Q$ from this derivation is:

$$Q = \frac{2m_g(m_i - m_a)}{(m_a + m_g)m_i}\left(E - (EE')^{\frac{1}{2}}\cos\theta\right) \tag{21}$$

This expression is very similar to the one derived from QDA which Neumann *et al.* used[3,4] in calculating $Q$. It is important to point out that had he employed equation (21), the variation of $Q$ with incident ion mass would be the same.

From equation (21), it is obvious that the maximum amount of energy available for transfer to internal energy, $Q_{max}$, is:

$$Q_{max} = \frac{4m_a m_g}{\left(m_a + m_g\right)^2} \cdot \frac{m_g}{m_i} \cdot \frac{\left(m_i - m_a\right)}{m_i} E \tag{22}$$

Since multiatomic ion collisions are being considered, expression (22) can be simplified to:

$$Q_{max} = \chi E_{cm} \tag{23}$$

where $\chi = \dfrac{4m_a m_g}{\left(m_a + m_g\right)^2}$ and $E_{cm}$ = centre-of-mass collision energy.

$\chi$ is known as the efficiency factor and will be discussed in more detail later.

The average internal energy uptake, $\langle Q \rangle$, is found by integrating and normalising. Its value is half that of $Q_{max}$.

$$\langle Q \rangle = \tfrac{1}{2} Q_{max} \tag{24}$$

The final important expression is for the highest scattering angle, $\theta_{high}$:

$$\theta_{high} = \arcsin(\mu - 1) \tag{25}$$

7.2.3. *Implications Of ICT Theory.* The first implication of ICT theory is that both the average translational energy loss, $\langle \Delta E \rangle$, and the average internal energy uptake, $\langle Q \rangle$, increase linearly with the incident ion energy, $E$. This is consistent with experiment.[31,32] From this, it follows that $\Delta E$ and $Q$ are related. High-energy losses result in high energy excitation and low-energy losses result in low energy excitation. The experiments of Alexander *et al.*[24] are in agreement with this.

The efficiency of conversion of centre-of-mass collision energy, $E_{cm}$, to internal energy of the ion increases when the gas atom and the atom involved in the collision are comparable in mass. The efficiency factor, $\chi$, has a value of one when the two atoms are equal in mass. This idea of efficiency was used by Bradley *et al.*[7,8] and will be discussed shortly.

Expression (25) suggests that, for a constant ion energy, scattering increases as the target gas mass is increased. This prediction is intuitive and has been borne out by molecular dynamics calculations.[10]

The final implication is that translational energy loss, $\Delta E$, (and therefore internal energy uptake, $Q$) are inversely proportional to incident ion mass. This is inconsistent with experiment.[3,4] These experimental results show that, at constant ion energy, increasing the incident ion mass, and therefore *decreasing* its velocity, results in a *greater* translational energy

232

loss. This observation is counter-intuitive and, as yet, no theory submitted has been able to explain it.

## 8. The Effect Of Collision Gas Pressure On Translational Energy Losses.

In a direct follow-up to the work of Sheil *et al.*[5], Bradley *et al.*[7,8] undertook an investigation into the effect of collision gas pressure on translational energy losses. The first study employed the peptide valinomycin and helium as the collision gas. This was then extended to include the larger peptide valine-gramicidin A, the caesium iodide cluster ion $(Cs_5I_4)^+$ and xenon as a collision gas.

As in the previous studies on collisional activation, the large-scale reverse geometry double-focusing mass spectrometer (MMM)[16] was employed with the collision cell sited at the focal point between the magnetic and electric sectors. The electric sector potential was swept to produce a MIKE spectrum. The ion beams were generated by field desorption.

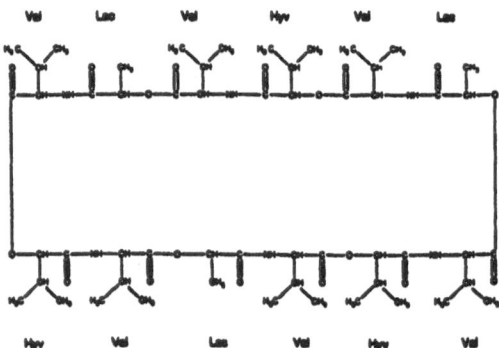

*Figure 22*[7]: The structure of valinomycin.

The structure of valinomycin[32] is shown in *figure 22*[7] and the CID spectrum of the valinomycin [M+Na]$^+$ ion (m/z 1133.6) with collision gas helium, incident ion energy 25.075 keV and a transmission of 40%, is shown in *figure 23*[7].

The fragment ions studied are as follows:

(a)    m/z 1090.6    corresponding to loss of $C_3H_7$ side chain from M+Na]$^+$

(b)    m/z 962.6    [(Hyv-Val-Lac-Val)$_2$-Hyv-Val+Na]$^+$

(c)    m/z 948.6    [(Hyv-Val-Lac-Val)$_2$-Hyv-Val+H+Na]$^+$ minus ·$CH_3$ from a side chain

(d)    m/z 936.5    [(Hyv-Val-Lac-Val)$_2$+2H+Na]$^+$

(e)    m/z 779.5    [(Hyv-Val-Lac-Val)$_2$-Lac+2H+Na]$^+$ minus ·$CH_3$ and ·$C_3H_7$ from a side chain

(f)    m/z 393.2    [Hyv-Val-Lac-Val+Na]$^+$

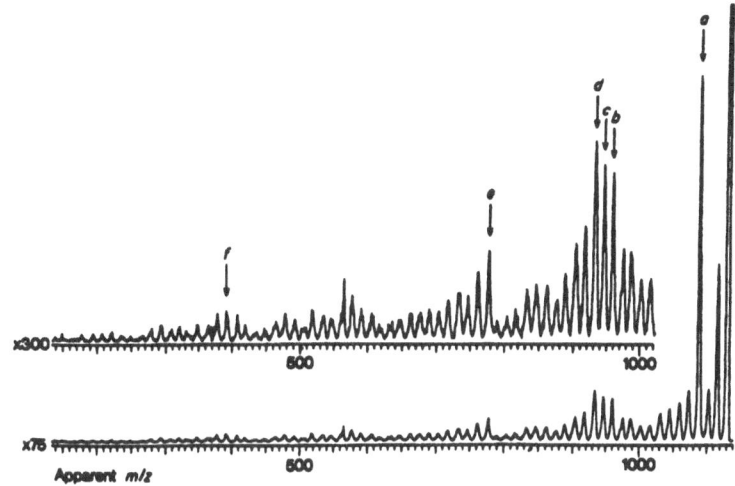

*Figure 23*[7]: **CID mass spectrum of valinomycin [M+Na]$^+$ ion (m/z 1133.6). Collision gas = helium. Incident ion energy = 25.075 keV. Transmission = 40%.**

*Figure 24*[7] shows plots of the translational energy losses, $\Delta E$, against incident ion transmission, for the series of fragments (a) to (f). The experimental results are shown with an estimation of error. The curves represent theoretically predicted relationships between $\Delta E$ and transmission obtained from Poisson statistics.

Fragments (a) to (d) show that $\Delta E$ does not undergo a significant increase on increasing the collision gas pressure. For fragment (e), $\Delta E$ does increase with increasing gas pressure and for fragment (f), this dependence is even stronger.

In principle, it is possible that this increase with $\Delta E$ at low transmissions is due to a trade-off between CID and metastable contributions. When transmission is high, metastable decomposition peaks would dominate and when transmission is low CID peaks would dominate. For the fragments used here, however, MIKE scans in the absence of collision gas showed the number of metastable ions to be negligible.

The theoretical curves plotted in (a) to (f) in *figure 24*[7] were calculated from Poisson statistics. Poisson statistics were mentioned earlier in the section on optimum pressure. As in these earlier calculations, the average number of collisions, l, for each transmission can be found and then substituted into equation (3) to find the probabilities $P(1)$, $P(2)$, $P(3)$... of one, two, three... collisions occurring at each transmission.

The fragmentation efficiency at 40% transmission was found to be 9% i.e. for every hundred incident ions entering the collision cell, nine fragment ions were detected. (Fragment ion losses due to charge transfer, ions colliding but not decomposing and scattering were discussed earlier). The overall probability $P$ of detecting *any* ion (incident or fragment) is then 0.4 +0.009. The value of $P_D$ (the probability of an ion being detectable) can then be found from the series:

234

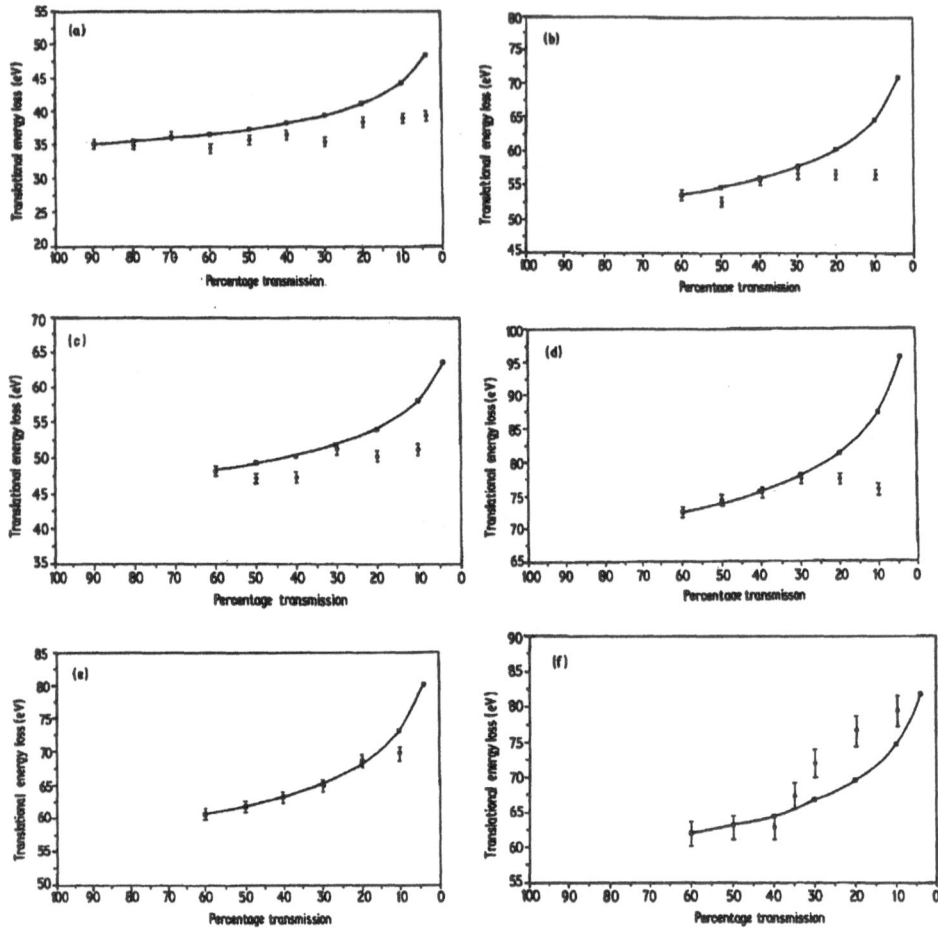

*Figure 24*[7]: Plots of translational energy losses against incident ion transmission for the fragments (a) to (f). Experimental data points are shown with an estimate of error. The curves represent theoretically predicted relationships between translational energy loss and transmission obtained by Poisson statistics.

$$\underline{P} - \sum_{x=0}^{x} P_D^x P(x) \tag{26}$$

$$\underline{P} - P(0) + P_D P(1) + \left(P_D\right)^2 P(2) + \left(P_D\right)^3 P(3) + .... \tag{27}$$

In this study $P_D$ was found to be 0.222. This value was used to calculate the number of ions in a fragment peak that were the result of one, two, three, etc. collisions. These numbers are given in *table 4* as percentages.

| Transmission (%) | 1 collision | 2 collisions | 3 collisions | 4 collisions |
|------------------|-------------|--------------|--------------|--------------|
| 4 | 69.56 | 24.43 | 5.81 | 1.03 |
| 10 | 76.66 | 19.55 | 3.32 | 0.42 |
| 20 | 83.23 | 14.83 | 1.76 | 0.157 |
| 30 | 87.26 | 11.63 | 1.03 | 0.069 |
| 40 | 90.19 | 9.15 | 0.62 | 0.03 |
| 50 | 92.52 | 7.10 | 0.36 | 0.01 |
| 60 | 94.45 | 5.34 | 0.2 | 0.0056 |
| 70 | 96.11 | 3.79 | 0.1 | |
| 80 | 97.55 | 2.41 | 0.04 | |
| 90 | 98.85 | 1.15 | | |

*Table 4*: **Percentage of fragment ions reaching the detector as a result of the incident ion colliding 1, 2, 3 and 4 times.**

In order to obtain the translational energy loss for each fragment ion, $\Delta E$, the percentages given in *table 4* were used to set up expressions for each transmission.

At 60% transmission the expression was as follows:

$$100\Delta E - 94.45\Delta E_1 + 5.34\Delta E_2 + 0.2\Delta E_3 + ... \tag{28}$$

where $\Delta E_1$, $\Delta E_2$, $\Delta E_3$ .. are the translational energy losses after 1, 2, 3. . . collisions. An assumption was made that, for each fragment, each collision resulted in the same energy loss i.e. $\Delta E_2 = 2\Delta E_1$, $\Delta E_3 = 3\Delta E_1$. Solving these expressions, allowed the theoretical relationship between $\Delta E$ and the transmission to be plotted.

Fragments (a) to (d) showed a weaker dependence of translational energy loss on transmission than theoretically predicted. Fragment (e) showed a similar dependence to that predicted and for fragment (f), there appeared to be a stronger dependence than expected from the theory.

It can be concluded, then, that the nature of the fragment ion affects the relationship between translational energy lost by that fragment and the transmission. This is due to the different methods of formation available. *Figure 25*[7] depicts these.

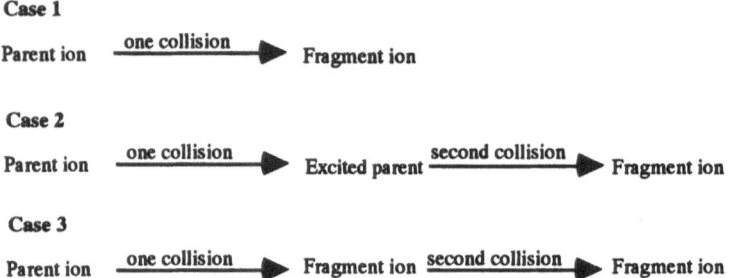

*Figure 25*[7]: **Schematic diagram of the number of collisions between the incident ions and collision-gas atoms which may be occurring prior to decomposition to the observed fragment ions.**

Case 1, where a single collision provides enough internal energy for fragmentation, describes the process by which fragments (a) to (d) were formed. In this case the value of $\Delta E$ would be expected to remain constant over a range of transmissions. Case 2 describes the situation for fragment (e). A fragment formed by case 2 is dependent on sufficient collisions taking place and enough internal energy being deposited for fragmentation. Since the average number of collisions rises with increasing gas pressure, then obviously $\Delta E$ will increase proportionally. It is possible that, at very high pressures, the incident ion will have gained so much internal energy that its immediate fragment will have enough energy to decompose further.

Case 3 represents the sequential decomposition process postulated by Sheil *et al.*[5] and described earlier. In this situation, a fragment ion may also undergo collision-induced decomposition to produce a further fragment and so on. This results in the $\Delta E$ values being very large for low mass fragments (see earlier) and explains the stronger than expected dependence of $\Delta E$ on transmission for the low mass fragment (f). *Figure 26*[7] shows a comparison of two CID spectra of valinomycin where one is recorded at 40% and the other at 10%. The two are normalised so that the intensities of the m/z 1118.6 peaks are equal in both. Decreasing the transmission from 40% to 10% results in higher relative intensities of low mass to high mass fragment ions. Similar trends have been shown for a range of gases.[31] This experimental finding is in agreement with the idea that for low transmissions (high pressures), high mass fragment ions are lost to low mass fragment ions as a result of sequential collisions.

In developing this work the systems shown in *table 5* were investigated[8].

Dependences of translational energy loss, $\Delta E$ on transmission for the valinomycin system are shown in *figure 27*[8]. Both fragments show insignificant variation of $\Delta E$ with transmission. These can be compared with *figure 24*[7] *(a)* and *(d)*. Whilst having the same *variation* of $\Delta E$, the *magnitude* of $\Delta E$ for the m/z 1090.6 fragment ion is less when using xenon (~ 15 eV). In the case of xenon, the m/z 936.5 ion has a larger translational energy loss than the m/z 1090.6 ion. These energy loss magnitudes can be accounted for using ICT theory, in particular equation (20).

| Parent ion | | Fragment ions | Collision gas |
|---|---|---|---|
| Valinomycin [M+Na] | m/z 1133.6 | 1090.6 | Xe |
| | | 936.5 | |
| $Cs_5I_4^+$ | m/z 1172 | 652.5 | He |
| | | 392.7 | |
| Valine-Gramicidin A [M+K] | m/z 1920 | 1876 | He |
| | | 1232.2 | |
| | | 933.6 | |

*Table 5:* Systems investigated in study of effect of collision gas pressure on translational energy losses.

Considering the m/z 1090.6 ion, $\Delta E$ is smaller when xenon is the target gas and larger when helium is used. The average value for $(\mu/2\varepsilon)$ however is greater when using xenon than helium. Bradley *et al.*[5] calculated $Q$, the internal energy uptake, to be 13 eV in both cases for the m/z 1090.6 ion. The fact that the translational energy loss for the m/z 936.5 ion is greater than that for the m/z 1090.6 ion is consistent with the idea that low mass fragments require more energy to form.

The invariance of the translational energy loss with transmission is explained by the fragment being the result of a single collision as in case 1, *figure 25*[7].

*Figure 28*[8] shows the variance of $\Delta E$ with transmission for the fragment ions m/z 652.5 $[Cs_3I_2]^+$ and m/z 392.7 $[CsI]^+$ from $[Cs_5I_4]^+$ (m/z 1172) on collision with helium. Both show a slight rise at low transmissions suggesting that more than one collision takes place prior to formation of the fragment.

238

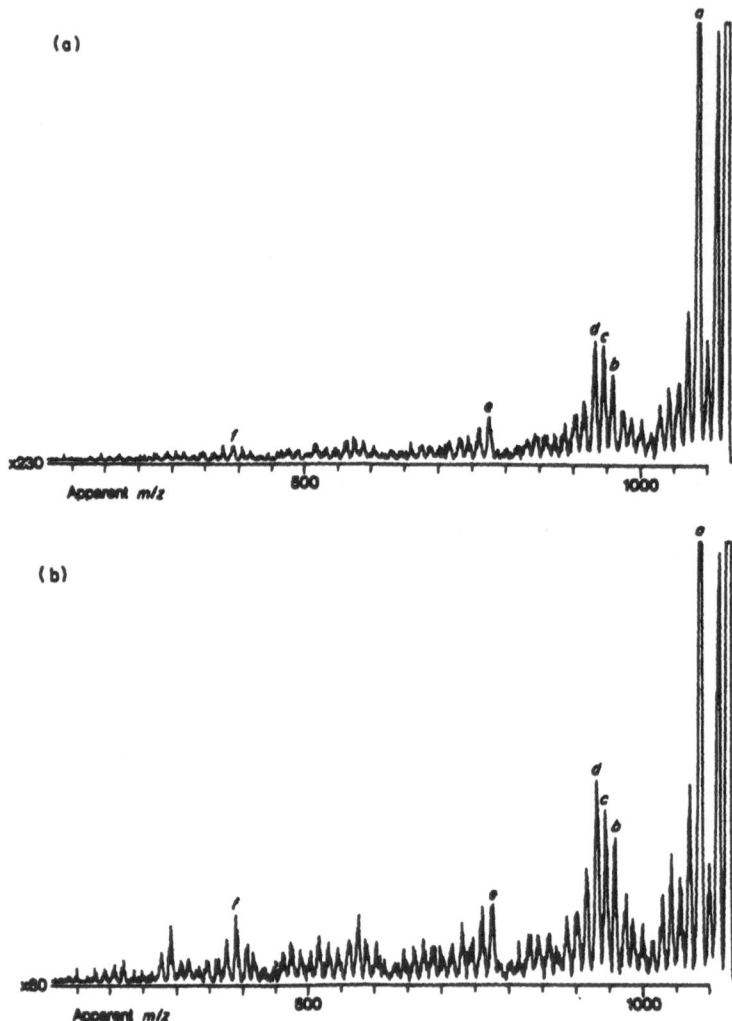

*Figure 26*[7]: CID spectra of valinomycin at (a) 40% and (b) 10% transmission. Incident ion energy = 14.92 keV. Collision gas = helium.

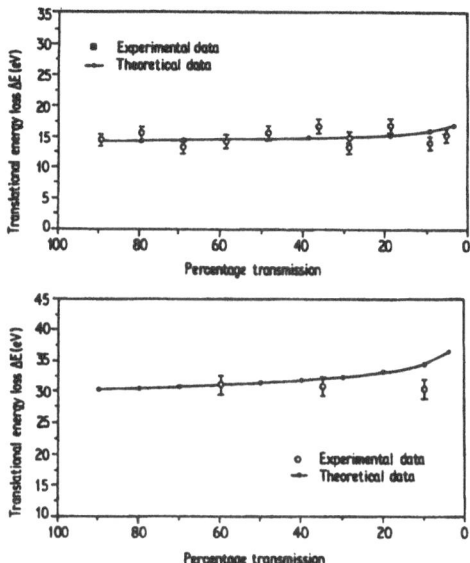

*Figure 27*[8]: Translational energy losses for [M+Na]+ of valinomycin , m/z 1133.6, on collision with xenon against percentage transmissions. Experimental points are obtained from the fragment ions m/z 1090.6 (top) and m/z 936.5 (bottom). Incident ion energy = 14.9 keV.

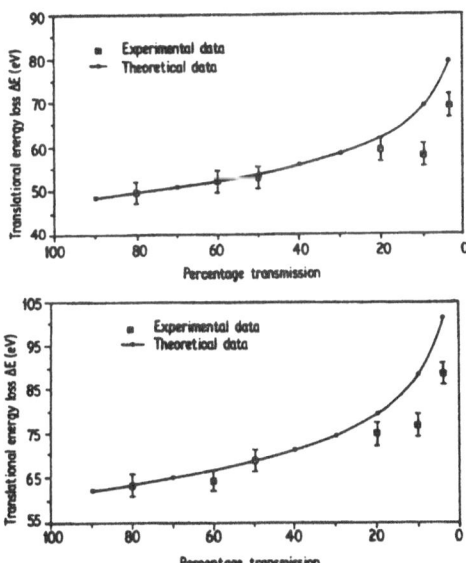

*Figure 28*[8]: Translational energy losses for (Cs$_5$I$_4$)+, m/z 1172, on collision with helium, against percentage transmission. Experimental points are obtained from the fragment ions m/z 652.5 (top) and m/z 392.7 (bottom). Incident ion energy = 10.4 keV.

240

The translational energy losses for the inorganic cluster ions are much the same as the translational energy losses for organic peptides of similar mass when using the same collision gas. This has also been noticed by others.[33] Using ICT theory, it is found that, in the case of the inorganic ion, $Q$ is an order of magnitude smaller than for the organic ion. This difference is a result of the different masses of the *atoms colliding*. With caesium iodide, the mass of the atom colliding with the helium is around 130 Da whereas for a peptide, atomic masses lie within the range 1 - 16 Da. This has an effect on $(\mu/2\varepsilon)$ and results in a lower value of $Q$ for the inorganic ion.

The results for valine-gramicidin A are depicted in *figure 29*[8]. The trends for this large peptide are more difficult to rationalise than those for valinomycin. The m/z 933.6 ion shows a very slight increase of $\Delta E$ with transmission. If there were no variation then the fragment would be said to be formed following a single collision. The slight increase in $\Delta E$, however , suggests the possibility of another fragmentation pathway that gives the m/z 933.6 ion.

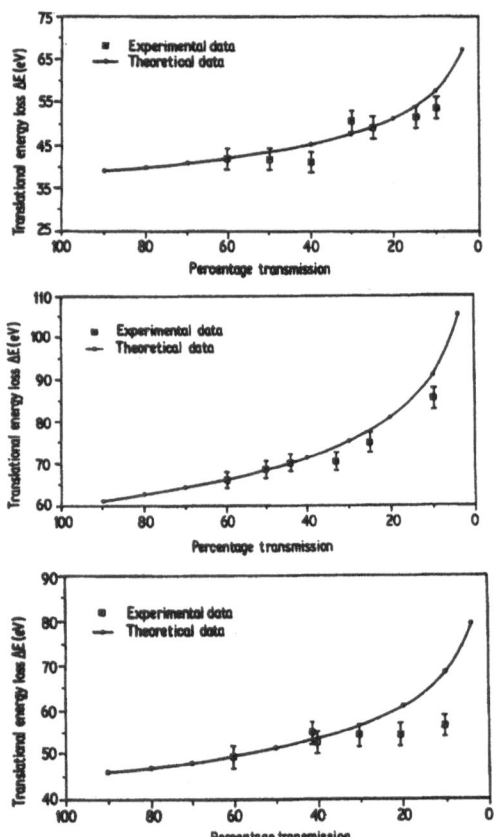

*Figure 29*[8]: Translational energy losses for [M+K]$^{+}$ of valine-gramicidin A, m/z 1920, on collision with helium. Experimental points are obtained from the fragment ions m/z 1876 (top), m/z 1232.8 (middle) and m/z 933.6 (bottom). Incident ion energy = 14.9 keV.

The m/z 1232.8 ion is consistent with the theoretical data predicted on the basis that multiple collisions give rise to the fragment. The results for the m/z 1876 ion are the most difficult to interpret. Bradley *et al.*[8] concluded that it was possible for the fragment to form as a result of multiple collisions and assigned the low energy losses (transmission 40-60%) to single collisions and the high energy losses (transmission 10%) to two or more collisions.

## 9. Centre-of-mass Collision Energy.

In order to increase the usefulness of tandem mass spectrometry as an analytical tool providing structural information, the efficiency of CID needs to be improved. Neumann *et al.*[4] suggested that a major factor in the difficulty in fragmenting large ions is that the centre-of-mass collision energy (equation 6) is too small. Bradley *et al.*[6] conducted a study into the effect of the centre-of-mass collision energy, the target gas and the incident ion energy on tandem mass spectra of various peptides.

The study employed two instruments: a Kratos Concept IHQ hybrid mass spectrometer which was fitted with an rf-only quadrupole to give EBqQ geometry and a Kratos Concept IIHH four sector mass spectrometer of EBEB geometry. In the case of the hybrid instrument, the incident ion energy could be set as high as 500 eV. This high ion energy (for a quadrupole) had to be accompanied by an increase in the rf-amplitude in the rf-only quadrupole.[34] The collision gas in the hybrid experiments was argon.

On the four sector, the incident ion energy was set between 500 eV and 4000 eV by floating the collision cell. The collision gases employed were helium and argon.

*Figure 30*[6] shows the tandem mass spectra of $[M+H]^+$ of dynorphin A fragment 1-9 (m/z 1137.5) measured on the hybrid at incident ion energy = 100, 400, 500 eV with and without argon collision gas. This peptide was chosen because, in previous studies[35], it gave a very poor CID spectrum when using a hybrid. It was reported that, even with multiple sample loadings, little information could be gleaned from the spectrum at either 30 eV or 100 eV incident ion energy.

At 400 eV and 500 eV, the collision gas has had a pronounced effect on fragmentation. At 100 eV, the dominant fragment ions are $b_2$, $y_3+2$, $b_6$, $b_7$ and $b_8+H_2O$. These reflect the position of the arginine residues in the peptide. The CID spectra at higher incident ion energies offer more sequence information and also show ions due to side chain losses: $w_5$ (m/z 654) and $d_7$ (m/z 737).

*Figure 31*[6] shows the CID spectra of the same peptide measured an the four sector using helium and argon collision gases. At high incident ion energy (4000 eV) with helium, the results are better than those obtained on the hybrid at 500 eV with argon, having the additional ion $w_2$ (m/z 243), $w_3$ (m/z 342), $w_4$ (m/z 498), $v_6$ (m/z 768). When both instruments use the same incident ion energy (500 eV) and argon, the hybrid results are superior to the four-sector results. The four-

242

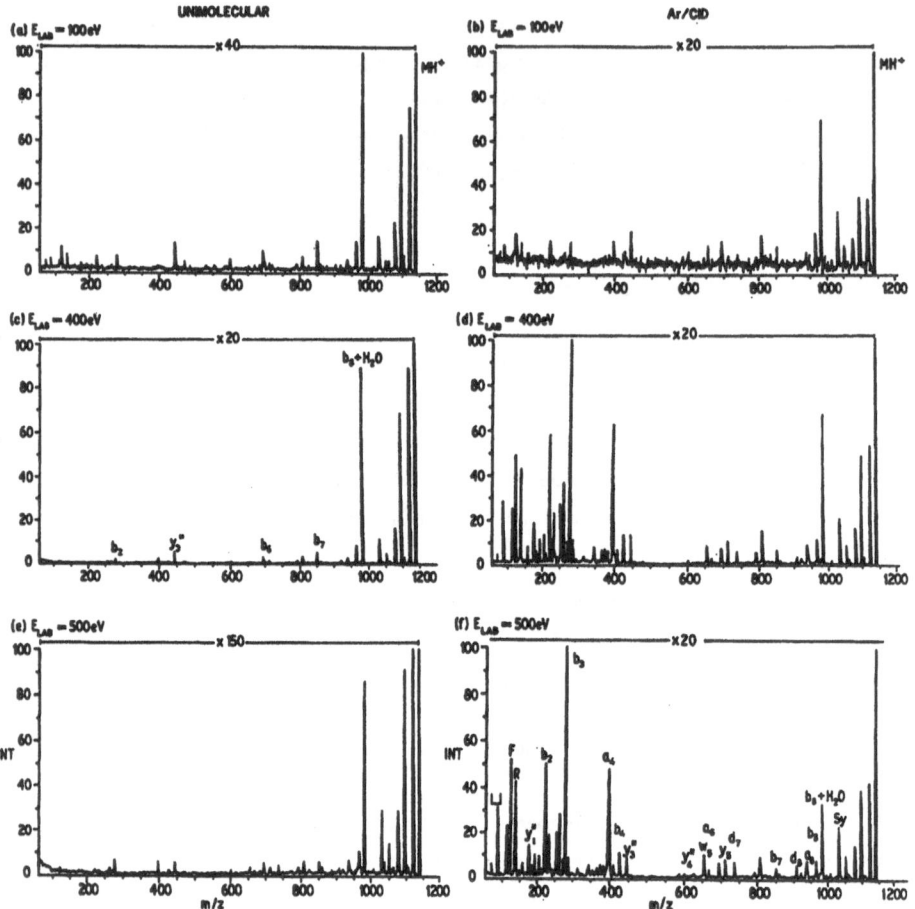

*Figure 30*[6]: **CID mass spectra of [M+H]⁺ of Dynorphin A fragment 1-9, m/z 1137.7, obtained on the hybrid at incident ion energies of 100, 400 and 500 eV. (a), (c) and (e) are without collision gas. (b), (d) and (f) are with collision gas = argon.**

sector spectrum has a number of artefacts contained within it. This is a result of transitions in the fourth field-free region,[36]which occur when the collision cell is floated close to the accelerating voltage.

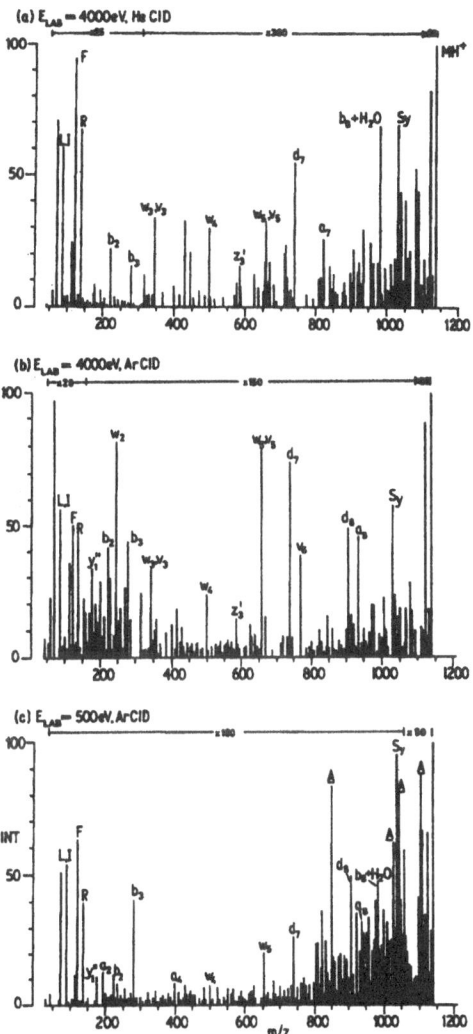

*Figure 31*[6]: CID mass spectra of [M+H]$^+$ of Dynorphin A fragment 1-9, m/z 1137.7, obtained on the four sector. (a) Incident ion energy = 4000 eV (He CID). (b) Incident ion energy = 4000 eV (Ar CID). (c) Incident ion energy = 500 eV (Ar CID). A = artefact from the fourth field free region.

Further illustrations of how increasing the incident ion energy results in more structural information are given in *figures 32*[6], *33*[6]*& 34*[6]. *Figure 32*[6] shows the CID spectrum of Gramicidin S [M+H]$^+$ , m/z 1141.7 , (cyclic Val-Orn-Leu-Phe-Pro), at ion energies 30 eV, 200 eV and 400 eV. Low mass fragments are shown in high abundance at 200 eV and 400 eV. These

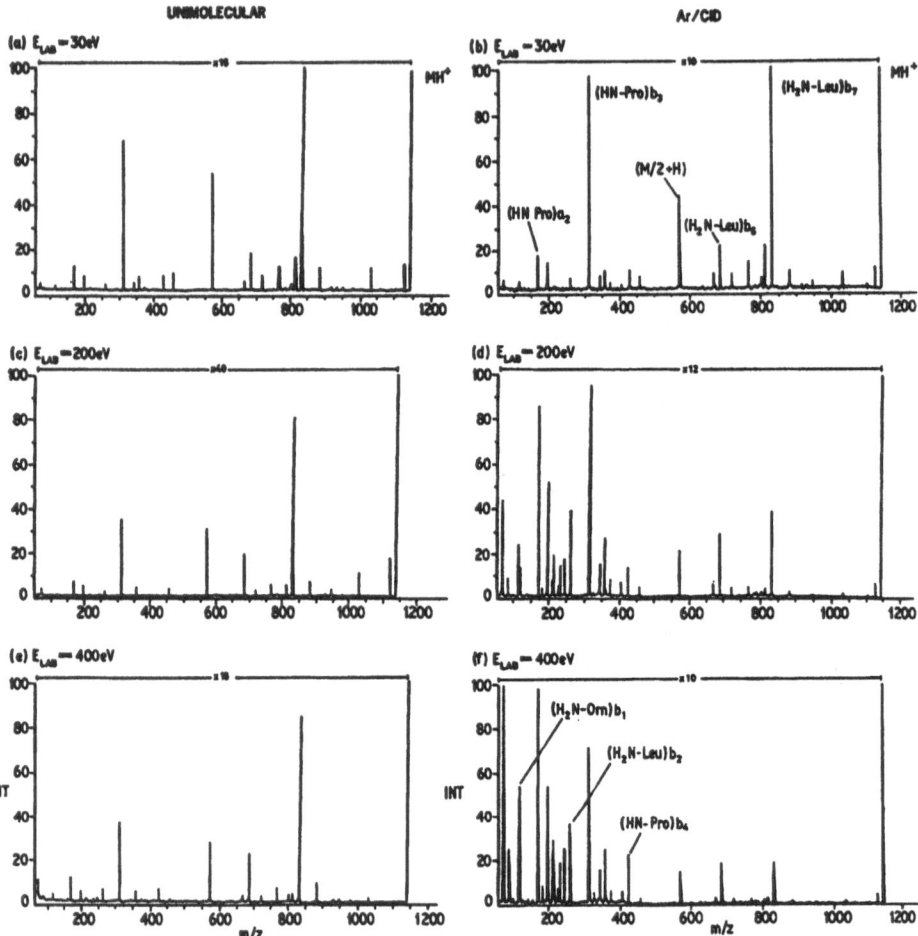

*Figure 32*[6]: **Gramicidin S [M+H]**[+], **m/z 1141.7, tandem mass spectrum obtained using the hybrid at incident ion energies 30-400 eV, unimolecular and argon CID.**

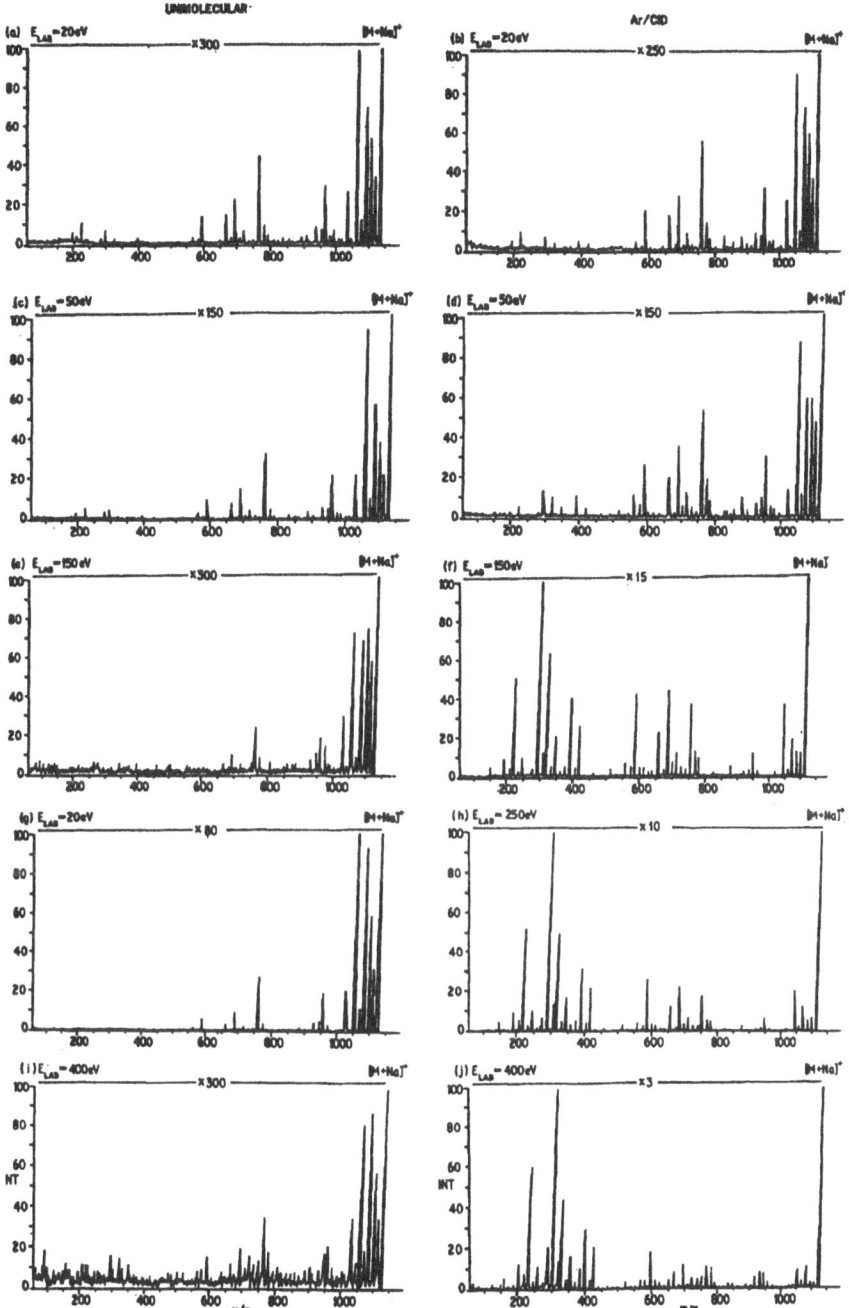

*Figure 33*[6]: Valinomycin [M+Na]+, m/z 1133.6, tandem mass spectrum obtained using the hybrid at incident ion energies 20-400 eV, unimolecular and argon CID.

fragment ions are predominantly the $a_n$ and $b_n$ series formed following ring opening between either Leu and Phe or between Phe and Pro.

*Figures 33*[6] & *34*[6] show valinomycin ((Val-Lac-Val-Hyv)$_3$) [M+Na]$^+$. As the incident ion energy is raised from 150 eV to 250 eV, fragmentation becomes more intense particularly at the low mass end of the spectrum. At incident ion energy 400 eV, the spectrum resembles the high energy four-sector mass spectrum.

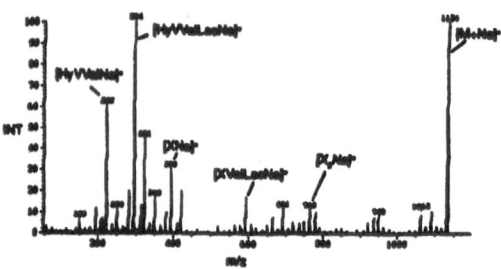

*Figure 34*[6]: Expanded version of *figure 33*[6] (j). X = [Val-Lac-Val-Hyv].

So far, the discussion has concerned the centre-of-mass collision energy, $E_{cm}$ being varied by changing the incident ion energy, $E_{lab}$, and the target gas mass. If the mass of the precursor ion is raised, then $E_{cm}$ will be decreased. This was investigated using the insulin A chain [M+H]$^+$ ( m/z 2531). The CID spectrum showed very little fragmentation with incident ion energy less than 300 eV but as the incident ion energy approached 500 eV useful fragmentation was recorded.

The above results can be summarised as follows: when the centre-of-mass collision energy and the incident ion energy are the same in both instruments a similar fragmentation pattern is seen. However, when $E_{cm}$ is kept constant but $E_{lab}$ is not i.e. on the four-sector $E_{lab}$ = 4000 eV, target gas = helium ($\sim$ 4 Da) and on the hybrid $E_{lab}$ = 400 eV, target gas = argon ($\sim$ 40 Da), more fragmentation is seen on the four-sector. The conclusion reached then is "that $E_{cm}$ is not the sole arbiter of collisionally activated dissociation processes".[6]

In order to examine this more closely, it is necessary to turn back to equations (23) and (24). These expressions combined give:

$$\langle Q \rangle = \tfrac{1}{2}\chi E_{cm} \tag{29}$$

The efficiency factor, $\chi$, has a value of one when the mass of the atom within the ion colliding, $m_a$, is equal o the mass of the gas atom, $m_g$. In the present situation, an atom within a peptide has a mass in the range 1 to $\sim$16 Da, in the absence of sulphur and metal atoms. The number of hydrogen atoms available for collision suggest that energy transfer is more likely to take place to an atom at the lower end of this mass range than to the atoms at the higher end. Helium is closer in mass to the mass of a constituent atom of a peptide than argon is. *At constant centre-of-mass collision energy*, energy transfer to a peptide will be more efficient when using helium. Leading on from this, it can be postulated that xenon will result in the most efficient energy transfer to a caesium iodide cluster since all the atoms involved have masses around 130 Da.[37]

## 10. Target Capture

Much current study into collisional activation is concerned with target capture. This was first predicted by Neumann *et al.*[4] and is the process whereby the incident ion and the collision gas do not separate following collision. For incident ion energy 8 keV and target gas helium, this phenomenon was predicted to occur at ~1500 Da.[4] The implication is that by tuning the mass of the target gas and the incident ion energy, it is possible for any incident ion to perform target capture. Whilst capture of a helium atom holds only inherent interest, the capture of a reactive molecular species opens up the possibility of chemical reactions within a complex. The complex has known composition, known internal energy and known lifetime and so is particularly useful.

The discovery of Buckminsterfullerene ($C_{60}$) in 1985[38] and the subsequent determination of its structure as that of a truncated icosahedron (soccer ball) opened up avenues for target capture. The fact that $C_{60}$ is a closed cage and that its constituent atoms are all of the same mass make it particularly suitable.

The inclusion of the target gas helium within the incident ion was first observed by Weiske *et al.*[39] They noticed that the daughter ions of $C_{60}$ had a second peak, four mass units higher, associated with them. The experiment was repeated using $^3He$ and, once again, peaks of three mass units higher were associated with the daughter ion peaks. In conclusion they wrote "there is no doubt that a helium atom is incorporated into the $C_x^+$ (x = 60, 70) cluster in the course of the collision". [39]

This was confirmed by Ross and Callahan[40] who used a hybrid tandem mass spectrometer of BEqQ geometry to identify $He@C_{60}^+$. (The notation $M@C_{60}^+$ to describe an atom trapped within $C_{60}^+$ was suggested by Chai *et al.*[41]). *Figure 35*[40] shows the results of Ross's work. The top spectrum is the 8 keV $C_{60}^+$/He CID/MIKE spectrum over a narrow energy range with the quadrupole set to transmit ions of m/z 720. The lower spectrum is the same as above but with the quadrupole set to transmit ions of m/z 724.

Later work by Mowrey, Ross and Callahan[42] showed that the optimum collision energy for production of $He@C_{60}^+$ is 5000 eV. This agrees qualitatively with Neumann's results.[4] If an ion of 1500 Da should undergo target capture at an incident ion energy of 8000 eV, then it follows that an ion of smaller mass (720 Da) will require a lower incident ion energy for target capture. A lower incident ion energy means a lower centre-of-mass energy, so in *figure 8*, the translational energy loss and centre-of-mass energy curves will cross at a lower mass. Hence $Q$, the internal energy uptake, will reach a maximum at a lower mass and target capture will take place at a lower mass.

Further studies have shown that $C_{60}^+$ can capture $D_2$, neon and argon to produce $D_2@C_{60}^+$, $Ne@C_{60}^+$ and $Ar@C_{55}^+$.[43] Schwarz *et al.* have shown that doubly- and triply-charged fullerene ions will incorporate helium and neon on collision.[44-46] The sequential inclusion of He within $C_{60}^+$ has also been observed.[47]

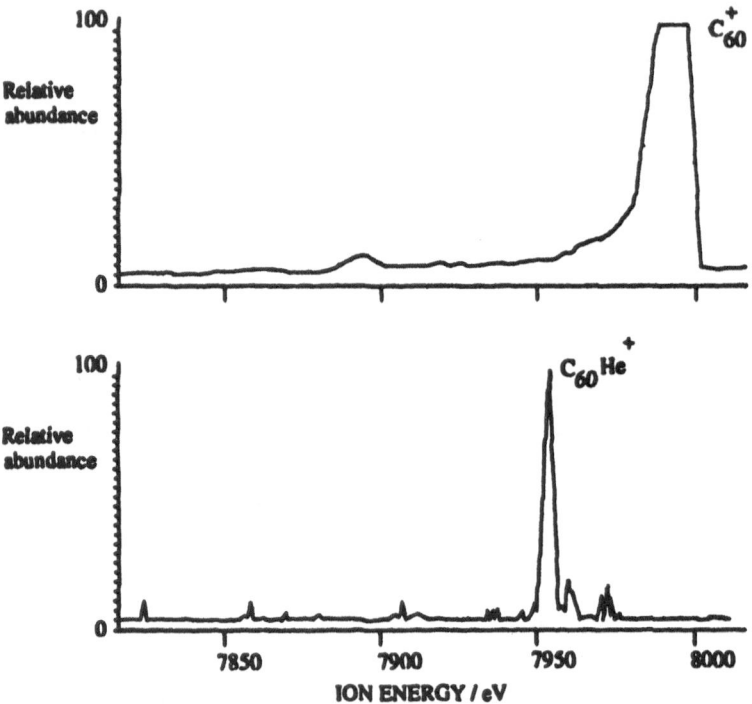

*Figure 35*[40]: **CID/MIKE spectra of $C_{60}^+$ in collision with helium over a narrow energy range. Quadrupole is set to transmit ions of m/z 720 (top) and m/z 724 (bottom). Incident ion energy = 8 keV.**

The study of the phenomenon of target capture has been extended to include capture by peptide ions[48], in particular renin substrate $[M+H]^+$ ions. A four-sector mass spectrometer of EBEB geometry was used and, as for most $C_{60}$ experiments, the CID spectra were recorded by means of a B/E = constant linked scan. The second electric sector slits were set to allow an energy band pass of $\pm 110$ eV. This energy band pass is vital - energy equal to the centre-of-mass collision energy is lost on forming target capture ions as stated earlier.

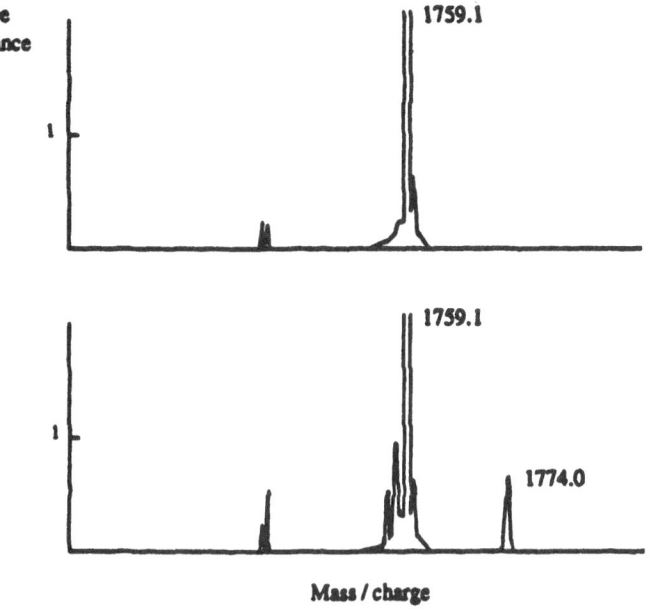

Renin substrate / NH₃

Collision energy 21 eV

Ion energy 2.2 keV

*Figure 36*[48]: Top: Spectrum of renin substrate [M+H]⁺, m/z 1759.1, in the absence of collision gas, incident ion energy = 2.2 keV. Bottom: CID spectrum of renin substrate [M+H]⁺, m/z 1759.1, incident ion energy = 2.2 keV, collision gas = ammonia (30% T).

The target gases used in these experiments were ammonia ($NH_3$) and methane ($CH_4$). *Figure 36*[48] shows the spectrum of renin substrate [M+H]⁺ ion, m/z 1759.1, without (above) and with (below) ammonia as the target gas, an incident ion energy of 2200 eV and pressure sufficient to attenuate the beam to 30% (single collision conditions) (lower figure). The introduction of ammonia results in the fragment peaks $[M+H-1]^+$, $[M+H-2]^+$, $[M+H-14]^+$ and $[M+H-15]^+$. These are standard CID fragments of renin substrate and are also observed when argon or helium are employed as the target gas. The most interesting feature of the CID spectrum, however, is the peak at 1774.0 Da, 15 Da higher than the incident ion. This has been assigned $[M+H+NH_3-2H]^+$. The greatest abundance of this ion is found under the conditions above i.e. incident ion energy = 2200 eV, centre-of-mass collision energy = 21.0 eV. When the incident ion energy was dropped below 2000 eV, a new peak appeared at 1776 Da. This corresponds to $[M+H+NH_3]^+$. The

250

abundance of this peak also varied with centre-of-mass collision energy. The maximum yield was found at centre-of-mass collision energy = 12.4 eV, incident ion energy = 1300 eV. *Figure 37*[48] shows the relationship between intensity and centre-of-mass collision energy for these two ions ([M+H+NH$_3$-2H]$^+$ and [M+H+NH$_3$]$^+$).

**Relative Intensity (%)**

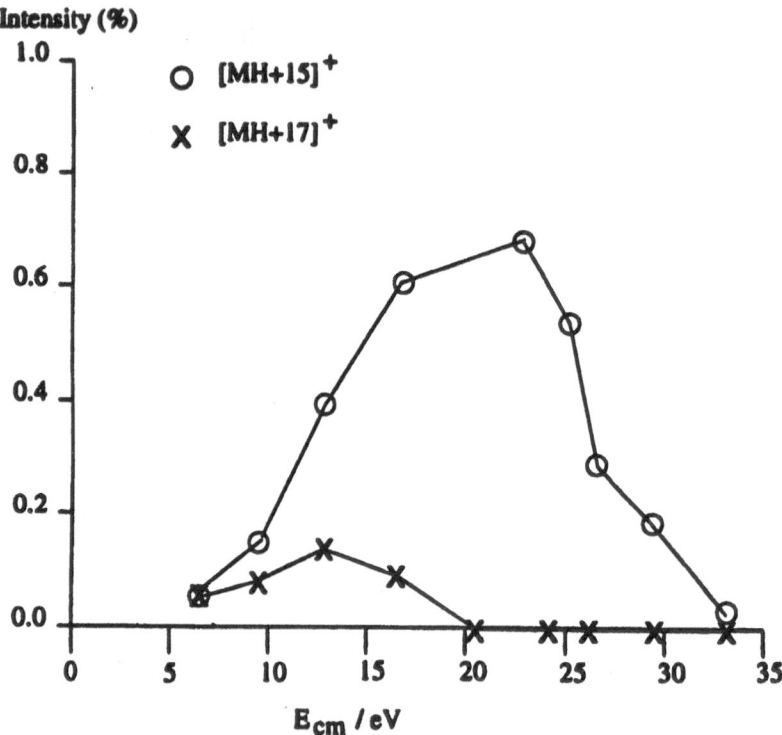

*Figure 37*[48]: **Intensity versus centre-of-mass collision energy for the ions [M+H+NH$_3$-2H]$^+$ and [M+H+NH$_3$]$^+$.**

The results for methane are very similar. The CID spectrum of [M+H]$^+$ of renin substrate with methane as the target gas, shows a peak at mass [M+H+14]$^+$. This is assigned to [M+H+CH$_4$-2H]$^+$ and has a maximum yield at incident ion energy 2400 eV and centre-of-mass collision energy 21.6 eV. At lower incident ion energies, the peak corresponding to [M+H+CH$_4$]$^+$ appeared.

Isotopic labelling was used in both cases to determine the origin of the lost hydrogens. It was found that one hydrogen came from the peptide and one from the target gas. This shows that chemical reactions can take place in high energy collision experiments.

Cheng and Fenselau[48] also studied target capture by other peptides and in each case found the [M+H+XH$_n$-2H]$^+$ peak. The optimal collision energy was measured for each peptide. The results showed higher mass peptides requiring higher collision energies. This trend is in accordance with the results of Neumann *et al.*[4] The theoretical predictions of Neumann *et al.*[4] do not perfectly fit the experimental data, however. The experimental collision energies are a little lower than those predicted. This can be accounted for as follows: the experimental results show the optimum collision energy at which the complex is *observed* rather than *formed*. It is easy then to visualise a large amount of the complex being formed but, the complex due to its high internal energy decomposes to produce fragment ions. This can also be used to explain the results in *figure 37*[48]. The greatest abundance of $[M+H+NH_3]^+$ appears at a lower energy than its counterpart $[M+H+NH_3-2H]^+$. At higher energies, although many $[MH+NH_3]^+$ ions may be formed, most of them will decompose to give $[M+H+NH_3-2H]^+$.

### 11. Recent Translational Energy Spectra Studies

### 11.1 ANGLE-RESOLVED TRANSLATIONAL ENERGY SPECTROSCOPY OF THE PARENT ION.

The question as to the fate of the energy lost by the precursor ion was addressed earlier. The suggestion that the major part of this energy was redeployed as electronic excitation of the target was made by Bricker and Russell.[23] The opposing view, put forward by Derrick *et al.*[3,4] and Thibault *et al.*[24], is that the recoil energy of the target gas is a major sink for the translational energy lost. Recently further work on this problem has been presented by Thibault *et al.*[49,50]

Measurements of translational energy losses of reactant ions that did not compose were made as a function of scattering angle. Very high beam collimation is required for angle-resolved mass spectrometry and so it is necessary to choose an abundant parent ion. $Cs_4I_3^+$ was used. The use of a caesium iodide cluster also removed the possibility of mis-assignment of daughter ions since there are several hundred Da between peaks. The third advantage of using a caesium iodide cluster is that there is no matrix interference.

Thibault *et al.*[49] derived expression (30) from the conservation of energy.

$$\Delta E = \left[ \left( U_R - U_R^\circ \right) + \left( U_g - U_g^\circ \right) + h\nu \right] + \Delta E_g = Q + \Delta E_g \qquad (30)$$

where $^\circ$ describes a pre-collision parameter, $U$ is internal energy, subscript $R$ denotes the reactant ion and subscript $g$ denotes the target gas. It should be noted that different symbols were used by Thibault *et al.*[49] The radiation $h\nu$ emitted by large ions is considered insignificant in accordance with experiment.[51,52,53]

To remove the possibility of target gas excitation, inert gases were employed. In this case, the parameter $\left( U_g - U_g^\circ \right)$ can only have a value corresponding to an excitation energy of the atom. These are well-known and by studying translational energy losses lower than the first excitation energy there is no question of target gas excitation occurring.

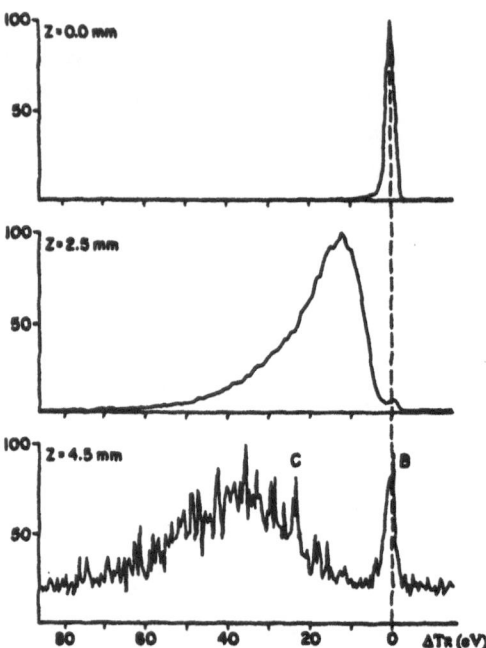

*Figure 38*[49]: **Translational energy spectra of Cs₄I₃⁺ ions, m/z 912, after collision with helium at z-displacements 0.0 mm, 2.5 mm and 4.5 mm. This corresponds to observations angles of 0˚, 0.14˚ and 0.26˚ respectively.**

As in the earlier experiments, a VG Analytical ZAB-EQ instrument was used. The ion beam was produced by FAB (fast-atom bombardment). Collimation was achieved as described earlier and angular dependence was measured using a movable z-slit (maximum displacement 5 mm). The maximum angle that could be measured was $0.29 \pm 0.03˚$ to the x-axis.

The translational energy spectra of the Cs₄I₃⁺ ion in collision with helium and neon are shown in *figures 38*[49] and *39*[49]. *Figure 40*[49] shows graphically the results for all the gases employed.

Two trends are evident from *figures 38*[49], *39*[49] and *40*[49]. The first is the relative intensities of the peaks labelled A and B. This can be explained in terms of the types of ion detected. They are:

1. Ions that have not collided.
2. Ions that have collided elastically i.e. there has been no internal energy uptake.
3. Ions that have collided inelastically.

It should be expected that ions of types 1 and 2 will be predominant in the translational energy spectra at low scattering angles. Ions of type 3, however, should become relatively more intense

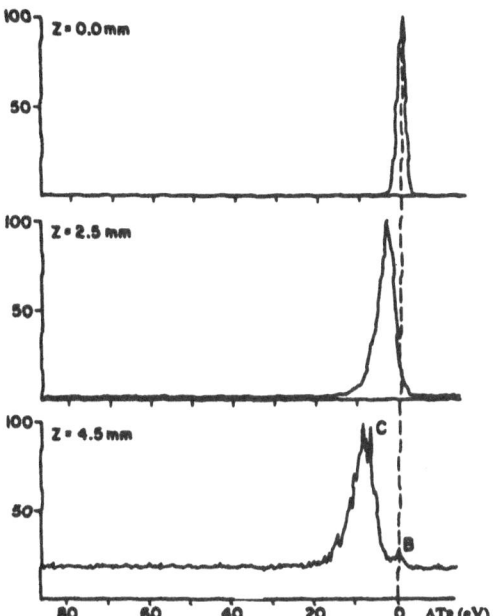

*Figure 39*[49]: Translational energy spectra of $Cs_4I_3^+$ ions, m/z 912, after collision with neon at z-displacements 0.0 mm, 2.5 mm and 4.5 mm. This corresponds to observations angles of 0°, 0.14° and 0.26° respectively.

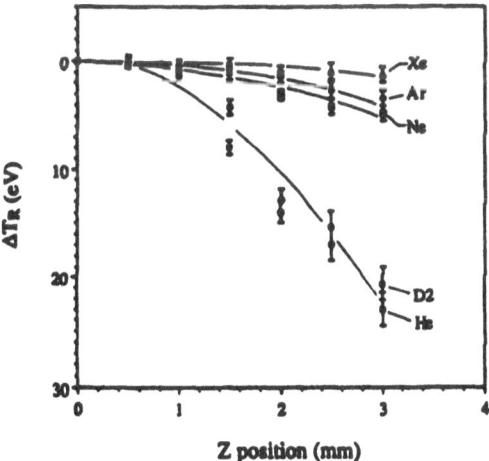

*Figure 40*[49]: Translational energy losses of $Cs_4I_3^+$ ions with various target gases, as a function of scattering angle.

at higher scattering angles showing larger translational energy losses. This is in qualitative agreement with *figures 38*[49] and *39*[49].

The second trend is that ions in collision with lighter target gases show greater translational energy losses for a given scattering angle. Since the possibility of excitation of the target gas has been removed, this trend needs to be studied in terms of a momentum transfer mechanism. Thibault *et al.*[49] compared the 'quasi-diatomic approximation' (QDA) with impulsive collision transfer (ICT) theory and found both able to account for the experimental observations qualitatively and semi-quantitatively.

In order to use the QDA, it is necessary to supply values of internal energy uptake, $Q$. Considerations of the energetics of caesium iodide clusters led to the range chosen for internal energy uptake being 1-3 eV. With ICT theory, $Q$ is not an input parameter as it is determined as a fixed fraction of the translational energy loss.

*Figure 41*[49] *shows* the results of the QDA calculations for the target gases helium and neon. *Figure 42*[49] shows the results of the ICT theory calculations for helium, neon, argon and xenon. (The experimental limit is indicated by a broken line).

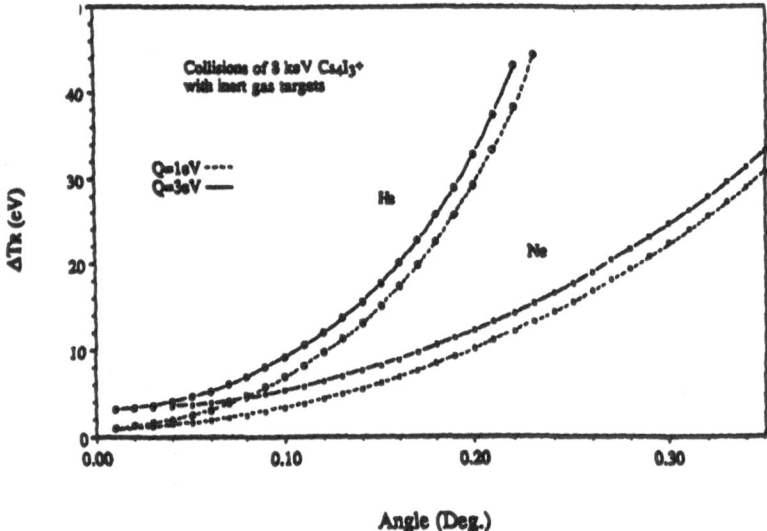

Angle (Deg.)

*Figure 41*[49]: **Translational energy losses as calculated by the QDA, as a function of scattering angle for collision gases helium and neon.**

The difference between the translational energy loss and the internal energy uptake is the translational energy gain of the target gas. This seems much larger for helium than for the other gases. For two collision gases and fixed internal energy uptakes and scattering angle, a small mass ratio requires a large velocity ratio in order for momentum to be conserved.

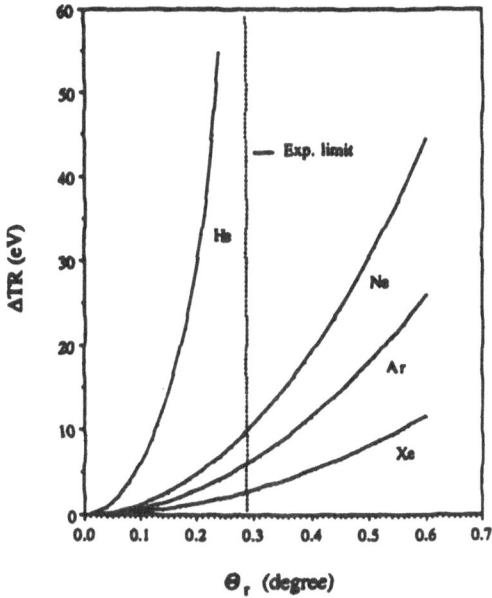

*Figure 42*[49]: Translational energy losses as calculated by ICT theory, as a function of scattering angle for collision gases helium, neon, argon and xenon.

Although both ICT theory and the QDA appear successful in predicting the above results, it is still not clear whether the translational energy loss suffered by the ion appears predominantly in the internal energy uptake of the ion or in the recoil energy of the target gas. It is clear, however, that excitation of the target gas is not involved.

## 11.2 INTERNAL ENERGY STUDIES

In order to clarify the question posed above, namely is the majority of the translational energy lost converted to internal energy of the ion or recoil energy of the target, Thibault *et al.*[50] looked at the delayed dissociation spectra of survivor ions form collisional activation. Collisionally activated, but intact, ions were selected and their internal energy characterised by looking at their delayed unimolecular decompositions. Choosing a particular translational energy loss, for a fixed initial internal energy of the reactant ion will specify the combined internal energy of the fragments. (See equation (30)). As before only inert target gases were used to remove the possibility of target gas excitation. The instruments used were a VG Analytical ZAB-EQ (BEqQ) and a VG Analytical ZAB-4F (BEEB).

Three precursor ions were used in this study: $C_{10}H_{14}^{+\cdot}$ of butyl-benzene, Leu-enkephalin and penta($d_3$)acetylgalactose. The fragmentation of the cation $C_{10}H_{14}^{+\cdot}$ of butyl-benzene has been studied extensively as a function of internal energy [52, 54-59] and the cation has a reasonably

256

large energy tail. The butyl-benzene system has had the intensity ratio of the fragment ions m/z 91 and 92 calibrated against internal energy. It is the only calibrated system of the three used in this study. The results of this system are presented below.

In order that an appropriate reaction time window was available, a four-sector instrument was used. Collisional activation took place between B1 and E1. E1 was used to select the survivor ions. A linked scan at constant E2/B2 monitored the unimolecular dissociations occurring in the field free region between E1 and E2. Since the four-sector was not equipped with a movable z-aperture, angular selection was impossible. The results correspond to an integration over the range of scattering angles. Translational energy losses of 0, 3, 6, 9 and 12 eV were chosen. The results for this system are shown in *figure 43*[50].

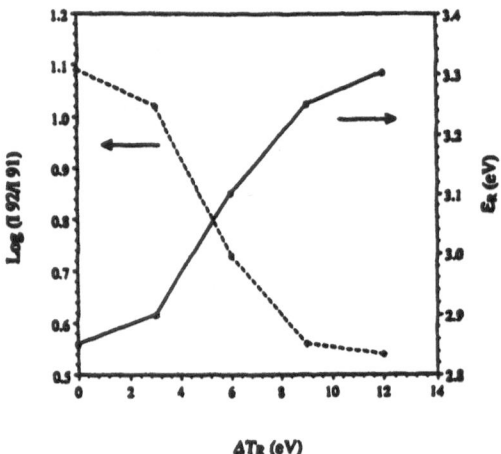

$\Delta T_R$ (eV)

*Figure 43*[50]: **Logarithm of ratio of intensity for m/z 92 to m/z 91 for butyl benzene system and the average internal energy of butyl benzene molecular ions as deduced from the intensity ratio against translational energy loss.**

The internal energy of the reactant ion at zero translational energy loss (no collision) is 2.85 eV. This leads to the conclusion that the maximum internal energy of the ion *prior to the collision* is 2.85 eV. At a translational energy loss of 12 eV, the internal energy is 3.3 eV. If it is assumed that the minimum internal energy prior to the collision is zero, then the maximum internal energy uptake, $Q$, is 3.3 eV (extending to a minimum of (3.3-2.85) = 0.45 eV). If the possibility of radiative emission is discounted then a maximum of only 25% of the translational energy lost is converted to internal energy of the ion. The only other destination for the translational energy loss is the recoil energy of the target gas.

Thibault *et al.*[50] used ICT theory to predict internal energy uptakes for translational energy losses in the range 7-9 eV and found the maximum values of Q to lie in the range 0.42-2.1 eV. These are in agreement with the experimental results.

The conclusions reached here are based on the underlying assumption that radiative cooling is negligible. For this emission to effect the results, emission of several electronvolts over a few micro-seconds would be required. Infra-red cooling of vibrationally excited ions is expected to be too slow[60] and UV-visible emission probabilities are not expected to compete effectively with internal conversion to the electronic ground state.[49]

## 12. Conclusion

In conclusion, it can be said that energy shifts accompanying collisional activation have a magnitude of tens of electronvolts for ions of m/z > 1000.

A direct momentum transfer mechanism explains the phenomenon of collisional activation better than electronic excitation. Of the two momentum transfer mechanisms postulated, ICT theory seems preferable, due to the essential idea of localised energy deposition. ICT theory, however, is still inadequate. In particular experimental results show that translational energy losses increase with the mass of the parent ion and this cannot be recreated by the ICT theory in its present form.

## 13. References

1. Jennings, K. R. (1968) *Int. J, Mass Spectrom. Ion Phys.*, 1, 227.

2. ' Metastable Ions', Cooks, R. G., Beynon, J. H., Caprioli, R. M., Lester, G. R.; Elsevier Scientific Publ. Co.; Amsterdam, 1973.

3. Neumann, G. M., Derrick, P.J. (1984) *Org. Mass Spectrom.*, 19, 165.

4. Neumann, G. M., Sheil, M. M., Derrick, P.J. (1984) *Z. Natursforsch*, 39a, 584.

5. Sheil, M. M., Derrick, P.J. (1988) *Org. Mass Spectrom.*, 23, 429.

6. Bradley, C. D., Curtis, J. M., Derrick, P. J., Wright, B. (1992), *Anal. Chem.*, 64, 2628.

7. Bradley, C. D., Derrick, P. J. (1991) *Org. Mass Spectrom.*, 26, 395.

8. Bradley, C. D., Derrick, P. J. (1993) *Org. Mass Spectrom.*, 28, 390.

9. Boyd, R. K., Kingston, E. E., Brenton, A.G., Beynon, J. H. (1984) *Proc. R. Soc. Lond.*, A392, 59.

10. Uggerud, E., Derrick, P. J. (1991) *J. Phys. Chem.*, 95, 1430.

11. Cooper, H. J., Derrick, P. J., Jenkins, H. D. B., Uggerud, E. (1993) *J. Phys. Chem.*, 97, 5443.

12. Desiderio, D. M., Sabbatini, J. Z. (1981) *Biomed. Mass Spectrom.*, 8, 565.

13. Puzo, G., Prome, J. C. (1980) *Adv. Mass Spectrom.*, 8, 1003.

14. Craig, A. G., Derrick, P. J. *unpublished results*.

15. Kim, M.S., McLafferty, F. W. (1978), *J. Am. Chem. Soc.*, 100, 3279.

16. Cullis, P. G., Neumann, G. M., Rogers, D. E., Derrick, P. J. (1980) *Adv. Mass Spectrom.*, **8**, 1729.

17. Neumann, G. M., PhD Thesis, La Trobe University, (1983).

18. Weber, R., Levsen, K. (1980) *Biomed. Mass Spectrom.*, **7**, 314.

19. Matsuo, T., Matsuda, H. Katakuse, I., Shimonishi, Y., Maruyama, Y., Higuchi, T., Kabota, E. (1981) *Anal. Chem.*, **53**, 416.

20. (a) J. Durup in ' Recent Developments in Mass Spectrometry', ed. K. Ogata & T. Hayakawa, p. 921, University Park Press, Baltimore, Maryland. (1969). (b) H. Ewald, W. Seibt, *ibid*, p. 39.

21. Todd, P. J., Warmack, R. J., McBay, E. H. (1983) *Int. J, Mass Spectrom. Ion Phys.*, **50**, 299.

22. Sheil, M. M., Gilbert, R. G., Derrick, P. J. in 'Advances in Mass Spectrometry 1985' ed. J. F. J. Todd, John Wiley and Sons Ltd., 1986, 1161.

23. Bricker, D. L., Russell, D. H. (1986) *J. Am. Chem. Soc.*, **108**, 6174.

24. Alexander, A. J., Thibault, P., Boyd, R. K. (1990) *J. Am. Chem. Soc.*, **112**, 2484.

25. Ouwerkerk, C. D., McLuckey, S. A., Kistemaker, P. G., Boerboom, A. J. H. (1984) *Int. J, Mass Spectrom. Ion Proc.*, **56**, 11.

26. Laramee, J. A., Cameron, D., Cooks, R. G. (1981) *J. Am. Chem. Soc.*, **103**,12.

27. Roepstorff, P., Fohlman, J. (1984) *Biomed. Mass Spectrom.*, **11**, 601.

28. Derrick, P. J. (1986) *Fres. Z. Anal. Chem.*, **324**, 486.

29. McLuckey, S. A., Ouwerkerk, C. D., Boerboom, A. J. H., Kistemaker, P. G. (1984) *Int. J, Mass Spectrom. Ion Proc.*, **59**, 85.

30. Gilbert, R. G., Sheil, M. M., Derrick, P.J. (1985) *Org. Mass Spectrom.*, **20**,430.

31. Sheil, M. M., PhD Thesis, University of New South Wales, Australia. (1987).

32. Sheil, M. M.,Uggerud, E., Derrick, P. J. (1989) *Adv. Mass Spectrom.*,**11**, 1012.

33. Derrick, P. J., Colburn, A. W., Sheil, M. M., Uggerud, E. (1990) *J. Chem Soc. Faraday Trans.*, **86**(13), 2533.

34. Alexander, A. J., Dyer, E. W., Boyd, R. K. (1989) *Rapid Commun. Mass Spectrom.*, **3**, 364.

35. Alexander, A. J., Thibault, P., Boyd, R. K., Curtis, J. M., Rinehart, K. L. (1990) *Int. J, Mass Spectrom. Ion Proc.*, **98**, 107.

36, Boyd, R. K., Harran, D. J., (1985) *Int. J, Mass Spectrom. Ion Proc.*, **65**, 273.

37. Cooper, H. J., Derrick, P. J., *unpublished results*.

38. Kroto, H. W., Heath, J. R., O' Brien, S. C., Curl, R. F., Smalley, R. E. (1985) *Nature*, **318**, 612.

39. Weiske, T., Bohme, D. K., Hrusak, J., Kratschmer, W., Schwarz, H. (1991) *Angew Chem. Int. Ed. Engl.*, **30**, 884.

40. Ross, M. M., Callahan, J. H. (1991) *J. Phys. Chem.*, **95**, 5720.

41. Chai, Y., Guo, T., Jin, C., Haufler, R. E., Chibante, L. P. F., Fure, J., Wang, L., Alford, J. M., Smalley, R. E. (1991) *J. Phys. Chem.*, **95**, 7564.

42. Mowrey, R. C., Ross, M. M., Callahan, J. H. (1992) *J. Phys. Chem.*, **96**, 4755.

43. Caldwell, K. A., Giblin, D. E., Hsu, C. S., Cox, D., Gross, M. L. (1991) *J. Am. Chem. Soc.*, **113**, 8519.

44. Weiske, T., Hrusak, J., Bohme, D. K., Schwarz, H. (1991) *J. Phys. Chem.*, **95**, 8451.

45. Weiske, T., Hrusak, J., Bohme, D. K., Schwarz, H. (1992) *Helv. Chim. Acta.*, **75**, 79.

46. Weiske, T., Hrusak, J., Bohme, D. K., Schwarz, H. (1991) *Chem. Phys. Lett.*, **186**, 459.

47. Weiske, T., Schwarz, H. (1992) *Angew Chem. Int. Ed. Engl.*,**31**, 605.

48. Cheng, X., Fenselau, C. (1993) *J. Am. Chem. Soc.*, **115**, 10327.

49. Thibault, P., Alexander, A. J., Boyd, R. K. (1993) *J. Am. Soc. Mass Spectrom.*, **4**, 835.

50. Thibault, P., Alexander, A. J., Boyd, R. K., Tomer, K. B. (1993) *J. Am. Soc. Mass Spectrom.*, **4**, 845.

51. Reference 29 from (49).

52. Baer, T., Dutuit, O., Mestdagh, H., Rolando, C. (1988) *J. Phys. Chem.*, **92**, 5674.

53. Laramee, J. A., Hemberger, P. H., Cooks, R. G. (1980) *Int. J, Mass Spectrom. Ion Phys.*, **33**, 231.

54. Griffiths, I.W., Mukhtar, E. S., March, R. E., Harris, F. M., Beynon, J. H. (1981) *Int. J, Mass Spectrom. Ion Phys.*, **39**, 125.

55. Griffiths, I.W., Mukhtar, E. S., Harris, F. M., Beynon, J. H. (1982) *Int. J, Mass Spectrom. Ion Phys.*, **43**, 283.

56. Harrison, A. G., Lin, M. S. (1983) *Int. J, Mass Spectrom. Ion Phys.*, **51**, 353.

57. Chen, J. H., Hays, J. D., Dunbar, R. C. (1984) *J. Phys. Chem.*, **88**, 4759.

58. Boyd, R. K., Harris, F. M., Beynon, J. H. (1985) *Int. J, Mass Spectrom. Ion Proc.*, **66**, 185.

59. Brown, P. (1970) *Org. Mass Spectrom.*, **3**, 1175.

60. Uechi, G. T., Dunbar, R. C. (1992) *J. Chem. Phys.*, **96**, 8897.

# HYBRID INSTRUMENTS OF UNUSUAL GEOMETRY

D. FRAVETTO and P. TRALDI
*CNR, Area della Ricerca,*
*Corso Stati Uniti 4, I-35020 Padova, Italy*

ABSTRACT. The principle of operation of tandem mass spectrometers where sectors and quadrupoles are combined with time-of-flight and ion traps is described.

## 1. Introduction

Tandem mass spectrometry has shown a constant growth of interest due to its high capabilities either in structural investigations or in the analytical field [1]. This interest resulted in the research and design of new tandem instruments, constituted by different arrangements of different analysers. Thus, aside the multisector and triple quadrupole well consolidated geometries, hybrid instruments, constituted by analysers based on different physical phenomena, have received a particular attention. The most common of these instruments combine, with different geometries, sectors and quadrupoles [2] : this approach is well established and many of such instruments are commercially available. Less common are those which combine sectors and quadrupoles with time of flight instruments and ion traps; their description will be the object of this work.

## 2. HYBRID TOF

The first paper on the linking between a magnetic sector mass spectrometer and a TOF instrument was due to Enke et al. [3].
The sector instrument was modified such as to provide measurements of ion flight time and this investigation resulted in a novel instrument, providing the same data obtained by more usual MS/MS instrumentation. The combination of ion momentum and velocity analysis provides unambiguous mass assignment for parent and daughter ions, while flight time comparison allows to establish parent-daughter relationship. Usually, without any time resolved detection, the fragment ions formed within the source and the magnetic sector of a single focus instrument appear as "metastable" peaks, with apparent mass $m^* = m_2^2/m_1$, where $m_2$ and $m_1$ are the masses of the parent and the daughter ion respectively. Such a relationship can not lead to unequivocal determination of $m_2$ and $m_1$. However, considering that ion current can result from both stable and daughter ions having the same momentum, this superimposed contribution to ion current can be evaluated by time resolution since daughter ions maintain the same velocity as their parent (figure 1).
For any selected momentum, the stable ions will arrive first, followed by daughter of successively heavier parents. The masses of the parent and daughter ions are determined by the related Bt

*R. M. Caprioli et al. (eds.), Mass Spectrometry in Biomolecular Sciences, 261–270.*
© 1996 *Kluwer Academic Publishers.*

product. This can be well understood looking at the plot magnetic field strength (Tesla) vs TOF (µs) presented in figure 2.

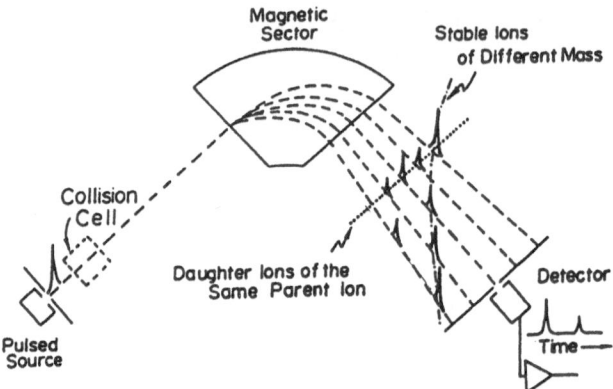

**Figure 1. Concept of time-resolved magnetic dispersion mass spectrometry**

Stable ions will lye on the line of unit slope, all ions of a given parent, maintaining the velocity of the parent, will occur at the same flight time. Parent and daughter ions of the same mass fall on a hyperbole of constant Bt. Thus, as an example, ions at m/z 150 in Figure 2 can be determined as originating by a parent at m/z 400; in fact its flight time does not correspond to 15 µs related to the B point on the stable ion line, but about 25 µs. This flight time allows to determine its precursors.

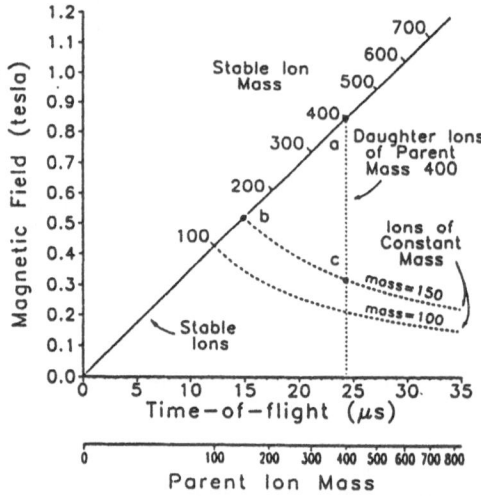

**Figure 2. Plot of field strength vs. flight time for stable and daughter ions. Points a and b are for stable ions of *m/z* 400 and 150. Point c is a daughter of *m/z* 150 from a parent of *m/z* 400**

A general problem can arise in the linking between sector instrument and OF, lying in the different ion current required by the two systems. In fact, while a sector (or Q) machine generates and analyses a continuous beam of ions, OF requires a pulse as short as possible of ions. For such reason, the research carried out by Damson and Gilhouse[4] is of high interest. It consists in the development of a new TOF mass analyser compatible with conventional continuous ionisation methods, and is based on the principle of orthogonal acceleration (oa). The TOF oa shown in Figure 3 and 4 is constituted by different components, more precisely:

    i)       a continuous ion source
    ii)     a collimating lens to form a parallel ion beam
            normal to the TOF direction
    iii)    a pulsed ion sampling (o.a. region)
    iv)    a short fixed field region
    v)     the main accelerator
    vi)    the drift region
    vii)   ion detector

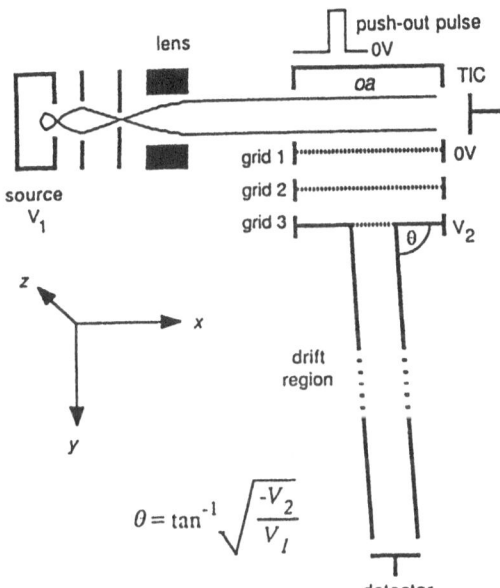

**Figure 3. Schematic diagram of TOF-oa mass spectrometer**

The orthogonal accelerator shown in figure 4 alternates between fill-up mode and push-out mode. In the former case, push-out plate and exit grid are grounded: the ions just traverse the oa region and reach the total ion monitoring. Driving the push-out plate to the suitable potential makes the ion beam reach the first grid producing a field for the first step of ion acceleration. When the last of the ions in the sample have passed through the first grid, the push-out potential can be relaxed. Such an approach can be validly employed to be mounted in a field free region of any multisector machine[5]. A possible design is reported in Figure 5

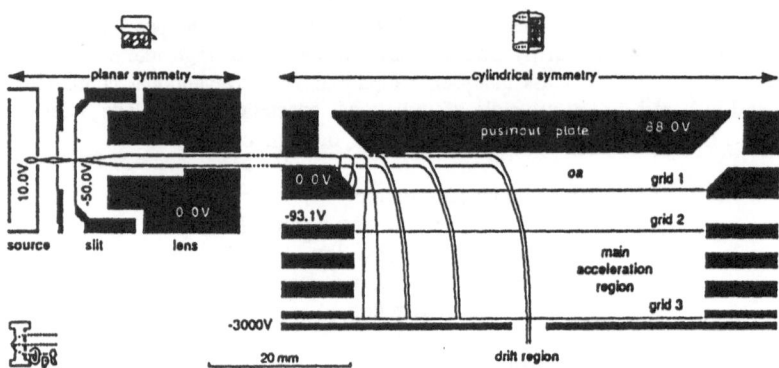

**Figure 4.** Selected ion trajectories from an EI source, through the collimating lens and into the oa. The orthogonal acceleration of the ions in the oa is shown at various points along the continuous ion beam. The trajectories were obtained with a three-stage simulation in a potential function having planar symmetry and a single-stage simulation in cylindrical symmetry. The slit width employed was 0.05 mm.

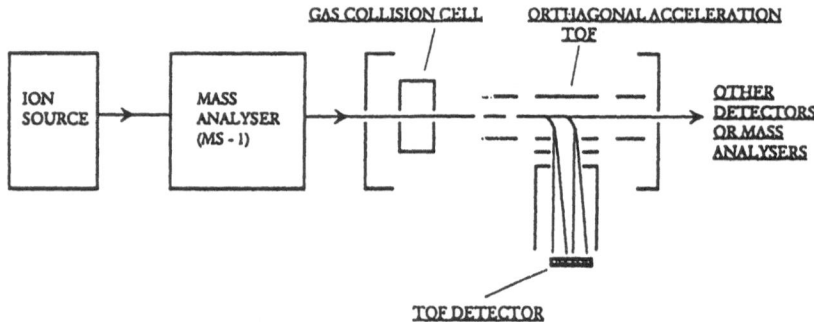

**Figure 5. Possible design of a TOF mass spectrometer**

reprinted with permission of reference [5]

Such an instrument, whilst retaining the capabilities of high resolution precursor ion selection, would be smaller, less expensive, less complex and, in particular, an order of magnitude more sensitive that a four sector instrument with array detector. A tandem quadrupole/TOF instrument has been developed and described by Glish and Goeringer[6]. The related scheme is shown in Figure 6. The sample is introduced into a thermal ion source developed at Oak Ridge. The ions are focused in a quadrupole which can work also in a r. f. only mode. The entrance and exit aperture of the quadrupole are 3 mm in diameter. After the quadrupole, the ions are focused by means of an einzel lens into a 10 mm collision cell. A four element lens accelerates the ions exiting the collision cell into a drift region, approximately 50 cm long. Since the quadrupole is operated at ground potential, the drift region needs to be electrically floated at the acceleration voltage (-700 to - 1000 V). A set of deflector is mounted on the last of the accelerating lenses to gate the ions for TOF analysis. The normal pulse width used in this study was 500 ns. Some data obtained by this approach are reported in Figure 7. The tandem mass spectra of tetraethylammonium cation is reported; a dependence on the laboratory collision energy and on the collision gas pressure is clearly evidenced.

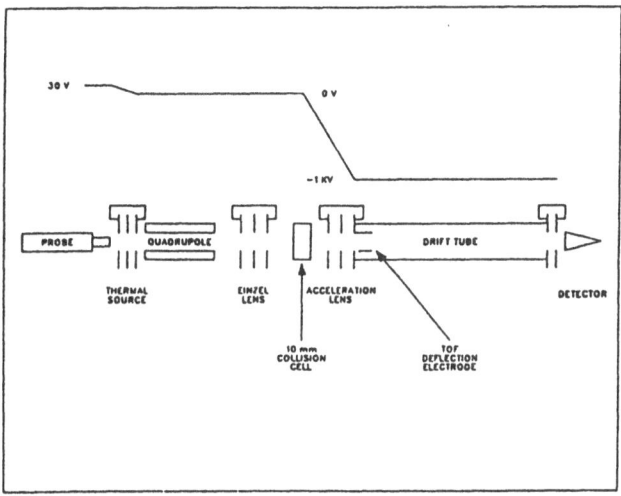

**Figure 6. Schematic of the tandem quadrupole/time-of-flight instrument and acceleration scheme used for positive ions** reprinted with permission of reference [6]

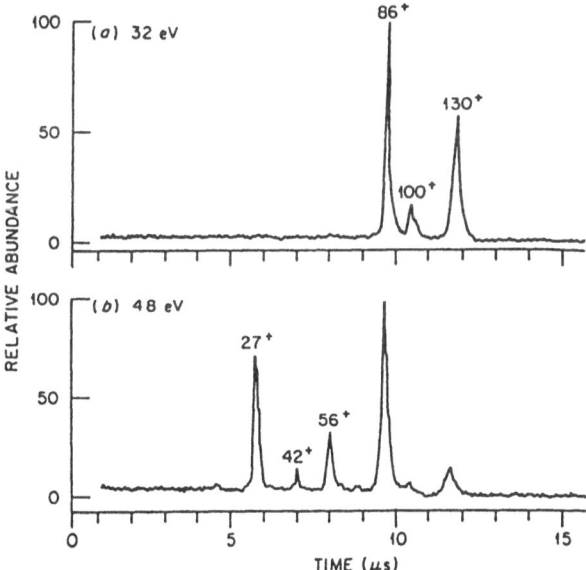

**Figure 7. MS/MS spectra of *m/z* 130 from tetramethylammonium bromide at a collision gas pressure of 5x10⁻⁵ torr.. (a) Collision energy of 32 eV; (b) collision energy of 48 eV**

reprinted with permission of reference [6]

In the same paper, the authors reported a discussion on comparison between B/TOF and Q/TOF instruments, reaching the following conclusions:

I.   Q/TOF is much more convenient for acquiring daughter ion spectra of few ions
II.  Q/TOF has a higher parent ion resolution (in B/TOF it is dependent from the kinetic energy release)
III. QTOF operates at lower CAD energies
IV. QTOF are easier to be computer-controlled.

## 3. HYBRID ION TRAPS

March and co-workers were the first to develop a EB coupling with Ion Trap[7]. In their earlier version the projectile and fragment ions were decelerated within the trap by multiple collisions with a pulsed collision gas, while in an improved version a 10 cm long collision chamber was inserted between the sector instrument and the ion trap. More recently, a hybrid mass spectrometry of BE(IT)Q geometry has been developed and described by Schlunegger et al.[8]. Its schematic representation is reported in Figure 8.The sector machine consist of a Finnigan MAT CH5 (double focusing, reverse geometry) fitted with an electron impact source with an acceleration voltage of 3Kv. The retardation system allows a deceleration from 3Kv to 5eV (laboratory frame-of-reference). The cylindrical axis of symmetry of ion trap-quadrupole combination is parallel to the incoming ion beam. The ions enter in the trap through a 1,5 mm hole in the centre of the top end-cap. Both trap and quadrupole control units are operated with a floating ground, at a potential identical to the retardation voltage. An einzel lens just before the trap is used for the effective focusing of the ion beam not the trap, while a second ion-optical element is placed between trap and quadrupole (see Figure 9 A and B)

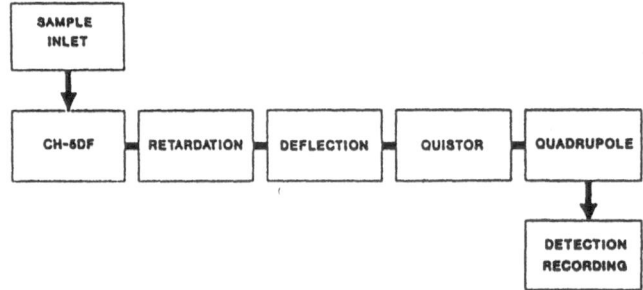

**Figure 8. Schematic representation of the BEQiQu instrument**

The deflection of the incoming ion beam, necessary to enable undisturbed ion storage in the trap, was achieved with a cylindrical electrode which was cut in half, with one half held at floating ground potential. A pulse of more than +60V was applied to the other half, which consequently deflected the ions. By such instrumental arrangement it is possible to store ions for up to 8 s after selecting the precursor with the double focusing section and reducing its translations energy to a few eV. The daughter ions formed in the trap can be easily identified. The same group developed a

Q(IT)Q system[9] which allows for complete flexibility in the choice of inlet technique, ionisation method and mass selection of ions to be stored. Once stored in the trap, ion can be subjected to further studies, e.g. ion molecule reactions, collisional induced dissociations, photodissociation, and the products of these experiments can be easily analysed by the second quadrupole. The schematic diagram of the instrument is reported in Figure 10.

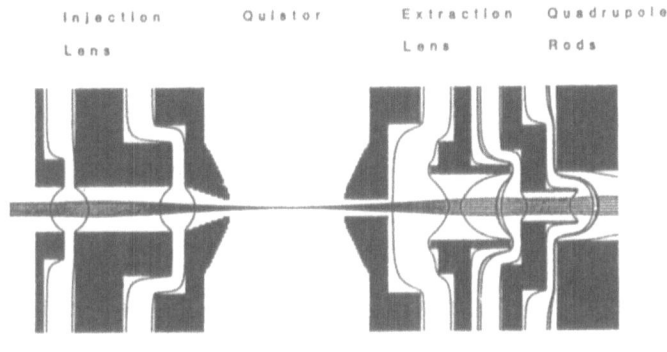

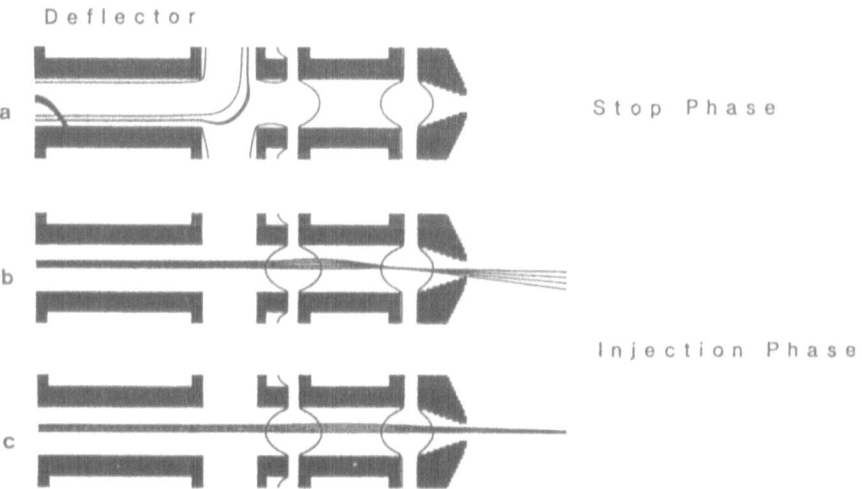

**Figure 9. (A): Injection and extraction lenses. (B): Deflection and injection phase: (a) deflection, (b) optimal storage conditions, (c) optimal transmission though the quistor**

Each part of the instrument is housed and pumped independently. Gas flow restrictions between the vacuum chambers allows a differential pressure exceeding 30000 between ion source and trap chamber to be maintained. The modularity of the system allows to perform numerous experiments. The more typical one is characterised by pulses of ionisation, storage delay and detection.

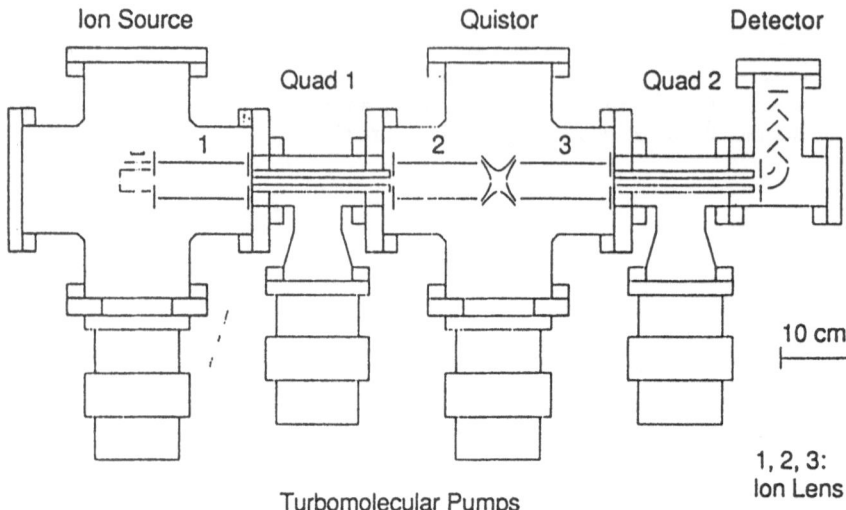

**Figure 10. Schematic diagram of vacuum system and ion optics of the QuaQuiQua tandem mass spectrometer**

During the first pulse, primary ions are generated in the ion source, selected by the first quadrupole and accumulated into the trap. The variable storage delay allows the formation of product ions in the trap. Finally, during the detection pulse, the product ions are extracted from the trap, mass analysed by the second quadrupole and detected. As a test for the instrument, the reaction $O_2^{+\cdot}$ + $CH_4$ was studied, because of its dependence on energy. It is known that for hot $O_2^{+\cdot}$ ions predominant formation of endothermic products $CH_3^+$ and $CH_4^{+\cdot}$ is observed, while for thermal ions the formation of $H_2COOH^+$ takes place. By the Q(IT)Q instrument, it was observed that the endothermic product ions ($CH_3^+$ and $CH4^{+\cdot}$) were present at short storage times only. A different arrangement of quadrupole and ion trap is that due to Cooks and co-workers[10]. Its schematic diagram is reported in Figure 11 A and was obtained by coupling a Finnigan model 4000Q with an r.f. only Q and custom built ion trap with internal radius 0.5 cm.

Ion beam focusing is achieved by three sets of einzel lenses. A typical scan sequence for spectra acquisition is shown in Figure 11 B (a) while in Figure 11 B (b) the sequence with the addition of a reaction time directly following the injection period is reported. This new hybrid instrument readily accesses to both collision activation and ion molecule reaction. The ion generation and selection external to the trap allows easy coupling with desorption ionisation techniques avoiding the interference from matrix ion. The sensitivity of the instrument was found particularly good. In comparison with the sector/trap hybrid instrument the tandem quadrupole/trap has the important advantage that ion does not need to be floated in the keV range.

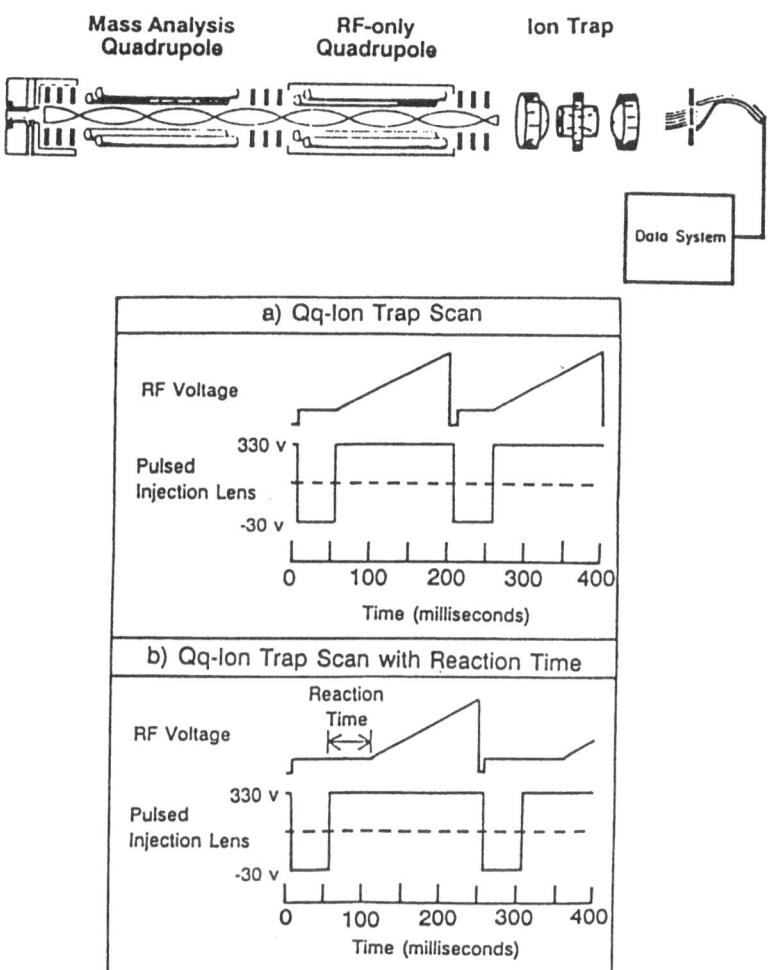

Figure 11. (A) Schematic diagram of quadrupole-ion-trap tandem mass spectrometer. (B) Scan functions used in recording mass spectra by mass-selective instability scans, without (a) and with (b) a reaction time.

**References**

1. K.L. Bush, G.L. Glish, S.A. McLuckey, *Mass Spectrometry/Mass Spectrometry. Techniques and Applications of Tandem Mass Spectrometry.* VCH Publishers, New York, 1988.
2. S.J. Gaskell, "Hybrid Tandem Mass spectrometers in Biological Research", *Biol. Mass Spectrom.* **21**, 413-419, (1992).
3. C.G. Enke, J.T. Stults, J.F. Holland, J.D. Pinkston, J. Allison, "MS/MS by Time Resolved Magnetic Dispersion Mass Spectrometry", *Int. J. Mass Spectrom. Ion Phys.* **46**, 229-232, (1983).
4. J.H.J. Dawson, M. Guilhaus, "Orthogonal-acceleration Time-of-flight Mass Spectrometer", *Rapid Commun. Mass Spectrom.* **3**, 155-159, (1989).
5. E. Clayton, R:H. Bateman, "Time-of-flight Mass Analysis of High-energy Collision-induced Dissociation Fragment Ions", *Rapid Commun. Mass Spectrom.* **6**, 719-720, (1992).
6. G.L. Glish, D.E. Goeringer, "Tandem Quadrupole/Time-of-flight Instrument for Mass Spectrometry/Mass Spectrometry", *Anal. Chem.* **56**, 2291-2295, (1984).
7. M. Ho, R.J. Hughes, E.M. Kazdan, P.J. Matthews. A.B. Young, R.E. March, in *35ᵗʰ ASMS Conference on Mass Spectrometry and Allied Topics 24-29.5.1987, Denver, Colorado,* Proceedings, 237 (1987).
8. M. J.F. Suter, H. Gfeller, U.P. Schlunegger, "A Novel Hybrid Mass Spectrometer", *Rapid Commun. Mass Spectrom.* **3**, 62-66, (1989).
9. P. Kofel, H. Reinhard, U.P. Schlunegger, "A Novel Quadrupole, Quistor, Quadrupole Tandem Mass Spectrometer", *Org. Mass Spectrom.* **26**, 463-467 (1991).
10. K.L. Morand, S.R. Horning, R.G. Cooks "A Tandem Quadrupole-Ion Trap Mass Spectrometer", *Int. J. Mass Spectrom. Ion Processes* **105**, 13-29, (1991).

# Part III
# Application to Biomolecules

CONTINUOUS-FLOW FAST ATOM BOMBARDMENT MASS SPECTROMETRY:
APPLICATIONS TO DYNAMIC SYSTEMS

RICHARD M. CAPRIOLI
*Analytical Chemistry Center and Dept. of Biochemistry & Molecular Biology*
*University of Texas Medical School*
*P.O. Box 20708*
*Houston, Texas 77225 (USA)*

ABSTRACT. Continuous flow fast atom bombardment (CF-FAB) mass spectrometry provides a means for particle bombardment analysis of aqueous solutions continuously infused into a mass spectrometer. Flow-injection analysis of biological reactions or systems are readily accommodated by this technique. This chapter describes the basic concepts and performance advantages of CF-FAB as compared to normal (standard) FABMS. Many specific applications are discussed, including direct injection of peptide mixtures obtained from protease digests. The coupling of CF-FAB for applications involving on-line liquid chromatography and capillary electrophoresis is described in some detail and is illustrated by the analysis of tryptic digest of proteins. Finally, the use of CF-FAB with microdialysis is described for the *in vivo* analysis of drugs and endogenous metabolites. In these applications, mass specific detection is used in drug pharmokinetic analysis and in the monitoring of the time-course of drugs crossing the blood/brain barrier. In conclusion, the outlook of the technique is briefly discussed.

## 1. Introduction

Continuous-flow fast atom bombardment (CF-FAB) ionization is a technique which permits the direct introduction of aqueous solutions of sample into the mass spectrometer [1]. Like FAB [2] itself, it has been employed in applications to biological molecules and has its greatest sensitivity with polar or charged molecules, especially peptides and small proteins. Indeed, the original impetus for the development of CF-FAB was to develop a means for the facile introduction of biological molecules in their natural (aqueous) environment without prior purification and derivatization. Today, a wide variety of applications in analytical biochemistry have been reported using CF-FAB including flow-injection analysis of isolated samples, mixture analysis using combined chromatography or electrophoresis and mass spectrometry, enzyme reaction monitoring, and on-line process monitoring such as *in vivo* microdialysis. This chapter will summarize the important concepts in the use of CF-FAB and will describe many different applications of the technique to analytical biochemical problems. This is meant as a tutorial chapter, not a review, and as such will illustrate concepts and

*R. M. Caprioli et al. (eds.), Mass Spectrometry in Biomolecular Sciences, 273–297.*
© 1996 *Kluwer Academic Publishers.*

applications with only a few examples selected from many publications in the literature. For more extensive discussion and references, the reader is referred to a more detailed monograph [3].

## 2. Basic Concepts

The term CF-FAB has now come to include several variations of design of a direct insertion probe, or similar device, where an aqueous-based solution is continuously pumped onto a sample stage in the source of a FAB mass spectrometer. The first use of the term CF-FAB, also sometimes called dynamic FAB, described a system for flow-injection analysis [1]. A second design, termed frit FAB, was first reported as an LC/MS interface [4]. A typical CF-FAB direct insertion probe is shown in Figure 1, equipped with an injection valve for direct sample introduction. The constant infusion of an aqueous solvent into a high vacuum

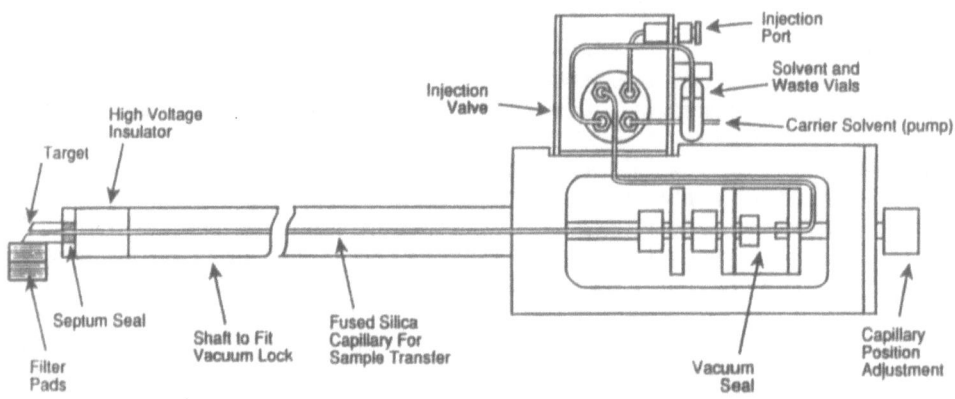

Figure 1. A continuous-flow FAB probe design with attached sample injection valve. (Reprinted with permission from Reference 25).

chamber must be done carefully to achieve slow, steady evaporation of the solvent without disruptive 'boiling', a situation that develops when a liquid droplet forms, as shown in Figure 2A. Stable operations can be achieved in several ways. First, in the case where the capillary (usually made of fused silica) is left open depositing liquid directly on the target, gentle heating of the target is used to balance flow and evaporation. In addition, a fiber wick can be used to absorb excess liquid from the surface which, because of its great surface area, provides a more even rate of evaporation (see Figure 2B). This design shows the greatest improvements in lowering ion suppression and background, and gives the least tailing of injected samples, but takes experience to use routinely. Another design terminates the capillary with a fine frit, which acts to disperse the liquid and provide a sample stage (see Figure 2C). Stable evaporation, and thus stable operation of the device, is easier to achieve, although peak tailing, higher background, and limitation to lower flow rates are generally the consequence. A third design, a compromise between the two, uses a coarse screen to give

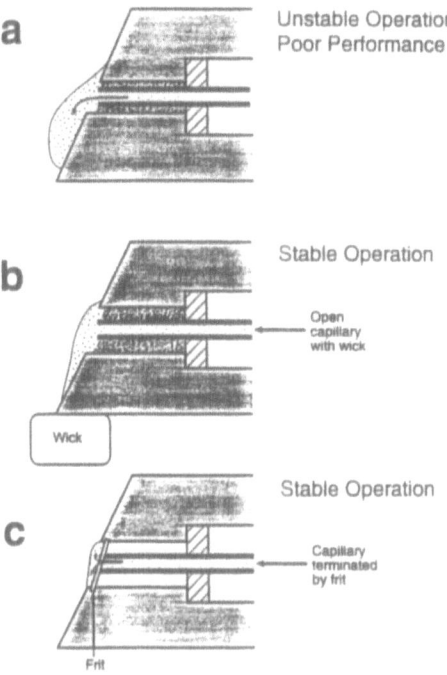

**a**      Unstable Operation
Poor Performance

**b**      Stable Operation

Open
capillary
with wick

Wick

**c**      Stable Operation

Capillary
terminated
by frit

Frit

Figure 2. Target conditions on the CF-FAB probe tip that gives (a) unstable, and (b) and (c), stable performance, Panel (b) illustrates the open capillary/wick design, and (c) the frit design.

some dispersion of the liquid but is sufficiently open to allow free flow of the capillary. All of these designs work well, enhancing the performance of FAB analysis, with the particular choice of one design over another mainly dictated by the availability from the manufacturer of the mass spectrometer. Collectively, these variations will simply be termed CF-FAB for purposes of this chapter and this is generally consistent with recent terminology in the literature.

Flow rates used with CF-FAB can vary depending on the particular design of the probe, but usually lie in the range of 0.5-10 µL/min. Fused silica capillaries of about 75 µm (i.d.) are used to deliver liquid and are chosen because of their flexibility and electrical insulation properties, essential for high voltage sources. Smaller inner diameter capillaries can be used, but suffer the disadvantage of being more problematic to handle and use because they become plugged easily. In applications to LC/MS, the flow rate range of CF-FAB couples best with packed capillary bore columns (300 µm i.d. or less). Microbore columns (1 mm i.d.) can also be used effectively but require splitting the effluent about 4:1 because these columns are typically eluted with at least 25-30 µL/min of eluant to achieve reasonable separation efficiencies. Applications of CF-FAB to LC/MS will be discussed in some detail later in this chapter.

Glycerol, thioglycerol, triethanolamine, nitrobenzyl alcohol, as well as other organic modifiers, are added to the carrier solvent at 1-5% (v/v) in order to obtain optimal

276

performance. The choice of one of these matrix materials over another depends upon the nature of the analyte, but most often glycerol or thioglycerol give satisfactory performance. These viscous liquid materials probably operate by providing a miscible hydrophilic surface layer on the metal target surface and also forming a liquid seal between the capillary and the target. The latter prevents solvent from becoming trapped and vapor from disruptively exiting from confined areas of the target tip as the solvent evaporates.

## 3. Performance Advantages

The use of CF-FAB, relative to normal (static) FAB, offers several important performance

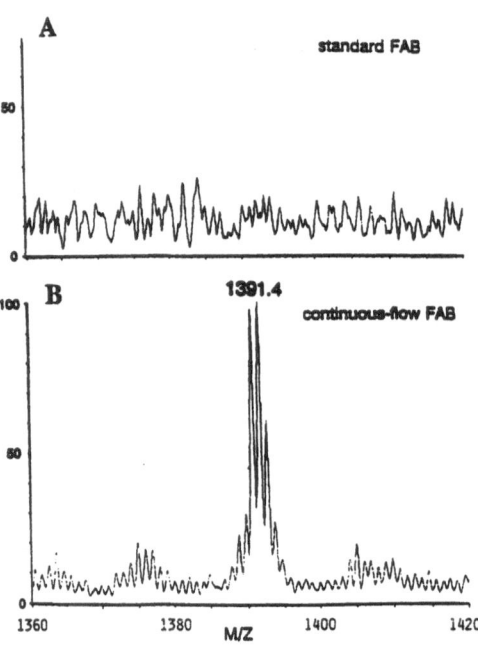

Figure 3. Comparison of the molecular ion $(M+H)^+$ regions from an octapeptide using standard FAB and CF-FAB.

advantages which can be extremely important in a given analysis. One of the major advantages is significantly better sensitivity [5]. For peptides, this has been shown to come largely as a result of a decrease in the background signals derived mostly from the high percentage matrix used for std-FAB, and to a lesser extent from increased ion yields from aqueous solvents. At analyte levels in the low picomole range or less, the increase in the limit of detection can be over 100 fold. In some cases where CF-FAB produces a high quality spectrum, no signal can be seen by static FAB. For example, Figure 3 shows mass spectra of an octapeptide where 20 pmol total was analyzed by CF-FAB and 30 pmol by std-FAB. Although this is an unusual case, this hydrophilic peptide was not present in the surface layers in the std-FAB glycerol droplet and therefore was not detected. In contrast,

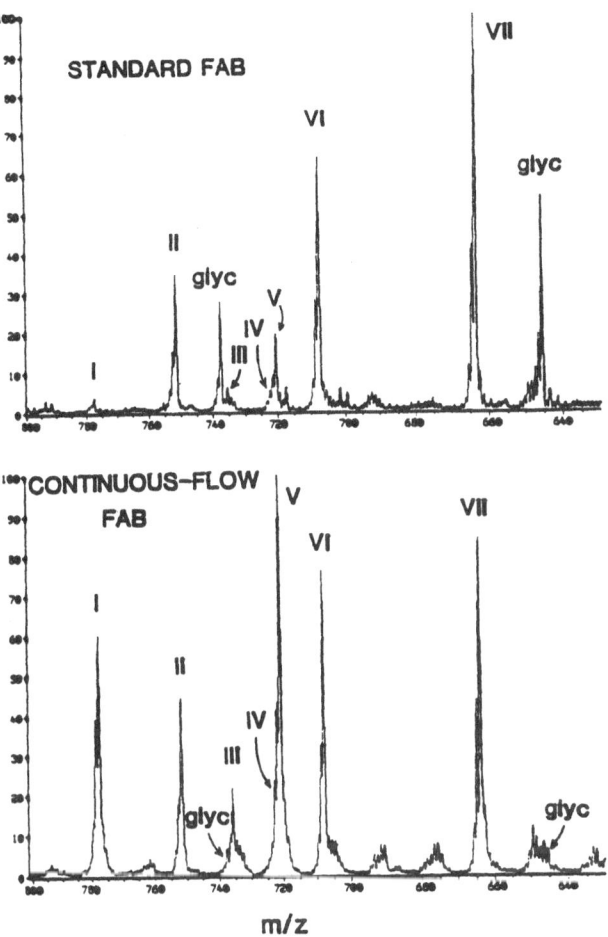

Figure 4. Comparison of spectra obtained from a mixture containing both hydrophilic and hydrophobic peptides using standard FAB and CF-FAB. (Reprinted with permission from Reference 6)

the mechanical mixing of the flowing solvent and high water content in CF-FAB provided the proper conditions to produce an $(M+H)^+$ ion of rather high intensity.

Similarly, compounds in a mixture can be analyzed more effectively by CF-FAB because the ion suppression effect is greatly diminished [6]. This effect results in the decreased sensitivity of one compound in the presence of one or more other compounds in the sample. Figure 4 shows the analysis of a few picomoles of a peptide mixture, where the peptide $(M+H)^+$ ions are labeled I-VII. For std-FAB, peptides I, III, IV, and V show significant suppression relative to the others, and two glycerol cluster ion peaks of high intensity are seen at m/z 645 and 737. For the analysis of the mixture by CF-FAB, all peptide $(M+H)^+$ ion signals are recorded at relatively high intensity, while the glycerol peaks are significantly reduced in intensity.

Another advantage of CF-FAB is the temporal relationship of the sample signal relative to the background level. A sample that is injected into the continuous-flow solvent appears as a peak in time, much like that of a chromatographic peak. This is clearly seen in Figure 5A for a peak produced from the flow-injection analysis of substance P. Mass spectra can be obtained before and after that of the sample, permitting efficient background subtraction, as shown in Figure 5B. In addition, the flow-injection of a series of samples from a batch

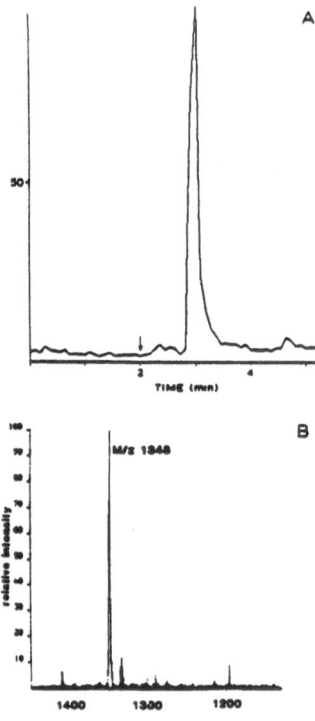

Figure 5. Analysis of substance P by flow-injection CF-FAB, showing (A) the injection profile and (B) the upper region of the mass spectrum containing the $(M+H)^+$ ion at m/z 1348. (Reprinted with permission from Reference 1).

process, such as an enzyme reaction, for example, provides much better quantitative comparison between samples than could be obtained by std-FAB. For example, Figure 6 shows consecutive injections of aliquots removed from the tryptic hydrolysis of angiotensin at m/z 1046. These peak areas can be directly compared to measure the kinetics of the hydrolysis, a feat not easily done with std-FAB. This is due to the fact that in CF-FAB, the position of the probe remains untouched and the fluid dynamics remain constant while consecutive samples are injected, while for std-FAB, repositioning of the probe from sample to sample is required and obtaining exactly the same sample droplet size and consistency is nearly impossible.

In comparison to newer ionization techniques such as electrospray and matrix-assisted laser

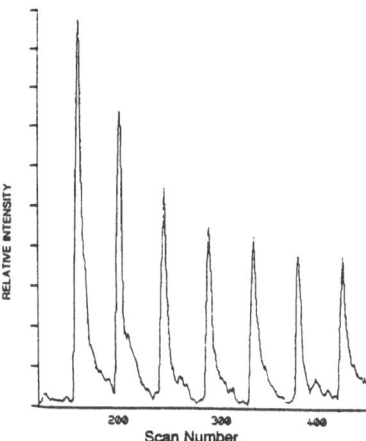

Figure 6. The injection profile for the time-course analysis of the digestion of angiotensin II with trypsin, monitoring the $(M+H)^+$ ion at m/z 1046.

desorption, FAB is useful only in a relatively low mass range, up to about m/z 5000-6000. Although several publications report successful analyses beyond this molecular weight, it cannot be considered routine. Generally, CF-FAB has similar limitations. In one study of the high mass performance of CF-FAB [7], it was shown that peptides having molecular weights in the 3000-4000 range gave about a ten-fold increase in signal-to-noise over that obtained by std-FAB. However, this is instrument dependent. Above approximately m/z 5000, CF-FAB, like std-FAB, is not generally effective.

## 4. Applications

One of the advantages of CF-FAB is that it is able to accept samples from dynamic systems i.e., where analyte concentrations are changing over time. These include devices such as liquid chromatographs and electrophoresis instruments where sample peaks are emerging in solvent flow. Such systems can also involve chemical and enzymic reaction processes, and integrated biological processes such as the *in vivo* monitoring of live animals using microdialysis. Solvent systems from pure aqueous to pure organic can be used, often together, for example, as needed in LC reverse phase gradient elution chromatography. Finally, the technique is compatible with moderate levels (about 50-100 mM) of many buffers, such as tris and acetate, although certain salts such as phosphates should be avoided.

The following sections describe some selected applications in specific modes of operation of the CF-FAB technique. The examples are illustrative and references are provided for those seeking additional details.

### 4.1. DIRECT ANALYSIS

Many samples are subjected to direct analysis by mass spectrometry because it is easy and less complicated than LC/MS. For single component samples and simple mixtures, direct analysis can be effective as well as fast. However, one must always keep in mind, especially in complex mixtures, that some compounds may be preferentially ionized, and the gas phase ion mixture may not reflect the liquid phase molecular mixture.

Typically, for the direct analysis of a sample by flow injection, 0.5-10 μL of sample solution in an aqueous or organic solvent is injected into a carrier solvent flowing at about 2-5 μL/min. For example, Figure 7 shows the CF-FAB direct injection analysis of the tryptic

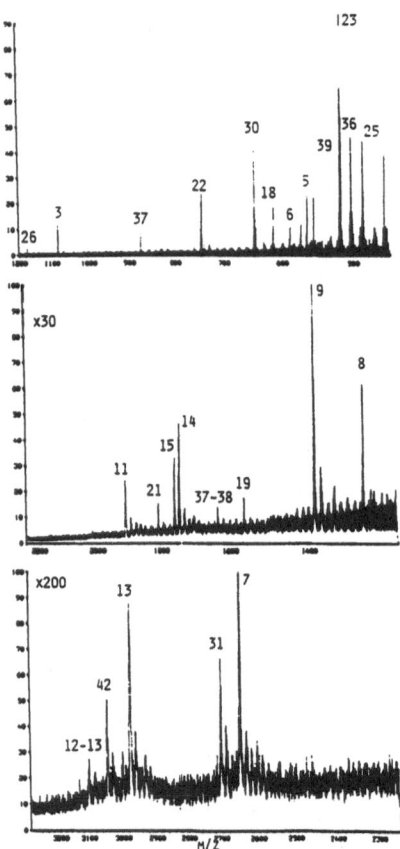

Figure 7. Direct injection of 24 pmol of the tryptic digest of the recombinase A protein and analysis using CF-FAB. The tryptic fragments identified, beginning from the N-terminus, are labeled for each (M + H)⁺ ion. Details are given in the text.

hydrolysis of 26 pmol of the recA protein (mw 37,000) [8]. Twenty-six tryptic fragments were quickly identified, representing about 70% of the fragments expected within the mass range recorded. The other fragments were either suppressed in the ionization process or were further hydrolyzed by contaminant protease activity. In a similar study of the tryptic hydrolysis of human recombinant growth hormone [9], direct injection CF-FAB analysis detected 10 of the 16 peptides expected in the mass range recorded. In contrast, the same sample analyzed by LC/CF-FAB methods identified 15 tryptic peptides. Clearly, if a complete identification of a complex sample is required, direct injection by CF-FAB, or any other ionization technique, is not recommended. However, if the analytical task involves a target compound or a survey analysis of a complex mixture, the ease of operation, time-saving and simplicity of direct injection makes the technique quite attractive when suitable analytical conditions are employed.

Direct analyses of tryptic digests of small peptides by CF-FAB are quite efficient. In one study [10], the hydrolysis of eel calcitonin (mw 3416) was followed with time in order to identify peptides formed during the course of the reaction, although some of these were only transient and were further digested. Peptides produced during the time-course of the hydrolysis can be identified as nearest neighbors because the sum of their molecular weights (obtained from the $(M+H)^+$ values measured) minus 18 mass units for water, add up to that of their precursor. Figure 8 shows a composite mass spectrum for this hydrolysis, obtained by adding spectra during the 15 min reaction. Peptides at m/z 2312 and 2718 are transient

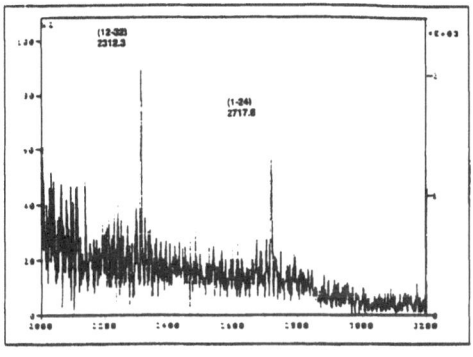

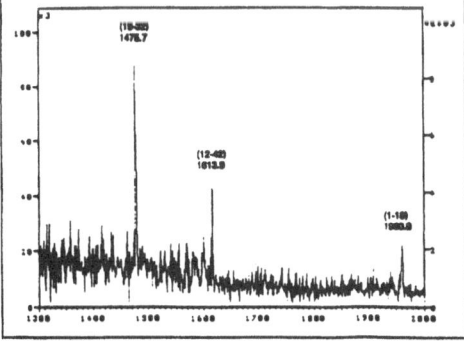

Figure 8. Analysis of the tryptic digest of eel calcitonin, a 42-residue polypeptide, by trypsin under conditions giving a low rate of reaction in order to observe transient peptide fragments. (Reprinted with permission from Reference 10).

in that they are further digested to smaller tryptic peptides, e.g., m/z 2312 → m/z 855 + m/z 1476. A series of these precursor/product correlations provides overlap information so that a nearest neighbor map of the final tryptic (limit digest) peptides can be constructed, giving the specific sequence order if one terminus is known. In this case, the four final tryptic peptides could be ordered from the N-terminus, (1123-855-778-717).

Many other types of compounds have been analyzed by direct injection, including nucleotides [11], bile acids [12], leukotrienes [13], drugs of abuse [14], therapeutic drugs [15], air and moisture sensitive compounds [16], as well as many others. Although most of these applications utilized aqueous carrier solvent, pure organic solvents such as toluene [16] can also be used.

A number of investigators have used CF-FAB with flow-injection analysis in experiments which have utilized MS/MS instrumentation. In one case, quantitative determination of the levels of platelet-activating factor (PAF) in cell extracts were performed through the use of stable isotope labeled standards and selected ion monitoring by MS/MS [17]. Collisionally-activated decomposition was employed and the principal fragmentations of hexadecyl-PAF and its tri-deutero analog (as internal standard) were measured, i.e., m/z 524 → 184 and m/z 527 → 185, respectively. The results showed that hexadecyl-PAF was present in amounts of about 2.6 nanograms per million cells. Other applications include detection of peptides in mixtures [18], carotenoids [19], and investigation of side-products of synthetic peptide production [20].

## 4.2. LC/MS

One of the most powerful techniques at the disposal of the bioanalytical chemist is the combination of mass spectrometry with separations techniques such as LC and CZE. Advantages accrue because the analysis is mass specific over a relatively wide mass range, provides facile sample handling to save both time and effort, allows very small amounts of sample (for example, femtomolar amounts) to be analyzed with minimal loss, and largely eliminates the need to derivatize compounds. Basically, combining LC and MS is truly a symbiotic process; the LC, in fractionating samples in time, allows a higher ionization efficiency for many compounds, and the MS provides a mass domain which allows compounds to be 'separated' even though overlapping in the time domain.

Commercially available CF-FAB probes operate best at flow rates below 10 μL/min and so coupling to LC is best when using capillary bore (100-300 μm i.d.) and microbore (1 mm i.d.) columns. The latter may require effluent splitting, of from 3:1 to 5:1, since microbore columns can be eluted at 25-50 μL/min and provide satisfactory performance. A full-bore column normally is eluted at 0.5-1 ml/min and therefore would require extensive effluent splitting (100:1) for use with CF-FAB and is practical only where sample amounts are not a limiting factor.

A typical LC/CF-FAB instrumental set-up is shown in Figure 9 based on combination with a microbore LC system [21]. Two splitters are used in this system, one before the sample injector so that solvent mixtures for gradient elutions can be run at effective levels (500 μL/min) while allowing only about 25 μL/min to flow into the microbore column, and a splitter after the column to allow about 5 μL/min of the effluent to flow onto the CF-FAB target. The remaining sample (20 μL/min) can be directed to a UV detector and, if desirable, into a fraction collector. The transfer line from the column to the CF-FAB interface should be of small i.d., typically 50-75 μm, to keep transfer volumes as low as

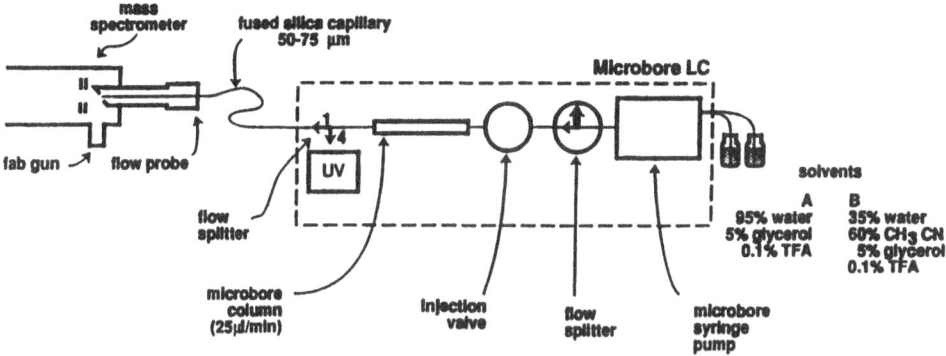

Figure 9. Instrumental set-up for microbore LC/CF-FAB mass spectrometry. Buffers are labeled for a typical analysis of a peptide mixture. (Reprinted with permission from Reference 21).

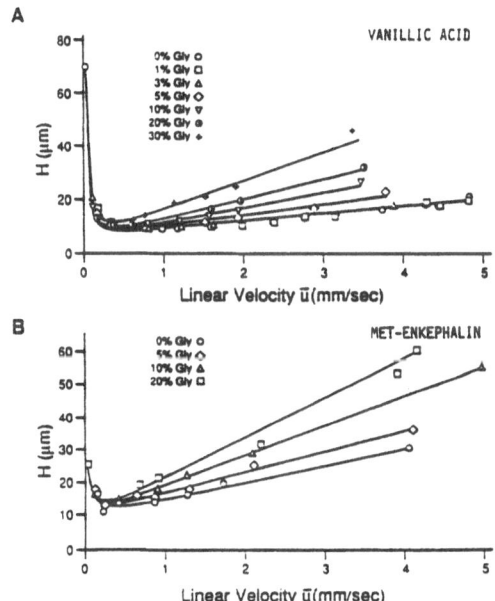

Figure 10. Effect of glycerol concentrations on the separation efficiencies of vanillic acid and methionine enkephalin using a microbore C-18 column. Theoretical plate height (H) is plotted versus linear velocity of the mobile phase. (Reprinted with permission from Reference 22).

possible. A transfer line of inner diameter smaller than 75 μm would be better in this regard, although plumbing connections and clogging of capillaries becomes significant problems. The viscous organic matrix, in many cases, can be added directly to the elution buffers without serious deterioration of chromatographic performance. This is the case, for example, for separations of small peptides in tryptic digests using reverse phase chromatography with a C-18 column and acetonitrile gradients. However, for some separations and column types, addition of matrix to the buffers can compromise separation efficiency. The effect of glycerol in this regard was investigated for several types of compounds, including vanillic acid, Met-enkephalin, and p-hydroxybenzoic acid [22]. The increase in the normalized peak width for met-enkephalin was about 15% for a glycerol content of 5% in the elution buffer (1-2% glycerol is normally used in CF-FAB). The increase in the plate height (H) versus linear velocity (ū) was measured for several glycerol concentrations and are shown in Figure 10. The decreased separation efficiency is attributed to the higher viscosity of the mobile phase.

The post-column addition of FAB matrix may be accomplished using a co-axial interface design [23]. This technique is depicted in Figure 11 and essentially consists of two concentric capillaries. The inner capillary transports the column effluent and the outer transports the matrix solution so that after mixing, a solution containing 1-5% of matrix is deposited on the target. The advantage of the coaxial interface is that chromatographic resolution is maintained, although at the cost of additional complexity in employing an additional pump and technical manipulation of the coaxial arrangement. For packed capillary columns, it has been shown that the capillary can be brought to within 2-3 mm of the probe tip inside the ion source before vacuum induced flow becomes a problem.

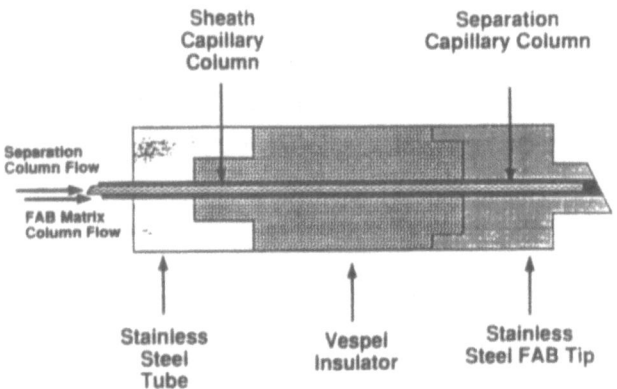

Figure 11. Coaxial design for the CF-FAB probe. (Reprinted with permission from Reference 23).

One of the continuing strengths of CF-FAB in the current era of electrospray ionization and matrix-assisted laser desorption ionization is the simplicity of the mass spectrum. Usually, $(M+H)^+$ or $(M-H)^-$ ions are produced although for positive ion spectra, cation addition products such as $(M+Na)^+$ or $(M+K)^+$ can be seen if these cations are present in sufficiently high concentrations in the sample. The molecular species is thus rather easy to identify in the spectrum since there is almost always one charge state, the singly-charged species. This is quite clearly seen in Figure 12 for the analysis of the tryptic digest of human apolipoprotein AI. The total ion chromatogram is shown in the upper panel, where over 40 peptides could easily be identified in this LC/MS analysis of the crude enzyme digest. The mass spectrum obtained from less than 24 pmol of peptide is shown for one fragment. The $(M+H)^+$ molecular species is clearly identified at m/z 1013 with excellent signal-to-noise in this crude and complex digest.

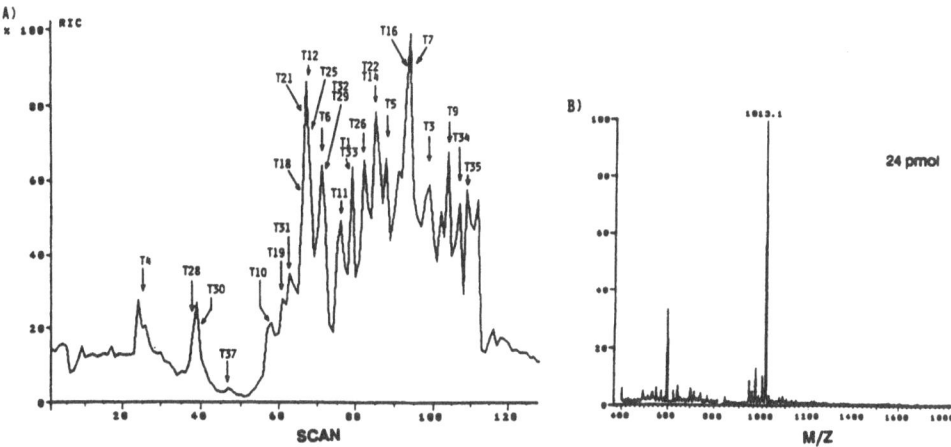

Figure 12. Analysis of the tryptic digest of human apolipoprotein AI by LC/CF-FAB. (A) total ion chromatogram with tryptic fragments labeled, and (B) mass spectrum of fragment nonapeptide with $(M+H)^+$ at m/z 1013.

Peptide mapping by LC/MS has become an extremely useful method and its speed and accuracy are major attributes. One example, nicely illustrating the power of LC/CF-FAB is the analysis of the tryptic digest of $\beta$-casein [24]. Figure 13 compares the UV chromatogram

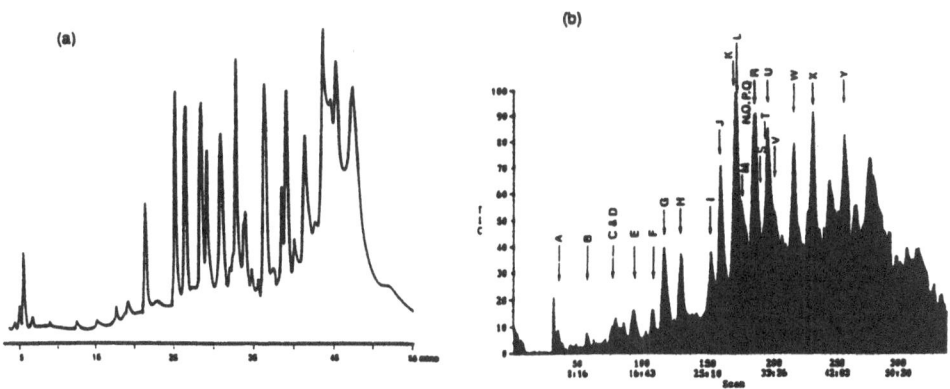

Figure 13. Analysis of the tryptic digest of $\beta$-casein by LC/CF-FAB MS. (A) UV and (B) total ion chromatograms. Peptide fragments are identified in Figure 14. (Reprinted with permission from Reference 24).

with the total ion chromatogram for this digest. As expected, although there is a general correlation, the intensities are distinctly different owing to the fact that the two detectors respond quite differently: the UV detector at 200 nm responds in a way so that the greater the number of peptide bonds the larger the signal, while the MS usually gives higher signal intensities at lower molecular weights. Figure 14 shows the selected ion chromatograms for this analysis, and the utility of LC/MS becomes immediately obvious. This particular analysis took about 45 min, and identified all of the fragments produced by their molecular weights.

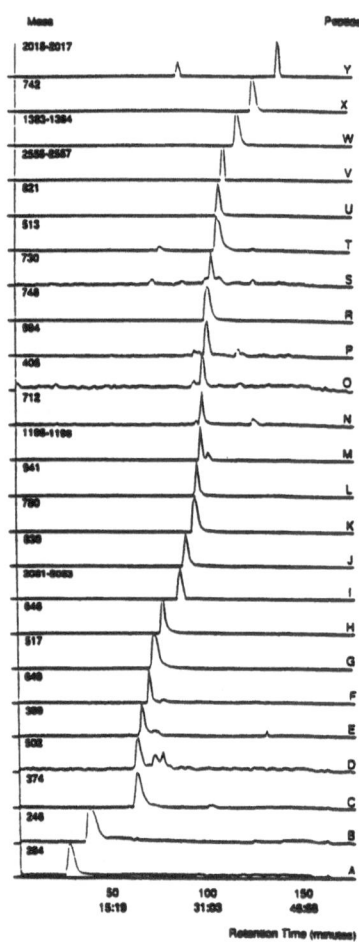

Figure 14. Selected ion chromatograms for the LC/MS analysis of β-casein. (Reprinted with permission from Reference 24).

Although peptides have been involved in many of the early LC/MS investigations, many other types of molecules have also been shown to be amenable to LC/CF-FAB analysis, including drugs and drug metabolites, glycopeptides, oligosaccharides, bile salts, glycosphingolipids and gangliosides, oligonucleotides, prostaglandins, food contaminants, and many more. The reader is referred to several more recent reviews for references and discussions of this work [25, 26]. In addition, other chapters in this volume cover topics relevant to LC/MS.

## 4.3. CE/MS

Capillary electrophoresis (CE) has recently emerged as one of the promising separations techniques, particularly well suited for combination with a mass spectrometer detector. The technique achieves separation of analytes through a combination of the charge on the molecule and its hydrodynamic volume, providing a separation process different from that of LC. CE is a nanoscale technique capable of very high separation efficiencies, often achieving theoretical plate estimates of greater than $10^6$. Briefly, an electric field is applied within a capillary filled with an electrolyte in which the analytes of interest migrate according to their electrophoretic mobilities. The charges species can cause an electroosmotic flow towards one of the electrodes, and the magnitude and direction of this flow depends on conditions such as pH, chemical nature of the buffer, and the charge on the capillary wall. Normally, with uncoated capillaries, the positive electrode (anode) is the sample loading end and the negative or ground electrode (cathode) is at the MS end. A typical instrumental set-up for CE/CF-FAB is shown in Figure 15. Samples can be loaded at the anode either hydrodynamically using slight pressure or electrophoretically. Electrophoretic loading is gentle and is capable of loading very small volumes but may discriminate against some charged analytes in the sample. Hydrodynamically all species are loaded, but the process is mechanical and is more prone to disrupting the smooth flow of the capillary.

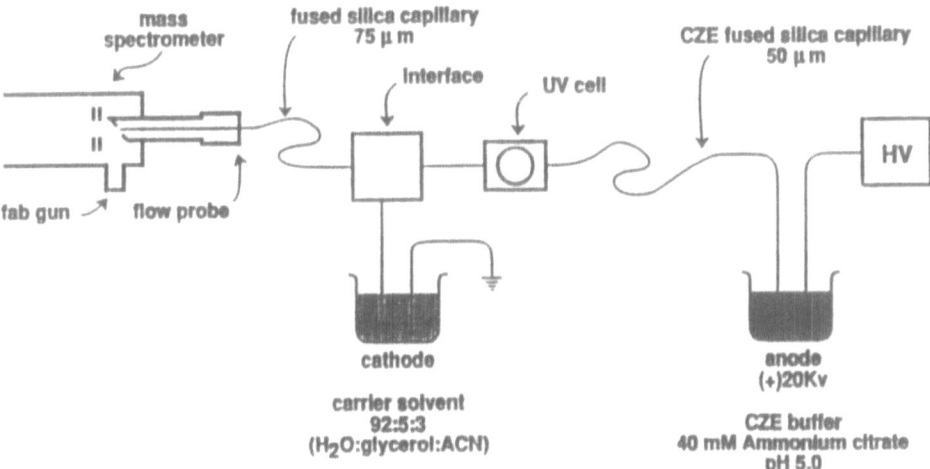

Figure 15. Instrumental set-up for CE/CF-FAB MS. (Reprinted with permission from Reference 28).

The interface between the CE capillary and the mass spectrometer source is critical and two different types have been shown to be successful, the coaxial and liquid junction interfaces. Schematic representations of these interfaces are shown in Figure 16. Each has particular advantages and drawbacks. Initial reports of the coupling of CF-FAB and CE utilized the liquid junction interface [27, 28]. It is relatively easy to set up, allows a UV

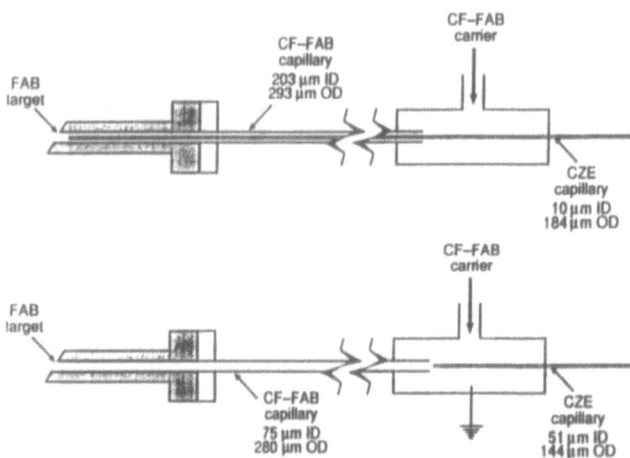

Figure 16. Interface arrangements for CE/CF-FAB MS; A) coaxial and B) liquid junction.

detector cell to be positioned at the end of the CE capillary, and easily accommodates discontinuous buffer systems (described below). A major disadvantage is that peak broadening occurs because the sample band from the CE capillary is transferred to the flow-probe capillary probe for delivery to the target. The coaxial interface more closely maintains the integrity of the sample band because the CE capillary terminates very close to the target. However, this interface is technically more difficult to set up and operate routinely, and does not allow a UV detector cell to be placed near the end of the CE capillary. A compromise design uses a liquid junction reservoir with a partial coaxial arrangement for more efficient band transfer [29]. The amount of coaxial overlap of the capillaries is adjustable and can be optimized for individual analytical tasks.

For applications involving mass spectrometry as the detection system for a CE separation, the use of discontinuous buffer systems can be advantageous, especially when using buffers having high concentrations $Na^+$, $K^+$, etc, (> 1 mM). This is shown nicely in Figure 17 for the CE/MS analysis of one of the tryptic fragments of cytochrome c [28]. The top part of the figure shows the mass spectrum of the peptide in a 10 mM tricine buffer containing 50 mM NaCl for both the CE and interface buffers. The spectrum shows $(M+H)^+$ at m/z 674, $(M+Na)^+$ at m/z 696, and $(M+Na_2)^+$ at m/z 718. In another experiment, a discontinuous buffer system was employed in which the CE buffer was composed of 10 mM tricine with 40 mM NaCl, and the interface 'buffer' was 5% aqueous glycerol with 0.5% heptafluorobutyric acid to provide electrical conductivity. In this case, only the $(M+H)^+$ ion was observed, simplifying the spectrum and improving signal-to-noise.

Many of the first applications of CE/MS involved the analysis of small peptides and mixtures of peptides using 20-50 μm i.d. fused silica capillaries. Buffer systems such as 50 mM acetate (pH 6) containing 50% acetonitrile have been shown to give good performance for protease digests of small proteins. Also, sensitivities are quite high, with sub-femtomole detection in the best cases. Peptides behave well in CE systems because of their different chemical properties, lack of adsorption to the capillary wall and elution with simple volatile

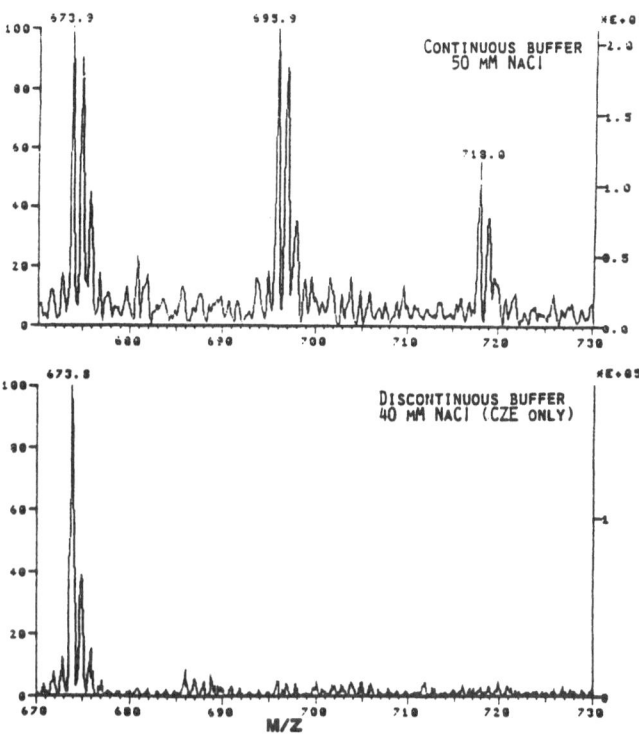

Figure 17. Molecular species recorded from the CE/CF-FAB MS analysis of a tryptic peptide using (A) continuous and (B) discontinuous buffer systems. Details are given in the text. (Reprinted with permission from Reference 28).

buffer systems. Although a number of reports of the CE/MS analysis of proteins have been published, these have employed electrospray mass spectrometry because FAB ionization is limited to compounds of molecular weight 5000 or less in its practical application.

With most current instrumentation, spectra are obtained by scanning a magnetic or quadrupole analyzer, i.e., one mass at a time is recorded. For many applications where unknowns are being analyzed, a wide mass range needs to be scanned. This presents some difficulty for current CE/MS instruments because electrophoretic peaks can be quite narrow in time, i.e., less than 1 sec wide. Mass spectrometer scan times of a few seconds per scan for a wide mass range is not unreasonable, especially when dealing with low levels of compounds. Simultaneous ion detection is therefore preferred. For magnetic analyzers, array detectors can be used although they are quite expensive. New types of analyzers that provide partial simultaneous ion detection such as time-of-flight and ion trap instruments may be preferable for use with high resolution CE separations.

### 4.4. MICRODIALYSIS / MS

Microdialysis (Md) was first reported about 10 years ago for application to *in vivo* brain chemistry [30]. Its major advantage is that it allows samples to be taken in relatively inaccessible places, e.g., a substructure of brain or other tissue, without loss of fluid from the tissue because in using isotonic perfusion fluids, there is only a net passage of analytes from the tissue into the Md probe. Once a sample is taken, a number of analytical instruments can be used to detect and identify compounds, such as liquid chromatography, electrophoresis, radioimmunoassay, radioreceptor assay, electrochemical detectors, etc. Recently, mass spectrometry has been employed [31]. The advantages of coupling Md and MS are much the same as those for combining MS with other separations processes, as described earlier in this chapter. The mass specificity advantage plays an even bigger role here because of the truly complex mixtures of metabolites present in the dialysates obtained from most tissues.

One of the common designs for a Md probe is a concentric arrangement, shown in Figure 18, which consists of a small hollow dialysis tube whose tip is enclosed with a dialysis membrane [30]. Typically, the o.d. of the probe is 0.5 mm or less and the length of the

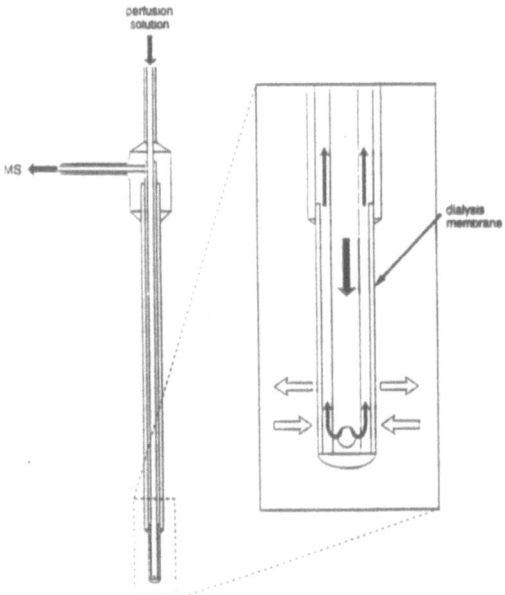

Figure 18. Microdialysis probe design. (Courtesy Carnegie Medicin, Uppsala, Sweden).

dialysis region is 1-5 mm. The analytes in the tissue can freely diffuse into, and out of, the dialysis tubing, with net flow down the concentration gradient. The inner reservoir of the Md probe is constantly perfused with an aqueous isotonic salt solution so that analytes are swept out of this chamber into a holding loop or other collection device. An advantage of this arrangement is that some analytes or metabolites that might be acted on by enzymes in the tissue are effectively isolated from these processes since the membranes have molecular weight cut-offs of 5000-10,000, or less, depending on the membrane material. Thus, one can obtain a sample for a time-slice out of a dynamic system in which concentrations of metabolites are changing over time. In another use of the Md probe, drug delivery to specific regions of tissues can be accomplished by injecting a dose of a drug or compound into the perfusate before it enters the Md probe.

In all experiments involving Md where even relative amounts or semi-quantitative analyses are required, it is important to precisely control the perfusion flow rate. The concentration of a compound in the perfusate is proportional to its concentration in the extracellular space of the tissue, and the amount of the compound recovered is termed 'relative recovery'. Several factors effect relative recovery, including perfusion flow rate, the surface area of the membrane, the diffusion rate of the substance, and the properties of the membrane. The relative recovery of small molecules increases as the perfusion rate decreases. Typically, one could expect a relative recovery of about 15% at perfusion rates of 5-6 $\mu$L/min and 30-50% or more at about 0.5 $\mu$L/min. In some experiments, the term 'absolute recovery' is used and refers to the total amount of substance that is removed by the perfusion medium during a specific time period. The absolute recovery of a substance increases with increasing perfusion flow rate up to about 5-10 $\mu$L/min and tends to remain constant thereafter.

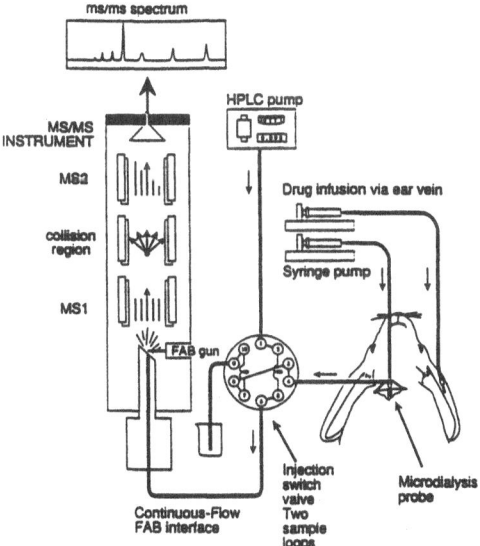

Figure 19. Experimental arrangement for on-line *in vivo* microdialysis/CF-FAB MS/MS analysis of drugs infused intravenously.

The use of Md/MS for the *in vivo* analysis of drugs and endogenous metabolites was first reported in experiments designed to measure the pharmacokinetics of penicillin in rats and rabbits [31]. In one case, a Md probe was inserted into the jugular vein of a rabbit and penicillin G was administered intravenously, as shown in Figure 19. The Md probe was perfused at about 2 µL/min with sterile water containing 5% glycerol (isotonic perfusate was not needed in this case) and the perfusate allowed to flow into a 5 µL loop of an injection valve, with samples taken at 10 min intervals. These samples were then flow-injected into a CF-FAB MS/MS system for analysis. The collision-induced dissociation fragment ion from penicillin G at m/z 192 was monitored to construct the pharmacokinetic curve, which agreed well with similar curves produced using classical methods. The benefit of MS/MS is considerable here because of the extremely high numbers of metabolites present in blood perfusate. The remarkable cleanup capability of MS/MS is shown in Figure 20, where the negative ion MS/MS spectrum of penicillin G obtained from blood dialysate from rabbit and rat are compared to that of a solution of pure drug. It should be pointed out that rigorous quantitative determination of the concentration of a drug in plasma is difficult because recovery must be accurately measured and only free drug, not protein bound drug, is able to diffuse into the Md probe. Several recent papers have examined the calibration and quantitative aspects of the use of Md in experiments requiring calculations of tissue levels of compounds [32, 33].

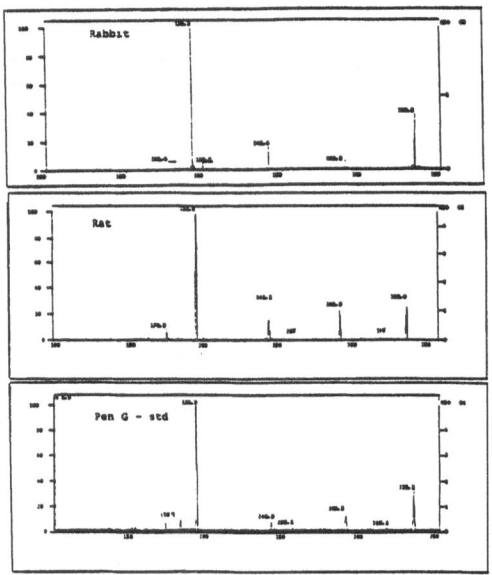

Figure 20. MS/MS spectra of penicillin G in dialysates from experimental *in vivo* drug infusion studies in rabbit and rat. The spectrum of the pure drug is shown in the bottom panel.

A second example of Md/MS methodology for *in vivo* studies utilized two probes simultaneously to monitor the kinetics of valproic acid (VPA), an anti-epileptic drug, in both brain and blood [34]. CF-FAB/MS was used with a switching valve to analyze dialysate from both probes. Figure 21 shows the single ion monitoring trace at m/z 143 ((M+H)$^+$ for VPA)

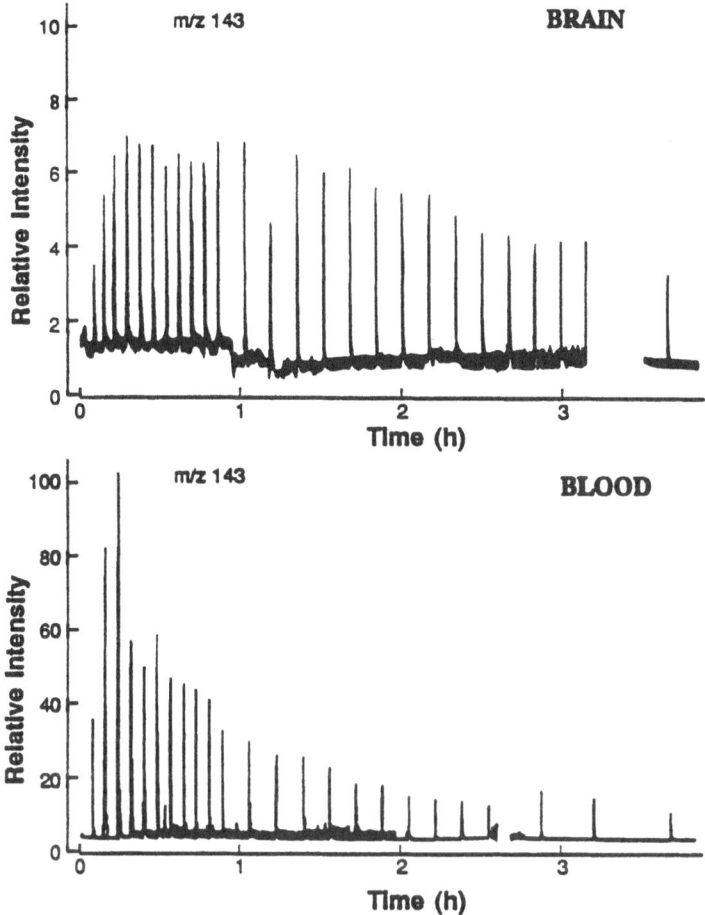

Figure 21. Relative blood and brain levels of valproic acid administered IV to a rat, using double-probe microdialysis/CF-FAB/MS. The ion at m/z 143 is $(M+H)^+$. (See text for details).

after injection of 50 mg/kg was administered intravenously to a 250 g rat. The data clearly show passage of the drug through the blood/brain barrier and its subsequent rise and fall in both dialysates. Brain levels were about one-tenth that of blood, and elimination of the drug from brain was considerably slower than that from blood.

Other reports of the use of Md/MS have appeared in the literature, and the reader is referred to a recent review of this subject for additional references and details [35].

## 5. Conclusions

CF-FAB is a sensitive mass spectrometric ionization technique that can be used for the analysis of polar and charged molecules up to a molecular weight of about 5000. It has been used effectively with many types of compounds and retains the rather wide diversity of application established by the FAB technique. However, relative to the standard FAB method, CF-FAB has several particular advantages in that it has significantly greater sensitivity, less ion suppression effects, better salt tolerance, improved background substraction capability, excellent relative quantitative comparison sample to sample, and high sample throughput. The technique is directly compatible with low flow rate techniques such as LC, CE, microdialysis, batch analysis through aliquot sampling, flow injection analyses, etc. These advantages come at the expense of a more complex technical set-up in that a low flow rate pump and capillary bore tubing are needed to maintain a constant flow of carrier liquid onto the target inside the mass spectrometer.

Although new ionization processes have been introduced over recent years and are quite powerful, FAB and CF-FAB ionization continue to play a major role in the analysis of biological samples. The ability of CF-FAB to directly analyze aqueous sample solutions remains one of its major assets.

## 6. Acknowledgement

The author wishes to thank the National Institutes of Health (grant #GM43783-05) for partial financial support.

## 7. References

1.  R. M. Caprioli, T. Fan, and J. S. Cottrell, "Continuous-Flow Sample Probe for Fast Atom Bombardment Mass Spectrometry", *Anal. Chem.* **58**, 2949-2954, (1986).

2.  M. Barber, R. S. Bordoli, R. D. Sedgewick, "Fast Atom Bombardment of Solids (F.A.B.): A New Ion Source for Mass Spectrometry", *J. Chem. Soc. Chem.* 325, (1981).

3.  R. M. Caprioli, "Continuous-Flow Fast Atom Bombardment Mass Spectrometry", in Biologically Active Molecules, (U. Schlunegger, ed.), Springer-Verlag, Berlin, 59-77, 1989.

4.  Y. Ito, T. Takeuchi, D. Ishii, and M. Goto, "Direct Coupling of Micro HPLC with Fast Atom Bombardment Mass Spectrometry", *J. Chromatogr.* **346**, 161-166, (1985).

5.  R. M. Caprioli and T. Fan, "High Sensitivity Mass Spectrometric Determination of Peptides: Direct Analysis of Aqueous Solutions", *Biochem. Biophys. Res. Commun.* **141**, 1058-1065, (1986).

6.  R. M. Caprioli, W. T. Moore, G. Petrie, and K. Wilson, "Analysis of Mixtures of Hydrophilic Peptides by CF-FAB MS", *Int. J. Mass Spectrom. Ion Proc.* **86**, 187-199, (1988).

7.  K. B. Tomer, J. R. Perkins, C.E. Parker, and L. J. Deterding, "Coaxial CF-FAB for Higher Molecular Weight Peptides: Comparison with Static FAB and ESI", *Biol. Mass Spectrom.* **20**, 783-788, (1991).

8. B. Whaley and R. M. Caprioli, unpublished work.

9. W. J. Henzel, J. H. Bourell, and J. T. Stults, "Analysis of Protein Digests by Capillary High-Performance Liquid Chromatography and On-Line Fast Atom Bombardment Mass Spectrometry", *Anal. Biochem.* **187**, 228-233, (1990).

10. B. Whaley and R. M. Caprioli, "Identification of Nearest-neighbor Peptides in Protease Digests by Mass Spectrometry for Construction of Sequence-ordered Tryptic Maps", *Biol. Mass Spectrom.* **20**, 210-214, (1991).

11. R. B. van Breemen, L. B. Martin, C. Le, "Continuous-Flow Fast Atom Bombardment Mass Spectrometry of Oligonucleotides", *J. Am. Soc. Mass Spectrom.* **2**, 157-163, (1991).

12. A. Lawson, "Atom Bombardment Method Offers Improved Analysis of Bile Acids", *C&EN*, 20-21, (1989).

13. M. J. Raftery, G. C. Thorne, R. S. Orkiszewiski and S. J. Gaskell, "Analysis of Leukotrienes Using Continuous Flow Fast Atom Bombardment/Tandem mass Spectrometry", *Proceedings of the 37th ASMS Conference on Mass Spectrometry and Allied Topics*, p. 760-761, Miami Beach, Florida, May 21-26, 1989.

14. W. E. Seifert, Jr., A. Ballatore, and R. M. Caprioli, "Direct Analysis of Drugs by Continuous-flow Fast-atom Bombardment and Tandem Mass Spectrometry", *Rapid Commun. Mass Spectrom.* **3**(4), 117-122, (1989).

15. M.P. Knadler, B. L. Ackermann, J. E. Coutant, and G. H. Hurst, "Metabolism of the Anticoagulant Peptide, MDL 28,050, in Rats", *Drug Metabolism and Disposition* **20**(1), 89-95, (1992).

16. M. Barber, D. J. Bell, M. Morris, L. W. Tetler, M. Woods, G. A. Gott, P. P. MacRory, C. A. McAuliffe, "Mass Spectrometric Studies of Air/Moisture Sensitive Compounds by CF-FAB", *Proceedings of the 36th ASMS Conference on Mass Spectrometry and Allied Topics*, p. 545-546, San Francisco, CA, June 5-10, 1988.

17. S. J. Gaskell and R. S. Orkiszewski, "Trace Analysis", in Continuous-Flow Fast Atom Bombardment Mass Spectrometry, (R. M. Caprioli, ed.) John Wiley & Sons Ltd., Sussex, UK, 29-43, 1990.

18. B. L. Ackermann, J. E. Coutant and T-M Chen, "Incorporation of Tandem Mass Spectrometric Detection to the Analysis of Peptide Mixtures by Continuous Flow Fast Atom Bombardment Mass Spectrometry", *Biol. Mass Spectrom.* **20**, 431-440, (1991).

19. R. B. van Breemen, H. H. Schmitz, and S. J. Schwartz, "Continuous-Flow Fast Atom Bombardment Liquid Chromatography/Mass Spectrometry of Carotenoids", *Anal. Chem.* **66**, 965-969, (1994).

296

20. W. T. Moore, "Integration of Mass Spectrometry into Strategies for Peptide Synthesis", *Biol. Mass Spectrom.* **22**, 149-162, (1993).

21. R. M. Caprioli, B. B. DaGue and K. Wilson, "Optimization of Chromatographic Conditions for Combined Microbore HPLC/Continuous-Flow Fast-Atom Bombardment Mass Spectrometry", *J. of Chromatographic Science* **26**, 640-644, (1988).

22. J-P Gagné, A. Carrier and M. J. Bertrand, "Source of Band Broadening in Liquid Chromatographic-Fast Atom Bombardment Mass Spectrometric Systems with Precolumn Addition of Viscous Matrix to the Mobile Phase", *J. of Chromatog.* **554**, 47-59, (1991).

23. M.A. Moseley, L. J. Deterding, J. S. M. de Wit, K. B. Tomer, R. T. Kennedy, N. Bragg, and J. W. Jorgenson, "Optimization of a Coaxial Continuous Flow Fast Atom Bombardment Interface Between Capillary Liquid Chromatography and Magnetic Sector Mass Spectrometry for the Analysis of Biomolecules", *Anal. Chem.* **61** 1577-1584, (1989).

24. D. S. Jones, W. Heerma, P. D. van Wassenaar and J. Haverkamp, "Analysis of Bovine β-Casein Tryptic Digest by Continuous-flow Fast-atom Bombardment Mass Spectrometry", *Rapid Commun. Mass Spectrom.* **5**, 192-195, (1991).

25. R. M. Caprioli and W. T. Moore, "Continuous-Flow Fast Atom Bombardment Mass Spectrometry", in Methods in Enzymology, (ed. J. A. McCloskey), Academic Press, Inc., San Diego, CA, 214-237, 1990.

26. R. M. Caprioli, "Design and Operation", in Continuous-Flow Fast Atom Bombardment Mass Spectrometry, (ed. R. M. Caprioli), John Wiley and Sons Ltd., Sussex, UK, 1-27, 1990.

27. R. D. Minard, P. D. Curry, Jr., and A. G. Ewing, *Proceedings of the 37th ASMS Conference on Mass Spectrometry and Allied Topics*, Miami Beach, FL, May 21-26, 1989.

28. R. M. Caprioli, W. T. Moore, M. Martin, B. B. DaGue, K. Wilson and S. Moring, "Coupling Capillary Zone Electrophoresis and Continuous-Flow Fast Atom Bombardment Mass Spectrometry for the Analysis of Peptide Mixtures", *J. of Chromatog.* **480**, 247-257, (1989).

29. M.J.-F. Suter and R. M. Caprioli, "An Integral Probe for Capillary Zone Electrophoresis/Continuous-Flow Fast Atom Bombardment Mass Spectrometry" *J. Am. Soc. Mass Spectrom.* **3**, 198-206, (1992).

30. U. Ungerstedt, "Measurement of Neurotransmitter Release by Intracranial Dialysis", in Measurement of Neurotransmitter Release In Vivo, (ed. C. A. Marsden), John Wiley & Sons Ltd., Sussex, UK, 81-105, 1984.

31.  R. M. Caprioli and S.-N. Lin, "On-line analysis of penicillin blood levels in the live rat by combined microdialysis/fast atom bombardment mass spectrometry", *Proc. Natl. Acad. Sci.* **87**, 240-243, (1990).

32.  J. B. Justice, Jr., "Quantitative microdialysis of neurotransmitters", *J. of Neuroscience Methods* **48**, 263-276, (1993).

33.  S. Menachery, W. Hubert, and J. B. Justice, Jr., "*In Vivo* Calibration of Microdialysis Probes for Exogenous Compounds", *Anal. Chem.* **64**, 577-583, (1992).

34.  S.-N. Lin, J. M. Slopis, P. Andrén-Johansson, S. Chang, I. J. Butler, and R. M. Caprioli, *Proceedings of the 39th ASMS Conference on Mass Spectrometry and Allied Topics*, p. 583-584, Nashville, TN, May 19-24, 1991.

35.  P. E. Andrén and R. M. Caprioli, "Microdialysis/Mass Spectrometry of Neuropeptides", in Biological Mass Spectrometry - Present and Future, (eds. T. Matsuo, R. M. Caprioli, M. L. Gross, and Y. Seyama), John Wiley & Sons Ltd., Sussex, UK, 355-367, 1994.

# HYBRID TANDEM MASS SPECTROMETRY OF PEPTIDES

SIMON J. GASKELL
*Department of Chemistry*
*University of Manchester Institute of Science and Technology*
*P.O. Box 88*
*Manchester M60 1QD*
*U.K.*

ABSTRACT. Tandem mass spectrometric analyses of peptides using hybrid sector/quadrupole instruments may incorporate low energy collisionally activated decomposition (CAD), in the rf-only multipole region, or high energy CAD, in a field-free region of the sector component. Low energy CAD of peptides is strongly influenced by the site of charge and the extended time scale is conducive to the observation of prominent rearrangement processes. High energy CAD provides data of complementary value (though such data are recorded with poorer sensitivity than may be achieved with four-sector instruments). Sequential MS ($MS^3$ and $MS^4$) may be performed with hybrid instruments, permitting the elucidation, for example, of peptide fragmentation pathways.

## 1. Introduction

This chapter reviews the application of hybrid tandem mass spectrometry to the structure elucidation of peptides. In this context, the meaning of "hybrid" will be restricted to those instruments in which a double-focusing mass spectrometer (of forward or reverse geometry) is coupled to an rf-only quadrupole (or other multipole) and a quadrupole mass filter. The principles and applications of hybrid instruments have been reviewed previously [1-3].

Figure 1 shows a schematic diagram of a hybrid tandem instrument of BEqQ configuration (where B = magnetic sector, E = electric sector, q = rf-only quadrupole, and Q = quadrupole mass filter). Source-formed ions are accelerated to kinetic energies of (typically) 6-10 keV and are separated in the magnetic and electric fields. Selected precursor ions are decelerated prior to entry into the quadrupole portion of the instrument, commonly to kinetic energies of 10-100 eV. The majority of tandem MS applications of hybrid instruments of this type involve collision-induced decompositions occurring in q so that collision energies (10 -100 eV in the laboratory frame-of-reference) are equivalent to those generally used on a triple quadrupole instrument (though higher collision energies have occasionally been used for collisional activation in q; see Section 4). Accordingly, collision energies in the center-of-mass frame of reference ($E_{cm} = E_{lab} \times m_t/(m_t + m_p)$) are typically < 4 eV for a peptide ion of m/z 1000, using argon as the collision gas. Fragmentation pathways of peptides observed under these conditions are discussed in Section 2.

*R. M. Caprioli et al. (eds.), Mass Spectrometry in Biomolecular Sciences, 299–315.*
© *1996 Kluwer Academic Publishers.*

Additional modes of MS/MS analysis may be implemented on hybrid instruments. The double-focusing portion of the instrument may be used alone to investigate decompositions in the field-free region between the ion source and the first sector. At this point the high ion kinetic energies allow collision energies (laboratory frame-of-reference) in the keV range. The familiar modes of linked scanning of B and E [4] may be used to generate spectra of product ions from a selected precursor (B/E fixed), precursor ions of a selected product ($B^2/E$ fixed) or product ions derived from various precursors by loss of a common neutral fragment ($B^2(1-E)/E^2$ fixed). All such scan modes suffer from poor effective resolution of precursor ions. On instruments incorporating a reverse geometry (BE) component, mass-analyzed ion kinetic energy spectrometry (MIKES) analyses may also be performed; in this instance, the decompositions observed are those occurring (spontaneously or following collisional activation) in the field-free region between B and E. Precursor ion resolution is satisfactory but the mass resolution of product ions is poor. The latter may, however, be improved by utilizing the mass resolving capabilities of the quadrupole filter. The specific attributes of hybrid instruments in the study of high energy CAD of peptide ions are discussed in Section 3.

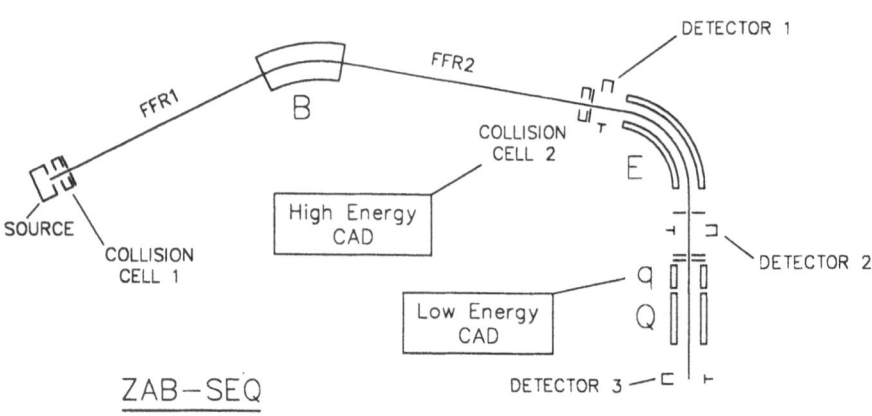

Figure 1. Schematic diagram of a hybrid mass spectrometer of BEqQ geometry (the VG ZAB SEQ): B, magnetic sector, E, electric sector; q, rf-only quadrupole; Q, quadrupole mass filter; FFR1, first field-free region; 2FFR, second field-free region.

The incorporation in hybrid instruments of several decomposition regions and of three ion analyzers (magnetic sector, electric sector and mass-resolving quadrupole) allows for the possibility of $MS^3$ and $MS^4$ experiments. These operational modes and their application to the elucidation of peptide fragmentation pathways are discussed in detail in Section 4.

## 2. Low energy decompositions of peptides

This Section reviews hybrid tandem MS analyses which have contributed to our understanding of the spontaneous and low energy collision-activated decompositions of peptide ions. Differences between hybrid and triple quadrupole instruments with respect to the selection of the precursor ions mean that decomposition processes occurring in the rf-only quadrupoles cannot be *assumed* to be identical in the two cases. Nevertheless, the gas phase ion chemistry is expected to be qualitatively similar and the peptide decompositions observed are indeed generally the same. The most prominent low energy fragmentation processes are within the following categories:

(i) Simple cleavages of the peptide backbone, with or without hydrogen migration.

(ii) Formation of "internal" fragment ions as a result of two cleavages of the peptide backbone; such ions include immonium ions derived from single amino acid residues.

(iii) Rearrangement ions (other than simple hydrogen migrations).

Although product ions in each of these categories are commonly observed, the understanding of the fragmentation processes remains insufficient to predict with confidence the relative significance of the different cleavages for particular peptide structures. Much of the data, however, is consistent with the general principle that low energy fragmentatons are directed by the site of charge. Bursey and coworkers evaluated the low energy decompositions of simple protonated peptides (observed using a hybrid instrument) and discussed their data in the context of the predicted proton affinities of sites within the peptide structure [5,6]. Thus, for example, substitution of a proline residue in pentaalanine promoted peptide bond cleavage N-terminal to the proline residue, consistent with the greater basicity of the proline N [6]. More complex considerations apply to homopolymers of amino acid residues lacking a basic side-chain [5]. The basicities of the amide bonds and of the N-terminal amino group are modified by hydrogen bonding and "internal solvation" [7].

Burlet et al. [8] explicitly evaluated the effects of intra-ionic interactions (reflecting specific gas-phase conformations) in a study of protonated peptides which incorporated C-terminal arginine and mid-chain cysteic acid residues. Clear series of Y"-type ions were observed. In contrast, the product ion spectra derived from the cysteine-containing analogues were dominated by the $Y_1$" ion (Figure 2). Removal of the C-terminal arginine residue abolished the difference in properties between cysteine- and cysteic acid-containing analogues. It was proposed that the presence of the cysteic acid residue resulted in an intra-ionic acid/base interaction which effectively reduced the propensity for charge sequestration by the arginine residue [8]. The consequent heterogeneity of the precursor ion population with respect to the site of protonation resulted in the observation of a multiplicity of cleavage sites. In the absence of the cysteic acid residue, charge localization on the arginine residue promoted proximal fragmentation affording the $Y_1$" ion.

Direct evidence for the influence of charge site on low energy peptide fragmentations has been obtained by study of the decompositions of pre-charged derivatives. Such derivatives have been advocated for the improvement of sensitivity in FAB MS analyses and useful high energy CAD product ion spectra have been reported [9]. Figure 3 shows a comparison of product ion spectra obtained following low energy CAD of a protonated heptapeptide and of a pre-charged cationic derivative [10]. The spectrum obtained for the derivative (Figure 3b) was uninformative, being dominated by a fragmentation proximal to the site of the charged derivative group. Useful fragmentation information was restored when the protonated pre-charged derivative (a doubly charged precursor ion) was analyzed by low energy CAD (Figure

302

3c); it was proposed that the proton provided an additional driving force for fragmentation and that the precursor ion population was heterogeneous with respect to the site of protonation [10].

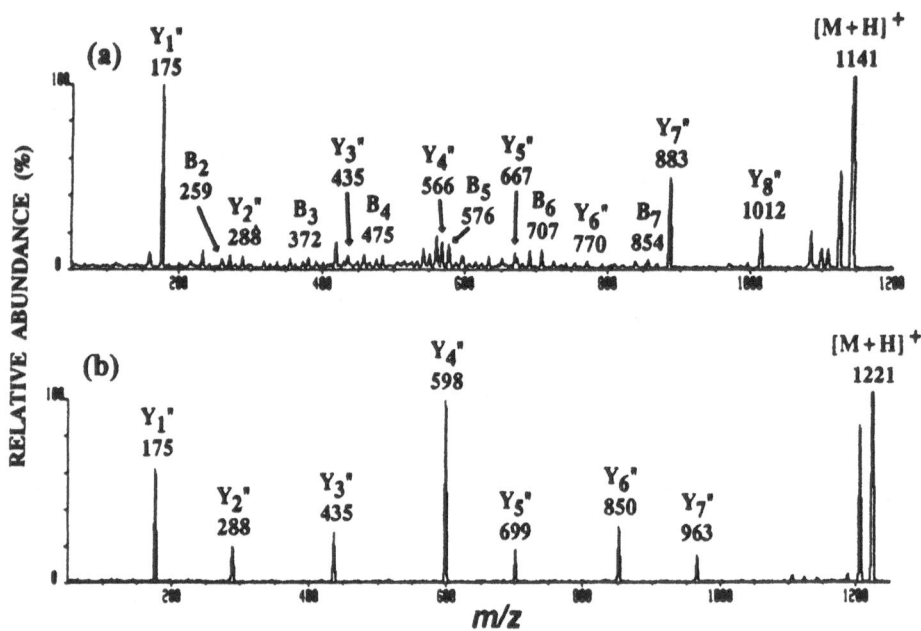

Figure 2. Low energy CAD product ion spectra of the [M+H]⁺ ions of (a) native EELCTMFIR at m/z 1141, and (b) EELC*TM*FIR at m/z 1221 (where the conventional single letter code is used to designate unmodified amino acids; C* is cysteic acid and M* is methionine sulfoxide). Reproduced with permission from reference [8].

More complex peptide fragmentation processes (categories (ii) and (iii), above) are promoted by the conditions of decomposition in the rf-only quadrupole. Thus, multiple low energy collisions promote multi-stage decompositions leading to internal fragment ions. These processes are effectively studied by sequential MS and are summarized in Section 4.

The time-scale of decompositions of low kinetic energy precursor ions occurring in rf-only quadrupoles is conducive to the observation of rearrangement processes. The most widely studied and reported rearrangement is that resulting in the loss of the C-terminal amino acid residue, with retention of a hydroxyl moiety from the C-terminal carboxyl group. This process was originally reported as a decomposition of alkali metal-cationized peptides [11-13] but is also a prominent fragmentation of some protonated peptides [14]. Though the mechanisms of rearrangement of the protonated and metal cationized peptides are presumably similar, some differences are suggested by different propensities to rearrangement shown by [M+H]⁺ and [M+Na]⁺ ions derived from the same peptide [14]. Detailed mechanistic investigations of C-terminal rearrangements have been performed using the sequential MS (MSⁿ) capabilities of a BEqQ hybrid instrument; they are discussed in Section 4.

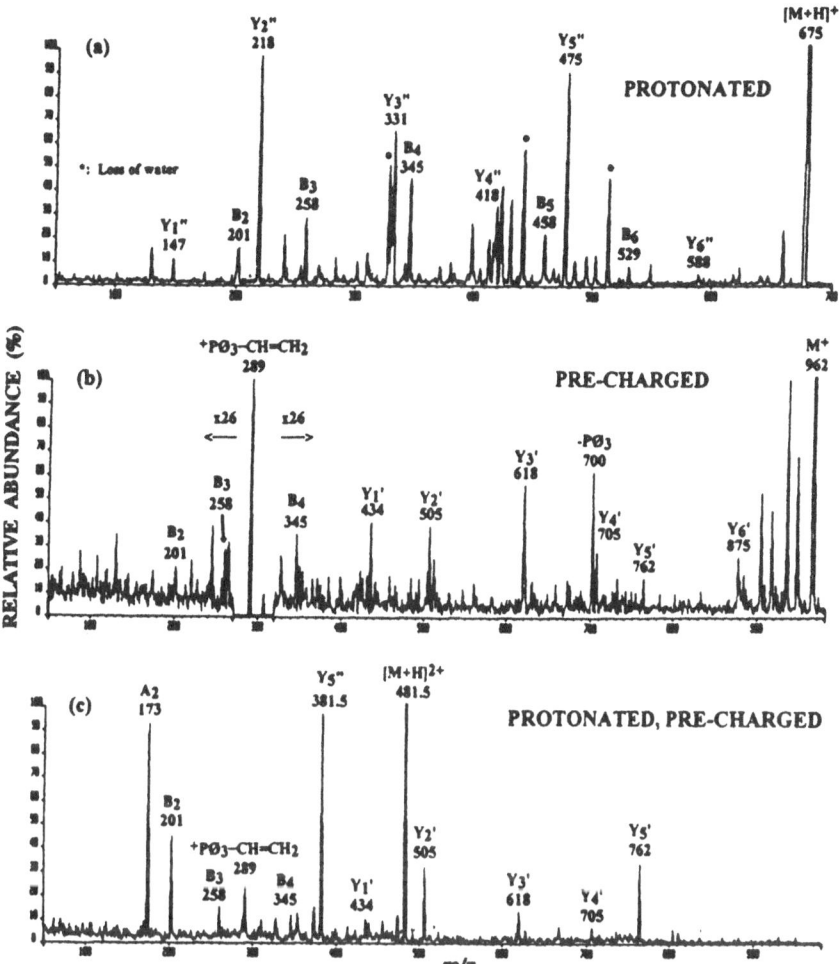

Figure 3. Product ion spectra recorded following low energy CAD of precursors derived from the heptapeptide, SIGSLAK. (a) the protonated peptide, [M+H]⁺, m/z 675; (b) the pre-charged C-terminal ethyl-triphenylphosphonium derivative of the amide analogue, M⁺, m/z 962; (c) the protonated, pre-charged derivative, [M+H]²⁺, m/z 481.5. m/z ratios are rounded down to the nearest integral or half-integral value. Reproduced with permission from reference [10].

## 3. High energy decompositions of peptides

High energy CAD ($E_{lab}$ in the keV range) of protonated peptides using four-sector instruments has demonstrated structurally informative fragmentations of amino acid residue side-chains, permitting, for example, the distinction between leucine and isoleucine [15]. These observations have provided the incentive for the exploration of the use of hybrid instruments in a manner in which the same fragmentations may be observed. Alexander et al. [16] pointed out that an increase in the rf amplitude in q enables efficient transmission of higher kinetic energy ions (as required for higher energy collisional activation in q). They demonstrated that w-type ions, resulting from amino acid side-chain cleavage, could be detected using $E_{lab}$ values of 300 eV (Figure 4). A similar approach was adopted by Bradley et al. [17] who compared CAD of a protonated peptide of m/z 1137 using hybrid and four-sector instruments. The use with the hybrid of argon collision gas and an $E_{lab}$ of 400 eV was equivalent, with respect to $E_{cm}$, to the four-sector experiment using helium and an $E_{lab}$ of 4 keV. Broadly similar data were obtained, including the observation of side-chain cleavage products. The four-sector data were, however, judged to be superior, perhaps reflecting a more efficient conversion of collision energy to internal energy when helium was used as the collision gas, but no doubt also resulting from the different time-scales of the decomposition experiments.

An alternative approach to the performance of high energy CAD on hybrid instruments is to make use of the collision cell in the first field-free region in both EBqQ and BEqQ instrument types, with the BEqQ providing the additional possibility of second field-free region CAD. The basic modes of operation (linked scanning and MIKES analyses) are derived from the double-focusing sector component of the hybrid instrument with the consequent, well-recognized limitations on precursor or product ion resolution. Mass resolution of product ions formed in the second field-free region may, however, be improved by simultaneous scanning of E and Q; the mass filtering of the energy-separated product ions provides mass resolution defined by Q. The experiment requires accurate tracking of the energy and mass scan functions, together with synchronized scanning of several float potentials. The practical difficulties may be somewhat alleviated either by superimposing a modulation on the electric sector scan [18] or by accumulating data through multiple scans of E with simultaneous but asynchronous scanning of Q [19]. These approaches have the further advantage of accommodating kinetic energy losses accompanying CAD.

E/Q linked scanning has been used to assess the contribution of low energy decomposition processes to product ions spectra recorded in a CAD/MIKES analysis [20]. The scan law typically applied during E/Q scanning assumes that the collision cell in the second field-free region is held at ground potential. If, alternatively, the collision cell is floated, then product ions formed in the cell will be shifted in kinetic energy and will not be transmitted through Q during the E/Q linked scan. The transmission of product ions formed outside the collision cell (predominantly via low energy processes) will be unaffected. Figure 5 shows a comparison of E/Q linked scan analyses of the decompositions of a protonated nonapeptide, with and without floating of the collision cell [20]. The near elimination of certain product ion signals from the spectrum recorded with the floated cell clearly identifies the high energy decomposition processes. Conversely, the continuing prominence of other product ions emphasizes the importance, in this instance, of low energy processes to the CAD/MIKES spectrum.

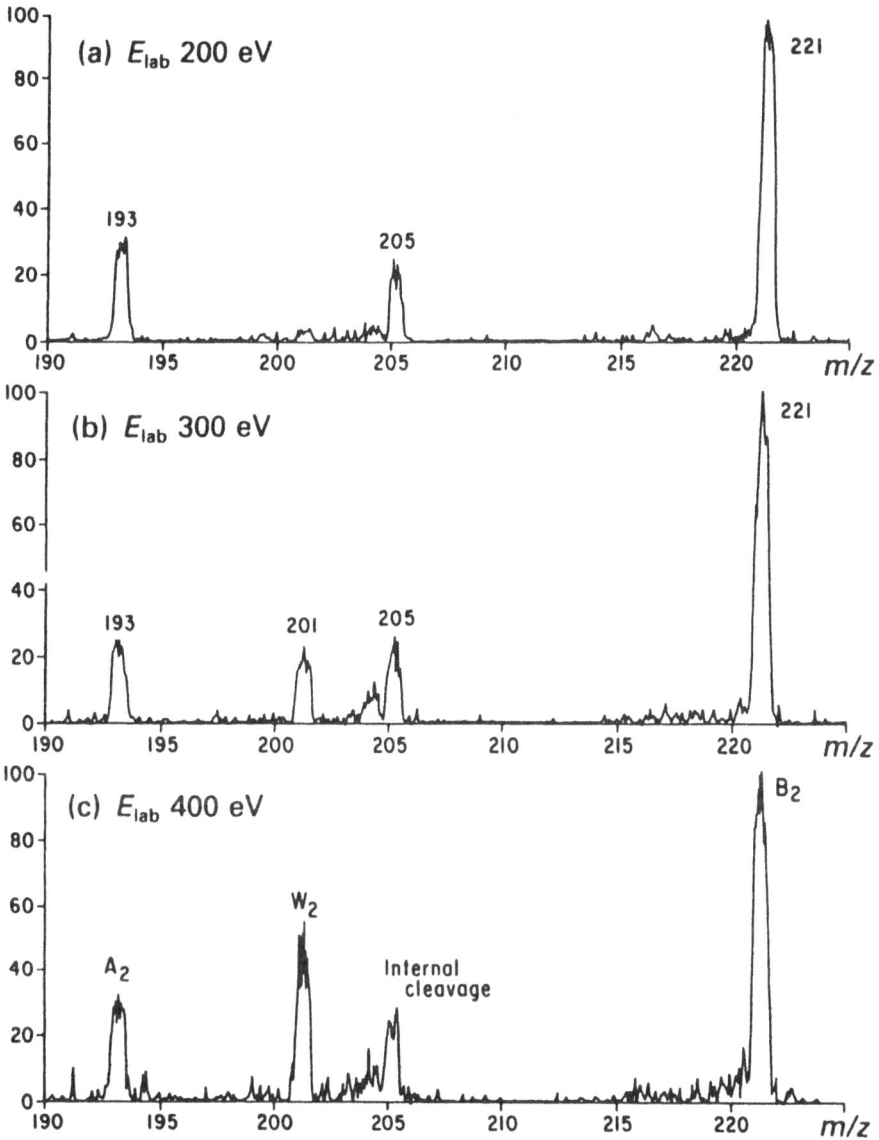

Figure 4. Partial product ion spectra recorded following CAD of protonated YGGFLK in the rf-only quadrupole (q) of an instrument of BEqQ geometry. Argon was used as collision gas, with 50% attenuation of the precursor ion beam. The $E_{lab}$ values quoted correspond to $E_{cm}$ values of 11-22 eV. (Reproduced with permission from reference [16].)

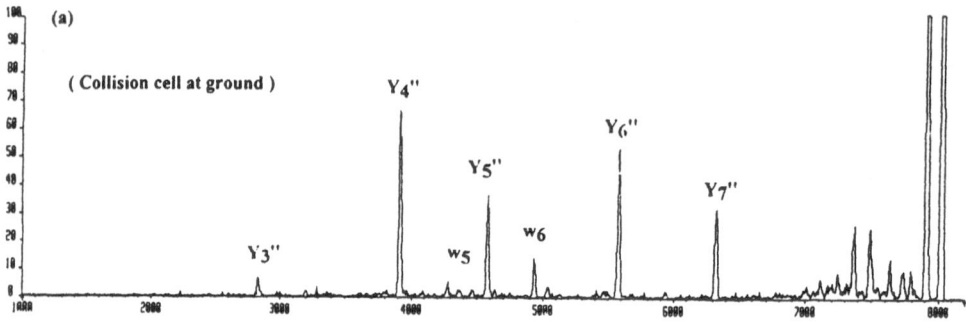

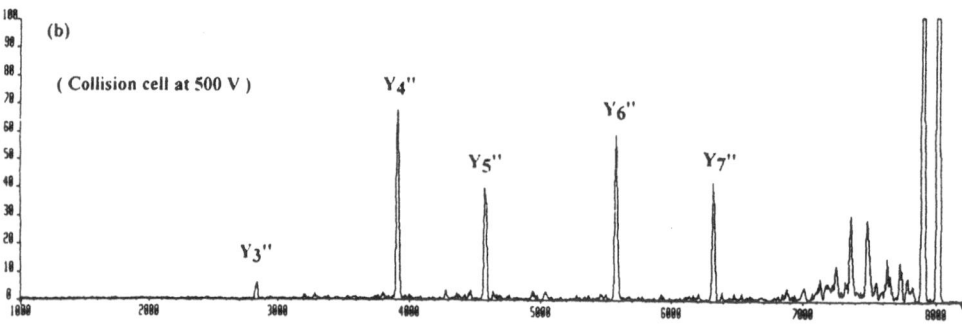

Figure 5. E/Q linked scans recorded with high-energy CAD in the second field-free region of the [M+H]+ ion, m/z 1221, of EELC*TM*FIR (where C* is cysteic acid and M* is methionine sulfoxide): (a) collision cell at ground potential; (b) collision cell floated at 500 eV. Reproduced with permission from reference [20].

For the purpose of defining specific high energy CAD products, limited range scanning of either E or Q with the other fixed may achieve the requisite mass assignment [20]. All such techniques exploiting the sequential energy-resolving and mass-resolving capabilities of E and Q, however, suffer from severely reduced transmission and should not be considered equivalent to analyses of high energy CAD product ions using a four-sector instrument. Furthermore, the use of the quadrupole filter as the final analyzer precludes the incorporation of array detection, which can provide a substantiialimprovement in sensitivity on four-sector instruments.

## 4. Sequential mass spectrometry of peptides

All tandem MS experiments are concerned with the establishment of "connectivity" between precursor and product ions. "Conventional" MS/MS analyses relate source-formed precursor ions to the detected product ion. In general, $MS^n$ experiments enable the definition of ion fragmentation pathways through (n-1) stages and higher order (n>2) experiments may be beneficial either for the study of fragmentation mechanisms or for the selective detection of analytes of related structure. As discussed in Section 1, hybrid instruments of the type discussed in this chapter incorporate three sequential ion analyzers; for reverse geometry (BEqQ) hybrids, both $MS^3$ and $MS^4$ experiments are possible. As the order of the tandem MS experiment increases, so does the number of possible scan modes [21]. Few $MS^3$ and $MS^4$ scan types have yet found practical applicability; sequential product ion scanning and reaction intermediate scanning are discussed here.

Sequential product ion scanning (MS3) may be achieved on the BEqQ hybrid in two distinct ways. (i) First generation product ions formed in the first field-free region from defined precursors are transmitted to the point of double-focus by appropriate selection of B and E. Further decomposition takes place in q, with scanning of Q to yield a spectrum of second generation product ions. As with linked B/E scanning of first field-free region decompositions, the effective mass resolution of precursor ions is poor. (ii) Alternatively, precursor ions may be selected with satisfactory resolution using B alone and fragmented in the second field-free region; E is then used as a low mass-resolution separation device to select first-generation product ions which again fragment further in q. These scan modes, together with reaction intermediate scan methods using the same decomposition regions, are summarized in Table 1, which is adapted from reference [22].

Both sequential product ion scanning and reaction intermediate scanning have been used to elucidate the relative importance of different fragmentation pathways leading to the formation of "internal" fragment ions from peptides [22]. Such product ions are derived from cleavage at two points in the peptide chain and may be formally described as $(A_rY_s)'_{(r+s-t)}$ or $(B_rY_s)'_{(r+s-t)}$ using the Roepstorff and Fohlman nomenclature [24]. In a study [22] of the formation of the $(B_4Y_4)'_3$ fragment from the protonated pentapeptide, leucine-enkephalin, an initial MIKES analysis (Figure 6a) provided a spectrum of product ions available for further fragmentation to yield the internal fragment; the $B_4$ single cleavage predominated. A reaction intermediate scan (with B set to transmit the protonated precursor and Q set to transmit $(B_4Y_4)'_3$, m/z 262), however, showed that both $B_4$ and $Y_4''$ were quantitatively significant intermediates (Figure 6b), suggesting a greater propensity of the $Y_4''$ ion to fragment yielding the internal fragment. Sequential product ion scans (Figure 6c,d) confirmed these fragmentation pathways. (The symbolism used in Figure 6 to describe the different scans is adopted from Louris et al. [25].

Reaction intermediate scanning (mode 2, Table 1) was demonstrated by Schey et al. [23] to aid in the interpretation of product ion information. For the example of YAGFM, the peptide $[M+H]^+$ was selected as the precursor ion and the $A_1$ fragment as the final product; reaction intermediate scanning yielded a spectrum including, as expected, only N-terminal fragments, facilitaing their discrimination from the C-terminal fragments also present in the conventional (MS/MS) product ion spectrum.

TABLE 1

Sequential MS scan modes (on a BEqQ hybrid) useful for the study of reaction mechanisms*

| Scan mode | Acquisition | | Decomposition[a] region | Effective mass resolution |
|---|---|---|---|---|
| | Static | Scanned | | |
| Second generation (precursor) product ion (intermediate) scan (mode 1 ) | B and E | Q | FFR1 (1) q (2) | Approx. 300 Approx. 800 Unit (products) |
| Second generation (precursor) product ion (intermediate) scan (mode 2) | B and E | Q | FFR2 (1) q (2) | > 1000 Approx 300 Unit (products) |
| Reaction (precursor) intermediate (intermediates) scan (mode 1) | Q | B and E | FFR1 (1) q (2) | Approx. 300 Approx 800 Unit (products) |
| Reaction (precursor) intermediate (intermediates) scan (mode 2) | B and Q | E | FFR2 (1) q (2) | > 1000 Approx. 300 Unit (products) |

* Reproduced, in modified form, from reference [22]

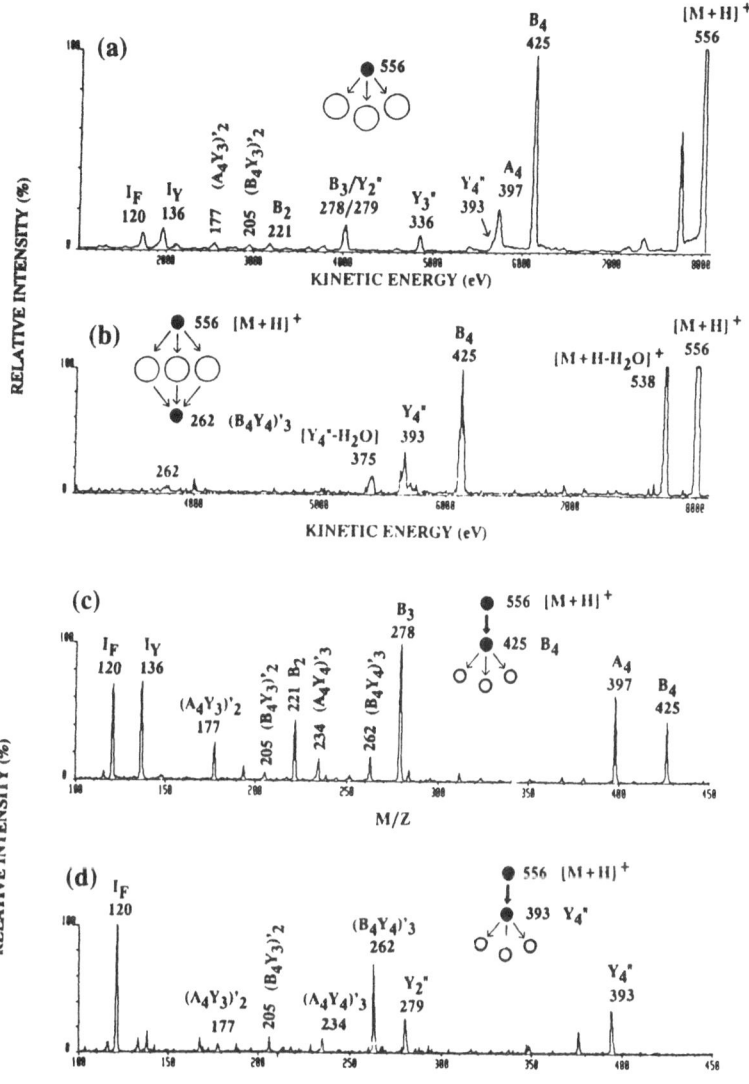

Figure 6. MS/MS and MS$^3$ analyses to probe the fragmentation of protonated YGGFL a) MIKES spectrum obtained under high energy (8 keV) CAD conditions. b) Reaction intermediate scan illustrating fragmentation pathways of the [M+H]$^+$ ion to yield the internal fragment, (B$_4$Y$_4$)'$_3$. c,d) Sequential product ion scans indicating the fragmentations of first generation product ions, m/z 425 and 393, derived from [M+H]$^+$. Reproduced with permission from reference [22].

It was noted in Section 2 that sequential MS has been used in the study of peptide ion rearrangement processes. [$^{18}$O]-labelling experiments established that fragmentation of protonated and metal cationized peptides can occur by loss off the C-terminal amino acid residue but with retention of an oxygen derived from the C-terminal carboxyl group [14]. Thus the product ion is formally designated as [B$_n$' + OH], in extension of the Roepstorff and Fohlman nomenclature [24]. Sequential product ion scanning analysis of protonated bradykinin (RPPGFSPFR) suggested that the rearrangement product ion structure was identical to that of protonated des-Arg$^9$-bradykinin [14]. In a number of examples it was observed that the [$^{18}$O$_1$]-rearrrangement product ion was accompanied by an [$^{18}$O$_2$]-labelled analogue whose genesis could not be explained by the simple rearrangement process [26]. Once again, sequential product ion scanning was emloyed, this time to establish the isotopic composition of the second generation rearrangement product ion. Figure 7 shows partial first and second generation product ion spectra recorded for [M+Na]$^+$ ions derived from [$^{18}$O$_2$]leucine-enkephalin (YGGFL). The first generation rearrangement product ion appeared as a doublet (Figure 7a), indicating species incorporating [$^{18}$O$_1$] and [$^{18}$O$_2$]. The second generation rearrangement ion derived via the [$^{18}$O$_1$]-labelled first generation product showed a near one-to-one doublet of singly labelled and unlabelled species (Figure 7b); this was consistent with the incorporation of the isotopic label in the C-terminal carboxyl group of the first generation product ion. The second generation rearrangement ion derived via the [$^{18}$O$_2$]-labelled first generation product showed predominantly incorporation of a single label, with a significant proportion of the double labelled analogue (Figure 7c); these data indicated that the incorporation of both $^{18}$O atoms in the first generation rearrangement product ion is in the C-terminal carboxyl group. These results suggested an isotope exchange process resulting in transfer of oxygen between the C-terminal carboxyl group and the C-terminal peptide bond. Additional evidence reported in the same study [26] indicated that the exchange process could occur independently of C-terminal rearrangement product ion formation. Evidence for both the C-terminal rearrangement and the isotope exchange process emphasized the significance of the gas-phase ion conformation in determining prominent low energy fragmentation pathways.

MS$^4$ sequential product ion scanning experiments may also be performed using a BEqQ hybrid [27]. The occurrence of a first field-free region decomposition in a single focussing magnetic sector instrument is detected as a broad peak in the conventional mass spectrum, generally at a non-integral m/z value. Such product ions may be transmitted in a BEqQ instrument by appropriate setting of B and further fragmented in the second field-free region. These second generation products may in turn be selected by appropriate setting of E and subsequently fragmented in q. Final scanning of Q yields a spectrum of third generation product ions derived from defined precursor and intermediate ions.

For most purposes, the components of the hybrid instrument which allow MS$^n$ experiments may be considered simply as alternative mass analyzers (albeit with differing mass resolution capabilities). In a few instances, however, analytical use has been made of the different ion-separating properties of the individual components. Thus, the magnet separates according to momentum/charge, the electric sector according to kinetic energy/charge, and the quadrupole (Q) according to mass/charge. These different separation principles were exploited by Ballard and Gaskell [28] in a study of the heterogeneity, with respect to kinetic energy, of protonated leucine-enkephalin ions produced in a fast atom bombardment ion source. Figure 8 shows mass/charge ratio-deconvoluted, energy-selected momentum spectra of the peptide [M+H]$^+$ ions. Experimentally, the spectra were recorded by setting of E and Q (the latter at a resolution to transmit a window of ions several m/z units wide) and scanning of B. Clearly these were not tandem MS experiments in the sense of establishing connectivity between precursor and product ions. The experiments served rather to define several related but distinct properties of source-formed ions. The data reproduced here (and substantiated by a variety of other data in the original publication [28]) indicate that a significant proportion of ions produced in a FAB source possess kinetic energies lower than that predicted from the accelerating potential. These ions may arise via desolvation and declustering processes in the acceleration region, or via FAB or chemical ionization in regions removed from the FAB target.

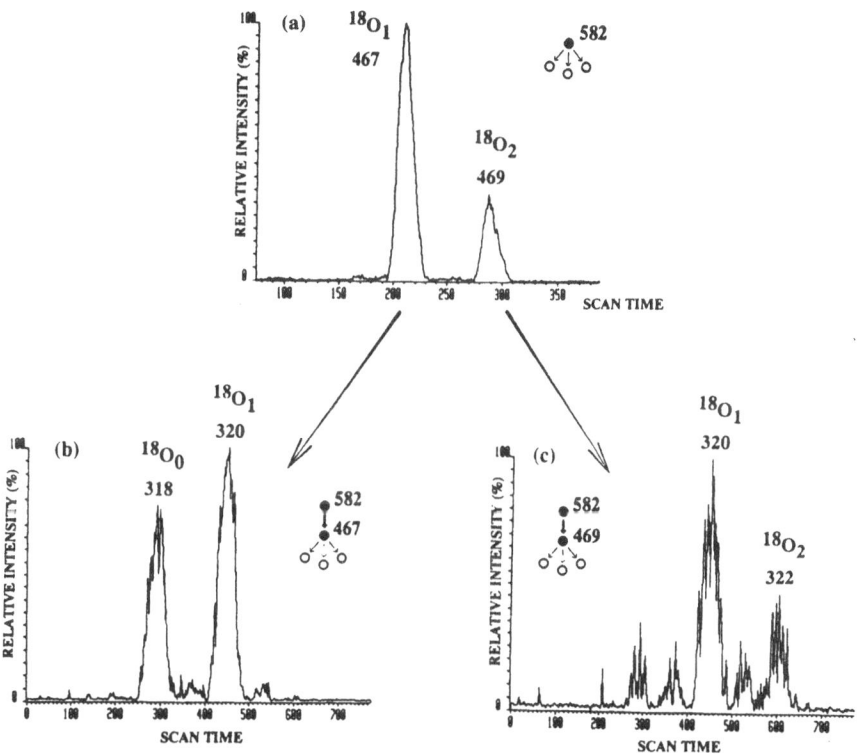

Figure 7. (a) Portion of the metastable first generation product ion spectrum of sodium cationized [$^{18}O_2$]leucine enkephalin, depicting the rearrangement ion region. (b) Second generation product ion spectrum of [$^{18}O_2$]leucine enkephalin, depicting the second generation rearrangement species derived through CAD of the [$^{18}O_1$]-labelled first generation rearrangement ion (m/z 467). (c) As in (b), depicting second generation rearrangement products derived from the [$^{18}O_2$]-labelled first generation rearrangement ion (m/z 469) from [$^{18}O_2$]leucine enkephalin. Reproduced with permission from reference [26].

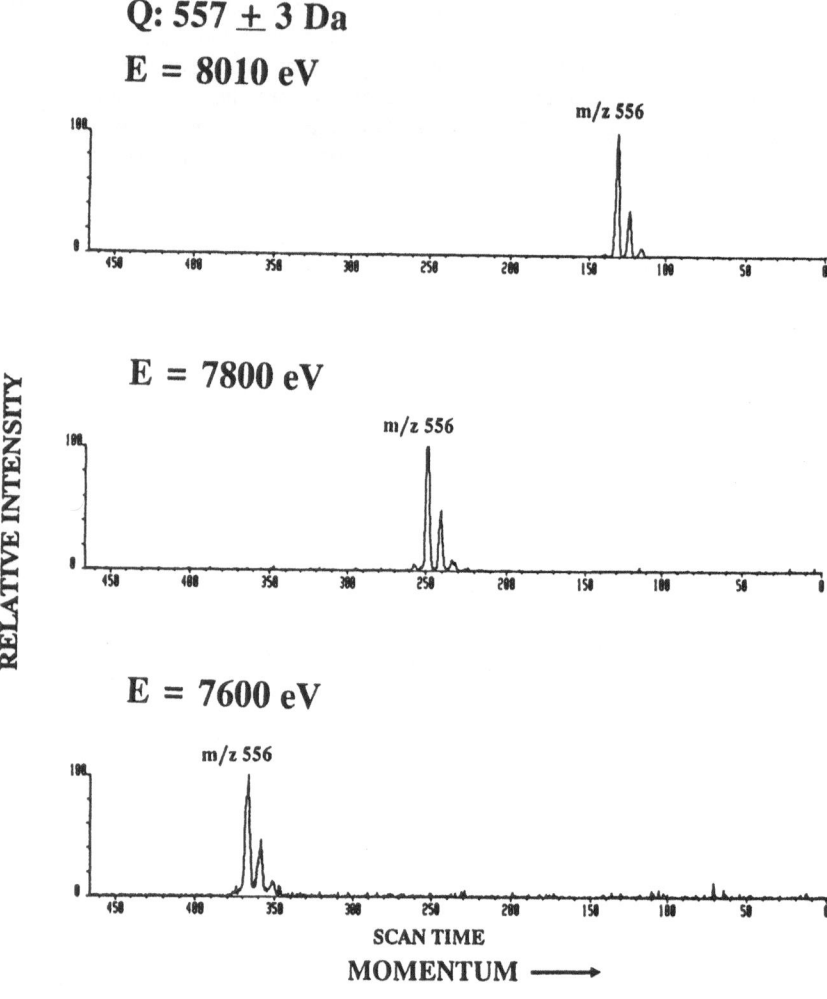

Figure 8. Mass/charge ratio-deconvoluted, energy-selected momentum spectra of leucine-enkephalin [M+H]⁺ ions. The ion source was operated at 8010 V. E was set to a selected kinetic energy value, and Q was set to transmit m/z 557±3. B was scanned to generate the momentum spectrum. Adapted from reference [28] and reproduced with permission.

## 5. Conclusions

Hybrid mass spectrometers have played a significant role in peptide analyses, though with respect to applications the contributions have been less substantial than those reported using four-sector instruments or triple quadrupoles. To an extent, this has been due to an under-appreciation of the capabilites of hybrid instruments. The great virtue of hybrids is their versatility. Thus, a single instrument may be used for accurate mass determination, low energy CAD and high energy CAD, albeit (in the last case) with some compromise in performance. Sequential MS (MS$^n$) capabilities further enhance the versatility of hybrid instruments. Some contributions to the continuing debate over the relative merits of low and high energy CAD have been made using hybrid instruments [29-31]. Recent data tend to suggest that improved understanding of the mechanisms of low energy fragmentation processes will facilitate the optimization of low energy CAD experiments. A continuing role for hybrid instruments is anticipated in the study of these and other issues.

## 6. Acknowledgements

Work reviewed here from the author's laboratory was performed at Baylor College of Medicine, Houston, Texas, U.S.A. The essential contributions of Dr. K.D. Ballard, Dr. O. Burlet and R.S. Orkiszewski are gratefully acknowledged. The research at Baylor was funded by the U.S. National Institutes of Health and by the State of Texas Advanced Technology Program.

## 7. References

1. R.A. Yost, R.K. Boyd, in Mass spectrometry: Methods in Enzymology, vol. 193, (J.A. McCloskey, Ed.) Academic Press, New York, 1990, pp. 154-200.

2. S.J. Gaskell, K.D. Ballard, in Mass Spectrometry in the Biological Sciences: a Tutorial, (M.L. Gross, Ed.) Kluwer Academic Publishers, Dordrecht, 1992, pp. 29-58.

3. S.J. Gaskell, "Hybrid tandem mass spectrometers in biological research", *Biol. Mass Spectrom.* **21**, 413-419, (1992).

4. K.R. Jennings, G.G. Dolnikowski, in Mass Spectrometry: Methods in Enzymology, vol. 193, (J.A. McCloskey, Ed. ) Academic Press, San Diego, 1990, pp. 37-61.

5. R.W. Yeh, J.M. Grimley and M.M. Bursey, "Collisionally induced fragmentation of protonated oligoalanines and oligoglycines", *Biol. Mass Spectrom.* **20**, 443-450, (1991).

6. B.L. Schwartz and M.M. Bursey, "Some proline substituent effects in the tandem mass spectrum of protonated pentaalanine", *Biol. Mass Spectrom.* **21**, 92-96, (1992).

7. D.F. Hunt, J.R. Yates, III, J. Shabanowitz, S. Winston and C.R. Hauer, "Protein sequencing by tandem mass spectrometry", *Proc. Natl. Acad. Sci. USA* **83**, 6233-6237, (1986).

8. O. Burlet, C.-Y. Yang and S.J. Gaskell, "The influence of cysteine to cysteic acid oxidation on the collisionally activated decomposition of protonated peptides: evidence for intra-ionic interactions", *J. Am. Soc. Mass Spectrom.* **3**, 337-344, (1992).

9. D.S. Wagner, A. Salari, D.A. Gage, J. Leykam, J. Fetter, R. Hollingsworth and J.T. Watson, "Derivatization of peptides to enhance ionization efficiency and control fragmentation during analysis by fast atom bombardment tandem mass spectrometry", *Biol. Mass Spectrom.* **20**, 419-425, (1991).

10. O. Burlet, R.S. Orkiszewski, K.D. Ballard and S.J. Gaskell, "Charge promotion of low energy fragmentations of peptide ions", *Rapid Commun. Mass Spectrom.* **6**, 658-662, (1992).

11. K. Tang, W. Ens, K.G. Standing and J.B. Westmore, "Daughter ion mass spectra from cationized molecules of small oligopeptides in a reflecting time-of-flight mass spectrometer", *Anal. Chem.* **60**, 1791-1799, (1988).

12. R.P. Grese, R.L. Cerny and M.L. Gross, "Metal ion-peptide interactions in the gas phase: a tandem mass spectrometry study of alkali metal cationized peptides", *J. Am. Chem. Soc.* **111**, 2835-2842, (1989).

13. D. Renner and G. Spiteller, "Linked scan investigation of peptide degradation initiated by liquid secondary ion mass spectrometry", *Biomed. Environ. Mass Spectrom.* **15**, 75-77, (1988).

14. G.C. Thorne, K.D. Ballard and S.J. Gaskell, "Metastable decomposition of peptide $[M+H]^+$ ions via rearrangement involving loss of the C-terminal amino acid residue", *J. Am. Soc. Mass Spectrom.* **1**, 249-257, (1990).

15. K. Biemann, in <u>Mass spectrometry: Methods in Enzymology, volume 193),</u> (J.A. McCloskey, Ed. ) Academic Press, New York, 1990, pp. 455-479.

16. A.J. Alexander, P. Thibault and R.K. Boyd, "Collision-induced dissociations of peptide ions. 2. Remote charge-site fragmentations in a tandem, hybrid mass spectrometer", *Rapid Commun. Mass Spectrom.* **3**, 30-34, (1989).

17. C.D. Bradley, J.M. Curtis, P.J. Derrick and B. Wright, "Tandem mass spectrometry of peptides using a magnetic sector/quadrupole hybrid - the case for higher collision energy and higher radio-frequency power", *Anal. Chem.* **64**, 2628-2635, (1992).

18. R.K. Boyd, E.W. Dyer and R. Guevremont, "A practical approach to linked E-Q scans over a wide mass range: partially linked scans with E-modulation", *Int. J. Mass Spectrom. Ion Processes* **88**, 147-160, (1989).

19. R. Guevremont and R.K. Boyd, "Operation of a hybrid tandem mass spectrometer of BEqQ configuration in scan modes using uncoupled analysers", *Int. J. Mass Spectrom. Ion Processes* **34**, 47-68, (1988).

20. O. Burlet, R.S. Orkiszewski and S.J. Gaskell, "Determination of high energy fragmentations of protonated peptides using a BEqQ hybrid mass spectrometer", *J. Am. Soc. Mass Spectrom.* **4**, 470-476, (1993).

21. J.C. Schwartz, A.P. Wade, C.G. Enke and R.G. Cooks, "Systematic delineation of scan modes in multidimensional mass spectrometry", *Anal. Chem.* **62**, 1809-1818, (1990).

22. K.D. Ballard and S.J. Gaskell, "Sequential mass spectrometry applied to the study of the formation of "internal" fragment ions of protonated peptides", *Int. J. Mass Spectrom. Ion Processes* **111**, 173-189, (1991).

23. K.L. Schey, J.C. Schwartz and R.G. Cooks, "Observation of sequence-specific peptide fragmentation using extended tandem mass spectrometry experiments", *Rapid Commun. Mass Spectrom.* **3**, 305-309, (1989).

24. P. Roepstorff and J. Fohlman, "Proposal for a common nomenclature for sequence ions in mass spectra of peptides", *Biomed. Mass Spectrom.* **11**, 601, (1984).

25. J.N. Louris, L.G. Wright, R.G. Cooks and A.E. Schoen, "New scan modes accessed with a hybrid mass spectrometer", *Anal. Chem.* **57**, 2918-2924, (1985).

26. K.D. Ballard and S.J. Gaskell, "Intramolecular [$^{18}$O] isotopic exchange in the gas phase observed during the tandem mass spectrometric analysis of peptides", *J. Am. Chem. Soc.* **114**, 64-71, (1992).

27. K.D. Ballard, S.J. Gaskell, R.C.K. Jennings, J.H. Scrivens and R.G. Vickers, "Sequential product ion spectra (MS$^3$ and MS$^4$) with array detection and reaction intermediate scanning on a four-sector mass spectrometer", *Rapid Commun. Mass Spectrum.* **6**, 553-559, (1992).

28. K.D. Ballard and S.J. Gaskell, "The origin of the tailing signal on the low energy side of the main beam in MIKES spectra", *J. Am. Soc. Mass Spectrom.* **3**, 644-655, (1992).

29. A.J. Alexander, P. Thibault, R.K. Boyd, J.M. Curtis and K.L. Rinehart, "Collision Induced Dissociation of Peptide Ions. Part 3. Comparison of results obtained using sector-quadrupole hybrids with those from tandem double-focusing instruments", *Int. J. Mass Spectrom. Ion Processes* **98**, 107-134, (1990).

30. L. Poulter and L.C.E. Taylor, "A comparison of low and high energy collisionally activated decomposition MS-MS for peptide sequencing", *Int. J. Mass Spectrom. Ion Processes* **91**, 183-197, (1989).

31. M.F. Bean, S.A. Carr, G.C. Thorne, M.H. Reilly and S.J. Gaskell, "Tandem mass spectrometry of peptides using hybrid and four-sector instruments: a comparative study", *Anal. Chem.* **63**, 1473-1481, (1991).

# Mass Spectrometry in the Study of Advanced Glycation End Products

A. LAPOLLA, D.FEDELE, S. CATINELLA*, P. TRALDI*
Institute of Internal Medicine - Division of Metabolic Disorder Via Giustiniani, 2-
35100 Padua (Italy)
*CNR Research Area, Corso Stati Uniti, 4 - 35100 Padua (Italy)

## Summary

The results obtained by EI MS, FAB MS, HPLC MS, Pyr/GC/MS and MALDI/MS
in the field of protein glycation are reported and discussed in detail in comparison
with those obtained by other analytical methodologies (fluorescence and absorbance spectroscopies, radioimmunoassay, enzyme-linked immunosorbent assay). The wide amount of data so obtained are discussed in terms of:
i) analysis on hydrolysis products of glycated proteins;
ii) direct analysis of undegraded glycated proteins;
iii) studies on the products arising from the reaction between protected lysine and
glucose. The general overwiew so achieved shows how mass spectrometry is a
particularly valid analytical approach in this field of research.

## 1.Introduction

The interaction between glucose and proteins is a topic of wide interest,
ranging from food chemistry to medicine (1, 2).

The reaction of amino groups of proteins with glucose, as well as other
reducing sugars, leads to highly reactive intermediates. These can react with
other reactive sites leading to intra or inter-molecular cross-linking (3).

Such kind of reactions, called Maillard reactions, proceeds classically
through early, intermediate and late stages. At the beginnig the sugars reacts with
free amino groups of proteins leading to labile Schiff bases; these bases then
undergoes an Amadori rearrangement (4,5) to form a stable adduct. In its turn this
is degraded into a series of carbonyl compounds (deoxyglucosones and sugar
fragmentation products) that act as propagators of the reaction (6). In the late

317

*R. M. Caprioli et al. (eds.), Mass Spectrometry in Biomolecular Sciences, 317–349.*
© 1996 *Kluwer Academic Publishers.*

stage these propagators react again with free aminogroups and through dehydration and rearrangement reactions lead to yellow-brown, insoluble and often fluorescent products, called advanced glycation end products (AGE) (7,8) (see Scheme 1).

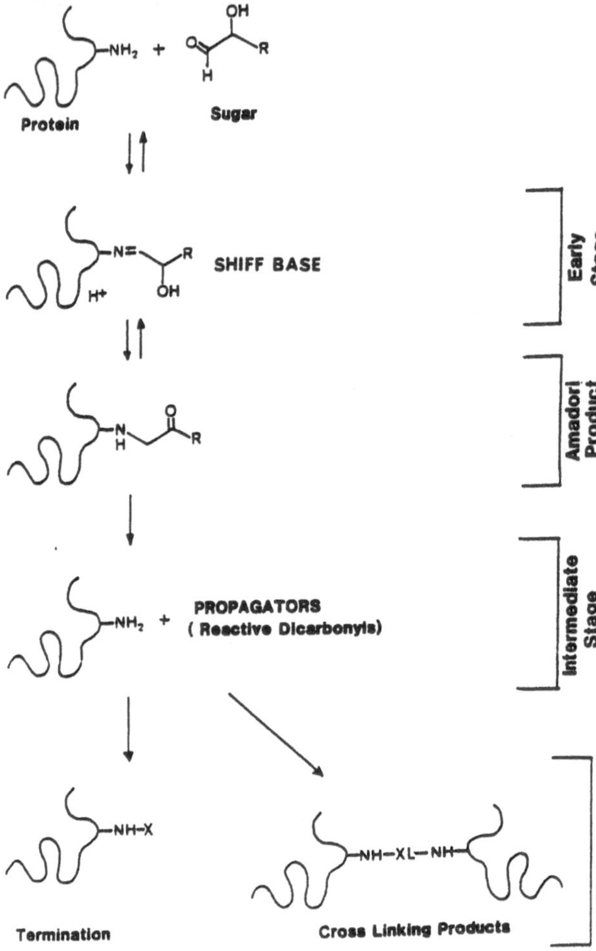

Scheme 1

In medicine such reactions are of wide interest. In particular, those leading to the formation of cross-linking advanced glycation products, are retained responsible for the long-term diabetic complications (nephropathy, retinopathy, neuropathy, macroangiopathy) (9-18). Consequently, the identification of AGE markers to be employed for the evaluation of tissutal modifications is of high interest, and 2-(2 furoyl)-4-(5)-(2-furanyl)1H-imidazole (FFI) and pentosidine were recently proposed as such (19-22). FFI, firstly identified by Pongor (19) among the hydrolysis products of glycated proteins, was proposed to originate from a condensation reaction between two glucose molecules and two amino groups of lysine, through cyclization and further autoxydation reactions. The in vitro glycated proteins and the synthetized FFI gave the same fluorescence spectra and this led to consider FFI a possible cross-link product.

Pentosidine was firstly identified in vivo on collagen samples by Sell (20) and further confirmed in vitro by the non enzymatic reaction of pentoses with lysine and arginine and it is considered an AGE responsible for protein cross-links (21, 22).

The analytical techniques usually employed in the protein glycation field are the spectroscopic ones, mainly fluorescence and absorption spectroscopies (7). In fact AGEs synthetized in vitro show a characteristic increase in fluorescence at 440 nm with excitation at 370 nm and an increase in absorbance between 300 and 400 nm with a shoulder at 320 nm. Furthermore brown pigments with fluorescence and absorbance properties very similar to those of in vitro AGE have been evidenced in vivo on collagen, lens proteins and other tissues of diabetic patients (1).

However spectroscopic techiques, also if exhibit a good sensitivity with respect to the analytical problem, lack of specificity, in particular considering the complexity of the natural substrate under investigation.

A more recently employed analytical approach with a surely higher specificity was based on RIA, i.e. on the development of antibodies specific vs substructures significative for glycation products (23). The authors, by developing an antibody reactivo against 4-furanyl-2-furoyl-1H imidazole-1hexanoic acid, reported different FFI levels in relation to different glucose concentrations. Unfortunately the antibody developed exhibits a not enough high specificity, showing a high binding affinity not only to FFI but also to different FFI-related compounds, all containing various furanyl and imidazole moieties (e.g. FFI-BSA, FFI-lysine, MeFFI, furoine, furoic acid) (23).

More recently an enzyme-linked immunosorbent assay (ELISA) was developed to detect glucose derived pyrroles and in particular 5-hydroxy-methyl-1-alkylpyrrole-2-carbaldehyde (pyrraline) (24).

Finally Makita et al. (25) using an ELISA method with a polyclonal antiserum to an AGE epitope, which is formed in vitro after the incubation of glucose with ribonuclease, demonstrated that this antiserum is able to dectect AGE in vivo from different tissues, as plasma and collagen. This antiserum can recognize cross-reactive epitopes which are formed from the reaction of different sugars with different carrier proteins, indicating that in vivo AGEs have a common immunological epitope which cross reacts with in vitro synthesized AGE. The fact that none of the known AGE (FFI, pyrraline, pentosidine, carboxymethyl lysine) were able to compete for binding to anti-AGE antibody could mean that these structurally defined AGE may be only a part of the products formed and it points the need of further structural studies of the advanced glycation pathways that really occur in vivo.

Mass spectrometry represents one of the analytical methods with higher specificity and sensitivity nowadays available. In fact, if on one hand detection limits of $10^{-12}$g are common, on the other hand the identification of molecular weight and the fingerprint obtained by the fragmentation patterns are highly diagnostic from the structural point of view (26). The development of "hyphenated methods", based on the coupling of a separative technique with mass spectrometry has furtherly increased the specificity of the analytical procedure, due to a sinergic effect of the coupling (26). With this respect the data obtained by Yost (27) in calculating the informing power ($P_{inf}$, as defined by Kaiser (28)) of different hyphenated mass spectroscopic techniques are worth of note. Thus, while $P_{inf}=$ 100 for a single quadrupole mass spectrometer with 1000 Da mass range and unitary resolution, $P_{inf}$ = 1000 for a HRGC/MS and $P_{inf}$= 10,000 for a MS/MS system. It follows that the most powerful method currently available results to be the MS/MS one. Furthermore, for the direct analysis of high molecular weight compounds, without any preliminary decomposition procedures, two different approaches can be nowadays employed. The first consists in the controlled thermal degradation (pyrolysis) of the biomolecule in a chamber directly connected to the analytical device (usually GC/MS) (29), the second consists in the ionisation/desorption (FAB (30) MALDI (31)) or desolvatation/ionisation (electrospray (32)) of the intact molecule.

All the above described mass spectrometric methods together with HPLC/MS, have been successfully applied in the study of protein glycation. Thus while either MS/MS or HPLC/MS have been used for the structural identification of the products arising from hydrolysis of glycated proteins, Pyr/GC/MS and MALDI have been employed for the direct characterization of glycated substrates, through the comparison with the behaviour of non-glycated ones.

## 1.1. MASS SPECTROMETRIC MEASUREMENTS IN THE STUDY OF HYDROLYSIS PRODUCTS OF GLYCATED PROTEINS

The first success in the study of glycation processes on proteic substrates was obtained by the comparison of the products arising from hydrolysis of glycated proteins and those derived from the non glycated ones. Among these hydrolysis products, the attention was firstly focalised mainly on three different products, i.e. FFI (19), furosine (33) and pentosidine (20-22). For their formation the mechanisms reported in Scheme 2 were suggested, and their structural identification was done on the basis of spectroscopical data .

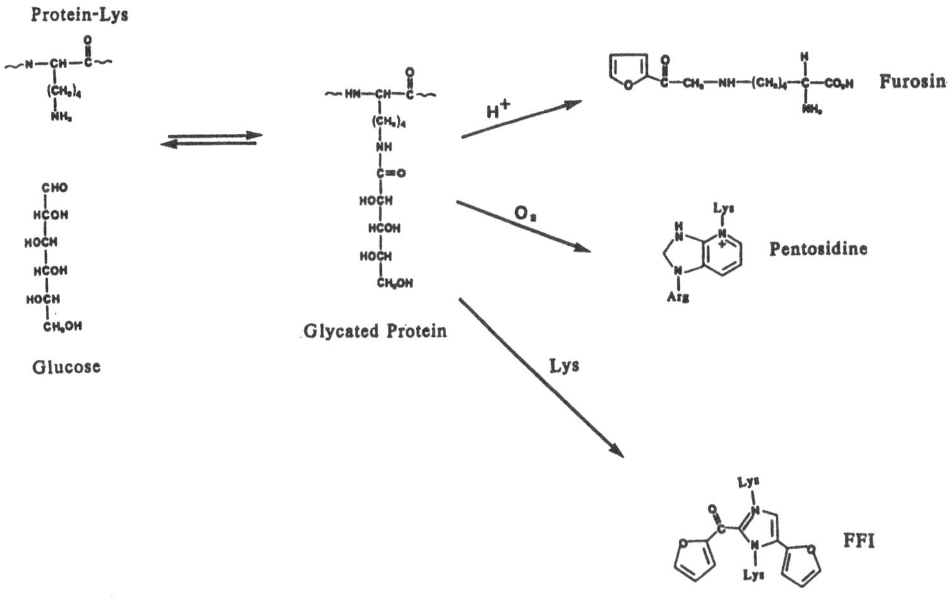

Protein-Lys

Furosin

Glucose

Glycated Protein

Pentosidine

Lys

FFI

Scheme 2

In particular FFI was proposed as an interesting AGE: the two nitrogen atoms were suggested to originate from the ε-amino groups of lysines present in two different proteins, while the furane moieties were proposed to originate from a rearranged glucose molecule. Hence it could give account not only to glycation but it could also be invoked as one of the cross-linking products, responsible for tissutal modifications. Consequently, FFI was employed as a possibly effective marker of advanced glycation: its structure was unequivocally proved by $^1$H NMR, mass spectrometry and the comparison with an "ad hoc" syntetized molecule. Successively it was dosed on the basis of fluorescence data and a specific antibody was developed to be employed in RIA. By such approach mean FFI levels (±ES) of 5.22 ± 0.25 pmol/mg and 8.4 ± 1.4 pmol/mg were respectively found for samples of globine and albumin of healthy controls, while they became 1744±274 pmol/mg for samples of in vitro glycated albumin (23).

Looking at the relevance of FFI in the determination of the end-glycation levels, it was thought of interest to employ a tandem mass spectrometric technique for its detection in the hydrolysis mixture; this was achieved by a double focusing, reverse geometry mass spectrometer. If the MS/MS spectrum of ionic species at m/z 228, isobaric to FFI molecular ion, is identical to that of M$^{+\cdot}$ of a pure sample of FFI, the FFI presence in the complex hydrolysis substrate is unequivocally proved.

In Figure 1 the MS/MS spectrum of M$^{+\cdot}$ of FFI is reported. The collisionally generated daughter ions are well related to the structure, as shown in Scheme 3.

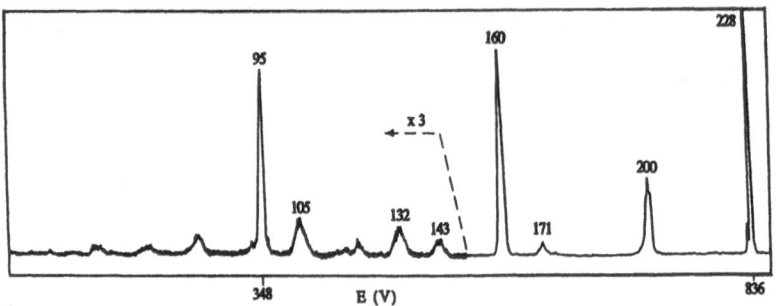

Figure 1: CAD MIKE spectrum of M$^{+\cdot}$ ion of FFI (m/z 228).

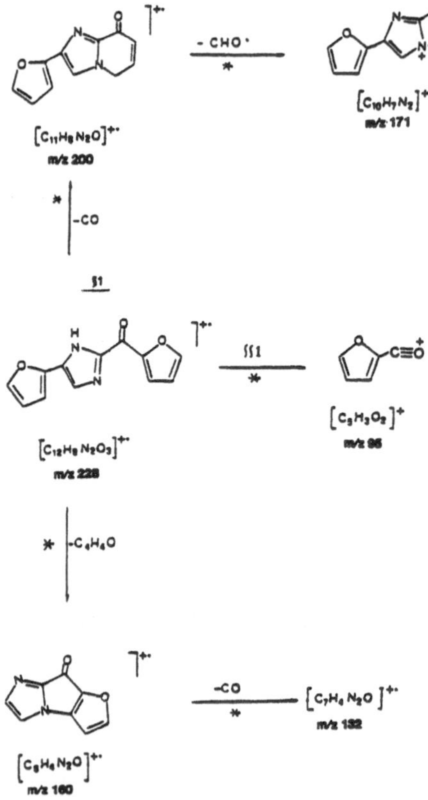

Scheme 3

The analysis of the ions at m/z 228 generated by EI on the hydrolysis mixture obtained by incubating polylysine and albumin (sodium phosphate buffer pH 7.5, 0.5 M Na, for 28 days at 37°C) without any glucose, led to spectra showing the presence of scarcerly abundant daughter ions completely unrelated to FFI structure. On the contrary the introduction of hydrolyzed glycated albumin and polylysine (obtained by 5g albumin, and 1g polylysine incubated with 50g glucose in sodium phosphate buffer at pH 7.5, 0.5 M Na, for 28 days at 37°C) into the EI source and the further MS/MS experiments on the ions at m/z 228, led to the spectra shown in Figure 2, practically identical to that achieved by pure FFI (see figure 1) (34).

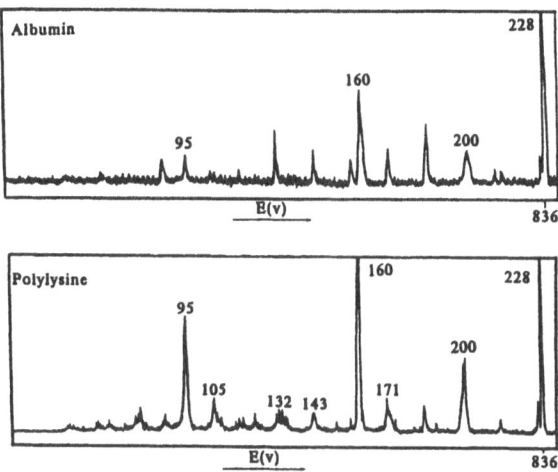

Figure 2: CAD MIKE spectra of ionic species at m/z 228 present in glycated albumin (up) and glycated polylysine (down).

In order to evaluate the sensitivity of the method, a hydrolyzed non glycated albumin sample was spiked with known amounts of pure FFI, allowing the evaluation of a detection limit in the range 50-100 pg. Considering that the detectable levels reported in literature correspond to the range from 912 to 6.1 x $10^4$pg, MS/MS resulted a more than suitable analytical method for FFI determination.

Such a judment seemed to be not valid when, instead of in vitro samples, in vivo glycated substrates were analyzed by MS/MS.

RIA data were in agreement with the fluorescence ones in determining high FFI levels in globine and in glycated proteins. Hence it was surprising how, by MS/MS, no FFI was evidenced in in vivo glycated samples. In an investigation on collagen of diabetic rats, while fluorescence and absorbance data were typical for the AGE presence, by MS/MS was impossible to detect any presence of FFI, also if, by spiking, the sensitivity was proved to be analogous to that obtained in in vitro samples. Looking at the high specificity of MS/MS, the obtained results could be explained not on the basis of the unvalidity of the analytical procedure, but for the real absence of FFI among the hydrolysis products of glycated collagen (35).

In the same period FFI was proved to be an artifact, originating from the NH$_4$OH neutralization after HCl hydrolysis (36). In fact, by using labelled $^{15}$NH$_4$OH, an increase of FFI molecular weight of 2 Da was found, proving that the imidazole nitrogen atoms do not originate from ε-aminogroups of lysine, but from the neutralization media. The absence of FFI in hydrolyzed collagen samples, as determined by MS/MS, is in agreement with such findings and clearly in contrast with the RIA data.

Such a discrepancy could be reasonably explained by a RIA positive response on molecules structurally related to FFI but with different structures and molecular weights. In order to investigate on such hypothesis, further mass spectrometric experiments were carried out. Looking at the fact that the RIA antibody was developed against 4-furanyl-2-furoyl-1H-imidazole-1-hexanoic acid, it was considered of interest to put in evidence, in the hydrolysis mixture, all the molecular species containing the furoyl moiety, which in principle could lead to a positive RIA response.

This could be easily obtained by B$^2$/E "linked scan" (37). By such approach in a hydrolyzed mixture from glycated albumin it was proved that, together with FFI, many other molecules containing the furoyl moiety, namely 2-(2-furoyl)-4-hydroxy-1H-imidazole (m/z 178), 2-(2-furoyl)-4-carboxy-1H-imidazole (m/z 206), furan-2,5-diacetyl (m/z 152), 2-furanglyoxal (m/z 124), 2-acetylfuran (m/z 110), and 2-furanaldehyde (m/z 96), were present (see Table 1 and Fig. 3) (38, 39).

The same method was applied for the identification of furoyl-containing advanced glycation products in collagen samples from diabetic and healty rats, by both acid and enzymatic hydrolysis of the proteinic substrate (40). The results so obtained are summarized in Table 2.

No great differences were found among the different samples, showing the presence of the same molecular species with differences only in relative abundances. 2-Furanaldehyde (responsible for the peak at m/z 96), 2 acetylfuran (m/z 110) and furanglyoxal (m/z 124) were already found in in vitro glycation of albumin and polylysine (38, 39). These molecular species were found in larger amounts in the samples originating from HCl hydrolysis, and this led to the conclusion that acidification and further neutralization are important for their production.

The molecular species at m/z 150 and 154 were not previously found among the hydrolysis products of in vitro glycated albumin (38, 39). Further mass spectrometric measurement devoted to their structural identification (accurate mass measurements and daughter ion spectroscopy) were succesfull only for the latter species, indicating for it the structure of 2-carboxy-2-acetylfurane.

325

TABLE I

Precursor ion spectra of furoyl cations (m/z 95) as obtained by $B^2/E$ linked scans [a]

| Compounds | Ionic species | | | | | | | |
|---|---|---|---|---|---|---|---|---|
| | 96 | 110 | 112 | 124 | 152 | 178 | 206 | 228 |
| HCl hydrolyzed NaOH neutralized glycated albumin | | ++ | ++ | ++ | ++ | ++ | ++ | ++ |
| Protease hydrolyzed glycated albumin | + | + | | + | | | | |
| HCl hydrolyzed NaOH neutralized glycated polylysine | | ++ | ++ | ++ | ++ | ++ | ++ | ++ |
| Protease hydrolyzed glycated polylysine | + | + | | + | | | | |
| Proposed structures | | | | | | | | |

[a] The ion abundances were calculated comparing the absolute signal obtained introducing comparable amounts of samples and measuring the peak intensities with respect to the most abundant one, kept equal to 100. Relative abundances: + 1–50%, ++ 50–100%.

reprinted with permission of reference [39]

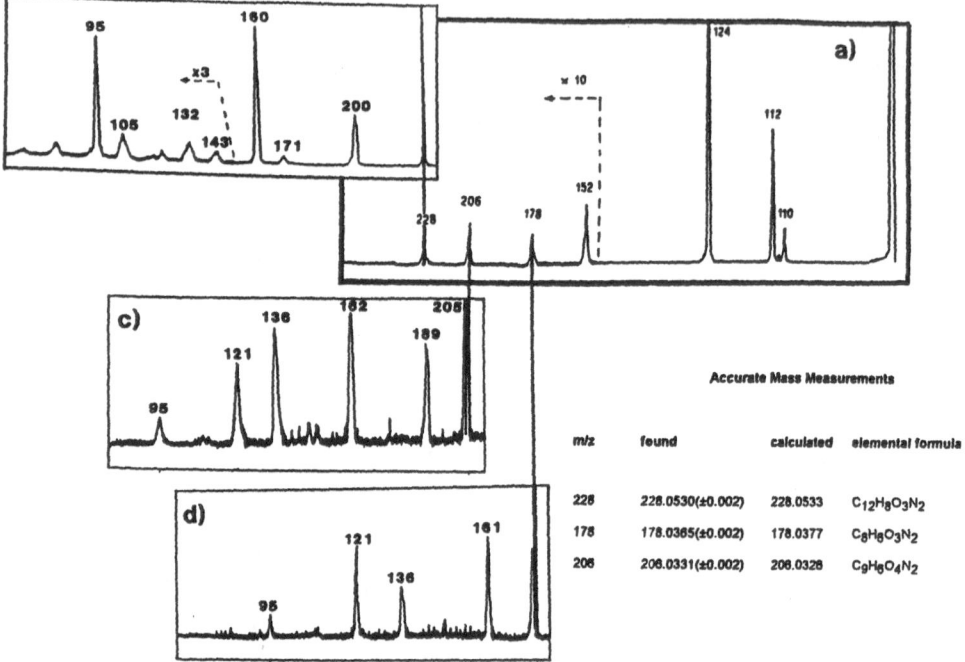

Figure 3: (a) Parent ion spectrum (linked scans at constant $B^2/E$) of ionic
species at m/z 95 generated by EI of glycated and HCl-hydrolyzed
albumin;
(b) CAD MIKE spectrum of the ion at m/z 228;
(c) CAD MIKE spectrum of the ion at m/z 206;
(d) CAD MIKE spectrum of the ion at m/z 178;
and accurate mass values of the same ionic species.

| m/z | Structures | m/z | Structures |
|---|---|---|---|
| 180 | Glucose ⟩+· | 270 | HOCH₂—(furan)—CH= (pyranone ring with OH, CH₂OH) + NH₄⁺  **d** |
| 189 | CH₃-C-NH-CH-COOH (CH₂)₄ NH₂  + H⁺ | | (pyrrole structure) + NH₄⁺  (CH₂)₄ NH₂-CH-COOH  **j** |
| 198 | Glucose + NH₄⁺ | 324 | (fused pyrano-pyrrole structure) ⟩+·  N (CH₂)₄ NH₂-CH-COOH  **n** |
| 162 | CH-C-CH₂-CHOH-CHOH-CH₂OH ⟩+·  **b**  COOH-CH₂-CH₂-O-C-CH-CH₃ ⟩+·  **i**  (pyranone with OH, CH₃) + NH₄⁺  **h**  (furanose ring) ⟩+·  **c**  (pyran ring with OH, CH₃) ⟩+·  **g** | 326 | (pyrrole dialdehyde) + NH₄⁺  (CH₂)₄ CH₃-C-NH-CH-COOCH₃  O |
| 222 | (pyranone with furan) CH=CH-(furan) ⟩+·  **f**  NH₂-CH-COOH (CH₂)₄ NH CH₂ COOH  + NH₄⁺  **k** | 342 | (pyrrole with CH₂OH, CHOH, CHOH chain) ⟩+·  (CH₂)₄ NH₂-CH-COOH  **m** |
| 240 | f + NH₄⁺ | 360 | m + NH₄⁺ |

TABLE 2: Possible structures detected by plasma-spray HPLC/MS

reprinted with permission of reference [49]

In conclusion, the data obtained by the above described investigations led to the following considerations:

i) glycated and not glycated hydrolyzed collagen samples contain the same furoyl-containing compounds;

ii) from a qualitative point of view, hydrolysis does not affect the production of such compounds, and it only slightly affects their relative abundance.

The mixtures arising from hydrolysis of proteins and glycated proteins are usually particularly complex and in order to obtain a their valid description, further investigations by HPLC and HPLC/MS techniques were carried out. The first study on this topic was devoted to the identification of the products arising from enzymatic digestion of advanced glycated albumin (41).

In Fig. 4 the HPLC chromatograms with UV detection (320 nm) of hydrolyzed albumin (up) and hydrolyzed glycated albumin (down) are reported. Their comparison shows that the glycated protein leads to a complex mixture, with a clear increase of the abundance of compounds with retention times between 11 and 18 min. Such molecules must necessarily contain a chromophoric moiety responsible for the absorption at 320 nm. However these chromatograms cannot exclude the presence of other compounds which do not contain such chromphores in their structure. This limitation of UV detection can be effectively overcome by using a HPLC/MS system.

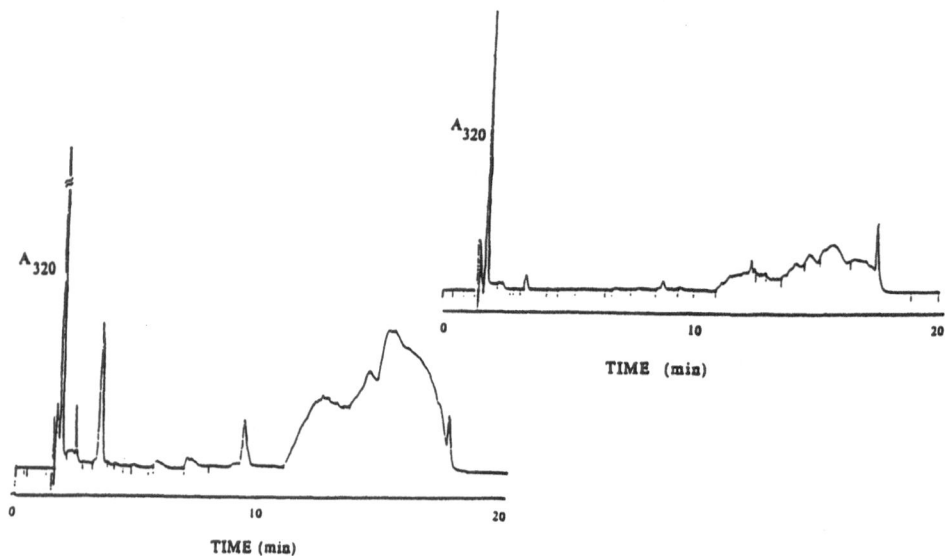

Figure 4: HPLC chromatogram obtained by detection at 320 nm of reaction mixtures:
- protease hydrolyzed albumin (up);
- protease hydrolyzed glycated albumin (down).

The measurements performed by this approach, using a plasmaspray inter-face, led to the results reported in Fig. 5.

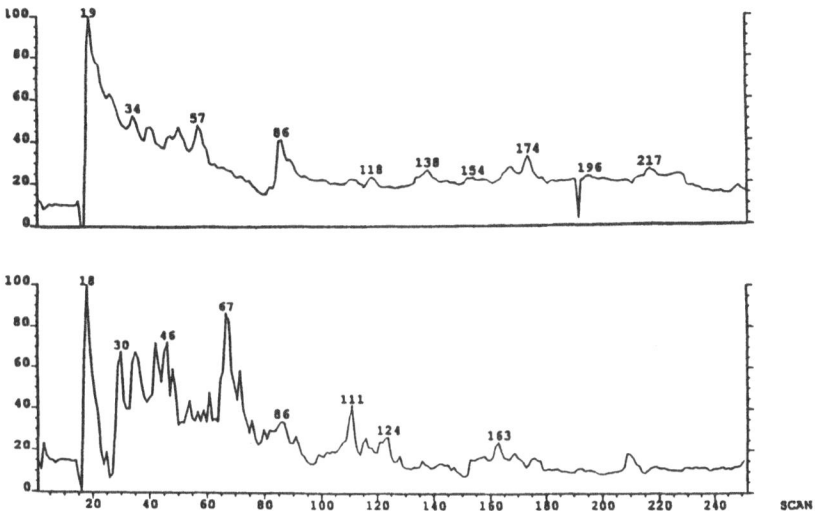

Figure 5: Reconstructed ion chromatogram of HPLC/MS run of reaction mix tures:
- protease hydrolyzed albumin (up);
- protease hydrolyzed glycated albumin (down).

The chromatograms strongly differ from those obtained by UV detection, proving the high effectiveness of MS as chromatographic detector. The mass spectra of the major component of the hydrolyzed mixture arising from glycated albumin led to the identification of the structures reported in Table 3.

In particular structure 1 can be proposed for ionic species at m/z 194: it can arise from the condensation reaction between ε-aminogroup of lysine and glucose, followed by dehydration and oxidation processes. Ionic species at m/z 206 could correspond to structure 2: an analogous compound was found by Olsson et al. (42) in the reaction mixture of glucose and methylamine. For the ion at m/z 132 a lactone structure, that could arise from an α-diketo cleavage from a 3-deoxyos-one, could be proposed (43). Two different structures (4 and 5), both originating from dehydration and further decomposition of 3-deoxyosone, could be assigned for ionic species at m/z 144 (44, 45). Also for ionic species at m/z 233 two different structures (6 and 7) both originating from a dimerization of 2-hydroxymethyl-pyrrolaldheyde, as suggested by Olsson et al. (42), could be assigned. A dimeri-zation of methyl-pyrrolaldehyde, a product of Strecker degradation, could deter-mine the formation of structure 8 (m/z 217). The structures 9 and 10 could be attributed to ions at m/z 168 and 316 respectively, as originating from the reaction between an aminoketose with 3-deoxyosone, as previously described (46).

330

Table 3 Possible structures identified by HPLC/MS

| m/z | Structures | | m/z | Structures | |
|---|---|---|---|---|---|
| 194 | 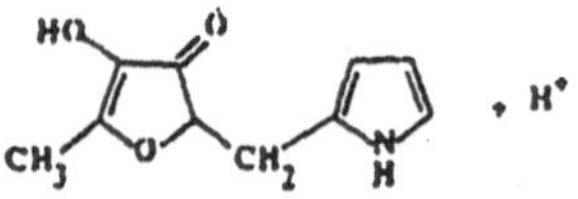 (structure 1) + H$^+$ | 1 | 217 | 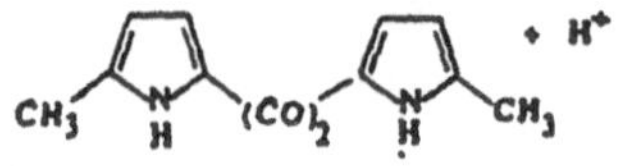 (structure 8) + H$^+$ | 8 |
| 206 | 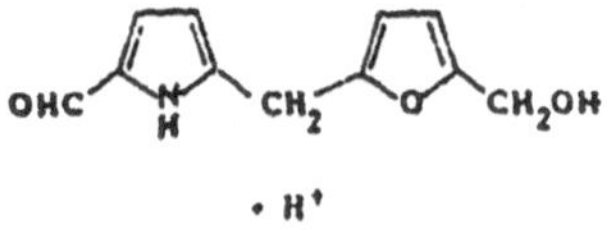 (structure 2) + H$^+$ | 2 | 168 | 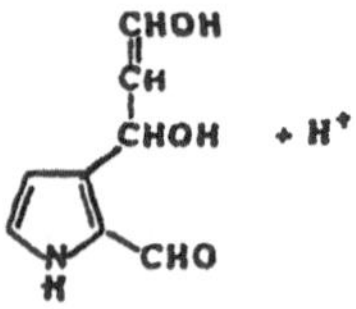 (structure 9) + H$^+$ | 9 |
| 132 | 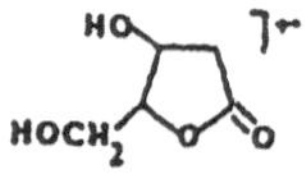 (structure 3) | 3 | 316 | 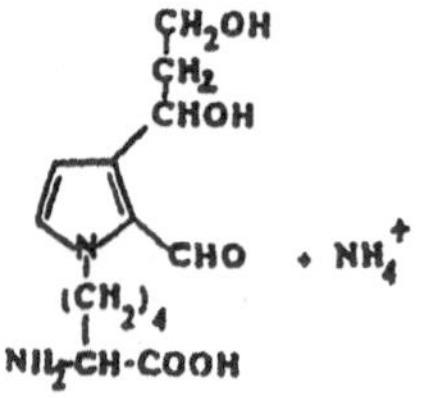 (structure 10) + NH$_4^+$ | 10 |
| 144 | 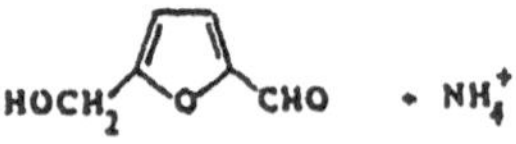 (structure 4) + NH$_4^+$ | 4 | 212 | 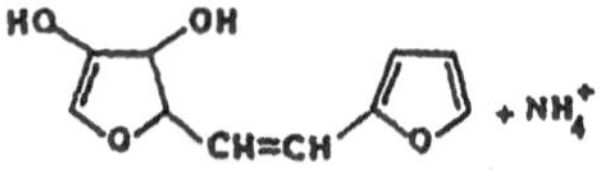 (structure 11) + NH$_4^+$ | 11 |
| 144 |  (structure 5) + NH$_4^+$ | 5 | 256 | 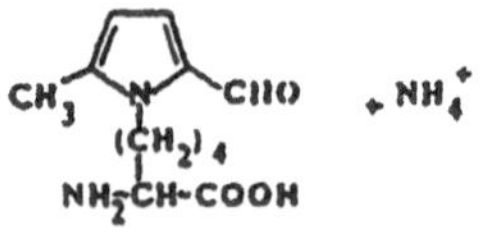 (structure 12) + NH$_4^+$ | 12 |
| 233 | 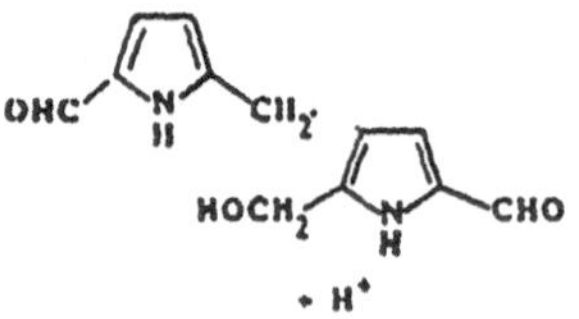 (structure 6) + H$^+$ | 6 | 300 | 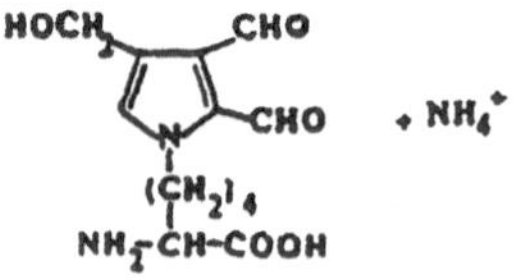 (structure 13) + NH$_4^+$ | 13 |
| 233 | 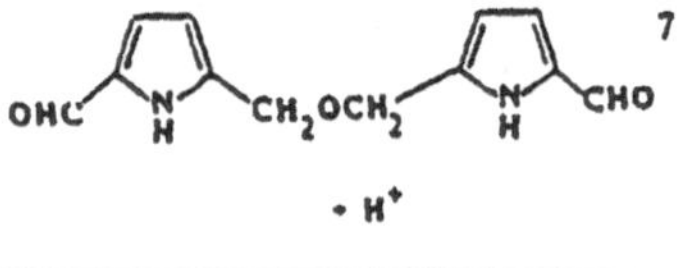 (structure 7) + H$^+$ | 7 | | | |

Table 3

A yellow compound, just described by Ledl et al. (44) as derived from the condensation of 1-deoxyosone with carbonyl compounds, could correspond to ionic species at m/z 212. Finally for ionic species at m/z 256 and 300 the structures 12 and 13 could be assigned: these pyrroles could arise from 3-deoxyosone via a Strecker degradation (46). Such structures must be considered propositive, being obtained just on the basis of molecular weight and literature data

Alternatively to hydrolysis procedures, experiments based on the controlled thermal degradation (pyrolysis) of glycated and not glycated polypeptidic molecules were carried out. As analytical method devoted to the identification of the pyrolysis products, GC/MS was choosen.

Firstly, in order to study a simple substrate, the pyrolysis of polylysine and glycated polylysine were studied (47). In Fig. 6 the gas chromatograms of the related pyrolysis products are reported. Clear differences are evidenced, with the presence of more numerous products in the case of glycated polylysine. Such compounds must necessarily arise from the interaction between polylysine and glucose.

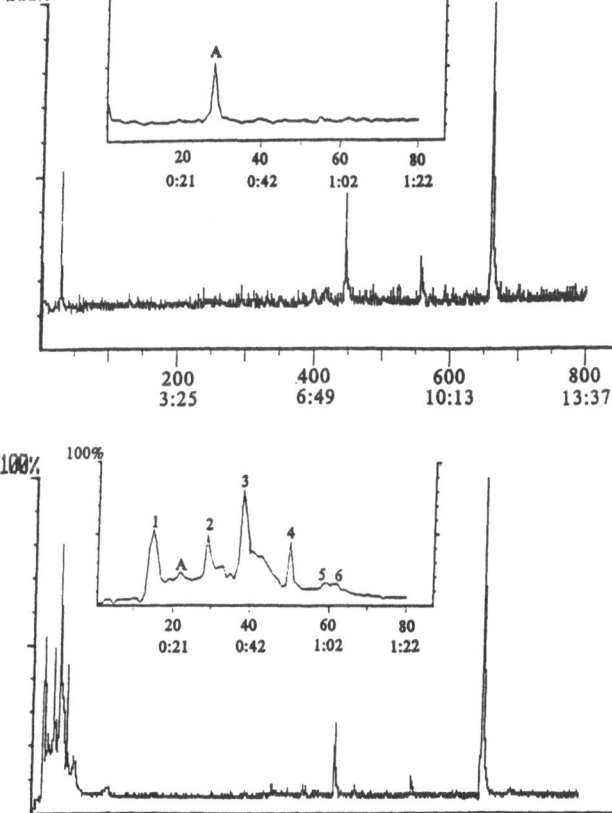

Figure 6: Reconstructed total ion chromatogram from pyrolysis/GC/MS of
- control poly-L-lysine (up);
- glycated poly-L-lysine (down).

In Table 4 the structural assignement, achieved on the basis of EI mass spectra of the various components, is resumed. The structure of propylacetate (m/z 102) could be attributed to compound 1, even if with a low fit value (719). Compound 2 (m/z 96), by library search was identified as furan-2-aldehyde. Compound 3 (m/z 110), the most abundant among the pyrolysis products arising from interaction between glucose and polylysine, was identified as 5-methylfuran-2-aldehyde.

Table 4 Retention times, relative molecular masses (RMM), molecular structures and fit values for compounds 1–6

| Compounds | Retention time (s) | RMM | Structure | Fit value |
|---|---|---|---|---|
| 1 | 15 | 102 | $CH_3COCH_2CH_2CH_3$ | 719 |
| 2 | 33 | 96 | | 800 |
| 3 | 36 | 110 | | 790 |
| 4 | 49 | 84 | | 837 |
| 5 | 59 | 126 | | 908 |
| 6 | 62 | 144 | | 823 |

Table 4

Then the structures of 2-methyl-4,5-dihydrofuran and 5-hydroxymethylfuran-2-aldehyde could be assigned to the components 4 (m/z 84) and 5 (m/z 126) respectively. These furane derivatives have been already reported among the acid and enzymatic hydrolysis products of glycated proteins (41, 44) and among the pyrolysis products of protein polysaccaride mixtures (48). Finally the structure of 2,3-dihydro-3,5-dihydroxy-6-methyl-4H-pyran-4-one (m/z 144), could be assigned to compound 6 even if with a relative low fit value (823): this compound was already described among the reaction products between protected lysine and glucose (49).

The same approach, when employed in the comparison of glycated and not glycated albumin (50), led to the chromatograms reported in Fig. 7.

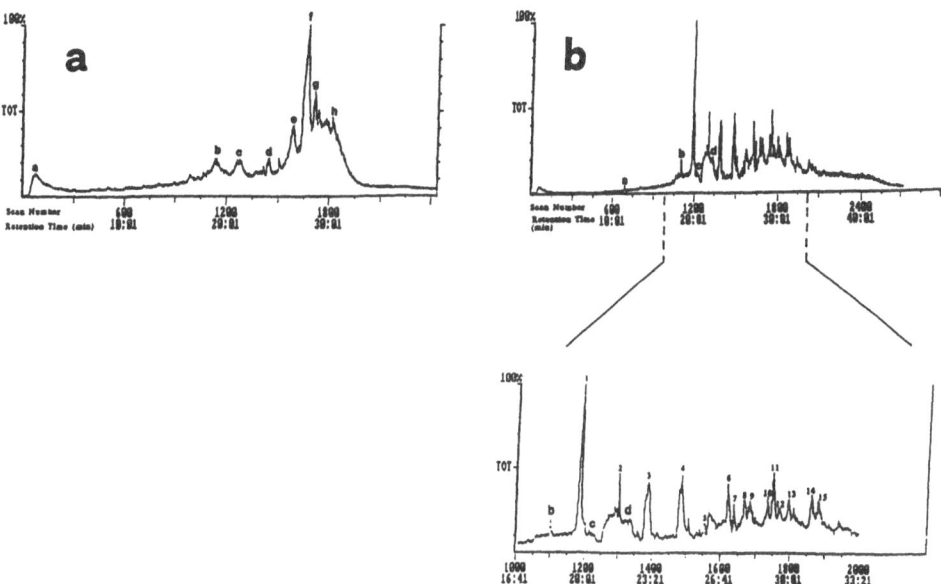

Figure 7:  Reconstructed total ion chromatogram of pyrolysis/GC/MS runs of:
(a) 0.3 mg of bovine serum albumin;
(b) 0.3 mg of glycated bovine serum albumin.

reprinted with permission of reference [50]

Again clear differences are present between the glycated and not glycated substrates, but, quite surprisingly, the molecular species characteristic for glycated polylysine are in the present case of very poor abundance, proving that the complexity of the proteinic substrate leads to a more complicated thermal decomposition pattern. However, diagnostic components for the glycation process are still present (see Table 5), proving that the Pyr/GC/MS approach can be validly employed for the evaluation of glycation levels. In fact, among the pyrolysis products identified by the mass spectra library search, those containg furane and pyrrole moieties, i.e. compounds 2, 4, 5, 6, 9, 11, 13, 14, are of interest, because often described as arising from glycation processes (44, 51).

TABLE 5
Possible structures identified at different retention times[a]

| Peak | Retention time (min) | Spectrum | Possible M.W. | Possible structure subunits | Remarks | FIT |
|------|------|------|------|------|------|------|
| 1 | 19:49 | 1188 | 136 | Furane, dihydrofurane, terminal methyl group | | |
| 2 | 21:43 | 1302 | 147 | 1-(2-furanylmethyl)1H-pyrrole | | 745 |
| 3 | 23:09 | 1388 | 503 | | Column bleeding | |
| 4 | 24:45 | 1484 | 236 | | | 803 |
| 5 | 25:55 | 1554 | 250 | | | 753 |
| 6 | 26:58 | 1620 | 236 | | | 720 |
| 7 | 27:18 | 1637 | 480 | Hexadecyl hexadecanoate | Unknown origin | 790 |
| 8 | 27:47 | 1666 | 450 | Tetradecyl 9-hexadecenoate | Unknown origin | 687 |
| 9 | 28:03 | 1682 | 323 | 1-Hydroxy-2,5-anisyl-3,4-dimethyl-pyrrole | | |
| 10 | 28:53 | 1732 | 504 | 9-Octadecenyl(Z,Z)-9-hexadecenoate | Unknown origin | 785 |
| 11 | 29:08 | 1747 | 299 | | | 478 |
| 12 | 29:25 | 1764 | 508 | Octadecyl hexadecanoate | Unknown origin | 711 |
| 13 | 29:55 | 1797 | 254 | 5-Dodecyl-dihydro-2(3H)-furanone | | 720 |
| 14 | 31:01 | 1810 | 296 | | | |
| 15 | 31:21 | 1880 | 534 | Eicosil Hexadecenoate | Unknown origin | 730 |

[a] FIT values are a measure of the degree to which the library spectrum is included in the unknown spectrum. An FIT value of 1000 indicate: all library peaks occur as peaks in the unknown; for those common peaks all intensities are exactly proportional.

Table 5

## 1.2. DIRECT MASS SPECTROMETRIC ANALYSIS OF UNDEGRADED GLY CATED PROTEINS

Until few months ago, only spectroscopical methods (i.e. fluorescence and absorbance spectroscopies) were employed for investigation on untreated proteins.

Such methods result effective for an evaluation of glycation levels, but they lack of the specificity necessary to give any structural information on the glycation products and this was the main reason for the employement of hydrolysis procedures, thus to investigate on smaller glycated moieties by different and more specific spectroscopical methods.

As shown above, mass spectrometry, and in particular MS/MS, demonstrated to be highly valid in this contest, but neither EI nor FAB MS were able to give any direct information on the whole glycated protein.

Recently new ionisation techniques have become available for the analysis of intact macromolecules, i.e. electrospray (ES) (32) and matrix assisted laser desorption/ionisation (MALDI) methods (31).

Both these techniques have been successfully applied in the field of proteins and for such reason the application of MALDI in the investigation of glycation processes was thought of interest. Until now the only data published on this argument pertain the study of in vitro glycation of albumin. In this work a parallel investigation on the basis of fluorescence and MALDI data was carried out (52).

| Samples | 1 | | 2 | | 3 | | 4 | | 5 | |
|---------|------|------|------|------|------|------|------|------|------|------|
| Time (days) | EU | ABS | EU | ABS | EU | ABS | EU | ABS | EU | ABS |
| 0 | 13.6 | 0.10 | 25.2 | 0.09 | 22.4 | 0.10 | 20.4 | 0.11 | 27.5 | 0.09 |
| 7 | 12.8 | 0.08 | 40 | 0.08 | 108 | 0.12 | 168 | 0.20 | 307 | 0.30 |
| 14 | 28 | 0.08 | 69.2 | 0.10 | 312 | 0.12 | 440 | 0.26 | 514 | 0.47 |
| 21 | 40 | 0.10 | 80 | 0.12 | 468 | 0.17 | 828 | 0.34 | 895 | 0.58 |
| 28 | 50 | 0.12 | 100 | 0.14 | 820 | 0.25 | 1104 | 0.54 | 1180 | 0.80 |

Table 6: Fluorescence and Absorbance data of samples 1-5. Fluorescence is expressed in Emission Unit (EU)/mg protein; absorbance is expressed in Absorbance (ABS)/mg protein

Table 6

Incubating a sample of albumin without any glucose (samples 1 of Table 6) (1.5 g bovine serum albumin in sodium phosphate buffer 0.05M at pH 7.5 at 37°C from 0 to 28 days), an increase in fluorescence is observed, while no increase in molecular weight is detected (see Table 6 and Fig. 8). Such a discrepancy can be explained by considering that the fluorescence data originate by intramolecular reactions of the protein without any detectable release of chemical moieties. In other words the practical constancy of molecular weight with respect to incubation time demonstrates that the elemental formula of BSA remains necessarily the same and that some structural modifications due to rearrangement processes are the responsible for the fluorescence data. Another possibility could be related to the production of particularly small amounts of different products not detectable by MALDI but, possibly due to their optical properties, easy detectable by fluorescence methods.

In presence of glucose (1.5 g BSA, glucose 2 M, in sodium phosphate buffer 0.05M at pH 7.5 at 37°C from 0 to 28 days) (samples 4 of Table 6) a clear increase in molecular weight with respect to incubation time is observed (see Fig. 9) proving that in such conditions condensation reactions of glucose on BSA take place. The yields in glycated products are influenced, as to be expected, by both glucose concentration and incubation time. An increment of molecular weight corresponding to a condensation of about 52 glucose units on BSA was detected.

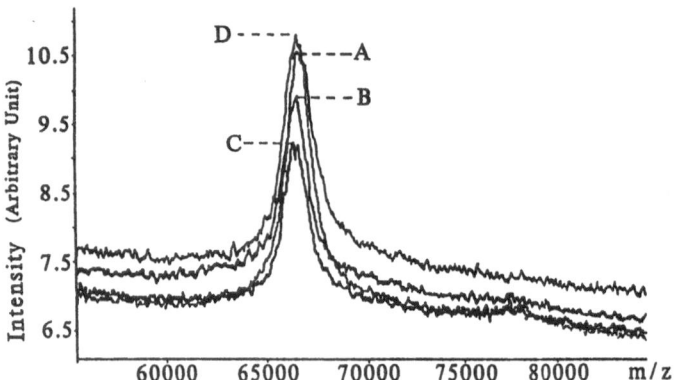

Figure 8:  MALDI spectra of BSA incubated with phosphate buffer without glucose (pH 7.5; 37°C) (samples 1) recorded at different incubation times:
A) Incubation time = 0 days (Molecular weight = 66429 Da)
B) Incubation time = 7 days (Molecular weight = 66449 Da)
C) Incubation time = 14 days (Molecular weight = 66434 Da)
D) Incubation time = 28 days (Molecular weight = 66464 Da)

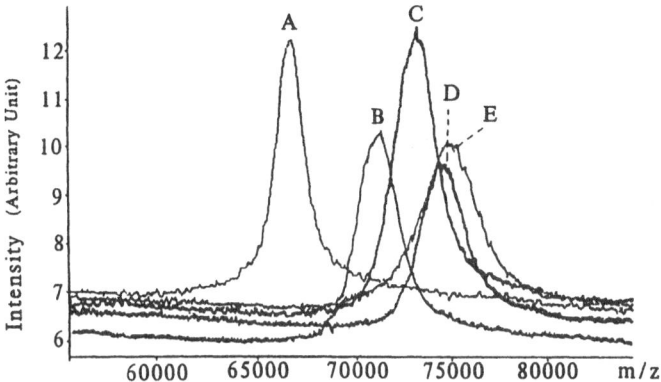

Figure 9: MALDI spectra of BSA incubated with glucose at 2M (pH 7.5; 37°C) (samples 4) recorded at different incubation times:
A) Incubation time = 0 days (Molecular weight = 66429 Da)
B) Incubation time = 7 days (Molecular weight = 71103 Da)
C) Incubation time = 14 days (Molecular weight = 73099 Da)
D) Incubation time = 21 days (Molecular weight = 74279 Da)
E) Incubation time = 28 days (Molecular weight = 74682 Da)

The data obtained by MALDI investigation suggest that, after the initial glucose addition to the ε-aminogroups, some further reactions must take place leading to the formation of more fluorescent moieties. Such reactions could be, in principle, either intramolecular or due to the reactivity of modified condensed glucose moieties vs free glucose.

## 1.3. MASS SPECTROMETRIC INVESTIGATION ON THE PRODUCTS ARISING FROM THE REACTION BETWEEN PROTECTED LYSINE AND GLUCOSE

Alternatively to the studies above described on the glycation products of proteins and mainly devoted to the identification of AGEs, in order to gain general information on the intimate mechanisms of the reaction between ε-amino group of lysine and glucose, it was thought of interest to undertake a study on the reaction between protected lysine and glucose. Such approach would give rise to an easier description of the final products as well as of possible intermediates of the glycation reactions; protected lysine should mimic the lysine contained in pro-

338

teinic chain, without all the problems related to the protein degradation.

In a first study (49) n-α acetyl-L-lysine methylester, prepared and purified according to Irwine and Gutman (53), was employed; 100 mg of it were incubated with D glucose (5g) in 5 ml of sodium phosphate buffer (pH 7.5, 0.05M Na) for 28 days at 37°C and then lyophilized. The HPLC chromatogram (UV detection) of the reaction mixture after 28 days is reported in Fig. 10: three main peaks are present with retention times of 6,11 and 12 min respectively.

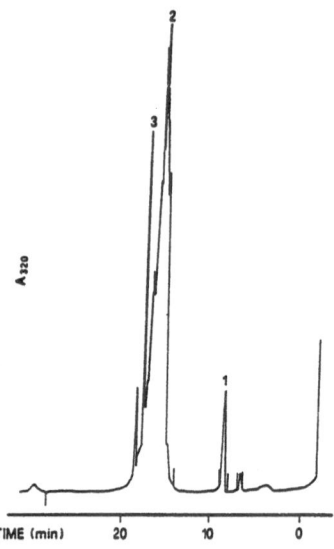

Figure 10: HPLC chromatogram of the reaction mixture of protected lysine and glucose, obtained by detection at 320 nm.

reprinted with permission of reference [49]

The same sample, when analyzed by HPLC/MS, gave rise to the chromatogram shown in Fig. 11. As it can be seen, instead of the well separated components obtained by UV detection at 320 nm, a practical continuous signal of total ion current with a maximum at 10 min, slowly decreasing, is present.

The direct analysis by FAB of the whole mixture gave a spectrum clearly different from that of the pure protected lysine (see Fig 12); the $M^{+\cdot}$ ion of protected lysine at m/z 203 is still present but with a low abundance. In the reaction mixture ions at m/z 285 and 263 are well detectable and, on the basis of their collisional spectra, structures 1 and 2 were respectively assigned (Scheme 4). The spectra obtained by HPLC/MS allowed to identify further possible molecular species at m/z 342, 326, 324, 270, 222 and 162, whose structure assignement is reported in Table 7. Furthermore, the presence of ions at m/z 189 demonstrate that a partial hydrolysis of the protected lysine takes place.

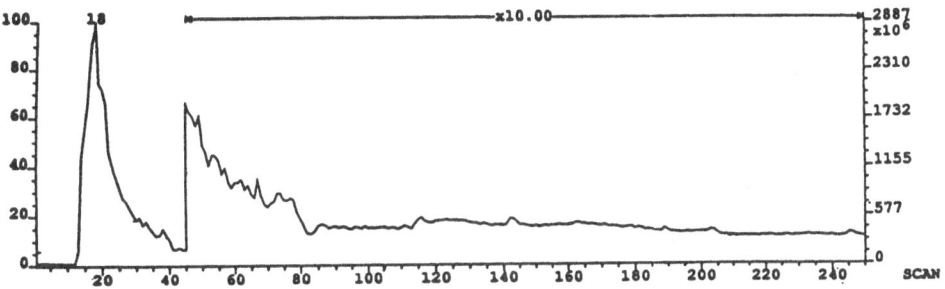

Figure 11: Reconstructed ion chromatogram for HPLC/MS analysis of the reac
tion mixture of protected lysine and glucose under plasma-spray con
ditions.

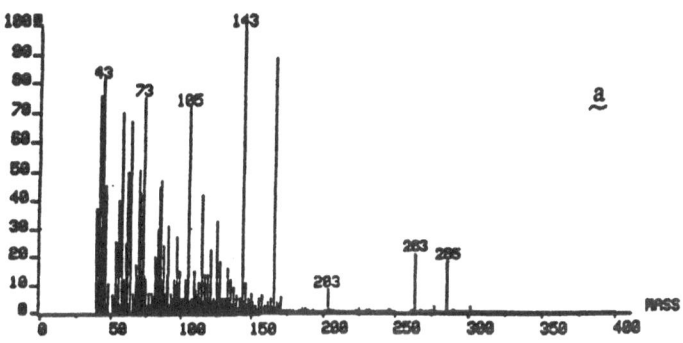

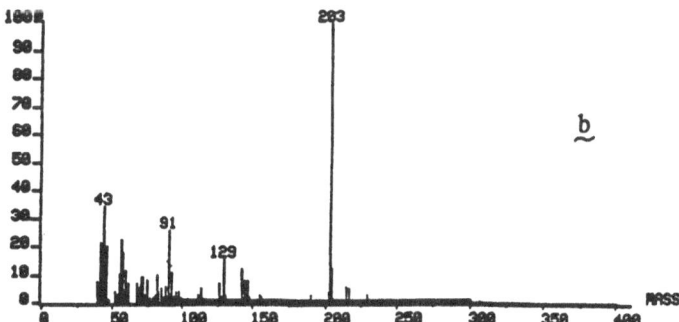

Figure 12: FAB mass spectra of:
(a) reaction mixture of protected lysine and glucose;
(b) α-protected lysine.

reprinted with permission of reference [49]

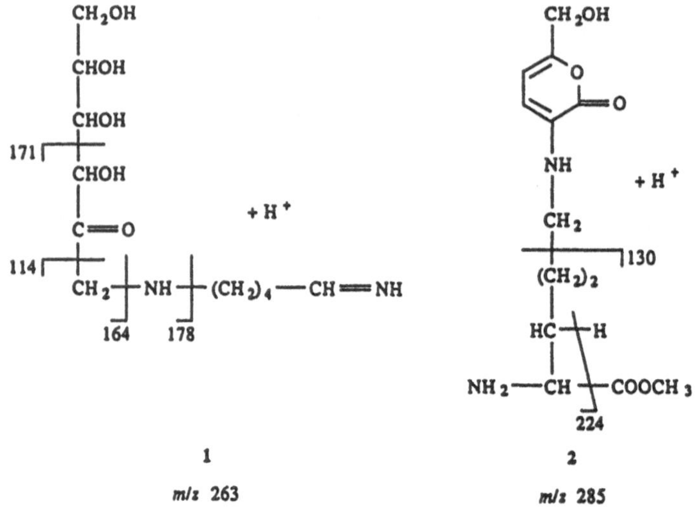

CH₂OH / CHOH / CHOH / 171 / CHOH / 114 / C=O / CH₂ — NH — (CH₂)₄ — CH = NH / 164 / 178

+ H⁺

CH₂OH / O / =O / NH / + H⁺ / CH₂ / 130 / (CH₂)₂ / HC — H / NH₂ — CH — COOCH₃ / 224

1

*m/z* 263

2

*m/z* 285

**Scheme 4**

reprinted with permission of reference [49]

For such reasons other investigation were performed using n-α-p-tosyl-lysine methylester hydrocloride; looking at the complexity of the reaction mixture, different separative and spectroscopical techniques were employed (54). 5 g of n-α-p-tosyl-lysine methylester hydrocloride were incubated with 2.57 g of anhydrous D-glucose in 6 ml of distilled water. The solution, kept at pH 7.2 was mantained at 37°C for 10 days under darkness. The free glucose was eliminated by an Extrelut cartridge and the acqueous solution was eluted with $CH_2Cl_2$. Different fractions were obtained by silical gel column chromatography. Thus, by eluting with $H_2O$, fraction A was obtained, while eluting firstly with $CH_2Cl_2$ the colorless fraction 1 was separated. Successively two different fractions were obtained with metanol: the brownish fraction 2 and the yellow fraction 3. The Extrelut cartridge was furtherly eluted with $CH_2Cl_2$ obtaining the lightly yellow fraction 4.

All the fractions were investigated by mass spectrometry and $^1H$ and $^{13}C$ nuclear magnetic resonance spectroscopy.

A portion of the reaction mixture just after glucose purification was analysed by gel permeation chromatography (GPC) in order to obtain reliable information on the molecular weight distribution of the reaction products. The results so obtained are shown in Fig. 13. A major peak with components corresponding to mean molecular weight of 870, 1228 and 1744 Da is evidenced. Other quite abundant components of the reaction mixture have mean molecular weights of 172 and 406, while a minor (but significant) component corresponds to a mean

molecular weight of 33454 Da, indicating the occurrence of extensive polimeriza-
tion processes.

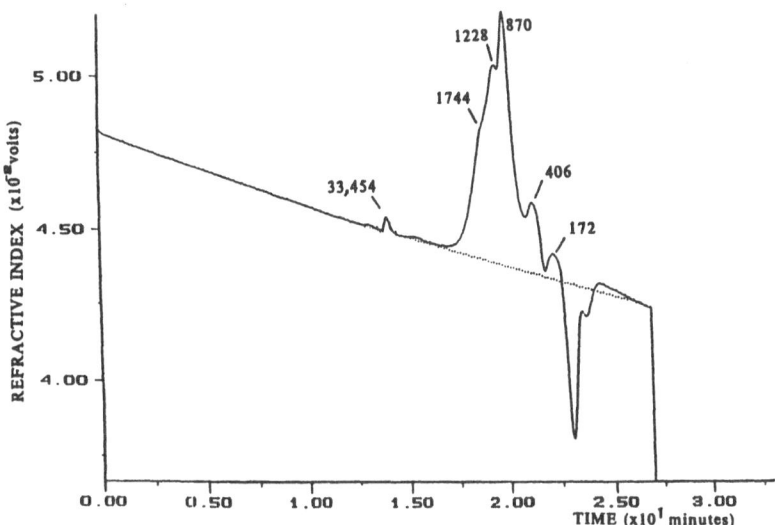

Figure 13: Gel permeation chromatogram of the whole reaction mixture obtained
by 10 days incubation of protected lysine and glucose (pH 7.2; 37°C).

This first screening allowed an estimation of the real complexity of the
reaction mixture and consequently the use of column chromatography was retained essential to obtain fractions with a more limitate number of components.

The ethereal fraction A and the four fractions 1-4 were analysed by mass
spectrometry and $^1$H and $^{13}$C NMR spectroscopy. While electron impact mass
spectrometry did not lead to any significative result, possibly due to the low
volatility of the reaction products, FAB mass spectrometry gave interesting analytical information. The FAB mass spectrum of the fraction A obtained by eluting
with diethyl-ether (to separate lower molecular weight organic products) evidenced the presence of a highly abundant ion at m/z 223 (see Fig 14). The ions
at m/z 337, 315, 283 and 238 present in the same spectrum do not have any
analytical value, being originated from protected lysine. Metastable ion studies
performed on the ion at m/z 223 led to the assignement of structure **b**, already
proposed by Ledl et al. (6).

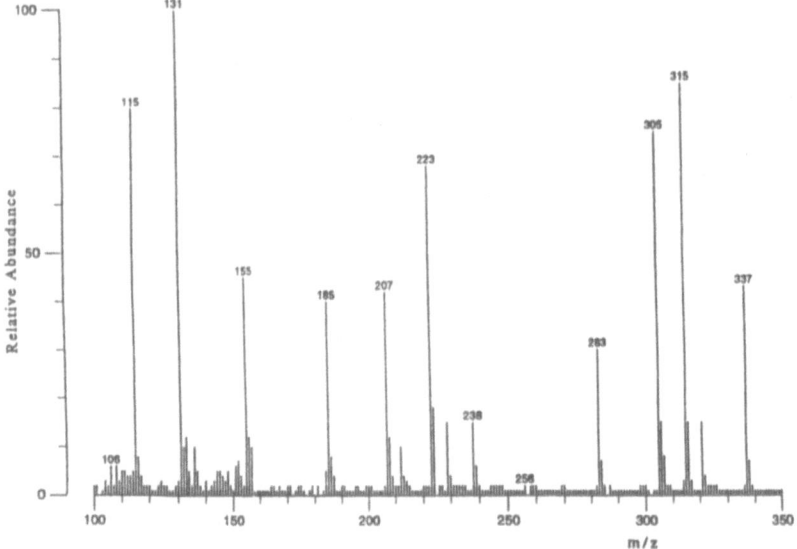

Figure 14: FAB mass spectrum of fraction A.

b

The FAB mass spectrum of fraction 1 showed the presence of highly abundant ions at m/z 191, 219, 315 and 531 with relative abundances of 13, 40, 25, and 8% respectively.

The molecular weight of 530 Da cannot be justified by the simple condensation of a glucose molecule on the protected lysine. In such case a molecular weight of 476 Da should be obtained. Hence the detected ionic species must necessarily originate by the reaction of two glucose units with the protected lysine and further skeletal rearrangement, leading to oxygen-containing heterocycles, always invoked in glycation processes. Either the fragment ions detected in the FAB spectrum or the MIKE data (showing the loss of $CH_3\cdot$ and the formation of the ions at m/z 315 and 222) are in agreement with structure **c** of figure 17. The $^1H$ NMR spectrum of the same fraction shows the signals related to the protected lysine (1.4-1.7 ppm, m, $CH_2$; 2.45 ppm, s, $CH_3$; 2.96 ppm, t, $\underline{CH_2}NH_2$; 3.44 ppm, s, $OCH_3$; 3.98 ppm, dd, CH; 7.45-7.79 ppm, m, Ph) together with an AA1 BB1 system centered at 4.11 ppm and a multiplet centered at 4.91 ppm, fully in agreement with the proposed structure **c**.

c

In the FAB mass spectrum of fraction 2 abundant ions at m/z 459 and 477 are present; on the basis of MIKE data and $^1H$ NMR spectrum (which does not show the presence of signals attributable to heterocyclic systems) the structure of a condensation product of glucose on the ε-amino group of lysine was assigned. Fractions 3 and 4 led to molecular species at m/z 783 and 1070 respectively; in both cases highly abundant ions at m/z 477 were detected; the two ions at higher mass can be considered originating from the addition to a protected lysine of three glucose units with losses of 4 $H_2O$ (m/z 783) and five glucose units with losses of 8 $H_2O$ (m/z 1070). The related $^1H$ NMR spectra show a very complex pattern. However integration data are in agreement with the proposed mechanism of multiple glucose addition on a lysine molecule and not with the possible polymerization of the Amadori products. Thus, even if fraction 3 and 4 are yellow (indicating the presence of browning products) the most abundant component in the fraction under study is the Amadori product itself. These features indicate that, under the experimental conditions employed, the late stage of the Maillard reaction is just initiated. Further confirmation of such data were also gained by $^{13}C$ NMR.

In conclusion  the studies until now developed in the field of advanced glycation end products and the results obtained by mass spectromeric metodologies have shown the great potentialities of such technique in this field and its more extensive use in the future will surely give an important contribution to the rationalization of Maillard reaction mechanism*S*.

References

1.  V.M. Monnier, A. Cerami "Non enzymatic glycosylation and browning of proteins in diabetes" *Clin. Endocrinol .Metab* . **11**, 431-52, (1982).

2.  L.C. Maillard "Synthèse des materies huniques par action des acid aminés sur le sucres reducteurs" *Ann. Chim. Sèr.* **9**, 258, (1916).

3.  L.C. Maillard "Action des acid amines sur les sucres; formation des mela noidines par voic methodique" *C R Acad. Sci.* **154**, 66-8, (1912).

4.  T.M. Reynolds "Chemistry of non-enzymatic browning" *Adv. Food. Res.* **12**, 1-52, (1963).

5.  T.M. Reynolds "Chemistry of non-enzymatic browing II" *Adv. Food. Res.* **14**, 167-283, (1965).

6.  Ledl F. "Chemical pathways in the Maillard reaction" In <u>The Maillard Reaction in Food processing. Human Nutrition and Physiology</u>, Finot P.A., Aesch bacher H.U., Hurrel R.F., Liarden E. (Eds.), Birkhäuser Verlag Basel, (1990).

7.  V.M Monnier, A. Cerami "Non-enzymatic glycosylation and browning of proteins in vivo" In <u>The Maillard Reaction in Foods and Nutrition</u>, G.R. Waller, F. Feather (Eds) . Am Chem Soc Symp Series 215, Washington D.C. The Am Chem Soc.(1983)

8.  M. Brownlee, H. Vlassara, A. Cerami "Nonenzymatic glycosylation and the pathogenesis of diabetic complications" *Ann. Intern. Med.* **101**, 527-37, (1984).

9.  V.M. Monnier, A. Cerami "The search for non enzymatic browning products" *Invest. Opthalmol .Vis. Sci.* **20**, 169-74, (1981).

10. T.S.Lyons, G. Silvestri, J.A. Dunn, D.G. Dyer, J.W. Baynes "Role of glycation in modification of lens crystallins in diabetic and non diabetic senile cataracts" *Diabetes* **40**, 1010-15, (1991).

11. H. Vlassara, M. Brownlee, A. Cerami "Accumulation of diabetic rats periph eral nerve myelin by macrophage increases with extent and duration of non enzymatic glycosylation" *Proc. Natl. Acad. Sci USA* **160**, 197-207, (1984).

12. N.J. Patel, V.P. Misra, P. Dandone, P.K. Thomas "The effect of non enzymatic glycation of serum proteins on their permeation into peripheral nerve in normal and speptozotocin-diabetic rats" *Diabetologia* **34**, 78-80, (1991).

13. S. Tanaka, G. Avigad, B. Brodsky, E.F. Enkenberry "Glycation induces expansion of the molecular packing of collagen" *J. Mol. Biol.* **203**, 495-505, (1988).

14. E.C. Tsilbary, A.S. Cheronis, L.A. Reger, R.M. Wohlhneter, L.T. Furcht "The effect of nonenzymatic glucosylation on the binding of the main noncolla genous NC1 domain to type IV collagen" *J. Biol. Chem.* **263**, 4302-8, (1988).

15. G. Lubec, A. Pollak "Reduced susceptibility of non enzymatic glucosylated glomerular basement membrane to protease: is thickening of diabetic glom erular basement membranes due to reduced proteolytic degradation?" *Renal Physiol* **3**, 4-8, (1980).

16. Z. Makita, S. Radolf, E.J. Rayfield, Z. Yang, E. Skolnik, V. Delany, E.A. Friedman, A. Cerami, H. Vlassara " Advanced glycosylation end products in patients with diabetic nephropathy", *N. Engl. J. Med.* **325**, 836-42, (1991).

17. M. Brownlee, H. Vlassara, A. Cerami A. "Advanced glycosylation end products in tissue and the biochemical basis of diabetic complications. *N. Engl. J. Med* **318**: 1315-21, (1988).

18. N.B. Ruderman, J.R. Williamson, M. Brownlee "Glucose and diabetic vascu lar disease" *FASEB J.* **6**: 2905-14, (1992).

19. S. Pongor, P.C. Ulrich, F.A. Bencsath, A. Cerami " Aging of proteins:isola tion and identification of a fluorescent chromophore from the reaction of pol ypeptides with glucose", *Proc. Natl. Acad. Sci. USA* **81**, 2684-4, (1984).

20. D.R. Sell, V.M. Monnier "Structure elucidation of a senescence crosslink from human extracellular matrix" *J. Biol. Chem.* **264**, 21597-602, (1989).

21. D.R. Sell, R.H. Nagaray, S.K. Grandhee, P. Odetti, A. Lapolla, J. Fogarty, V.M. Monnier "Pentosidine: a molecular marker for the cumulative damage to proteins in diabetes, aging, and uremia" *Diabetes Metabolism Review* **7** (4), 239-51, (1991).

22. D.R. Sell, A. Lapolla, P. Odetti, J.Fogarty, V.M.Monnier "Pentosidine forma tion in skin correlates with severity of complications in individuals with long-standing IDDM" *Diabetes* **41**, 1286-92, (1992).

23. J.C.F. Chang, P.C. Ulrich, R. Bucala, A. Cerami " Detection of an advanced glycosylation product bound to protein in situ", *J. Biol. Chem.* **13**, 7970-4, (1985).

24. S. Miyata, V.M. Monnier "Immunocytochemical detection of advanced glyco-sylation end products in diabetic tissues using monoclonal antibody to pyr-raline" *J. Clin. Invest.* **89**, 1102-12, (1992).

25. Z. Makita, H. Vlassara, A. Cerami, R. Bucala " Immunochemical detection of advanced glycosylation end products in vivo" *J. Biol. Chem.* **267** (8), 5133-8, (1992).

26. E. Gelpi "Trends in biochemical and biomedical applications of mass spec trometry" *Int J Mass Spectrom Ion Processes* **118/119**: 683-721, (1992).

27. R.A. Yost "MS/MS Tandem Mass Spectrometry" *Spectra* **4**: 2-6, (1983).

28. H. Kaiser "Foundations for the critical discussion of analytical methods." *Spectrochim Acta Part. B* **33B**: 551-76, (1978).

29. J.J. Boon "Analytical pyrolysis mass spectrometry: new vistas opened by temperature resolved in-source PY MS" *Int J Mass Spectrom Ion Processes* **118/119**: 755-88, (1992).

30. H. Barber, R.S. Bordoli, R.D. Sedgwick, A.N. Tyler "Fast atom bombardment as an ion source in mass spectrometry" *Nature* **293**: 270-5, (1981).

31. H. Karas, D. Bachmann, F. Hillenkamp "Influence of the wavelenght in high irradiance ultraviolet laser desorption mass spectrometry of organic mole cules" *Anal Chem* **57**: 2935-9, (1985).

32. R.D. Smith, J.A. Loo, C.G. Edmonds, C.J. Barinaga, H.R. Udset "New devel-opments in biochemical mass spectrometry: electrospray ionization" *Anal Chem* **62**: 822-899, 1990.

33. E. Schleicher, O.H. Wieland "Specific quantitation by HPLC of protein (lysine) bound glucose in human serum albumin and other glycosylated proteins" *J Clin Chem Clin Biochem* **19**: 81-7, (1981).

34. B. Pelli, A. Sturaro, P. Traldi, A. Lapolla, T. Poli, D. Fedele, G. Crepaldi "Collisional spectroscopy as a screening procedure for the determination of 2-(2-furoyl)-4(5)-(2-furanyl)-1H-imidazole from acid hydrolysis of β-poly (L-lysine) and β-albumin" *Biomed. Environ. Mass Spectrom.* **13**, 7-11, (1986).

35. A. Lapolla, C. Gerhardinger, B. Pelli, A. Sturaro, E. Del Favero, P. Traldi, G. Crepaldi, D. Fedele "Absence of brown product FFI in non diabetic and diabetic rat collagen" *Diabetes* **39**, 57-61, (1990).

348

36. F.G. Njoroge, A.A. Fernandes, V.M. Monnier "Mechanism of formation of putative advanced glycosylation end product and protein cross-link 2-(2-furoyl)-4(5)-(2-furanyl)-1H-imidazole", *J. Biol. Chem.* **263**, 10646-52, (1988).

37. K.L. Busch, G.L. Glish, S.A. McLuckey In: <u>Mass Spectrometry/Mass Spec trometry. Techniques and applications of Tandem Mass Spectrometry</u>, K.L. Busch, G.L. Glish, S.A. McLuckey (Eds), VCH New York, (1988).

38. A. Lapolla, T. Poli, C. Gerhardinger, D. Fedele, G. Crepaldi, D. Chiarello, E. Ghezzo, P. Traldi "Parent ion spectroscopy in the identification of advanced glycation products" *Biomed. Environ. Mass Spectrom.* **18**, 713-8, (1989).

39. C. Gerhardinger, A. Lapolla, G. Crepaldi, D. Fedele, E. Ghezzo, R. Seraglia, P. Traldi "Evidence of acid hydrolysis as responsible for 2-(2-furoyl)-4(5)-(2-furanyl)-1H-imidazole production" *Clin. Chim. Acta* **189**, 353-40, (1990).

40. A. Lapolla, C. Gerhardinger, E. Ghezzo, R. Seraglia, A. Sturaro, G. Crepaldi, D. Fedele, P. Traldi "Identification of furoyl containing advanced glycation products in collagen samples from diabetic and healthy rats" *Biochem. Biophys. Acta* **1033**, 13-8, (1990).

41. A. Lapolla, C. Gerhardinger, L. Baldo, G. Crepaldi., D. Fedele, C.J. Porter, R. Seraglia, D. Favretto, P. Traldi "Investigation of products arising from enzymatic digestion of advanced glycated albumin by high-performance liquid chromatography/mass spectrometry" *Rap. Commun. Mass Spectrom.* **5**, 624-28, (1991).

42. K. Olsson, P.A. Ternemalm, T. Popoff, O. Teanter "Formation of aromatic compounds from carbohydrates. V. Reaction of D-glucose and methylamine in slightly acid acqueous solution" *Acta Chem. Scan.* **B31**, 469-74, (1977).

43. M. Sengl "Identifizierung niedermolekularer, polaler Zuckerumwandlungs produkte sowie Nachweis eines proteingebundenen produkts aus der Spätphase der Maillard-Reaction" Dissertation University of Munic (1988).

44. F. Ledl, G. Fritsch, I. Hiebel, O. Parchmayr, T. Severin "Degradation of Mail-lard products" In: <u>Amino-Carbonyl Reactions in Food and Biological System</u>, M. Fuyimaki, M. Namiki, Kato H. (Eds), Elsevier, Amsterdam, Dev. Food Sci, (1986).

45. S.E. Hodge "The Amadori rearrangement" *Adv Carbohyd Chem* **10**: 169-205, (1955).

46. F.G. Njoroge, L.M. Sayre, V.M. Monnier "Detection of D-glucose-derived pyr-

role compounds during Maillard reaction under physiological conditions "*Car bohydr. Res.* **167**, 211-20, (1987).

47. A. Lapolla, C. Gerhardinger, L. Baldo, D. Fedele, D. Favretto, R. Seraglia, P. Traldi "Pyrolysis/gas chromatography mass spectrometry in the analysis of glycated poly-L-lysine" *Org. Mass Spectrom.* **27**, 183-7, (1992).

48. A.P. Snyder, J.H. Kremer, H.L. Meuzelaar, W. Winding "Curie point pyrolysis atmosperic pressure chemical ionization mass spectrometry as a probe on the effect of sodium chloride on biopolymers" *J Anal Appl Pyrrol* **13**: 77-88, (1988).

49. A. Lapolla, C. Gerhardinger, G. Crepaldi, D. Fedele, M. Palumbo, D. Dal Zoppo, C.J. Porter, E. Ghezzo, R. Seraglia, P. Traldi "Mass spectrometric approaches in structural identification of the reaction products arising from the interaction between glucose and lysine" *Talanta* **38**, 405-12, (1991).

50. A. Lapolla, C. Gerhardinger, L. Baldo, G. Crepaldi, D. Fedele, D. Favretto, R. Seraglia, O. Curcuruto, P. Traldi "Pyrolysis-gas chromatography mass spec trometry in the characterization of glycated albumin" *J. Anal. Appl. Pyr.* **24**, 87-103, (1992).

51. H. Paulsen, K.W. Pflughaupt "Glycosylamines" In: Carbohydrates IB, W. Pig man, D. Horton (Eds.) Academic Press New York, (1980).

52. A. Lapolla, C. Gerhardinger, L. Baldo, D. Fedele, A. Keane, R. Seraglia, S. Catinella, P. Traldi "A study on in vitro glycation processes by Matrix Assisted Laser Desorption Ionization Mass Spectrometry" Submitted to *Biochem Bio phis Acta*.

53. C.C. Irving, H.R. Gutmann "Preparation and properties of N-α-acyl-lysine ester." *J. Org. Chem.* **24**: 1979-82, (1959).

54. A. Lapolla, C. Gerhardinger, L. Baldo, D. Fedele, R. Bertani, G. Facchin, E. Rizzi, S. Catinella, R. Seraglia, P. Traldi "The lysine glycation. 1. A preliminary investigation on the products arising from the reaction of protected lysine and D-glucose" Submitted to *Aminoacids*.

# NUCLEIC ACIDS: OVERVIEW AND ANALYTICAL STRATEGIES

PAMELA F. CRAIN
*Department of Medicinal Chemistry*
*University of Utah*
*Salt Lake City, UT 84112*
*U.S.A.*

ABSTRACT. Mass spectrometry is finding increased utility for the analysis of nucleic acid constituents at all structural levels, from the heterocyclic bases to intact small nucleic acids. Problems of structure elucidation of modified nucleosides in nucleic acids and their quantification, and sequence determination of oligonucleotides are major areas of investigation. This review emphasizes the most practical approaches to solutions of these types of problems. In particular, the analysis of nucleic acid constituents in mixtures, and the techniques of tandem mass spectrometry (MS/MS) for structure elucidation and oligonucleotide analysis using electrospray and matrix-assisted laser desorption ionization are emphasized.

## 1. Introduction

Nucleic acids and their constituents have historically presented a substantial challenge for mass spectrometric analysis because of their high polarity. The goal of this review is to provide an overview of the biologically-relevant issues in nucleic acid structural biochemistry, and to provide an introduction to mass spectrometry-based analytical strategies useful for their solution. This review is not intended to be comprehensive or to present a historical overview, and citations that contain fundamental studies or that illustrate tested approaches will be emphasized over those that simply demonstrate feasibility. The literature cited may be consulted for leading references. Applications have exploited advances in ionization methods and instrumentation, and as newer methods are introduced, older methods, although useful at that point in time have since been displaced. Accordingly, older methods of historic importance, notably [$^{252}$Cf]plasma desorption-MS and field desorption-MS for their ability to ionize polar molecules, and pyrolysis-EI-MS for nucleoside isomer distinction using MIKES, will not be covered here, but have been earlier reviewed [1]. Approaches that facilitate analysis of nucleic acid constituents in mixtures will be emphasized, along with applications of tandem mass spectrometry for structure elucidation and methods that exploit the ability of both electrospray (ESI) and matrix-assisted laser desorption (MALDI) to generate ions of the highly polar oligonucleotides and nucleic acids.

Reflecting the growing importance of mass spectrometry as a tool for analysis of nucleic acids and their constituents, practical aspects of their analysis are reviewed in the recent volume in the Methods in Enzymology series devoted to biomolecular mass spectrometry [2–8]. Other recent

351

*R. M. Caprioli et al. (eds.), Mass Spectrometry in Biomolecular Sciences, 351–379.*
© *1996 Kluwer Academic Publishers.*

reviews of a general nature are refs. 1, 2 and 9–12.

Fundamental questions in nucleic acid structure problems (in common with those of all other biopolymers) can be grouped into three general categories [1]. What (is present)? How much (is there)? Where (is it located)? The first question comprises use of mass spectrometry to screen nucleic acids for the presence of modified nucleosides and to accomplish structure elucidation when unknown structures are encountered. The second question is addressed through the use of methods for quantification of compounds of interest, typically at the base level. Finally, the primary sequence is perhaps the most fundamental property of nucleic acids, and there is currently great interest in the application of mass spectrometry to sequence determination.

## 2. Basic features of nucleic acid structure

The nucleic acid polymer backbone is constructed by the joining of the nucleosides (Fig. 1) through a phosphodiester bond from the 3' hydroxyl of the sugar of one nucleoside to the 5' hydroxyl of the adjacent nucleoside. By convention the leftmost nucleoside is position 1 and the bond is described as 3' → 5'. The internucleotide (phosphate) linkage possesses one negative charge at pH > 1, so nucleic acids are polyanions and present a considerable challenge for mass spectrometric analysis at the nucleotide and larger level. The glycosidic bond connects the base

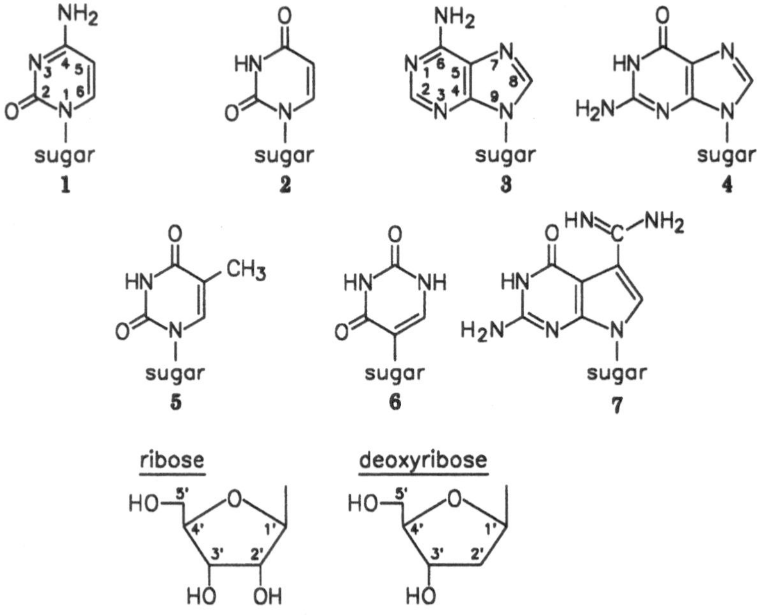

Figure 1. Structures of the major nucleosides in RNA (1–4; ribose sugar) and DNA (1, 3–5; deoxyribose sugar). Ring numbering systems are shown for a pyrimidine (1, cytosine) and a purine (3, adenine) base, and both sugars. Pseudouridine (6, 5-ribosyluridine) and archaeosine (7, a modified nucleoside unique to organisms in the archaeal kingdom) are common modified RNA nucleosides.

to C–1' of the sugar through N–1 (pyrimidines) or N–9 (purines). The alternative C–C glycosidic linkage is present in pseudouridine (6) and its derivatives, while 7 has the 7-deazaguanine (pyrrolo[2,3-*d*]pyrimidine) ring system. These modified structural features have important consequences for mass spectral features (Section 3.1.). Ref. 13 may be consulted for basic information on the chemical properties and structural and conformational properties of nucleic acids and their constituents.

## 3. Structure determination of modified nucleosides in nucleic acids

Both DNA and RNA contain modified nucleosides, and their structure elucidation historically represents the most successful application of mass spectrometry to problems in nucleic acid studies. The two nucleic acids differ structurally only in the nature of the sugar in the backbone. There are, however, substantial differences in the nature and degree of difficulty of structure problems to be addressed for each. RNA contains at least 93 reported naturally-occurring modified nucleosides [14], while mammalian DNA, for example, contains only one, 5-methyl-deoxycytidine (present at about 1 mole %). Sites of naturally-occurring modification are highly conserved and the extent of modification generally is complete for both RNA and DNA. As a consequence of the number of nucleosides in nucleic acid (76 in the smallest transfer RNA up to 4786 in the largest ribosomal RNA; $10^9$ in mammalian DNA) a single modified nucleoside in the largest RNA occurs at the 0.02 % level, while for DNA it occurs to an extent of $10^{-7}$ %. Adventitious modification by environmental and dietary sources is generally taken to be of no consequence to RNA function, but damaged DNA, if it escapes repair, is implicated in the processes of carcinogenesis and aging. Thus, in the case of RNA, the naturally-occurring modifications are of interest and their structure elucidation is generally tractable [2], while for DNA, it is the "unnatural" modifications that are of greatest interest, and their structural elucidation can represent a formidable challenge [15]. Nonetheless, mass spectrometry has considerable utility for structural studies of modified nucleosides from both RNA and DNA and is widely used in this area.

### 3.1. BASIC FEATURES OF NUCLEOSIDE MASS SPECTRA

Nucleosides, which are recovered from nucleic acids by digestion to the nucleotide (nucleoside monophosphate) level with nucleases followed by dephosphorylation with phosphatase, are the most readily analyzed integral structural unit in nucleic acids. The nomenclature utilized to describe fragment ions in the first report of nucleoside spectra by Biemann and McCloskey [16] refers to the base fragment as "B" and the sugar as "S" (Scheme I) is widely used, and will be

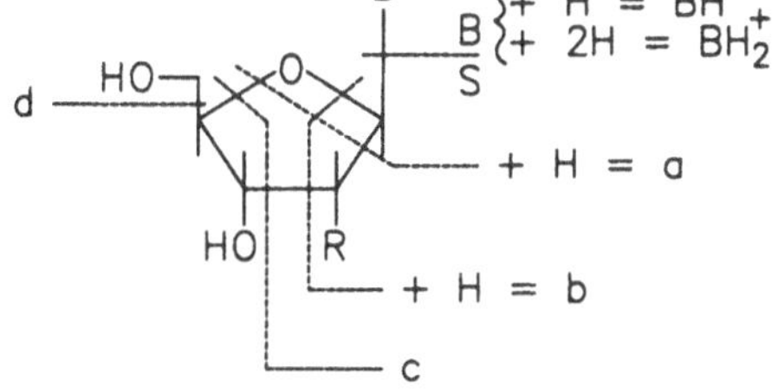

Scheme I

employed here. As a general rule, the major fragment ion seen in mass spectra is cleavage of the glycosidic bond and elimination of the sugar to give a base fragment accompanied by one hydrogen ($BH^{+\cdot}$ from EI [16]) or two hydrogens ($BH_2^+$ from chemical ionization (CI) [17], fast atom bombardment (FAB) [18,19] or electrospray ionization (ESI) [20]) in the positive ion detection mode. The C–C glycosidic bond in **6** is stronger than the normal C–N bond, so $BH^{+\cdot}$ and $BH_2^+$ ions are not produced. The substitution of C for N in the deazapurine nucleoside **7** leads to strengthening of the glycosidic bond, and the protonated B ions are likewise absent.

Fast atom bombardment is undoubtedly the most widely used ionization method for nucleoside mass spectra at present, owing to its simplicity, availability and suitability for examination of polar charged species. Nonetheless, ESI will find increased use as a much more sensitive option than FAB for analysis of nucleic acid substituents at the monomer level [20,21], but especially so when analysis at the nucleotide level is advantageous [22].

### 3.1.1. *Electron ionization.*

The chemistry of gaseous ions from EI of bases and nucleosides has been extensively reviewed [23-25] and will not be discussed in detail here. In addition to the two B-related ions, other fragment ions (*a–c*, Scheme I) result from cleavage through the sugar ring. The source of rearranged hydrogens in the $BH^{+\cdot}$, $BH_2^+$ and B + 30 (ion *a*) ions was shown by deuterium labeling to be largely from sugar hydroxyl groups and the latter two ions are reduced in abundance in spectra of deoxynucleosides. Ion *b* (B + 44, R is OH or B + 28, R is H) is also observed in spectra of free nucleosides and when the 2′ OH is methylated (a common RNA modification) ion *b* shifts by 14 u, making this ion diagnostic for this modification. EI with negative ion detection was used to study several nucleosides [26] and decreased fragmentation was reported, compared with (+)EI. The spectra contained essentially only the $B^-$ ion, and the observation was rationalized in terms of enhanced stability arising from the possibility of extensive charge delocalization in the heterocyclic base moieties. This factor most likely accounts for similar observations regarding stabilization of $B^-$ in mass spectra with other ionization modes.

### 3.1.2. *Desorption ionization.*

A comprehensive report of CI mass spectra of free nucleosides [17] used a variety of reagent gases to ionize a set of nucleosides which were sufficiently volatile to be evaporated from the direct insertion probe. The difference in proton affinities between the analyte and reagent gas was reported to correlate directly with abundance of the $MH^+$ ion. As reagent gas basicity increases, the protonation reaction becomes less exothermic, and so less energy is available for decomposition of $MH^+$. Much (if not all) of the ion chemistry is directly relevant to interpretation of nucleoside spectra produced from other "soft" ionization methods. Negative ion CI spectra were reported for several ribonucleosides and bases [27]; molecular ions for nucleosides were weak or non-existent and formation of $B^-$ was a dominant process. The most prominent ion in base spectra was $(M - H)^-$, formally equivalent to the $B^-$ ion from nucleosides.

Major ions in FAB mass spectra of nucleosides consist of the $MH^+$ and $BH_2^+$ ions ((+)FAB), and $(M - H)^-$ and $B^-$ ions ((–)FAB) [18,19,28]. The use of a labeled matrix permits the number of exchangeable hydrogens to be determined [29], as well as aids in ion assignment. Deuterated glycerol and thioglycerol are commercially available, and *m*-nitrobenzylalcohol-*d* is readily prepared [30]. Side reactions from analyte-matrix reactions have been reported [reviewed in ref. 1]. Nonetheless, FAB remains a popular and versatile technique, particularly in combination with tandem mass spectrometry (Section 3.3). Aspects of FAB-MS of nucleic acid constituents have been reviewed [1,31,32], and ref. 31 is especially recommended because it compares FAB-derived mass spectra with those from EI and CI, with emphasis on common ion chemistry and structures.

## 3.2. DERIVATIVES FOR MASS SPECTROMETRY

While FAB (particularly in combination with tandem mass spectrometry (MS/MS) is extensively utilized for nucleoside structure problems (Section 3.3.), microscale synthesis of volatile derivatives remains a useful tool for structural studies of nucleic acid constituents for several reasons. First, there is an absolute requirement for derivatization for gas chromatography-mass spectrometry (GC/MS). Second, EI-MS of volatile derivatives will remain a useful technique for specialized applications to structural studies (Section 3.1.1). Finally, trimethylsilylation produces derivatives that are highly hydrophobic and surface active, and allows for highly sensitive detection in FAB-MS of nucleosides [33,34] and nucleotides [33,35]. Permethyl derivatives share these features and have unique utility for MS/MS studies of modified guanines [36,37] (Section 3.3.).

Trimethylsilyl (TMS) derivatives were the first to be employed for EI-MS of nucleosides and nucleotides [38], and the ion chemistries for TMS derivatives of nucleosides [39] and nucleotides [40] have been thoroughly explored. These derivatives are readily prepared [4], and labeled derivatives synthesized from deuterated silylating reagents are useful for assignment of ion structures [41]. Side reactions have been reported [42–44], and while generally undesirable, they can be diagnostic for certain structural features, e.g., 7-alkylation of guanine residues [42].

Other derivatives that are useful in mass spectrometry of nucleosides include permethyl [45,46], trifluoroacetyl [47], and $2',3'-O$-isopropylidene [45]. Ion assignments for permethyl [46] and trifluoroacetyl [47] derivatives are the best documented.

## 3.3. METASTABLE AND COLLISION-INDUCED DISSOCIATION STUDIES

"Soft ionization" methods such as FAB and thermospray (in directly combined liquid chromatography-mass spectrometry (LC/MS; Section 4.1.2.) are widely used for the detection and characterization of modified nucleosides. Typically, besides the protonated (or deprotonated) molecular ion, only the $BH_2^+$ (or $B^-$) ion is produced, making the site of modification difficult, if not impossible, to determine if the structure is unknown. Both metastable- and collision-induced dissociation (CID) have considerable utility in structural studies of base-modified nucleosides in nucleic acids. In the case of FAB-MS, CID provides the additional benefit of reducing chemical noise from the matrix [18].

Several important concepts are relevant to utilizing MIKES or CID for structure studies of nucleoside constituents. It has been shown that the B-related ion from glycosidic bond cleavage in primary (ion source-produced) spectra of nucleosides, and the M-related ion of the corresponding free base, have the same structure in the gas phase [48,49]. As a consequence, when a comparison of the nucleoside of interest with a reference nucleoside is desired, and the latter is not readily available, the corresponding base may be substituted. In addition, as shown in Figure 2, product ions from CID of the $MH^+$ ion (from FAB) of adenine are the same ion types (shifted 1 u higher from the proton of ionization) as are present in the EI spectrum of the free base (Fig. 2A). The nucleic acid bases are well-studied, and consideration of EI mass spectra of nucleic bases may provide useful insights for interpretation of CID spectra from the corresponding protonated species. (For convenient access to the early literature, see refs. 23, 25 and 50). Finally, CID of the B-related ion is generally more informative than CID of the M-related ion [1,51,52] because in the latter case energy will be dissipated into the already favored glycosidic bond cleavage instead of into fragmentation of the base moiety.

Several experimental parameters that may influence the mass spectrum are ionization method (e.g. FAB vs. EI), ion detection mode (positive vs. negative) and collision energy (keV vs. eV).

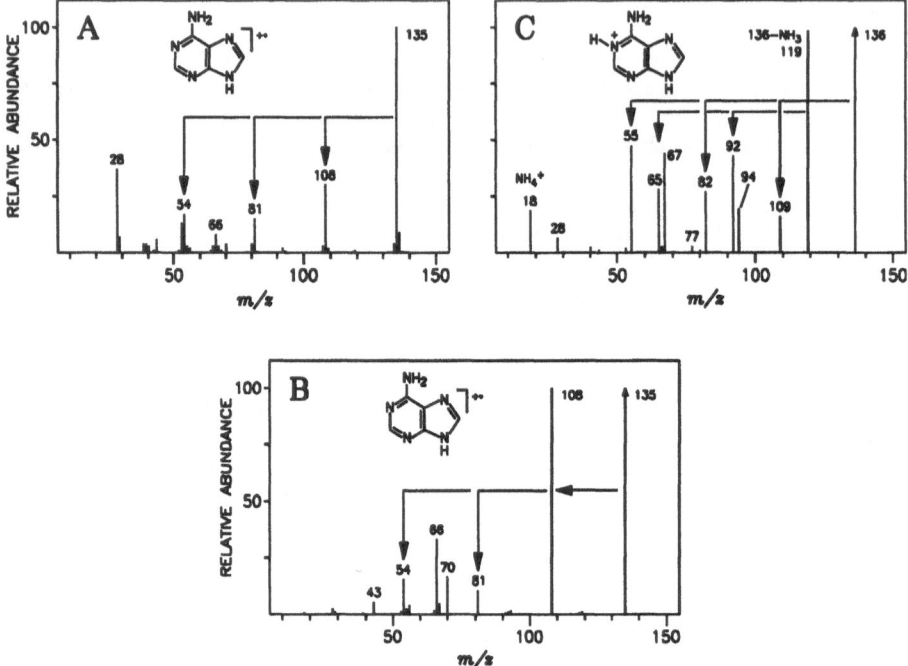

Figure 2. Mass spectra of adenine ($M_r$ 135) from: A. 70 eV electron ionization; B. CID of the $M^{+\cdot}$ ion from EI, $m/z$ 135; C. low energy CID of $MH^+$ from FAB, $m/z$ 136. Reproduced from [21] with permission.

Although few systematic studies have been carried out, some generalizations are appropriate. Differentiation of 1-, 2- and $N^6$-methyladenine isomers may be effected using CI and high energy CID with either positive [53] or negative [54] ions detected. In this example, the detection mode was not critical. On the other hand, high energy CID of $(M - H)^-$ ions of uridine and cytidine (1 and 2, ribose sugar; Fig. 1) resulted in loss of HNCO, but the corresponding loss was not observed for CID of $MH^+$ of 1 or 2 [18,52]. In contrast, $N^2$-methylguanosine loses methylene-imine ($N^2$ and the methyl) from $MH^+$, but not from $(M - H)^-$ [18]. These observations suggest that in some cases the utility of information derived from CID would be influenced by the choice of detection mode (+ vs. –). Other informative studies directed toward the issue of determining sites of substitution on the heterocyclic ring include the use of FT-MS/MS (negative ions) for isomer distinction among a series of methylguanosines [55] and the use of low energy CID (positive ions) for a set of methylated adenines [56]. The latter study placed heavy reliance on the use of isotopically labeled bases to dissect the fragmentation pathways. A rapidly developing trend is the use of hydrophobic derivatives for FAB-MS/MS studies [34,36,37,57,58] because of the enhanced sensitivity of detection they provide.

Applications of tandem mass spectrometry to the analysis of modified nucleic acid constituents have been directed largely toward structure elucidation of DNA adducts and related model compounds [57,59–64] (reviewed in refs. 12 and 15). API-MS/MS was used to characterize oxidized bases in irradiated polyadenylic acid [51]. Three unusual naturally occurring modified nucleosides in tRNA from hyperthermophilic organisms were characterized recently [36,58], and in the case of archaeosine (7, Fig. 1), from archaeal tRNA, among all of the mass spectrometric techniques applied to the structure elucidation, only MS/MS of the permethyl derivative provided the information that led to the structure assignment [36]. As the gene sequence of the tRNA suggested that archaeosine was a guanosine derivative, the study of isomeric methylguanosines by low-energy CID was undertaken as a separate exercise, but unambiguous isomer distinction could not be made. When the monomethylated guanosines were alkylated using deuteromethyl iodide, and examined by FAB-MS/MS, however, structurally significant ions could be discerned for the isomers. Figure 3 is the spectrum from CID of the $BH_2^+$ ion of per(deuteromethyl)-guanosine, showing the structures of ions whose shifts in mass reflect substitution at a particular site on the guanine moiety. Assignments were supported by $^{15}N$ and $^{18}O$ labeling. Given the frequency with which deoxyguanosine is a target for xenobiotic adduct formation, this study should prove useful for structure elucidation of carcinogen-modified DNA.

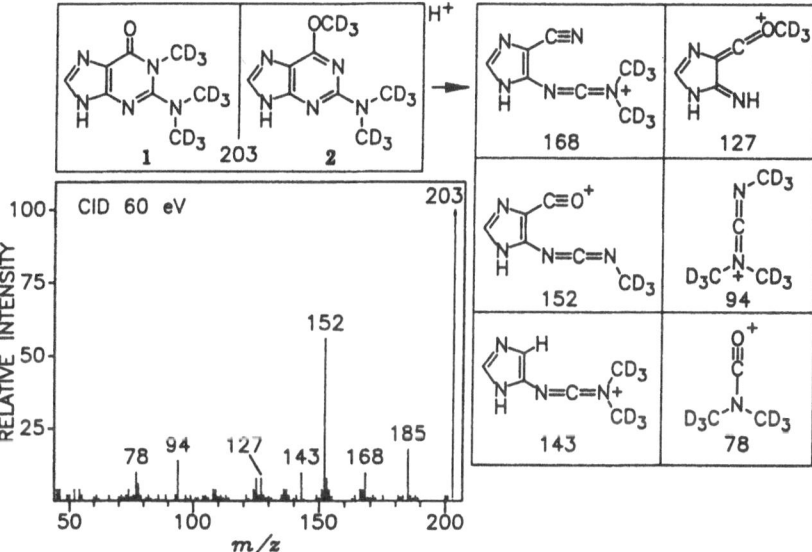

Figure 3. Product ions from CID of the $BH_2^+$ ion from perdeuteromethylated guanosine. Proposed structures for six diagnostic fragment ions are shown. Guanosine tautomers are 6-oxo (1) and 6-hydroxy (2). Reprinted from [37] with permission.

## 4. Analysis of nucleosides and bases in mixtures

Methods that allow for the analysis of modified nucleosides that occur at low levels in nucleic acids, without the necessity of isolation and its attendant sample losses at each stage, provide a powerful tool for structure studies. Two general approaches include combined chromatography-mass spectrometry and tandem mass spectrometry. While LC/MS and GC/MS require more effort than a straightforward MS/MS measurement, it should be noted that a retention timed gained from chromatographic sample introduction constitutes an additional parameter of identity. MS/MS can provide structurally useful information and has an advantage in speed and simplicity, but its application to complex mixtures is problematic if more than one isomer is present, which is generally not a problem with chromatographic sample introduction. The combination of both methodologies in LC-MS/MS, represents an especially powerful approach to structure characterization of nucleic acid constituents in mixtures [51].

4.1. DIRECTLY COMBINED CHROMATOGRAPHY/MASS SPECTROMETRY

The two most widely used techniques for sample introduction with fractionation are GC/MS and LC/MS and each method has attributes that would make it preferable to the other, depending on the type of information sought. As a generalization, GC/MS analysis is more sensitive than LC/MS, but has limited utility for the analysis of highly polar compounds even when volatile derivatives can be synthesized.

4.1.1. *Gas chromatography/mass spectrometry.* GC/MS analysis is most readily carried out at the base level, with some successful applications reported at the nucleoside level [65,66]. Bases can be cleaved from DNA by treatment with acid, typically 88% formic acid to effect complete digestion [7]. Volatile derivatives of purine and pyrimidine bases have favorable chromatographic properties, and in a well-characterized system, analysis at the base level is appropriate. An issue regarding base analysis is the assignment of origin once the base has been cleaved from the sugar. RNA is highly modified [14], and the sensitivity of GC/MS is typically sufficient to detect a modified base at the 1 in $10^5$ level, an amount easily present from RNA contamination to the extent of a few percent.

The analysis of modified bases in DNA produced by reaction with hydroxyl radical (Figure 4) represents an example of the most highly successful application of GC/MS to nucleic acid structure problems [7]. A variety of compounds have been identified by mass spectrometry of the trimethylsilyl derivatives [67] of bases (and acid-resistant deoxynucleosides) released from DNA subjected to treatment with agents that produce hydroxyl radicals.

The chromatographic separation of oxidized bases shown in Figure 5 illustrates the high resolving power of capillary gas chromatography. Synthetic standards were detectable at the 10 femtomole level. Treatment of oxidized DNA with acid, however, results in deamination of certain oxidized cytosine derivatives to give the corresponding uracil compound, e. g., $6 \rightarrow 4$ (Figure 5). These analyses are carried out on a simple quadrupole instrument and demonstrate the impressive problem solving capabilities of relatively unsophisticated instruments in favorable cases.

Other derivatives that have proven useful for base analysis by GC/MS are t-butyldimethylsilyl [68,69], and a variety of electrophoric derivatives (e.g. pentafluorobenzyl) for use in combination with electron capture negative ion chemical ionization [70,71].

Nucleosides are likewise amenable to GC/MS analysis, and this approach may be preferred, in combination with EI of TMS derivatives, because their spectra are well-understood [39]. When nucleosides are already present in the biological matrix, e. g., as in the analysis of modified nucleosides in cancer urine [66], a simple clean-up is sufficient prior to derivatization. When analysis of RNA constituents is desired, enzymatic digestion to the nucleoside level is recommended because the bases are not readily liberated from the biopolymer. Salts are usually specified in protocols for nucleic acid digestion for molecular biological studies, but the presence of salts can quench the silylation reaction, and a procedure for nucleic acid digestion using volatile buffers has been described [3]. The digest may then be dried down and directly treated

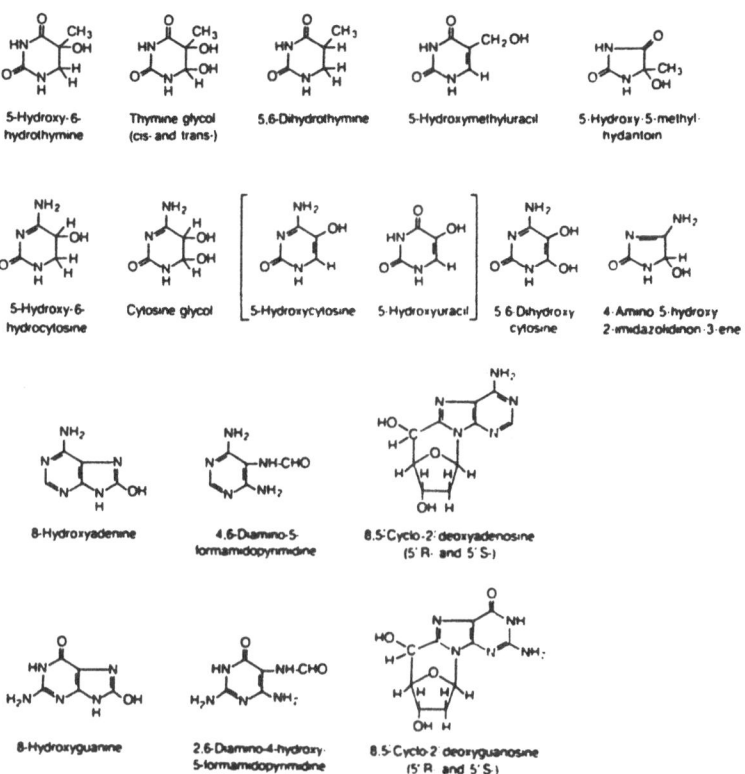

Figure 4. Structures of oxidized bases from γ-irradiated calf thymus DNA. Reproduced from [7] with permission.

with derivatizing reagent(s). EI-GC/MS is advantageous because EI can provide more structural information from fragmentation reactions (compared with "soft" ionization methods such as FAB or CI) for comparison with reference mass spectra. TMS derivatives of cytidine do not chromatograph well; however, conversion to the mixed N-TFA, O-TMS derivative [72] greatly decreases adsorption and improves peak shapes.

The utility of EI-GC/MS is illustrated by the structure determination of two unknown nucleosides (A* and U*) in the cap structure (the 5′-terminal hexanucleotide) in trypanosomal mRNA [65]. Severe limitations in sample availability (a total of about 200 pmol isolated over a two-year period) precluded isolation of the individual components, and all analyses were carried out using combined chromatography-MS methods. Preliminary mass spectral data suggested that one of the unknowns was an adenosine derivative with one methyl group in the sugar and two methyl groups in the base. Comparison of U* and A* with synthetic models was required for final structure assignment. There are six stable dimethyladenines, but biological precedent for the $N^6,N^6$-dimethyl isomer made it the first choice to synthesize for comparison with A*. There was no assurance that there was sufficient sample for FAB-MS/MS analysis of A*, or that the spectrum

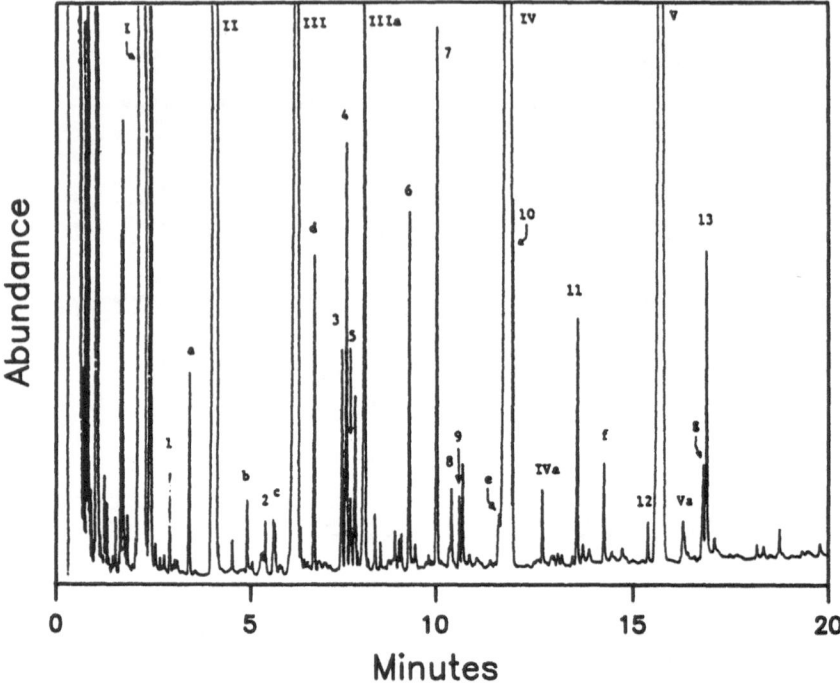

Figure 5. Chromatogram from capillary GC fractionation of a γ-irradiated DNA sample following acid digestion and trimethylsilylation. Peaks are trimethylsilyl derivatives of: I, phosphoric acid; 1, uracil; II, thymine; 2, 5,6-dihydrothymine; III, cytosine; d, 5-methylcytosine,; 3, 5-hydroxy-6-hydrothymine; 4, 5-hydroxyuracil; 5, 5-hydroxy-6-hydrouracil; IIIa, cytosine; 6, 5-hydroxycytosine; 7 and 8, *cis*- and *trans*- thymine glycol; 9, 5,6-dihydroxyuracil; IV and IVa, adenine; 10, 4,6-diamino-5-formamidopyrimidine; 11, 8-hydroxyadenine; 12, 2,6-diamino-4-hydroxy-5-formamidopyrimidine; V and Va, guanine; 13, 8-hydroxyguanine. Peaks a–g were present in control DNA samples. Adapted from [7] with permission.

would permit unequivocal assignment of the structure *de novo*. Synthesis of all dimethyladenine isomers would be required for structure assignment for comparison with A* by FAB-MS/MS. Because the favored $N^6,N^6$-dimethyl isomer contains no active hydrogens in the base, there was a high probability of success for trimethylsilylation of the enzymatic digestion solution for analysis by EI-GC/MS. An additional bonus with the choice of EI would be observation of an ion derived from loss of methyleneimine from the $BH_2^+$ ion, e.g. $m/z$ 164 → $m/z$ 134, a fragmentation pathway that is diagnostic for the presence of the $N,N$-dimethylamino function in heterocyclic compounds [73]. As shown in Figure 6, the EI mass spectra of the TMS derivatives of authentic A* and synthetic $N^6,N^6,-O-2'$-trimethyladenosine, acquired in back-to-back capillary GC/MS analyses, are essentially identical. The retention times, also an important criterion for establishing the mutual identity of two compounds, differed by 0.5 sec, well within the 1.5 sec scan cycle time, so the structure of A* was established as $N^6,N^6,-O-2'$-trimethyladenosine [65].

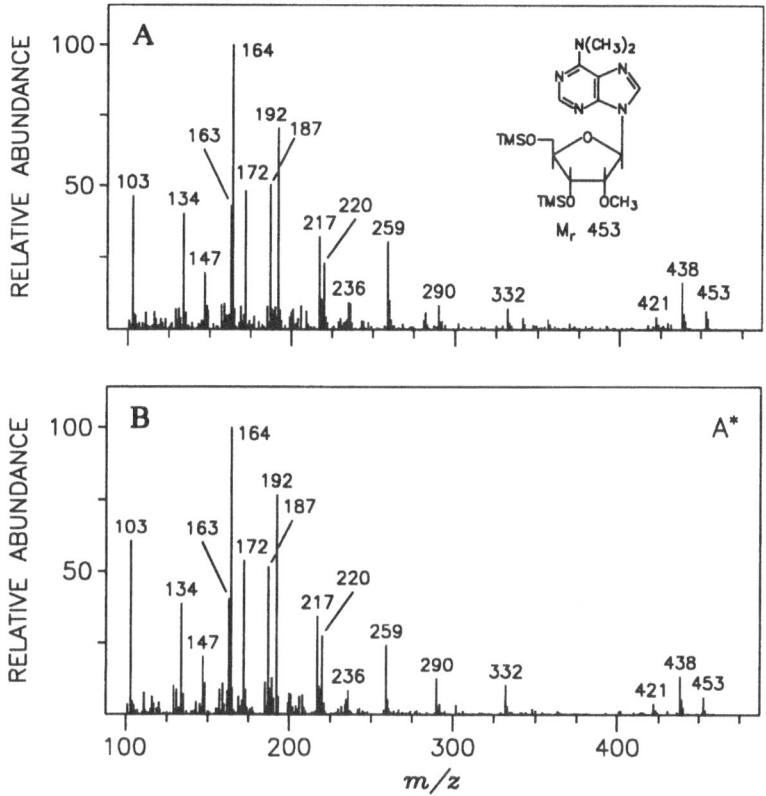

Figure 6. EI mass spectra from back-to-back capillary GC/MS analyses of the TMS derivatives of: (A), authentic $N^6,N^6,-O$-2′-trimethyladenosine, and (B) an uncharacterized trimethyladenosine from a trimethylsilylated digest of 5 pmol of *T. brucei* mRNA CAP 4 hexanucleotide. Reprinted from [65] with permission.

4.1.2. *Liquid chromatography/mass spectrometry.* Among the several different LC/MS interfaces [74] thermospray LC/MS has been the most extensively utilized for screening of RNA hydrolysates [5,75], and has proven to be a powerful tool for detection of new nucleosides in tRNA [11], and is equally suitable for DNA analyses [76]. Compared with GC/MS, LC/MS accommodates a much wider range of sample polarities, does not require derivatization, and is generally more tolerant to the presence of salts. Another advantage of LC/MS is the capability for determining on-line the number of replaceable hydrogens for every nucleoside in the mixture by using $D_2O$ instead of $H_2O$ in the HPLC eluants [77]. Detailed protocols for thermospray reversed phase LC/MS have been described [5]. The amount of sample required depends upon the levels of occurrence of the modified nucleoside of interest, but is in the range of 3–5 μg at the 1% level for a tRNA (76-mer). The thermospray mass spectra of $N^2$- and 2′-$O$-methylguanosine, two modified nucleosides that occur in RNA are shown in Figure 7. Both spectra show the characteristic MH$^+$ and BH$_2^+$ ion types from "soft" ionization methods. The site of methylation is easily

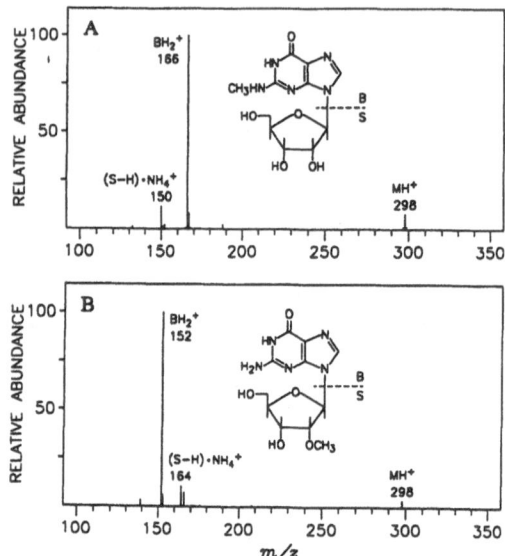

Figure 7. Thermospray mass spectra of: A, $N^2$-methylguanosine, and B, 2'-$O$-methylguanosine. Reproduced from [78] with permission.

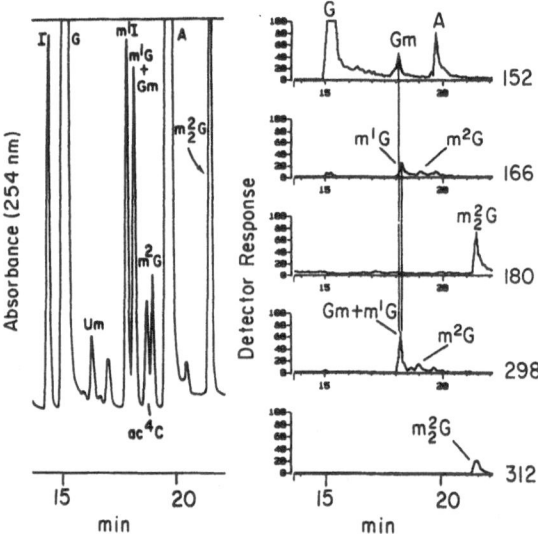

Figure 8. Thermospray RP-LC/MS of a digest of tRNA from *Haloferax volcanii.* Left panel: partial chromatogram from UV absorbance. Right panel: ion chromatograms for detection of methylguanosine isomers. Ion identities: $m/z$ 152, $BH_2^+$ of 2'-$O$-methylguanosine; $m/z$ 166, $BH_2^+$ of base-methylated guanosines; $m/z$ 180, $BH_2^+$ of base-dimethylated guanosines; $m/z$ 298, $MH^+$ of methylguanosines; $m/z$ 312, $MH^+$ of dimethylguanosines. Reproduced from [78] with permission.

determined from the difference in mass between the two ions, 132 u for normal ribose and 146 u for methylribose, which corresponds to loss of a neutral sugar moiety. Figure 8 shows the mass spectrometric detection of both 1- and 2'-O-methylguanosine, which are not visibly separated in the HPLC chromatogram, but which are revealed from the reconstructed ion chromatograms.

## 4.2. TANDEM MASS SPECTROMETRY

Tandem mass spectrometry-based techniques have some especially attractive features for analysis of nucleic acid constituents in mixtures. Typically used with FAB for polar molecules (e.g., [34,79–83]), or EI in the case of bases [84,85], it is convenient and rapid well suited for the study of polar molecules which would not be suitable for GC/MS or LC/MS. The different scan modes are all useful for nucleoside analysis in mixtures. For example, the presence of a nucleoside is apparent from a constant neutral loss scan for 116 u (deoxynucleosides), 132 u (ribonucleosides) or 146 u (sugar-methylated nucleosides). These losses would be unlikely from a non-nucleoside component. CNL scans are convenient to assess the purity of partially-purified fractions [82,83]: two (non-isobaric) nucleosides would yield two different $MH^+$ ions, although a single nucleoside can produce multiple ions due to cationization. These situations are readily distinguishable, however. Product ion scans give information about the structure of the nucleoside (Section 3.3.), while precursor scans indicate the origin(s) of the ion of interest.

Mixtures of nucleosides produced from enzymatic digestion of nucleic acids can be examined by MS/MS, without the necessity of sample cleanup [79–81]. The presence of the enzymes and ancillary salts does not interfere with the analysis, making this approach a rapid one when a simple screen is desired. Shown in Figure 9 is the product ion spectrum of $m/z$ 166, the $BH_2^+$ ion of a base-methylated methylguanosine, acquired directly from a digest of E. coli Ala-tRNA. The amount of sample loaded represents about 200 pmol, corresponding to about 25 ng of nucleoside. Comparison of the spectrum from the tRNA-derived material with product ion spectra of authentic 7-methylguanosine and $N^2$-methylguanosine permitted the assignment of the 7-methyl isomer as the methylguanosine present in this tRNA. It should be recognized that if other species with the same precursor mass are also present, overlapping spectra will result, whose interpretation may not be straightforward. Nonetheless, this approach is preferred over LC/MS, for example, for highly polar nucleosides, which do not give good thermospray mass spectra.

An innovative solution to the problem of the presence of isobaric components in a mixture utilizes H/D exchange to shift the mass from the analyte of interest away from the interfering component [85]. In this application, urine, following a one-pass cleanup on a reversed phase cartridge, was to be screened for the presence of modified guanines representing the products of repair of damaged DNA. Evidence from preliminary mass spectra indicated the presence of a component tentatively assigned as an ethylated or dimethylated guanine. The product ion spectrum indicated that there might be two or more isobaric $m/z$ 180 species. $N^2,N^2$-dimethylguanine is a known urinary constituent that originates from RNA turnover, and would interfere with the search for the ethylated component. The isobaric components were "separated" by treating the partially purified urine extract with 2 M DCl and adding the acidified extract directly to the FAB matrix. A putative ethylguanine would contain one more replaceable hydrogen than $N^2,N^2$-dimethylguanine. Product ion spectra of the fully deuterated $m/z$ 184 urinary component and authentic 7-, $O^6$- and $N^2$-ethylguanine permitted the identification of $N^2$-ethylguanine as the urinary constituent.

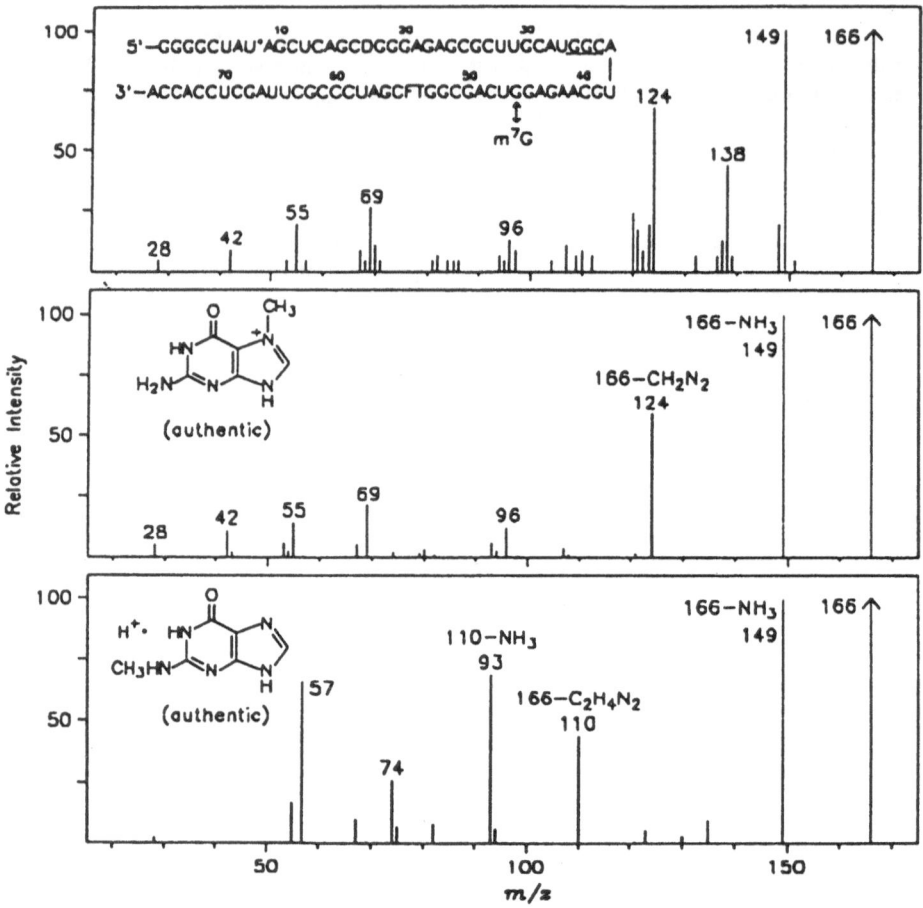

Figure 9. Product ions from CID of *m/z* 166 from: upper panel, 5 μg of digested Ala-tRNA from *E. coli*; middle panel, 7-methylguanosine; lower panel, $N^2$-methylguanosine. Reprinted from [80] with permission.

## 5. Quantification of nucleic acid constituents

A number of mass spectrometry-based strategies have been applied toward the problem of quantifying the amounts of naturally and xenobiotically modified bases in nucleic acids [1]. Among the most widely used are those based on capillary GC/MS, using comparison with weighed standards [7] or by isotope-dilution [8]. These methodologies are straightforward and reliable and will not be discussed in detail.

An emerging trend is the increased activity directed toward mass spectrometry-based methods for quantifying oxidized bases from DNA damaged as a consequence of interaction with reactive oxygen species, e. g. hydroxyl radical. Standard isotope dilution assays have been described for thymine glycol [86,87] and a number of other oxidized species shown in Figure 4 [87], released from DNA as the base. An alternative assay for thymine glycol at the parts-per-million level has been described, based on isotope-dilution GC/MS of a ring-opened derivative [88]. Labeled thymine glycol is added to the DNA, and treatment with alkali causes ring-opening of both standard and analyte to 2-methylglycerate. Following esterification the t-butyldimethylsilyl derivatives are analyzed by EI-GC/MS. A number of derivatives and mass spectrometric techniques (including MS/MS) were examined, but offered no improvement over the one chosen. Owing to the increased interest in the potentially deleterious role of DNA oxidation in a number of pathological processes, increasing use of mass spectrometry-based protocols quantification of oxidized bases from DNA can be anticipated.

## 6. Mass spectrometric analysis of oligonucleotides and nucleic acids

Although the first technique used to ionize and detect intact oligonucleotide molecular ion species was [$^{252}$C]PD-TOF-MS [89], the technique is little used currently because of its poor ability to ionize free oligonucleotides larger than about four residues [90], compared with FAB [91], MALDI [92,93] and ESI [94,95]. Although sample requirements for FAB-MS are greater than for MALDI- and ESI-MS (on the order of tens of nanomoles vs. tens of picomoles, respectively), it is useful for the analysis of synthetic oligonucleotides, where larger amounts of material are typically available, compared with more limited amounts of material available from natural nucleic acids, for which MALDI and ESI are better suited. Two important problems of nucleic acid structure which are amenable to solution at the oligonucleotide level are determination of molecular weight (hence verification of composition), and sequence determination. For successful application of mass spectrometry to oligonucleotide analysis, sample preparation is a crucial factor.

### 6.1. PREPARATION OF OLIGONUCLEOTIDES FOR ANALYSIS

Extraction of nucleic acids from cells, as well as electrophoretic or column chromatographic separation of oligonucleotides, generally requires high concentrations of non-volatile salts. If not removed, cations (e.g., $Na^+$, $K^+$) will replace H resulting in mass shifts that both make determination of molecular weight difficult and spread the ion signal over a wider mass range thereby diminishing signal intensity. Analysis is therefore facilitated by conversion of the oligonucleotide to a volatile (e. g., $NH_4^+$, $Et_3NH^+$) salt. Oligonucleotides are preferably analyzed by FAB-MS as triethylammonium salts [91]. Replacement of $Na^+$ by $NH_4^+$ is readily effected by precipitation from 2.5 M ammonium acetate solutions with ethanol [96]. An especially simple procedure for preparing salts for MALDI-MS is achieved by simply adding several cation-exchange resin beads ($NH_4^+$ form) directly to the analyte/matrix solution on the sample target [92].

Both FAB- and MALDI-MS generally yield predominantly the singly-charged $(M - H)^-$ ion. For an oligonucleotide with $n$ internucleotide bonds, $n - 1$ protons must be gained, a process which may be facilitated by formation of an ammonium salt and subsequent loss of neutral ammonia during sample vaporization, leaving the proton behind. In contrast, the generation of multiply-charged ions in ESI-MS does not require as great an extent of protonation, but the presence of ammonium salts in the analyte solution, although not detrimental, results in a small shift of the ion envelope to lower charge state [96]. An example of electrospray mass spectra before and after precipitation is shown in Figure 10 and illustrates these concepts.

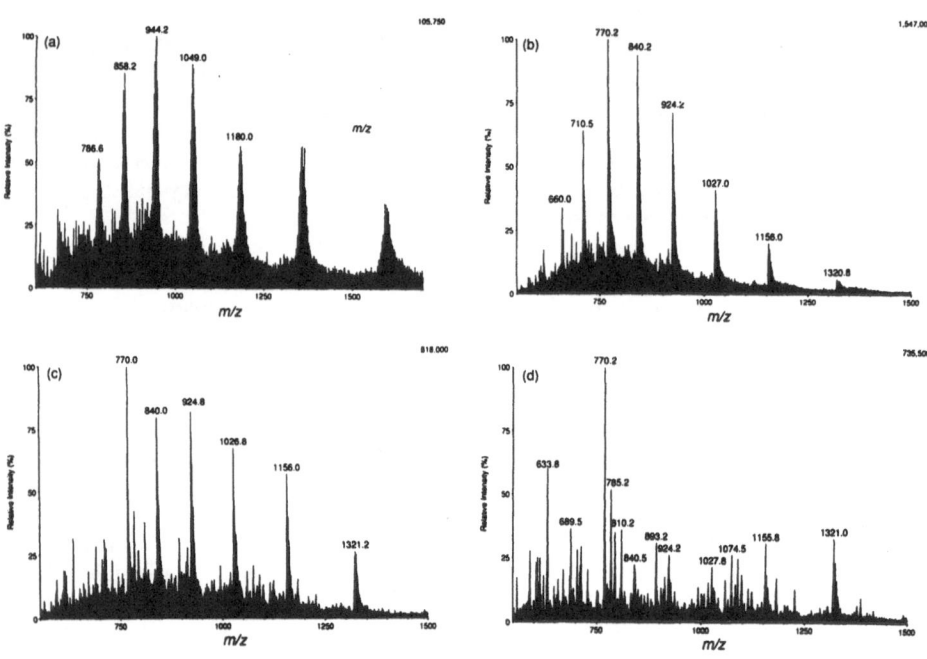

Figure 10. Electrospray mass spectra of an oligodeoxynucleotide 30-mer (a) before, and (b–d) after ethanol precipitation from 2.5 M ammonium acetate. Orifice potentials were (a,b) –70V, (c) –100V, (d) –130V. Adapted from [96] with permission.

## 6.2. DETERMINATION OF MOLECULAR WEIGHT AND OLIGONUCLEOTIDE MAPPING

The ability to determine molecular weight accurately has important consequences for verification of oligonucleotide composition. Clearly, if confirmation of identity of a synthetic oligonucleotide is desired, FAB-, ESI- and MALDI-MS are all suitable methods, subject to the constraints of sample purity. Caution must be taken using glycerol as the FAB matrix. Under these conditions, the spectrum of an octameric oligonucleotide yielded a mixture of molecular ion species consisting of 3% $(M - 3H)^-$, 12% $(M - 2H)^-$, 36% $(M - H)^-$, 27% $M^-$, 15% $(M + H)^-$, and 6% $(M + 2H)^-$ [97], making determination of identities of the different molecular species difficult to achieve by inspection.

Determination of molecular weight cannot, however, provide the composition of an uncharacterized oligonucleotide larger than about a tetramer, in the absence of structural constraints, owing to the large number of compositions accommodated [98]. It was shown that when the number of one of the four normal DNA or RNA nucleosides can be fixed at any known value, determination of molecular weight to within ± 0.01% allows assignment of a *unique* composition for oligonucleotides up to 14 residues [98]. This finding is especially advantageous for RNA structure analysis because there are nucleoside-specific nucleases that digest RNA to oligonucleotides with a unique nucleotide at one end (e.g. RNase $T_1$, which yields oligonucleotides terminating with Gp at the 3′-terminus). A method has been described whereby the presence of modified nucleosides within oligonucleotide sequences can be determined by comparison of molecular weights of oligonucleotides, determined by ESI-MS, with those predicted from the gene [99,100]. A discrepancy between expected and measured molecular weights allows the assumption that a modified residue is present, which can then be further characterized by other mass spectrometric methods.

## 6.3. DETERMINATION OF NUCLEIC ACID SEQUENCE

There is presently great interest in mass spectrometry-based methods for sequencing nucleic acids. Two approaches have been demonstrated: sequence derivation from fragmentation in the mass spectrometer, typically following CID, and use of the mass spectrometer to determine the masses of oligonucleotides produced enzymatically from the nucleic acid.

*5.2.1. Direct mass spectrometric sequencing.* Sequence information is readily inherent in oligonucleotide spectra acquired by FAB-MS, representing the largest number of reported examples of direct sequencing [91, 101–103], as extensively reviewed recently [1]. Scheme II shows a dinucleotide structure representing the smallest oligonucleotide structural unit. Ions resulting from backbone cleavages at every internucleotide phosphate, between the phosphate–C-3′ bond (ion *e*, scheme II) or the phosphate–C-5′ bond (ion *f*, scheme II) are produced from FAB-MS.

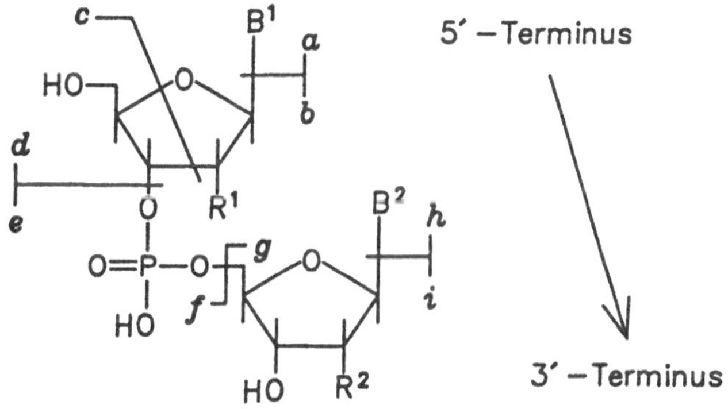

Scheme II

It was initially concluded that the *e*-type ions were always more abundant than the *f*-type ions, permitting deduction of the sequence. This observation was soon shown not to be universally true [101,104], leading to ambiguity in assigning the direction of the derived sequence (3′ → 5′ *vs.* 5′ → 3′). This uncertainty does not exist where the termini are not symmetrical, e.g. when a blocking group is present at one end or the other [102]. MS/MS was applied as a means to provide definitive sequence information, but with mixed success [105,106]. Cleavage of singly-charged oligonucleotide ions in MALDI analyses has also been reported [107,108], but extensive use for the purpose of oligonucleotide sequencing is not yet evident.

In contrast with the limited success obtained with MS/MS of singly-charged oligonucleotides, the spectra of the dinucleotides display remarkable ability to unambiguously assign the sequence.

Two studies are relevant here. The earliest utilized high-energy CID to examine the 12 heterodimeric deoxyribodinucleotides and 10 of the 12 heterodimeric ribodinucleotides [109]. The CU and UC pair differ in mass by only one $m/z$ u and so were not examined owing to low resolution in MS-2 (EBE instrument configuration). More recently, the complete set of 12 heterodimeric ribodinucleotides was examined using low energy CID (EBqQ instrument) [110]. Both studies indicated that the sequence of the dinucleotides could be unambiguously determined. In the case of high energy dissociations [109] the ion resulting from loss of the 5' base (ion $b$, Scheme II) was always more abundant than that from loss of the 3' base (ion $i$). In contrast, with low dissociation energy [110] sequence assignment is derived from the observation that the B$^-$ ion of the 5' residue (ion $a$, Scheme II) is always more abundant than that of the 3' nucleoside (ion $h$). An important finding from the low-energy study [110] was that it is the interaction between the 3' base and the phosphate group that directly influences fragmentation of the dinucleotides.

The use of collisional activation MS/MS of multiply-charged oligonucleotide ions in a quadrupole ion trap has been reported [111,112] and yields sequence information to a greater or lesser extent of completeness, depending on the composition of the oligomer. The first decomposition is loss of a base followed by cleavage of the internucleotide bond at the 3' position of the corresponding sugar. Preferential loss of Ade$^-$ and Thy$^-$ relative to Cyt$^-$ and Gua$^-$ can make sequence determination of the C,G pair difficult, however.

Despite the greater utilization of triple quadrupole mass analyzers, compared with ion traps, there are no reports of MS/MS for oligonucleotide sequence analysis of multiply charged ions using this type of instrument. Increasing production of fragment ions from multiply-charged oligonucleotides from ESI as a consequence of increased voltage drop between the needle and sampling orifice has been reported, and such an example is shown above [compare Figures 10b and 10c]. Very limited sequence information could be derived, and the spectra were described as not yielding any useful information [96]. MS/MS of multiply-charged ions has apparently not been utilized with tandem analyzer instruments (as opposed to trapped-ion analyzers), so it cannot be predicted whether the correlations derived for sequence determination using CID with the ion trap [111,112] will hold for analyses with the triple quadrupole analyzer. This is an area worthy of further investigation.

6.2.2. *Enzyme-based mass spectrometric procedures.* Routine sequencing of DNA on a large scale is accomplished using Sanger's method [113], which generates a nested set of oligonucleotide fragments that are subsequently electrophoretically separated and analyzed with the aid of fluorescence detection or $^{32}$P autoradiography. The underlying principle of the method is the use of a polymerase to copy the strand to be sequenced, and including in the mixture of the four deoxynucleoside triphosphate substrates a small amount of the corresponding 2',3' di-deoxy triphosphate analogues that are not substrates for the enzyme and thus terminate the elongation reaction. The solution to be sequenced consists of a mixture of oligonucleotides and while traditional sequencing methods require that the mixture be fractionated, use of mass spectrometry to determine molecular weights of the oligonucleotides in the mixture has been offered as a rapid and (potentially) more accurate means of determining the sequence.

MALDI-MS has features that currently render it the most straightforward to apply. Ideally, given an appropriate matrix/laser combination such that singly-charged ions are the predominant species produced, the problem reduces to determination of molecular weights of oligonucleotides in a mixture, for which multiply-charged ions resulting from ESI would not be suitable. The difference in mass between adjacent ions represents the last (chain terminating) residue added to the growing oligonucleotide, and permits the sequence to be read directly from the spectrum.

There are a number of issues remaining to be resolved before application of MALDI-based mass spectrometric techniques will become competitive with currently-available automated DNA sequencers. The sequencing mixture will contain protein and salts, which have the potential to interfere with analysis (Section 6.1.). Ion production efficiency that does not reflect composition must be achieved; presently, poly-dT oligomers are readily ionized while mixed-base oligomers present a challenge. Finally, the generation and detection of large ions remains to be optimized. There is much activity in this area of research and the prospects are intriguing.

Although there remain difficulties associated with the sequence analysis of large oligonucleotides, smaller ones are more tractable to mass spectrometric analysis. For example, the use of MALDI-TOF-MS to determine molecular weights of oligonucleotides resulting from sequential removal of nucleotides from the 3' or 5' termini by exonucleases, directly in the enzymatic digest has been reported [114]. The time course of conversion of larger to smaller oligonucleotides is reflected in the molecular weights of the components of the mixture. Consequently, determination of mass differences between adjacent ions reflects the identity of the nucleotide removed. New features described include use of a non-ionic matrix and of matrix additives to diminish the salt effect, which may prove generally useful.

## 7. Conclusions

The advent of electrospray and matrix-assisted laser desorption for the production of gaseous biopolymer ions has significantly enhanced the prospects for structure and sequence studies to be carried at the polynucleotide level. Both ionization methods have proven useful for the analysis of "small" oligomers, but the prospects for successful application to large-scale sequencing efforts (e.g. at the 100-mer level and beyond) are not certain. While its ability to produce large ions may be more exciting, electrospray ionization is also applicable to analysis of nucleic acid monomer subunits, where it is capable of greater sensitivity than FAB-based techniques, which are nonetheless extensively applied. Structure elucidation of modified nucleosides, both those that occur naturally and those that are xenobiotically produced, will continue to be the major contribution of mass spectrometry to nucleic acid-related problems. Sample introduction via microbore chromatographic or electrophoretic interfaces improve the ability to perform high sensitivity analyses directly on enzymatically generated mixtures, both at the nucleoside and oligonucleotide levels.

## 8. Acknowledgments

I thank NATO for financial support of the workshop and the National Institutes of Health (grant GM 21584) for partial support of preparation of this manuscript.

## 9. References

1.    P. F. Crain, "Mass spectrometric techniques in nucleic acid research", *Mass Spectrom. Rev.* **9**, 504–554, (1990).

370

2.    J. A. McCloskey, "Constituents of nucleic acids:   overview and strategy", *Methods Enzymol.* **193**, 771–781, (1990).

3.    P. Crain, "Preparation and enzymatic hydrolysis of DNA and RNA for mass spectrometry", *Methods Enzymol.* **193**, 782–790, (1990).

4.    K. H. Schram, "Preparation of trimethylsilyl derivatives of nucleic acid components for analysis by mass spectrometry", *Methods Enzymol.* **193**, 791–796, (1990).

5.    S. C. Pomerantz and J. A. McCloskey, "Analysis of RNA hydrolyzates by liquid chromatography–mass spectrometry", *Methods Enzymol.* **193**, 796–824, (1990).

6.    J. A. McCloskey, "Electron ionization mass spectra of trimethylsilyl derivatives of nucleosides", *Methods Enzymol.* **193**, 825–842, (1990).

7.    M. Dizdaroglu, "Gas chromatography–mass spectrometry of free radical-induced products of pyrimidines and purines in DNA", *Methods Enzymol.* **193**, 842–857, (1990).

8.    P. F. Crain, "Analysis of 5-methylcytosine in DNA by isotope dilution gas chromatography–mass spectrometry", *Methods Enzymol.* **193**, 857–865, (1990).

9.    B. Mompon, "Structure elucidation of nucleic acid components", in CRC Handbook of Chromatography:  Nucleic Acids and Related Compounds, Vol. I, Part A, A. Krstulovic, Ed., CRC Press, Boca Raton, 3–39, 1987.

10.   K. H. Schram, "Purines and pyrimidines", in Clinical Biochemistry: Principles-Methods-Applications. Mass Spectrometry, A. M. Lawson, Ed., Walter deGruyter, Berlin, 509–570, 1989.

11.   J. A. McCloskey, "Structural characterization of natural nucleosides by mass spectrometry", *Accts. Chem. Res.* **24**, 81–88, (1991).

12.   J. A. McCloskey and P. F. Crain,  "Progress in mass spectrometry of nucleic acid constituents:  analysis of xenobiotic modifications and measurements at high mass", *Int. J. Mass Spectrom. Ion Processes* **118/119**, 593–615, (1992).

13.   W. Saenger, Principles of Nucleic Acid Structure, Springer-Verlag, New York, 1984.

14.   P. A. Limbach, J. A. McCloskey and P. F. Crain, "The modified nucleosides in RNA", *Nucleic Acids Res.* **22**, 2183–2196, (1994).

15.   M. P. Chiarelli and J. O. Lay, "Mass spectrometry for the analysis of carcinogen DNA adducts", *Mass Spectrom. Rev.* **11**, 447–493, (1992).

16.   K. Biemann and J. A. McCloskey, "Application of mass spectrometry to structure problems. VI. Nucleosides", *J. Am. Chem. Soc.* **84**, 2005–2006, (1962).

17. M. S. Wilson and J. A. McCloskey, "Chemical ionization mass spectrometry of nucleosides. Mechanisms of ion formation and estimations of proton affinity", *J. Am. Chem. Soc.* **97**, 3436–3444, (1975).

18. F. W. Crow, K. B. Tomer, M. L. Gross, J. A. McCloskey and D. E. Bergstrom, "Fast atom bombardment combined with tandem mass spectrometry for the determination of nucleosides", *Anal. Biochem.* **139**, 243–262, (1984).

19. J. Eagles, C. Javanaud, and R. Self, "Fast atom bombardment mass spectrometry of nucleosides and nucleotides", *Biomed. Mass Spectrom.* **11**, 41–46, (1984).

20. D. M. Reddy and C. R. Iden, "Analysis of modified deoxynucleosides by electrospray ionization mass spectrometry", *Nucleosides Nucleotides* **12**, 815–826, (1993).

21. P. F. Crain and J. A. McCloskey, "Naturally modified nucleosides from nucleic acids", in Biological Mass Spectrometry: Present and Future, T. Matsuo, R. M. Caprioli, M. L. Gross, and Y. Seyama, Eds., John Wiley & Sons, Ltd., Chichester, 509–537, 1994.

22. J. H. Gommers-Ampt, F. Van Leeuwen, A. L. J. de Beer, J. F. G. Vliegenthart, M. Dizdaroglu, J. A. Kowalak, P. F. Crain and P. Borst, "β-D-Glucosyl-hydroxymethyluracil: a novel modified base present in the DNA of the parasitic protozoan *Trypanosoma brucei*", *Cell* **75**, 1129–1136, (1993).

23. J. A. McCloskey, "Mass spectrometry", in Basic Principles in Nucleic Acid Chemistry, Vol. I, P. O. P. Ts'o, Ed., Academic Press, New York, 209–309, 1974.

24. S. J. Shaw, D. M. Desiderio, K. Tsuboyama and J. A. McCloskey, "Mass spectrometry of nucleic acid components. Analogs of adenosine", *J. Am. Chem. Soc.* **92**, 2510–2522, (1970).

25. C. Hignite, "Nucleic acids and derivatives", in Biochemical Applications of Mass Spectrometry, First Supplemental Volume, G. R. Waller and O. C. Dermer, Eds., Wiley-Interscience, New York, 527–566, 1980.

26. D. L. Smith, K. H. Schram and J. A. McCloskey, "The negative ion mass spectra of selected nucleosides", *Biomed. Mass Spectrom.* **10**, 269–275, (1983).

27. H. Seto, T. Okuda, T. Takesuke and T. Ikemura, "Negative chemical ionization mass spectrometry of nucleosides", *Bull. Chem. Soc. Japan* **31**, 197–203, (1983).

28. D. L. Slowikowski and K. H. Schram, "Fast atom bombardment mass spectrometry of nucleosides, nucleotides and oligonucleotides", *Nucleosides Nucleotides* **4**, 309–345, (1985).

29. S. K. Sethi, D. L. Smith and J. A. McCloskey, "Determination of active hydrogen content by fast atom bombardment mass spectrometry following hydrogen-deuterium exchange", *Biochem. Biophys. Res. Commun.* **112**, 126–131, (1983).

30. A. M. Reddy, V. V. Mykytyn and K. H. Schram, "Deuterium-labeled 3-nitrobenzyl alcohol as a matrix for fast atom bombardment mass spectrometry", *Biomed. Environ. Mass Spectrom.* **18**, 1087–1095, (1989).

31. D. L. Slowikowski and K. H. Schram, "Fast atom bombardment mass spectrometry of nucleosides. Comparison with electron impact and chemical ionization mass spectra", *Nucleosides Nucleotides* **4**, 347–376, (1985).

32. G. Sindona, "Perspectives of desorption ionization methodologies in nucleic acid chemistry", *NATO ASI Ser., Ser. C.* **280**, 321–345, (1989).

33. K. H. Schram and D. L. Slowikowski, "Fast atom bombardment of trimethylsilyl derivatives of nucleosides and nucleotides", *Biomed. Mass Spectrom.* **13**, 263–264, (1986).

34. M. S. Bryant, J. O. Lay, Jr. and M. P. Chiarelli, "Development of fast atom bombardment mass spectral methods for the identification of carcinogen-nucleoside adducts", *J. Am. Soc. Mass Spectrom.* **3**, 360–371, (1992).

35. Q. M. Weng, W. M. Hammargren, D. Slowikowski, K. H. Schram, K. Z. Borysko, L. L. Wotring and L. B. Townsend, "Low nanogram detection of nucleotides using fast atom bombardment-mass spectrometry", *Anal. Biochem.* **178**, 102–106 (1989).

36. J. M. Gregson, P. F. Crain, C. G. Edmonds, R. Gupta, T. Hashizume, D. W. Phillipson and J. A. McCloskey, "Structure of the archaeal transfer RNA nucleoside G*-15: 2-amino-4,7-dihydro-4-oxo-7-β-D-ribofuranosyl-1*H*-pyrrolo[2,3-*d*]pyrimidine-5-carboximidamide (archaeosine)", *J. Biol. Chem.* **268**, 10076–10086, (1993).

37. J. M. Gregson and J. A. McCloskey, "The dissociation chemistry of permethylated guanosines as articulated by MS/MS", *Tetrahedron Lett.*, **34**, 6665–6668, (1993).

38. J. A. McCloskey, A. M. Lawson, K. Tsuboyama, P. M. Krueger and R. N. Stillwell, "Mass spectrometry of nucleic acid components. Trimethylsilyl derivatives of nucleotides, nucleosides and bases", *J. Am. Chem. Soc.* **90**, 4182–4184, (1968).

39. H. Pang, K. H. Schram, D. L. Smith, S. P. Gupta, L. B. Townsend and J. A. McCloskey, "Mass spectrometry of nucleic acid constituents. Trimethylsilyl derivatives of nucleosides", *J. Org. Chem.* **47**, 3923–3932, (1982).

40. A. M. Lawson, R. N. Stillwell, M. M. Tacker, K. Tsuboyama and J. A. McCloskey, "Mass spectrometry of nucleic acid components. Trimethylsilyl derivatives of nucleotides", *J. Am. Chem. Soc.* **93**, 1014–1023, (1971).

41. J. A. McCloskey, R. N. Stillwell and A. M. Lawson, "Use of deuterium labeled trimethylsilyl derivatives in mass spectrometry", *Anal. Chem.* **40**, 233–236, (1968).

42. D. L. von Minden, R. N. Stillwell, W. A. Koenig, K. J. Lyman and J. A. McCloskey, J. A., "Mass spectrometry of 7-methylpurine nucleosides. Studies of a unique oxygen incorporation reaction which occurs during trimethylsilylation", *Anal. Biochem.* **50**, 110–121, (1972).

43. J. A. Kelly, M. M. Abbasi and J. A. Beisler, "Silylation-mediated oxidation of dihydropyrimidine bases and nucleosides", *Anal. Biochem.* **103**, 203–213, (1980).

44. S. K. Sethi, P. F. Crain and J. A. McCloskey, "Formation of a new derivative of secondary amines during trimethylsilylation with *N,O*-bis(trimethylsilyl)trifluoroacetamide. *N*-(aminomethylene)-2,2,2-trifluoroacetamides", *J. Chromatogr.* **254**, 109–116, (1983).

45. J. J. Dolhun and J. L. Wiebers, "Mass spectra of nucleoside derivatives", *Org. Mass Spectrom.* **3**, 669–681, (1970).

46. D. L. von Minden and J. A McCloskey, "Mass spectrometry of nucleic acid constituents. N,O-permethyl derivatives of nucleosides", *J. Am. Chem. Soc.* **95**, 7480–7490, (1973).

47. W. A. Koenig, L. C. Smith, P. F. Crain and J. A. McCloskey, "Mass spectrometry of trifluoroacetyl derivatives of nucleosides and hydrolysates of deoxyribonucleic acid", *Biochemistry* **10**, 3968–3979, (1971).

48. G. Puzo and J. L. Wiebers, "Identification of modified nucleosides in intact transfer ribonucleic acid by pyrolysis - electron impact - collisional activation mass spectrometry", *Nucleic Acids Res.* **9**, 4655–4667, (1981).

49. S. K. Sethi, S. P. Gupta, E. E. Jenkins, C. W. Whitehead, L. B. Townsend and J. A. McCloskey, "Mass spectrometry of nucleic acid constituents. Electron ionization spectra of selectively labeled adenines", *J. Am. Chem. Soc.* **104**, 3349–3353, (1982).

50. C. Hignite, "Nucleic acids and derivatives", in Biochemical Applications of Mass Spectrometry, G. R. Waller, Ed., Wiley-Interscience, New York, 427–447, 1972.

51. A. J. Alexander, P. Kebarle, A. F. Fuciarelli and J. A. Raleigh, "Characterization of radiation-induced damage to polyadenylic acid using high-performance liquid chromatography/tandem mass spectrometry", *Anal. Chem.* **59**, 2484–2491, (1987).

52. T. Sakurai, T. Matsuo, A. Kusai and K. Nojima, "Collisionally activated decomposition spectra of normal nucleosides and nucleotides using a four-sector tandem mass spectrometer", *Rapid Commun. Mass Spectrom.* **3**, 212–216, (1989).

53. A. E. Schoen, R. G. Cooks and J. L. Wiebers, "Modified bases characterized in intact DNA by mass-analyzed ion kinetic energy spectrometry", *Science* **203**, 1249–1251, (1979).

54. C. H. Hocart and U. P. Schlunegger, "Negative-ion mass spectrometry of substituted adenine bases and adenosine nucleosides", *Rapid Commun. Mass Spectrom.* **3**, 249–254, (1989).

374

55.  R. L. Hettich, "The differentiation of methylguanosine isomers by laser ionization Fourier transform mass spectrometry", *Biomed. Environ. Mass Spectrom.* **18**, 265–277, (1989).

56.  C. C. Nelson and J. A. McCloskey, "Collision-induced dissociation of adenine", *J. Am. Chem. Soc.* **114**, 3661–3668, (1992).

57.  R. S. Annan, R. W. Giese and P. Vouros, "Detection and structural characterization of amino polyaromatic hydrocarbon-deoxynucleoside adducts using fast atom bombardment and tandem mass spectrometry", *Anal. Biochem.* **191**, 86–95, (1990).

58.  D. M. Reddy, P. F. Crain, C. G. Edmonds, R. Gupta, T. Hashizume, K. O. Stetter, F. Widdell and J. A. McCloskey, "Structure determination of two new amino acid-containing derivatives of adenosine from tRNA of thermophilic bacteria and archaea", *Nucleic Acids Res.* **20**, 5607–5615, (1992).

59.  E. G. Rogan, E. L. Cavalieri, S. R. Tibbels, P. Cremonisi, C. D. Warner, D. L. Nagel, K. B. Tomer, R. L. Cerny and M. L. Gross, "Synthesis and identification of benzo[*a*]pyrene-guanine nucleoside adducts formed by electrochemical oxidation and by horseradish peroxidase catalyzed reaction of benzo[a]pyrene with DNA", *J. Am. Chem. Soc.* **110**, 4023–4029, (1988).

60.  E. Van den Eeckhout, J. Coene, J. Claereboudt, F. Borremans, M. Claeys, E. Esmans and J. E. Sinsheimer, "Comparison of the isolation of adducts of 2′-deoxycytidine and 2′-deoxyguanosine with phenylglcidyl ether by high-performance liquid chromatography on a reversed-phase column and a polystyrene-divinylbenzene column", *J. Chromatogr.* **541**, 317–331, (1991).

61.  N. V. S. RamaKrishna, E. L. Cavalieri, E. G. Rogan, G. Dolnikowski, R. L. Cerny, M. L. Gross, H. Jeong, R. Jankowiak and G. J. Small, "Synthesis and structure determination of the adducts of the potent carcinogen 7,12-dimethylbenz[*a*]anthracene and deoxyribonucleosides formed by electrochemical oxidation:  models for metabolic activation by one-electron oxidation", *J. Am. Chem. Soc.* **114**, 1863–1874, (1992).

62.  N. V. S. RamaKrishna, F. Gao, N. S. Padmavathi, E. L. Cavalieri, E. G. Rogan, R. L. Cerny and M. L. Gross, "Model adducts of benzo[*a*]pyrene and nucleosides formed from its radical cation and diol epoxide", *Chem. Res. Toxicol.* **5**, 293–302, (1992).

63.  S. M. Wolf, R. S. Annan, P. Vouros and R. W. Giese, "Characterization of aminopoly-aromatic hydrocarbon-DNA adducts using continuous-flow fast atom bombardment and collision-induced dissociation:  positive and negative ion spectra", *Biol. Mass Spectrom.* **21**, 647–654, (1992).

64.  S. M. Wolf, P. Vouros, C. Norwood and E. Jackim, "Identification of deoxynucleoside-polyaromatic hydrocarbon adducts by capillary zone electrophoresis-continuous flow-fast atom bombardment mass spectrometry", *J. Am. Soc. Mass Spectrom.* **3**, 757–761, (1992).

65. J. D. Bangs, P. F. Crain, T. Hashizume, J. A. McCloskey and J. C. Boothroyd, "Mass spectrometry of mRNA Cap 4 from trypanosomatids reveals two novel nucleosides", *J. Biol. Chem.* **267**, 9805–9815, (1992).

66. J. I. Langridge, T. D. McClure, S. El-Shakawi, A. Fielding, K. H. Schram and R. P. Newton, "Gas chromatography/mass spectrometric analysis of urinary nucleosides in cancer patients; potential of modified nucleosides as tumour markers", *Rapid Commun. Mass Spectrom.* **7**, 427–434, (1993).

67. E. White, V, P. M. Krueger and J. A. McCloskey, "Mass spectra of trimethylsilyl derivatives of pyrimidine and purine bases", *J. Org. Chem.* **37**, 430–438, (1972).

68. P. F. Crain and J. A. McCloskey, "Analysis of modified bases in DNA by stable isotope dilution gas chromatography-mass spectrometry", *Anal. Biochem.* **132**, 124–131, (1983).

69. W. G. Stillwell, H. X. Xu, J. A. Adkins, J. S. Wishnok and S. R. Tannenbaum, "Analysis of methylated and oxidized purines in urine by capillary gas-chromatography-mass spectrometry", *Chem. Res. Toxicol.* **2**, 94–99, (1989).

70. R. S. Annan, G. M. Kresbach, R. W. Giese and P. Vouros, "Trace detection of modified DNA bases via moving belt liquid chromatography-mass spectrometry using electrophoric derivatization and negative chemical ionization", *J. Chromatogr.* **465**, 285–296, (1989).

71. M. Saha, G. M. Kresbach, R. W. Giese, R. S. Annan and P. Vouros, "Preparation and mass spectral characterization of pentafluorobenzyl derivatives of alkyl and hydroxyalkyl-nucleobase DNA adducts", *Biomed Mass Spectrom.* **18**, 958–972, (1989).

72. U. I. Krahmer, J. G. Liehr, K. J. Lyman, E. A. Orr, R. N. Stillwell and J. A. McCloskey, "Use of a mixed derivative for the gas chromatography and mass spectrometry of cytidine and its analogs", *Anal. Biochem.* **82**, 217–225, (1977).

73. D. L. von Minden, J. G. Liehr, M. H. Wilson and J. A. McCloskey, "Mechanism of electron impact induced elimination of methylenimine from dimethylamino heteroaromatic compounds", *J. Org. Chem.* **39**, 285–289, (1974).

74. A. L. Yergey, C. G. Edmonds, I. A. S. Lewis and M. L. Vestal, "LC/MS of nucleic acid constituents", in Liquid Chromatography/Mass Spectrometry: Techniques and Applications, Plenum Press, New York, 89–125, 1990.

75. C. G. Edmonds, M. L. Vestal and J. A. McCloskey, "Thermospray liquid chromatography-mass spectrometry of nucleosides and of enzymatic hydrolysates of nucleic acids", *Nucleic Acids Res.* **13**, 8197–8206, (1985).

76. J. Serrano, D. W. Kuehl and S. Naumann, "Analytical procedure and quality assurance criteria for the determination of major and minor deoxynucleosides in fish tissue DNA by liquid chromatography ultraviolet spectroscopy and liquid chromatography-thermospray mass spectrometry", *J. Chromatogr. Biomed. Appl.* **615**, 203–213, (1993).

376

77. C. G. Edmonds, S. C. Pomerantz, F. F. Hsu and J. A. McCloskey, "Thermospray liquid chromatography/mass spectrometry in deuterium oxide", *Anal. Chem.* **60**, 2314–2317, (1988).

78. C. G. Edmonds, R. Gupta, J. A. McCloskey and P. F. Crain, "Ribonucleic acid modification in microorganisms", in Mass Spectrometry for the Characterization of Microorganisms, C. Fenselau, Ed., ACS Symp. Ser. 541, American Chemical Society, Washington, D.C., 147–158, 1994.

79. C. C. Nelson and J. A. McCloskey, "Structural characterization of nucleosides and nucleotides: the emerging role of LC/MS and tandem mass spectrometry", *Adv. Mass Spectrom.* **11A**, 260–261, (1989).

80. T. Hashizume, C. C. Nelson, S. C. Pomerantz and J. A. McCloskey, "Applications of LC/MS and tandem mass spectrometry to the characterization of nucleosides in mixtures", *Nucleosides Nucleotides.* **9**, 355–360, (1990).

81. P. F. Crain, T. Hashizume, C. C. Nelson, S. C. Pomerantz and J. A. McCloskey, "The determination of RNA post-transcriptional modifications by mass spectrometry", in Biological Mass Spectrometry, A. L. Burlingame and J. A. McCloskey, eds., Elsevier, New York, pp. 509–525, 1990.

82. J. Claereboudt, E. L. Esmans, E. G. Van den Eeckhout and M. Claeys, "Fast atom bombardment and tandem mass spectrometry for the identification of nucleoside adducts with phenylglycidyl ether", *Nucleosides Nucleotides* **9**, 333–344, (1990).

83. P. P. Fu, D. W. Miller, L. S. Von Tungeln, M. S. Bryant, J. O. Lay, Jr., K. Huang, L. Jones and F. E. Evans, "Formation of C8-modified deoxyguanosine and C8-modified deoxyadenosine as major DNA adducts from 2-nitropyrene metabolism mediated by rat and mouse liver microsomes and cytosols", *Carcinogenesis* **12**, 609–616, (1991).

84. P. B. Farmer, "Tandem mass spectrometry study of urinary alkylated purines", *Biomed. Environ. Mass Spectrom.* **17**, 143–145, (1988).

85. J. R. Cushnir, S. Naylor, J. H. Lamb and P. B. Farmer, "Deuterium exchange studies in the identification of alkylated DNA bases found in urine, by tandem mass spectrometry," *Rapid Commun. Mass Spectrom.* **4**, 426–431, (1991).

86. H. Faure, M. F. Incardona, C. Boujet, J. Cadet, V. Ducros and A. Favier, "Gas chromatographic-mass spectrometric determination of 5-hydroxymethyl uracil in human urine by stable isotope dilution", *J. Chromatogr. Biomed. Appl.* **616**, 1–7, (1993).

87. M. Dizdaroglu, "Quantitative determination of oxidative base damage in DNA by stable isotope-dilution mass spectrometry", *FEBS Lett.* **315**, 1–6, (1993).

88. S. P. Markey, C. J. Markey, T. C. L. Wang and J. B. Rodriguez, "Gas chromatographic mass spectrometric method for the assessment of oxidative damage to double-stranded DNA by quantification of thymine glycol residues", *J. Am. Soc. Mass Spectrom.* 4, 336–342, (1993).

89. C. J. McNeal and R. D. Macfarlane, "Observation of a fully protected oligonucleotide dimer at m/z 12637 by 252Cf-plasma desorption mass spectrometry", *J. Am. Chem. Soc.* 103, 2132–2139, (1986).

90. R. D. Macfarlane, "252Californium plasma desorption mass spectrometry (252Cf-PDMS)", in Soft Ionization Biological Mass Spectrometry, H. R. Morris, Ed., Heyden & Son Ltd., London, 110–119, 1981.

91. L. Grotjahn, R. Frank and H. Blöcker, H., "Ultrafast sequencing of oligodeoxyribo-nucleotides by FAB-mass spectrometry", *Nucleic Acids Res.* 10, 4671–4678, (1982).

92. E. Nordhoff, A. Ingendoh, R. Cramer, A. Overberg, B. Stahl, M. Karas, F. Hillenkamp and P. F. Crain, "Matrix-assisted laser desorption/ionization mass spectrometry of nucleic acids with wavelengths in the UV and IR", *Rapid Commun. Mass Spectrom.* 6, 771–776, (1992).

93. E. Nordhoff, R. Cramer, M. Karas, F. Hillenkamp, F. Kirpekar, K. Kristiansen and P. Roepstorff, "Ion stability of nucleic acids in infrared matrix-assisted laser desorption/ionization mass spectrometry", *Nucleic Acids Res.* 21, 3347–3357, (1993).

94. T. R. Covey, R. F. Bonner, B. I. Shushan and J. Henion, "The determination of protein, oligonucleotide and peptide molecular weights by ion-spray mass spectrometry", *Rapid Commun. Mass Spectrom.* 2, 249–256, (1988).

95. R. D. Smith, J. A. Loo, C. G. Edmonds, C. J. Barinaga and H. Udseth, "New developments in biochemical mass spectrometry. Electrospray ionization", *Anal. Chem.* 62, 882–899, (1990).

96. J. T. Stults and J. C. Marsters, "Improved electrospray ionization of synthetic oligonucleotides", *Rapid Commun. Mass Spectrom.* 5, 359–363, (1991).

97. L. Grotjahn, "Oligonucleotide sputtering from liquid matrices", *Springer Proc. Phys.* 9, 118–125, (1986).

98. S. C. Pomerantz, J. A. Kowalak and J. A. McCloskey, "Determination of oligonucleotide composition from mass spectrometrically measured molecular weight", *J. Am. Soc. Mass Spectrom.* 4, 204–209, (1993).

99. E. Bruenger, J. A. Kowalak, Y. Kuchino, J. A. McCloskey, H. Mizushima, K. O. Stetter and P. F. Crain, "5S rRNA modification in the hyperthermophilic archaea *Sulfolobus solfataricus* and *Pyrodictium occultum*", *FASEB J.* 7, 196–200, (1993).

100. J. A. Kowalak, S. C. Pomerantz, P. F. Crain and J. A. McCloskey, "A novel method for the determination of posttranscriptional modification in RNA by mass spectrometry", *Nucleic Acids Res.*, **21**, 4577–4585, (1993).

101. L. Grotjahn, H. Blöcker and R. Frank, "Mass spectrometric sequence analysis of oligodeoxyribonucleotides", *Biomed. Mass Spectrom.* **12**, 514–524, (1985).

102. A. M. Hogg, J. G. Kelland, J. C. Vederas and C. Tamm, "Investigation of ribo- and deoxyribonucleosides and nucleotides by fast atom bombardment mass spectrometry", *Helv. Chim. Acta* **69**, 908–917, (1986).

103. C. R. Iden and R. A. Rieger, "Structure analysis of modified oligodeoxyribonucleotides by negative ion fast atom bombardment mass spectrometry", *Biomed. Mass Spectrom.* **18**. 617–619, (1989).

104. M. Panico, G. Sindona and N. Uccella, "Fast atom-bombardment-induced zwitterionic oligonucleotide quasi-molecular ions sequenced by MS/MS", *J. Am. Chem. Soc.* **105**, 5607–5610, (1983).

105. R. L. Cerny, K. B. Tomer, M. L. Gross and L. Grotjahn, "Fast atom bombardment combined with tandem mass spectrometry for determining structures of small oligonucleotides", *Anal. Biochem.* **165**, 175–182, (1987).

106. J. J. Dino, C. R. Guenat, K. B. Tomer and D. G. Kaufman, "Analysis of carcinogen-modified oligonucleotides by fast atom bombardment/tandem mass spectrometry", *Rapid Comm. Mass Spectrom.* **1**, 69–71, (1987).

107. R. Hettich and M. Buchanan, "Structural characterization of normal and modified oligonucleotides by matrix-assisted laser desorption Fourier transform mass spectrometry" *J. Am. Soc. Mass Spectrom.* **2**, 402–412, (1991).

108. E. Stemmler, R. L. Hettich G. B. Hurst and M. V. Buchanan, "Matrix-assisted laser desorption/ionization Fourier-transform mass spectrometry of oligodeoxyribonucleotides", *Rapid Commun. Mass Spectrom.* **7**, 828–836, (1993).

109. R. L. Cerny, M. L. Gross and L. Grotjahn, "Fast atom bombardment combined with tandem mass spectrometry for determining structures of dinucleotides", *Anal. Biochem.* **156**, 424–435, (1986).

110. D. R. Phillips and J. A. McCloskey, "A comprehensive study of the low energy collision-induced dissociation of dinucleoside monophosphates", *Int. J. Mass Spectrom. Ion Processes* **128**, 61–82 (1993).

111. S. A. McLuckey, G. J. Van Berkel and G. L. Glish, "Tandem mass spectrometry of small, multiply charged oligonucleotides", *J. Am. Soc. Mass Spectrom.* **3**, 60–70, (1992).

112. S. A. McLuckey and S. Habibi-Goudarzi, "Decompositions of multiply-charged oligonucleotide anions", *J. Am. Chem. Soc.* **115**, 12085–12095, (1993).

113. F. Sanger, S. Nicklen and A. R. Coulson, "DNA sequencing with chain-terminating inhibitors", *Proc. Natl. Acad. Sci. USA* **74**, 5463–5467, (1993).

114. U. Pieles, W. Zürcher, M. Schär and H. E. Moser, "Matrix-assisted laser desorption ionization time-of-flight mass spectrometry: a powerful tool for the mass and sequence analysis of natural and modified oligonucleotides", *Nucleic Acids Res.* **21**, 3191–3196, (1993).

# MONITORING OF HUMAN EXPOSURE TO XENOBIOTICS;
## Identification and Quantification of Cancer Initiators in Vivo

M. TÖRNQVIST
*Department of Environmental Chemistry*
*Wallenberg Laboratory*
*Stockholm University*
*S-106 91 Stockholm*
*Sweden*

## 1. Introduction

The term xenobiotics, i.e. chemicals foreign to the biologic system, covers a wide range of chemical compounds, which differ with regard to stability and other chemical properties. One extreme example is polychlorinated biphenyls, which are very persistent and e.g. do not even degrade in fuming sulfuric acid. They accumulate in adipose tissues and could be determined in biological samples with high analytical sensitivity by gas chromatography (GC) or gas chromatography/mass spectrometry (GC/MS), due to the content of chlorine atoms. On the other hand, so called genotoxic agents are xenobiotics which are reactive or form reactive intermediates in vivo and which usually have too short life-time to be analysed in biological samples. One further complication is that some compounds of this type also could occur as natural compounds in vivo at the same time as they occur as xenobiotics in e.g. urban air pollution.

These two examples of xenobiotics illustrate that the requirements on analytical methodologies for monitoring xenobiotics in biological samples could be very different. Genotoxic compounds require in many respects a completely different approach than xenobiotics in the more classical sense. This paper will deal with methods for monitoring biological samples with respect to exposure to genotoxic compounds, or more generally, compounds that through their reactivity cause chemical changes in cellular components.

## 2. Genotoxic (Reactive) Compounds or Intermediates

Most known chemical carcinogens which are genotoxic (i.e. cancer initiators, which are also mutagens) possess electrophilic reactivity, or are metabolized to electrophiles, most of which are alkylating (1). An electrophilic compound (RX) is reactive towards nucleophilic centers ($Y^-$) such as oxygen, nitrogen and sulfur atoms, and gives rise to reaction products (RY):

$$RX + Y^- \xrightarrow{k} RY + X^- \qquad \text{(Eqn. 1)}$$

Parallell to reactions with nucleophilic sites in DNA, some of which could lead to genotoxic effects, reaction products with nucleophilic sites in other biomacromolecules are formed.

*R. M. Caprioli et al. (eds.), Mass Spectrometry in Biomolecular Sciences, 381–395.*

Reaction kinetic studies and studies of mutagenicity of monofunctional alkylating agents have shown that the frequency of mutations per unit of dose (see Eqn. 2) of the reactive compound is proportional to the rate of reaction towards DNA oxygens (2). It has later been demonstrated that alkylation of guanine-$O^6$ is a premutagenic event. Genotoxic compounds of other classes induce mutation by reaction with other atoms in DNA bases.

Dose (D) is defined as the time integral of the concentration of the reactive compound (3):

$$D \ = \ \int_t [RX] \, (t) \, dt \qquad \qquad \text{(Eqn. 2)}$$

The in vivo dose of electrophilic reagents can be calculated from the levels of reaction products of biomacromolecules provided the reaction kinetic constant k in Eqn. 1 and the life-time of the monitor molecule are known. Such reaction products of, or adducts to, sufficiently long-lived biomacromolecules can be used for monitoring of exposure to genotoxic compounds.

## 3.    Basis for Monitoring Methods

### 3.1.    REQUIREMENTS ON ANALYTICAL TECHNIQUES

Genotoxic effects (mutation, cancer) are stochastic with a probability which is proportional to the dose of the causative factor, without any no-effect threshold. This means that even very small absorbed doses will lead to a risk increment. This calls for an analytical sensitivity, which is sufficiently high to permit determination of adducts from genotoxic compounds or their reactive intermediates, within the whole range of doses where associated risks are unacceptable. The analytical methods should further be able to identify  adducts (exposures are often complex mixtures, with partly unknown components). Furthermore, in order to use the method for risk estimation it should be possible to quantify adduct levels with good reproducibility. The only analytical technique that fulfills these requirements is mass spectrometry (MS).

### 3.2.    CHOICE OF MONITOR MOLECULE

It would seem that determination of products of reaction with DNA would be the most natural basis for monitoring of genotoxic compounds. However, DNA adducts are subject to repair at rates which vary between chemicals, tissues, cells and DNA-regions and therefore it is at present difficult to use DNA adducts for quantitative determination of in vivo doses. On the contrary, adducts to proteins are mostly stable.

The blood protein hemoglobin (Hb) accumulates adducts over a long and well-defined life-span (4 months in humans) and is accessible in large amounts. These properties are both prerequisites for high sensitivity. Furthermore, the reaction rates for the formation of adducts to DNA and Hb are proportional and therefore Hb adducts present an indirect measure of reactions with DNA. For these reasons Hb has been found to be a suitable monitor molecule for reactive compounds in vivo (4). From kinetic aspects serum albumin may have certain advantages as monitor molecule (5).

3.3.     METHODS FOR PROTEIN ADDUCT DETERMINATION

For many simple alkylating agents N-terminal valine, as well as cysteine-$S$ and histidine-$N$ are major reaction sites in Hb, the two latter being initially used for Hb adduct determination (6). For $N$-nitrosamines and -amides, oxygens in carboxyl groups are predominant reaction sites (7). Aromatic amines undergo a specific reaction with cysteines in Hb with the formation of cysteine sulfinamides (8). Specific reaction sites, probably due to configuration properties, have been identified in Hb or serum albumin for some other compounds (9, 10). It has to be realized that electrophilic compounds have different reaction patterns, with an influence on the method to be chosen for the determination of adducts.

It has been shown that adducts to biomacromolecules, particularly from low-molecular weight compounds, can be formed also as artefacts during preparation of biological samples (11). In accordance therewith the analysis of adducts to reactive atoms in residues of internal amino acids in the globin chains (e.g. cysteine-$S$, histidine ring-$N$ etc.), which requires hydrolysis of the protein, results in risks of artefact formation, high background and low sensitivity (12). In most cases it is important to find ways to separate adducts from the macromolecules by a mild treatment of the sample. This is achieved by the "N-alkyl Edman procedure" which is a mild procedure, based on a modified Edman degradation, for specific cleavage of $N$-alkylated N-termini, valines, of the four globin chains in Hb (13). Other mild procedures are the analysis of aromatic amines from cysteine sulfinamides (8) and adducts to carboxyl groups (14, 15) which could be hydrolysed off under mild conditions.

Protein adducts have been, following suitable derivatization, identified and quantified by mass spectrometrical methods. Various procedures for mass spectrometric analysis of protein adducts have recently been described in detail (16). Immunological (10, 17) and fluorimetric (15, 18) methods have also, to some extent, been applied for protein adduct monitoring. About thirty different protein adducts have been determined in occupationally exposed humans, in smokers or in humans without exposure (summarized in ref. 19). Protein adducts from a much larger number of compounds have been studied in animals.

4.     The N-alkyl Edman Method

4.1.     REACTION PRINCIPLE

With the N-alkyl Edman method alkylated N-terminal valines are specifically split off and extracted from the neutral solution in which the globin is treated with Edman reagents, phenyl isothiocyanates. In contrast, unsubstituted N-terminal valine requires acidification to be cleaved after coupling to an Edman reagent (20). Current studies indicate that the derivatives formed, phenylthiohydantoins (PTHs) of alkylated valines, are obtained without intermediate formation of the thiazolinones observed in the corresponding reaction of unalkylated N-termini. Particularly $N$-(2-hydroxyethyl)valine (HOEtVal), the adduct from ethylene oxide, has been used as a model compound in the method development. The corresponding derivative formed with the N-alkyl Edman procedure, where pentafluorophenyl isothiocyanate (PFPITC) has been chosen as the reagent, is a pentafluorophenylthiohydantoin (PFPTH), abbreviated HOEtVal-PFPTH (Fig. 1). In the developmental work the parameters examined were the sensitivity, reproducibility and simplicity of the method (reviewed in ref. 21).

**Fig. 1.** Reaction principle for the N-alkyl Edman method. The N-terminal valine of a globin chain is alkylated by an electrophilic reagent. The globin is treated with pentafluorophenyl isothiocyanate (PFPITC) in neutral solution in which the derivative (PFPTH) of the *N*-substituted valine is split off. Alkylation of Hb is exemplified by the reaction of ethylene oxide, the reactive metabolite of ethene.

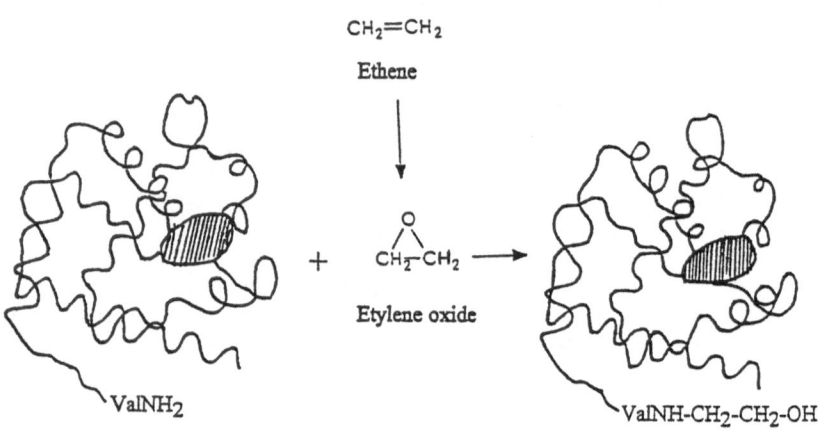

$CH_2{=}CH_2$

Ethene

$CH_2{-}CH_2$

Etylene oxide

ValNH$_2$ + → ValNH-CH$_2$-CH$_2$-OH

PFPITC

HOEtVal-PFPTH

## 4.2.    ACHIEVEMENT OF SUFFICIENT SENSITIVITY

The sensitivity required may be defined as the
- ability to detect current exposures, or
- ability to detect cancer risks at the limit of acceptability.
According to the International Commission on Radiological Protection a cancer mortality, due to one factor such as radiation, in the range 1 in $10^6$ - 1 in $10^5$ per year to members of the public may be considered acceptable (22). The steady-state level of adducts to N-terminal valine from ethylene oxide and compounds with similar reaction pattern corresponding to this risk is 1 - 10 pmol/g Hb. (The risk is estimated, from the dose of the genotoxic compound calculated from the adduct level, according to an approach described below.)

Replacement of the standard Edman reagent, phenyl isothiocyanate, by halogenated compounds, leads to an increase of the detectability of the derivatives. The pentafluorinated derivatives, pentafluorophenylthiohydantoins, obtained by using PFPITC as the reagent, give high sensitivity by analysis with negative ion chemical ionization mass spectrometry (NICI/MS). By switch from MS-analysis of PTHs by electron impact to PFPTHs by NICI approximately a factor of 100 was gained in sensitivity in analysis of standard compounds (23).

Typical fragments in NICI of some classes of *N*-substituted valine-PFPTHs which could be used for monitoring are given in Table 1. An example of an electrophile giving rise to a *N*-alkylvaline is methyl bromide and corresponding examples for *N*-(2-hydroxyalkyl)valines are ethylene oxide, propylene oxide etc., with ethene, propene etc. as precursors. PFPTHs of *N*-alkylvalines with unsubstituted alkyl groups (methyl, ethyl etc.) give fragments [M-1]⁻, but with *N*-benzylvaline the whole substituent is lost (corresponding to a fragment m/z 323). When the alkyl is 2-hydroxylated the major fragment is mostly [M-20]⁻ (loss of HF). In addition a fragment m/z 318 is found corresponding to additional loss of R'CHO from the alkyl group, where R' = H for the model compound, HOEtVal-PFPTH (see Fig. 2a). (For further details; see refs. 21 and 24.) Other chemical characteristics of PTHs and PFPTHs of a few *N*-substituted valines, as well as of the corresponding *N*-substituted valines, have been described by Rydberg et al. (24).

**Table 1.** Typical mass fragments, obtained by analysis with GC/MS, NICI, with methane as reagent gas, used for monitoring of PFPTHs of a few *N*-substituted valines.

| *N*-substituted valine | Adduct | Major mass fragments |
|---|---|---|
| Valine | -H | [M-1]⁻, [M-28]⁻ |
| *N*-alkylvaline | -$C_nH_{2n+1}$ | [M-1]⁻, [M-28]⁻ |
| *N*-(2-hydroxyalkyl)-valine | -CH₂CH-R'<br>        OH<br>R' = $CH_{2n+1}$ or<br>$C_nH_{2n-1}$ | [M-20]⁻, [M-R'CH₂O]⁻ (= m/z 318) |

Other steps important for achieving high sensitivity was e.g. reduction of by-products by introduction of formamide as the solvent during the derivatization to PFPTH and mild alkaline washing of the extract from the derivatized globin sample.

With analysis by GC/MS, NICI, selected ion monitoring, of the model compound HOEtVal-PFPTH the upper limit of the sensitivity aimed at, 10 pmol/g Hb, was reached. In the procedure 50 mg globin (corresponding to less than 0.5 ml blood) is derivatized and an amount corresponding to ca. 1 mg globin from the sample extract is injected on the GC/MS. Due to the by-products formed in the derivatization the sensitivity could not be increased by increasing the sample size.

This problem was overcome by the introduction of analysis by tandem mass spectrometry (GC/MS/MS). In the analysis of $N$-(2-hydroxyalkyl)valines the quantitation is based on selected reaction monitoring of the daughter ion m/z 318 of the fragment ion [M-HF]$^-$ (m/z 348 for HOEtVal-PFPTH) (see Fig. 2b). The sensitivity in the analysis has been optimized through studies of the effect of source temperature on the abundance of selected ions in the mass spectrum (Fig. 3a) and the effect of collision energy and collision pressure (argon as collision gas) on abundance of the daughter ion (Fig. 3b).(For details; see ref. 25.)

By this technique and without increasing the sample size the sensitivity was increased by a factor of 10 thus reaching a detection level below 1 pmol/g Hb, which corresponds to an injected amount of about 1 fmol (see Fig. 4). Since in this case the background noise is eliminated a further increase of the sensitivity is possible by using larger samples.

4.3.    REPRODUCIBILITY

The reproducibility requested is of course dependent on the problem to be solved, although generally a high reproducibility is desirable. It is mostly suitable to add the internal standard, such as a globin alkylated in vitro with ($^2$H$_4$)ethylene oxide (cf. Fig. 4) already before derivatization, thus correcting for yield of PFPTH in the derivatization step. Conditions for reproducible calibration have been studied, it is e.g. critical that the internal standard is very similar to the adduct to be quantified (26). Furthermore, as the derivatives could be unstable or stick to glass (especially those containing hydroxyl groups) it has been shown to be important to use a GC-column exclusively for analysis of PFPTH derivatives, to use a pre-column, to be expedient to use on-column injection or injection by program temperature vaporizer (on-column mode), and to be careful about "memory effects" (21, 26). Contamination or artefact formation, against which precautions have to be taken, could endanger the reproducibility particularly at low adduct levels; see 4.4. below. With regard to contamination the N-alkyl Edman procedure has the advantage of having a "tag", valine, from Hb left in the derivative to be determined.

Despite the existence of many pitfalls the interlaboratory variation in quantitative adduct determination by GC/MS with the N-alkyl Edman method has been shown to be acceptably small (coefficient of variation = 28 %) (27). Table 2 summarizes current status on the variations in analysis by GC/MS/MS due to specific steps in the procedure (25). In order to maintain a high reproducibility it is important to carefully calibrate on each analytical occasion and to quantify samples to be intercompared on the same occasion and preferably in the same derivatization series. By these precautions it is possible to keep the analytical error around 10 % and by repeated injections below 10 %.

Fig. 2. **a.** Mass spectrum of the PFPTH of *N*-(2-hydroxyethyl)valine (HOEtVal-PFPTH; M=368) obtained by NICI, with methane as reagent gas.
  **b.** Daughter ion spectrum of m/z 348 of HOEtVal-PFPTH obtained by MS/MS/NICI, with methane as reagent gas and argon as collison gas.

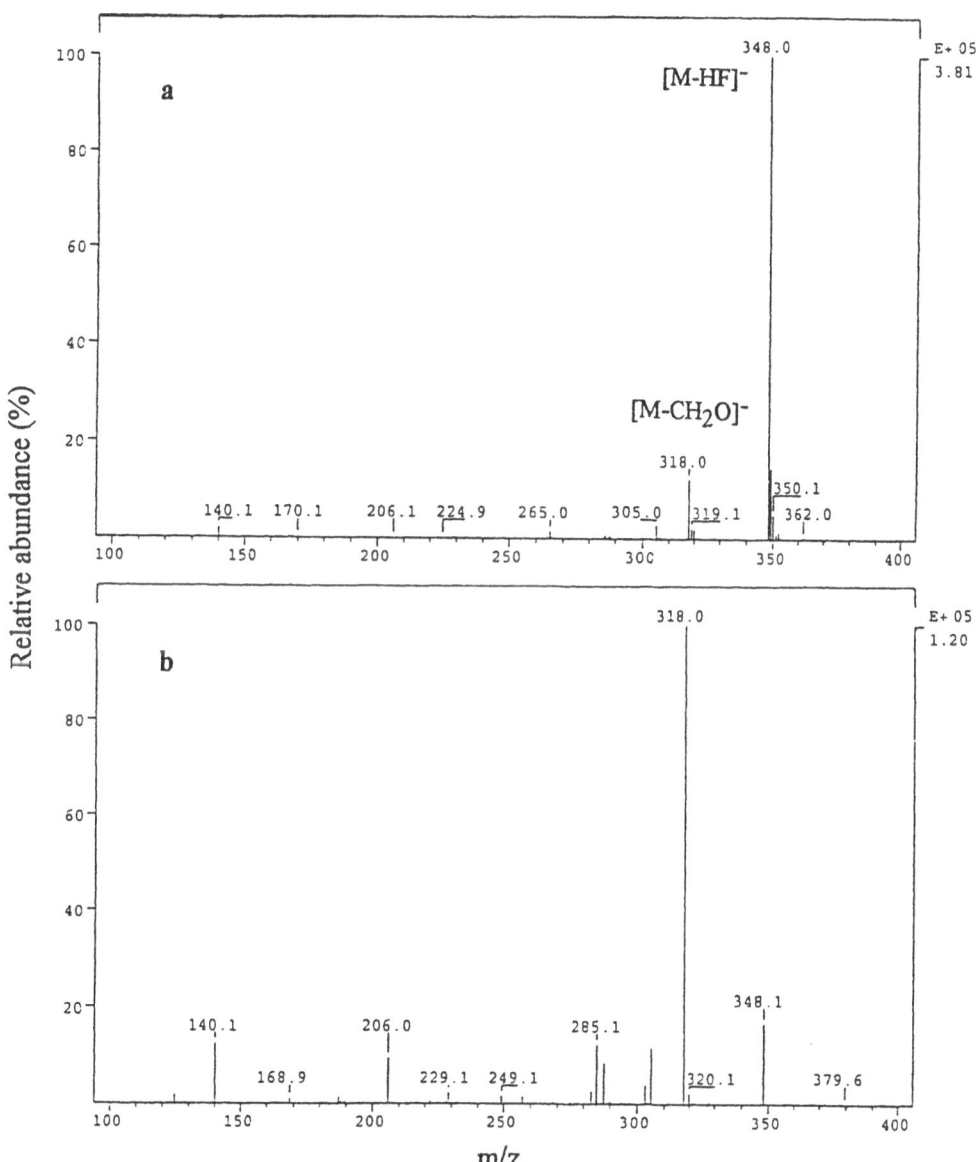

**Fig. 3.** **a.** Studies by MS/NICI of dependence on ion source temperature of abundance of m/z 348 and m/z 318 of HOEtVal-PFPTH.

**b.** Studies by MS/MS/NICI of dependence on collison gas pressure and collision energy of abundance of m/z 318 (daughter ion of m/z 348) of HOEtVal-PFPTH.

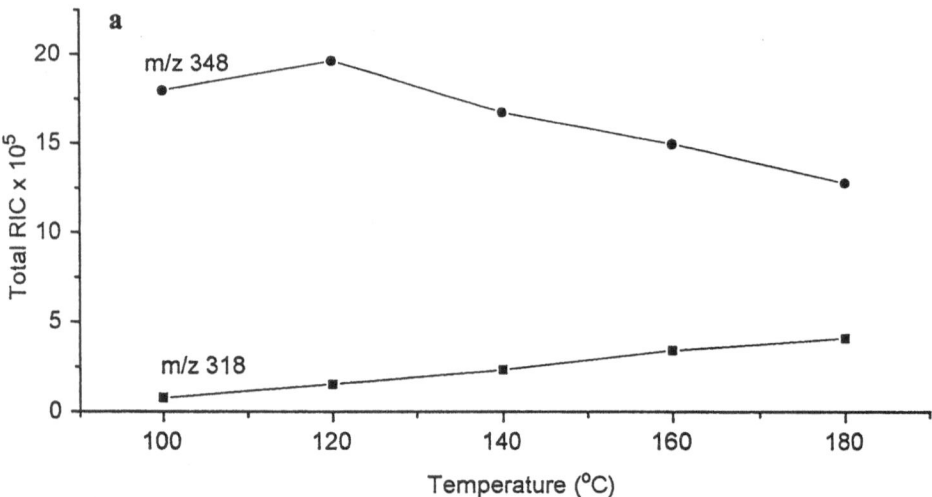

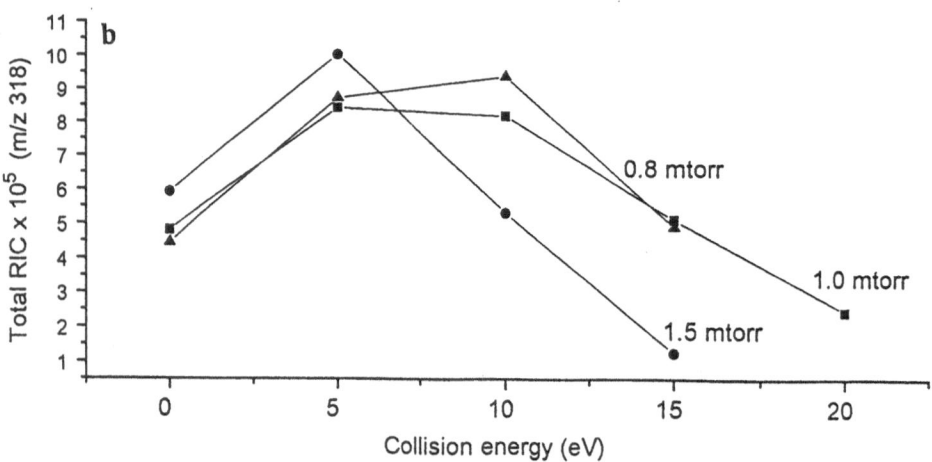

Fig. 4. Analysis of background adducts from endogenously formed ethylene oxide (HOEtVal-PFPTH; monitoring of m/z 318) and propylene oxide (HOPrVal-PFPTH, giving diastereomers; monitoring of m/z 318) in a derivatized blood sample from a non-exposed person. Globin alkylated with ($^2H_4$)ethylene oxide is used as internal standard [($^2H_4$)HOEtVal-PFPTH; monitoring of m/z 320]. Analysis of daughter ions by MS/MS/NICI (selected reaction monitoring).

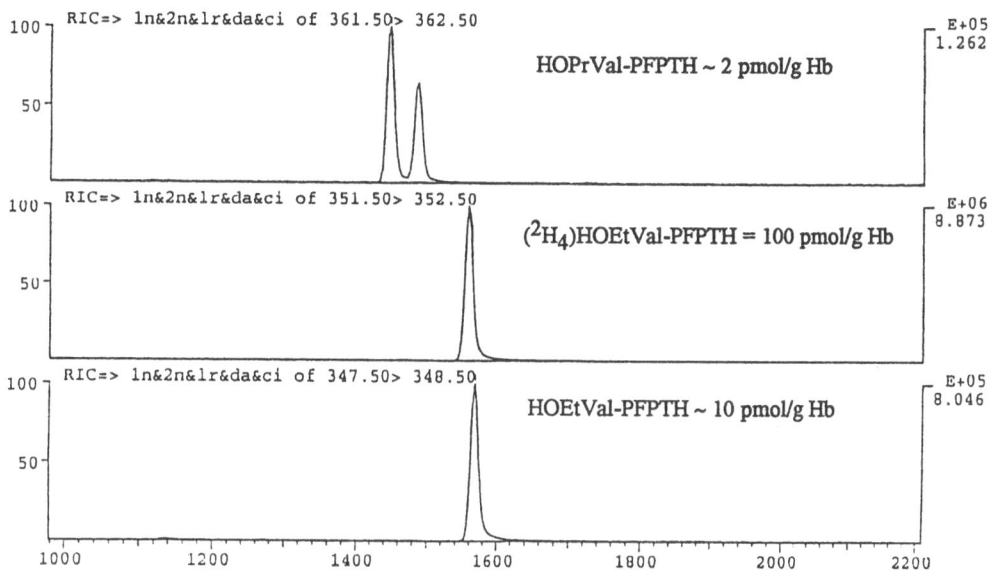

Table 2. Detection level and sources of analytical errors in mass spectrometrical analysis of PFPTHs from adducts to valine in Hb obtained by the N-alkyl Edman method.

| | GC/MS/MS[a,b] | GC/MS[c] |
|---|---|---|
| Detection level: | | |
| Globin sample (50 mg) | < 1 pmol/g | ≈ 10 pmol/g |
| Standard compound | 1 fmol | 10 fmol |
| Coefficient of variation: | | |
| Range of measurement | 2-20 pmol/g | around 20 pmol/g |
| Analytical occasion | 25 % | — |
| Derivatization | 10 % | — |
| Analytical error[d] | ≈ 10 % | 10 % |

[a] Present status of work under development (25).
[b] Analysis carried out on a Finnigan TSQ 700 triple stage quadrupole instrument.
[c] Analysis carried out on a Finnigan 4500 quadrupole instrument.
[d] After correction (by calibration) for occasion of analysis and derivatization.

## 4.4. PROCEDURE

The standard N-alkyl Edman procedure permits at least ten samples to be collected and analysed within four days. Samples are prepared and analysed according to the following (for details of the procedure, see refs. 25 and 26):
- Collection of blood sample (1 ml or more).
- Isolation, washing and hemolysation of red blood cells.
- Precipitation of globin.
- Derivatization: Dissolution of globin samples in formamide, addition of internal standard globin and the reagent, PFPITC. Isolation of the PFPTH-derivative by extraction and purification by washing the extract with mild alkaline solution.
- Analysis of the sample by GC/MS or GC/MS/MS, NICI, using methane as reagent gas.

It was set as a goal in the development of the method that the isolation and purification of globin samples and the derivatives should be simple in the standard procedure, e.g. time-consuming column separations should be avoided. Furthermore, treatment of samples in many steps increase the risk of artefact formation and contamination. It has been shown that even storage under longer periods of whole blood or red cells (intact or hemolysed), as well as lyophilization, may lead to formation of artefacts of low-molecular weight adducts and should be avoided. So far, it has been shown that blood samples are most safely stored as globins (non-dialysed, non-lyophilized) (11). Contamination has been avoided by using dedicated laboratory area and equipment during preparation of samples for the analysis of low adduct levels.

## 5. Monitoring of Exposure: Risk Identification and Risk Estimation

### 5.1. SURVEILLANCE OF EXPOSURES

The N-alkyl Edman method has been used for surveillance of work environments, particularly with respect to ethylene oxide. It has been shown that monitoring by means of Hb adducts is more sensitive, by orders of magnitude, than monitoring of cytogenetic endpoints (28). Furthermore, it has also been shown that measurements of in vivo doses by Hb adducts give more reliable values of the time-weighted average concentration than measurement of the exposure in an occupational environment. The method has been applied for the determination of adducts from for instance ethene (metabolite ethylene oxide), propene (propylene oxide) and butadiene (butadiene monoxide) in exposed humans and smokers (summarized in ref. 19). Adducts from aromatic amines have mostly been studied in smokers (8).

### 5.2 IDENTIFICATION OF RISK FACTORS

In animals treated with precarcinogens, reactive metabolites could be identified through the determination of Hb adducts. In individuals without known exposure "background adducts", particularly from many low-molecular weight adducts, have been found. The identification of these adducts and their sources is a possible way of discovering factors that could act as "background" cancer initiators/mutagens (see Fig. 5) (29). Among the compounds observed in this way is ethene produced endogenously, according to animal experiments associated with the intestinal flora and with unsaturated fatty acids in the diet (30). A number of aldehydes have been observed following reduction of Schiff bases to alkylvalines (31).

5.3.    ESTIMATION OF CANCER RISKS

Metabolism, with regard to activation to, or detoxification of reactive intermediates could be quantitatively studied in animals and humans through the determination of Hb adducts. Individuals with a variation in sensitivity due to polymorphism in bioactivating or detoxification enzymes may be identified in this way. In such studies a high reproducibility of the method is important (32).

Metabolic studies, in animals and humans, are at the same time studies of the relationship between exposure dose and blood dose (in vivo dose). This relationship gives a basis for risk estimation (see Fig. 5). In animal experiments the relationship between target dose, the dose of reactive intermediates in different organs, and blood dose could then be measured. Target dose could be obtained through measurement of DNA adducts after acute exposures with radiolabelled compounds. The relative genotoxic potency of reactive intermediates are then measured in mutation experiments using $\gamma$-radiation as reference standard (3, 33). $\gamma$-Radiation has been chosen as a reference standard for this approach as it has a rather well-known cancer risk per unit of in vivo dose in humans.

Through this approach cancer risks associated with exposure to ethylene oxide and ethene from different sources have been estimated (Table 3).


6.    **Discussion**

As said above the goal was reached to achieve a detection level for single compounds corresponding to 1 pmol valine adduct/g Hb, i.e. the level associated with a cancer risk of about 1 in $10^6$ per year. However, since exposure to humans are often mixed further improvement of sensitivity in the analysis of adducts from individual components will be required. This concerns both exogenous exposures such as air pollutants, food components etc. and the endogenous production of reactive intermediates. Although a major part of current cancer incidence can be ascribed to environmental factors (in a broad sense) (35) we have at present some knowledge of the cancer initiators in less than 10 % of the cancer cases (21). It is quite possible that adducts from endogenously produced compounds reflect the existence of unknown initiators responsible for a large part of the cancers occurring.

In order to tackle these problems analytical techniques will have to be developed for the identification of essential contributors to risk as well as measurement of total loads of whole classes of compounds. This requires development both of methods for the isolation and derivatization of adducts and of the analytical procedures. It seems that the principle of MS/MS-techniques is the only technique with a potential possibility to solve these problems.

Surprisingly high background adduct levels of DNA adducts have been found in for instance human lymphocytes (19). This indicates that long-lived cells with little or no DNA repair may exist in human tissues. These cells could become a tool for long-term dose monitoring provided the adducts are identified chemically and quantified.

The problems discussed in this paper require a multidisciplinary approach with MS-techniques playing a central role.

**Fig. 5.** Principles for identification of genotoxic risk factors and for estimation of cancer risks, by means of determination of in vivo dose through measurement of Hb adducts.

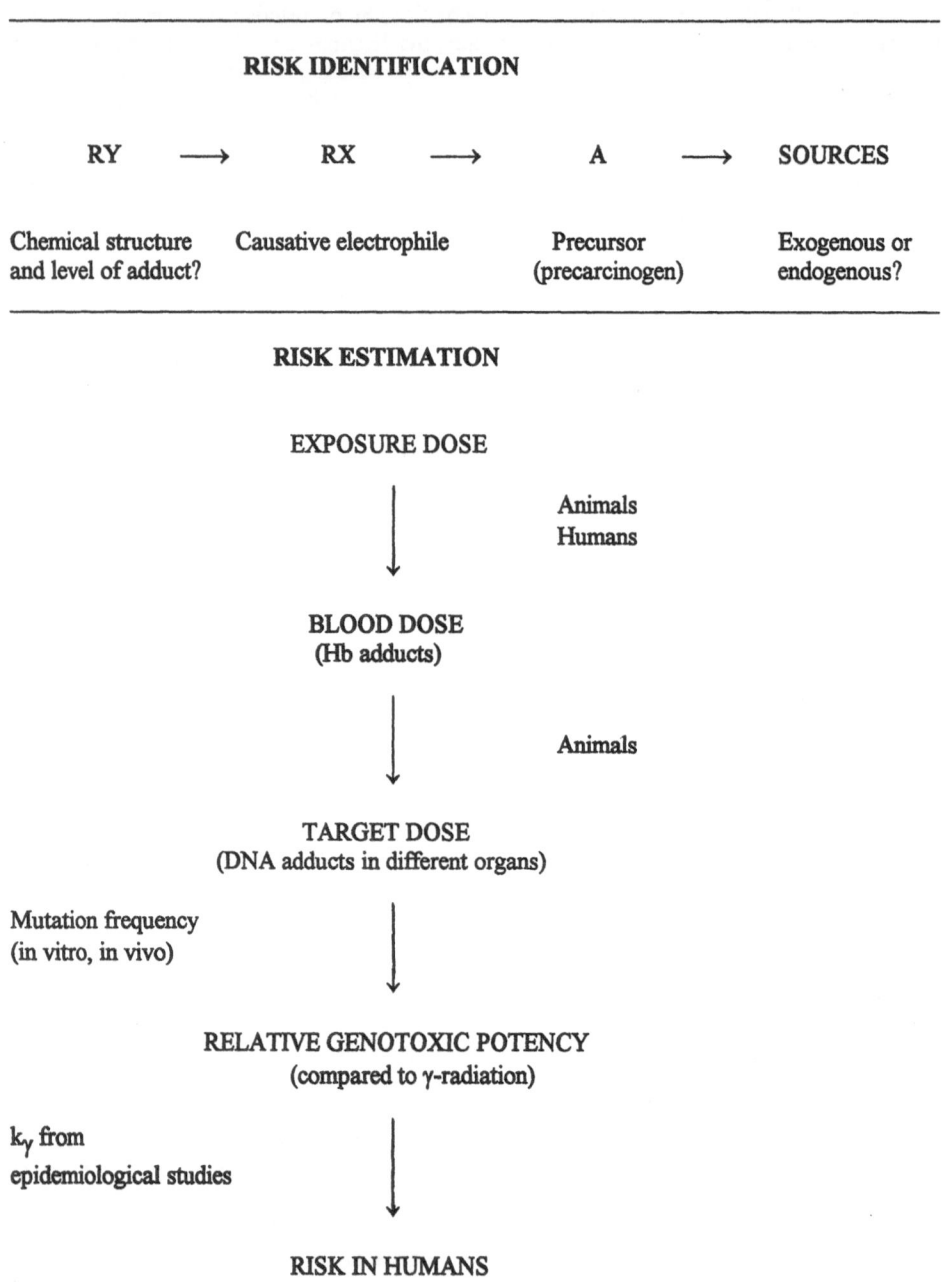

RISK IDENTIFICATION

RY ⟶ RX ⟶ A ⟶ SOURCES

Chemical structure and level of adduct?    Causative electrophile    Precursor (precarcinogen)    Exogenous or endogenous?

RISK ESTIMATION

EXPOSURE DOSE

Animals
Humans

BLOOD DOSE
(Hb adducts)

Animals

TARGET DOSE
(DNA adducts in different organs)

Mutation frequency
(in vitro, in vivo)

RELATIVE GENOTOXIC POTENCY
(compared to γ-radiation)

$k_\gamma$ from
epidemiological studies

RISK IN HUMANS

**Table 3.** Estimation of cancer risks associated with exposure to ethylene oxide and ethene from different sources (34).

| Exposure | HOEtVal steady-state level (pmol/g) | Rad-equivalent annual dose | Associated lifetime risk of cancer death from 1 year's exposure |
|---|---|---|---|
| Ethylene oxide, 1 ppm, 40 h/week | 2400 | 22 | $400 \cdot 10^{-5}$ |
| Ethene, 1 ppm, 40 h/week | 120 | 1 | $20 \cdot 10^{-5}$ |
| Ethene from smoking 10 cig./day | 85 | 0.8 | $16 \cdot 10^{-5}$ |
| Ethene in urban air 10 ppb, 168 h/week | ~ 4 | 0.036 | $\sim 1 \cdot 10^{-5}$ |
| Background | 20 ( 6-25) | 0.18 | $3.6 \cdot 10^{-5}$ |

## 7.    Acknowledgements

Mr. Vlado Zorcec is gratefully acknowledged for assistance with analyses by tandem mass spectrometry. These studies have been financially supported by the Swedish Environmental Protection Agency.

## 8.    References

1.  E.C.Miller and J.A.Miller, "Mechanisms of chemical carcinogenesis: Nature of proximate carcinogens and interactions with macromolecules", *Pharmacol. Rev.* **18**, 805-838 (1966).
2.  I.Turtóczky and L.Ehrenberg, "Reaction rates and biological action of alkylating agents preliminary report on bactericidal and mutagenic action in *E. coli*", *Mutat. Res.* **8**, 229-238 (1969).
3.  L.Ehrenberg, E. Moustacchi, S.Osterman-Golkar and G.Ekman, "Dosimetry of genotoxic agents and dose-response relationships of their effects", *Mutat. Res.* **123**, 121-182 (1983).
4.  S.Osterman-Golkar, L.Ehrenberg, D.Segerbäck and I.Hällström, "Evaluation of genetic risks of alkylating agents. II. Haemoglobin as a dose monitor", *Mutat. Res.* **34**, 1-10 (1976).
5.  F.Granath, L.Ehrenberg and M.Törnqvist, "Degree of alkylation of macromolecules in vivo from variable exposure", *Mutat. Res.* **284**, 297-306 (1992).
6.  C.J. Calleman, L.Ehrenberg, B.Jansson, S.Osterman-Golkar, D.Segerbäck, K.Svensson and C.A.Wachtmeister, "Monitoring and risk assessment by means of alkyl groups in hemoglobin

394

in persons occupationally exposed to ethylene oxide, *J. Environ. Pathol. Toxicol.* **2**, 427-442 (1978).

7.  D.Segerbäck, "Reaction products in hemoglobin and DNA after *in vitro* treatment with ethylene oxide and *N*-(2-hydroxyethyl)-*N*-nitrosourea", *Carcinogenesis* **11**, 307-312 (1990).

8.  M.S.Bryant, P.L.Skipper, S.R.Tannenbaum, and M.Maclure, "Hemoglobin adducts of 4-amino-biphenyl in smokers and nonsmokers", *Cancer Res.* **47**, 602-608 (1987).

9.  S.R.Tannenbaum, P.L.Skipper, J.S.Wishnok, W.G.Stillwell, B.W.Day and K.Taghizadeh, "Characterization of various classes of protein adducts", *Environ. Health Perspect.* **99**, 51-55 (1993).

10. L.Gan, P.L.Skipper, X.Peng, J.D.Groopman, J.Chen, G.N.Wogan and S.R.Tannenbaum, "Serum albumin adducts in the molecular epidemiology of aflatoxin carcinogenesis: correlation with aflatoxin B1 intake and urinary excretion of aflatoxin M1", *Carcinogenesis* **9**, 1323-1325 (1988).

11. M.Törnqvist, "Formation of reactive species that lead to hemoglobin adducts during storage of blood samples", *Carcinogenesis* **11**, 51-54 (1990).

12. C.J.Calleman, L.Ehrenberg, S.Osterman-Golkar and D.Segerbäck, "Formation of *S*-alkyl-cysteines as artifacts in acid protein hydrolysis, in the absence and in the presence of 2-mercaptoethanol", *Acta Chem. Scand. B* **33**, 488-494 (1979).

13. M.Törnqvist, J.Mowrer, S.Jensen, and L.Ehrenberg, "Monitoring of environmental cancer initiators through hemoglobin adducts by a modified Edman degradation method", *Anal. Biochem.* **154**, 255-266 (1986).

14. S.G.Carmella, S.S.Kagan, M.Kagan, P.G.Foiles, G.Palladino, A.M.Quart, E.Quart and S.S.Hecht, "Mass spectrometric analysis of tobacco specific nitrosamine hemoglobin adducts in snuff dippers, smokers, and non-smokers", *Cancer Res.* **50**, 5438-5445 (1990).

15. A.Weston, M.L.Rowe, D.K.Manchester, P.B.Farmer, D.L.Mann and C.C.Harris, "Fluorescence and mass spectral evidence for the formation of benzo[*a*]pyrene *anti*-diol-epoxide-DNA and -hemoglobin adducts in humans", *Carcinogenesis* **10**, 251-257 (1989).

16. J.Everse, R.M.Winslow and K.D.Vandergriff (eds) Hemoglobins Part B: Biochemical and Analytical Methods, Methods in Enzymology, vol. 231, Academic Press, New York, 1994 (in press).

17. M.J.Wraith, W.P.Watson, C.V.Eadsforth, N.J.van Sittert, M.Törnqvist and A.S.Wright, "An immunoassay for monitoring human exposure to ethylene oxide", In: H.Bartsch, K. Hemminki and I.K. O'Neill (eds.), Methods for Detecting DNA Damaging Agents in Humans: Applications in Cancer Epidemiology and Prevention, IARC Sci. Publ. 89, pp. 271-274, Lyon, 1988.

18. A.Umemoto, Y.Monden, S.Grivas, K.Yamashita and T. Sugimura, "Determination of human exposure to the dietary carcinogen 3-amino-1,4-dimethyl-5*H*-pyrido[4,3-*b*]indole (Trp-P-1) from hemoglobin adduct: the relationship to DNA adducts", *Carcinogenesis* **13**, 1025-1030 (1992).

19. M.Törnqvist, "Current research on hemoglobin adducts and cancer risks: an overview", In: C.C. Travis (ed.) Use of Biomarkers in Assessing Health and Environmental Impacts of Chemical Pollutants, pp. 17-30, Plenum Press, New York, 1993.

20. P.Edman and A.Henschen, "Sequence determination", In: Needleman, S.B. (ed.), Protein Sequence Determination, 2nd ed., p. 232, Springer-Verlag, Berlin and New York 1975.

21. M.Törnqvist, "Monitoring and cancer risk assessment of carcinogens, particularly alkenes in urban air ", Ph.D. Thesis, University of Stockholm, Stockholm (1989).

22. ICRP, Publication No. 26, Recommendations of the International Commission on Radiological Protection, Annual ICRP Vol. 1. No. 3, Pergamon Press, Oxford, 1977.

23. J.Mowrer, M.Törnqvist, S.Jensen and L.Ehrenberg, "Modified Edman degradation applied to hemoglobin for monitoring occupational exposure to alkylating agents", *Toxicol. Environ. Chem.* **11**, 215-231 (1986).

24. P.Rydberg, B.Lüning, C.A.Wachtmeister and M.Törnqvist, "Synthesis and characterization of *N*-substituted valines and their phenyl- and pentafluorophenyl-thiohydantoins", *Acta Chem. Scand.* **47**, 813-817 (1993).

25. M.Törnqvist, V.Zorcec and F.Granath, "Tandem mass spectrometry for monitoring of environmental carcinogens by macromolecule adducts", (manuscript) (1993).

26. M.Törnqvist, "Epoxide adducts to N-terminal valines", In: J. Everse, R.M.Winslow and K.D. Vandergriff (eds), Hemoglobins Part B: Biochemical and Analytical Methods, Methods in Enzymology, vol. 31, Academic Press, Inc, New York (in press), 1994.

27. M.Törnqvist, A.-L.Magnusson, P.B.Farmer, Y.-S.Tang, A.M.Jeffrey, L.Wazneh, G.D.T. Beulink, H.van der Waal and N.J van Sittert, "Ring test for low levels of *N*-(2-hydroxyethyl)valine in human hemoglobin", *Anal. Biochem.* **203**, 357-360 (1992).

28. A.D.Tates, T.Grummt, M.Törnqvist, P.B.Farmer, F.J.van Dam, H.van Mossel, H.M. Schoemaker, S.Osterman-Golkar, Ch.Uebel, Y.S.Tang, A.H.Zwinderman, A.T.Natarajan and L.Ehrenberg ,"Biological and chemical monitoring of occupational exposure to ethylene oxide", *Mutat Res* **250**, 483-497 (1991).

29. M.Törnqvist, "Search for unknown adducts: Increase of sensitivity through preselection by biochemical parameters", In: H.Bartsch, K. Hemminki and I.K. O'Neill (eds.), Methods for Detecting DNA Damaging Agents in Humans: Applications in Cancer Epidemiology and Prevention, IARC Sci. Publ. 89, pp. 378-383, Lyon, 1988.

30. M.Törnqvist and A. Kautiainen, "Adducted proteins for identification of endogenous electrophiles", *Environ. Health Perspect.* **99**, 39-44 (1993).

31. A.Kautianen, M.Törnqvist, K.Svensson and S.Osterman-Golkar, "Adducts of malonaldehyde and a few other aldehydes to hemoglobin", *Carcinogenesis* **10**, 2123-2130 (1989).

32. M.Törnqvist, M.Svartengren and C.H.Ericsson, "Methylations in hemoglobin from monozygotic twins discordant for cigarette smoking: hereditary and tobacco-related factors", *Chem.-Biol. Interactions* **82**, 91-98 (1992).

33. A.Wright, T.K.Bradshaw and W.P.Watson, "Prospective detection and assessment of genotoxic hazards: A critical appreciation of the contribution of L. Ehrenberg", In: H.Bartsch, K. Hemminki and I.K. O'Neill (eds.), Methods for Detecting DNA Damaging Agents in Humans: Applications in Cancer Epidemiology and Prevention, IARC Sci. Publ. 89, pp. 237-248, Lyon, 1988.

34. L.Ehrenberg and M.Törnqvist, "Use of biomarkers in epidemiology: quantitative aspects", *Toxicol. Lett.* **64/65**, 485.492 (1992).

35. J.Higginson and C.S.Muir, "Environmental carcinogens: misconceptions and limitations to cancer control", *J. Natl. Cancer Inst.* **63**, 1291-1298 (1979).

# THE ROLE OF MASS SPECTROMETRY IN BIOMONITORING EXPOSURE TO CARCINOGENS

## N. SANNOLO
*Sezione di Medicina Occupazionale e Igiene Industriale*
*Dipartimento di Biochimica e Biofisica*
*Seconda Università di Napoli, Piazza Miraglia 3*
*I-80138 Napoli, Italy*

## V. CARBONE, P. FERRANTI, I. FIUME, D. SANTORO and A. MALORNI
*Servizio di Spettrometria di Massa del C.N.R.*
*c/o Facoltà di Medicina dell'Università di Napoli Federico II*
*Via Pansini 5*
*I-80131 Napoli, Italy*

## 1. Introduction

It should be known that the well-being and effective functioning of present societies are grounded on chemicals. In fact, the quality of life, at both social and individual levels, would be severely limited if we had to live without petroleum products, synthetic drugs, plastics and textiles and man-made mineral fibres or without the synthetic materials necessary to produce modern electronic devices.

It has been evaluated that about 2000 new chemicals are synthesised annually and about 60.000-70.000 chemicals are used daily in the industrialised societies. Since the use and the production of chemicals imply not only the desired benefit but also some adverse effect, we are compelled to compromise between costs and benefits and some people have wondered how long our ecosystem can survive and how long the biosystem, man included, can tolerate the doubling of the consumption of chemicals that occurs every seven years.

In order to sustain the production of goods and services in each industrialised country, a large amount of the labour force is exposed to chemicals, many of which have been proved to be, while other are suspected to be, genotoxic. At present one of the most relevant problems in occupational medicine deals with the biomonitoring exposure to carcinogens in workplace and with the assessment and management of risk due to the exposure to genotoxic agents. From a historical point of view Bernardino Ramazzini, an italian physician, was the first to realise in 1713 a chemical causation of cancer pointing out the high incidence of brest cancer in nuns (1). But the first clear recognition of an occupational cancer was done by the english surgeon Pott in 1775, who described the common occurrence of cancer of the scrotum in chimney sweepers (2). He recognised the

397

*R. M. Caprioli et al. (eds.), Mass Spectrometry in Biomolecular Sciences, 397–415.*
© 1996 *Kluwer Academic Publishers.*

malignant nature of the disease and correctly attributed it to prolonged contact of the skin with soot. The first identification of carcinogenic chemicals, the polycyclic aromatic hydrocarbons, by Cook *et al.* 1933 (3) can be clearly traced back to the Pott's observations.

Over a century after Pott's publication, occupational cancer of the urinary bladder was recognised in a small group of german dyestuff workers by Rehn in 1895 (4), and this in turn led to the identification in 1954 of the agent responsible for the disease: the aromatic amines (5-7). In any case, since occupational cancers have occupational causes, this implies that such tumours at least should be preventable and this view is widely accepted by the scientific communities.

At present the range of known occupational carcinogens covers many chemicals of organic and inorganic types; unfortunately, however, we don't know the agents responsible for several processes associated with cancer risk, as well as we have not been able to obtain, up to now, firm evidence in the cases of chemicals under suspicion of having caused occupational cancers. This lack of knowledge is the ground for the different policies in the western countries both in the prevention of occupational carcinogenesis and in compensation of workers affected by cancer suspected of occupational origin.

## 2. Exposure to genotoxic agents

From a scientific point of view, workplaces represent an invaluable source of knowledge for chemical carcinogenesis since workers are generally exposed to relatively high doses of only one, or at least few genotoxic chemicals that are present in well known concentrations in workplaces. On the contrary, the genotoxic risk for the whole population, associated with the presence of mutagens and carcinogens in the environment, is difficult to evaluate because of possible multi-exposure, i.e. the exposure to several man made genotoxic agents at the same time, even if present at lower concentrations. On the other hand, it must not be forgotten that in the environment many genotoxic chemicals, released from natural sources, are present too. These substances are, generally, less considered in screening programmes for chemical carcinogenesis since they are focused primarily on synthetic industrial compounds according to the "chemicalization" point of view (8), although there is similarity in toxicology of synthetic and natural toxins (9).

Whatever the origin of the xenobiotic to which man is exposed (workplace, food, tap water, urban air, etc.) the amount of it, which enters the organism, is distributed among several compartments according to its lifetime (Fig. 1). Its biotransformation gives rise to active or inactive metabolites that, in turn, may be distributed in the body according to their lifetime and, like the parent compound, may be excreted in different ways. In fact, many carcinogens are pro-carcinogens that require enzymatic activation. During the last decade, it has been recognised that i) many carcinogens require multiple enzymatic steps to become activated to their ultimate carcinogenic form; ii) the activity of the rate-limiting enzymatic reaction, responsible for activation of carcinogen, differs among people and influences their cancer risk. Acting directly, or indirectly through its active metabolites, the xenobiotic binds to critical or non critical target molecules and a proportionality exists between the binding processes. Repair mechanisms release degradation products that may

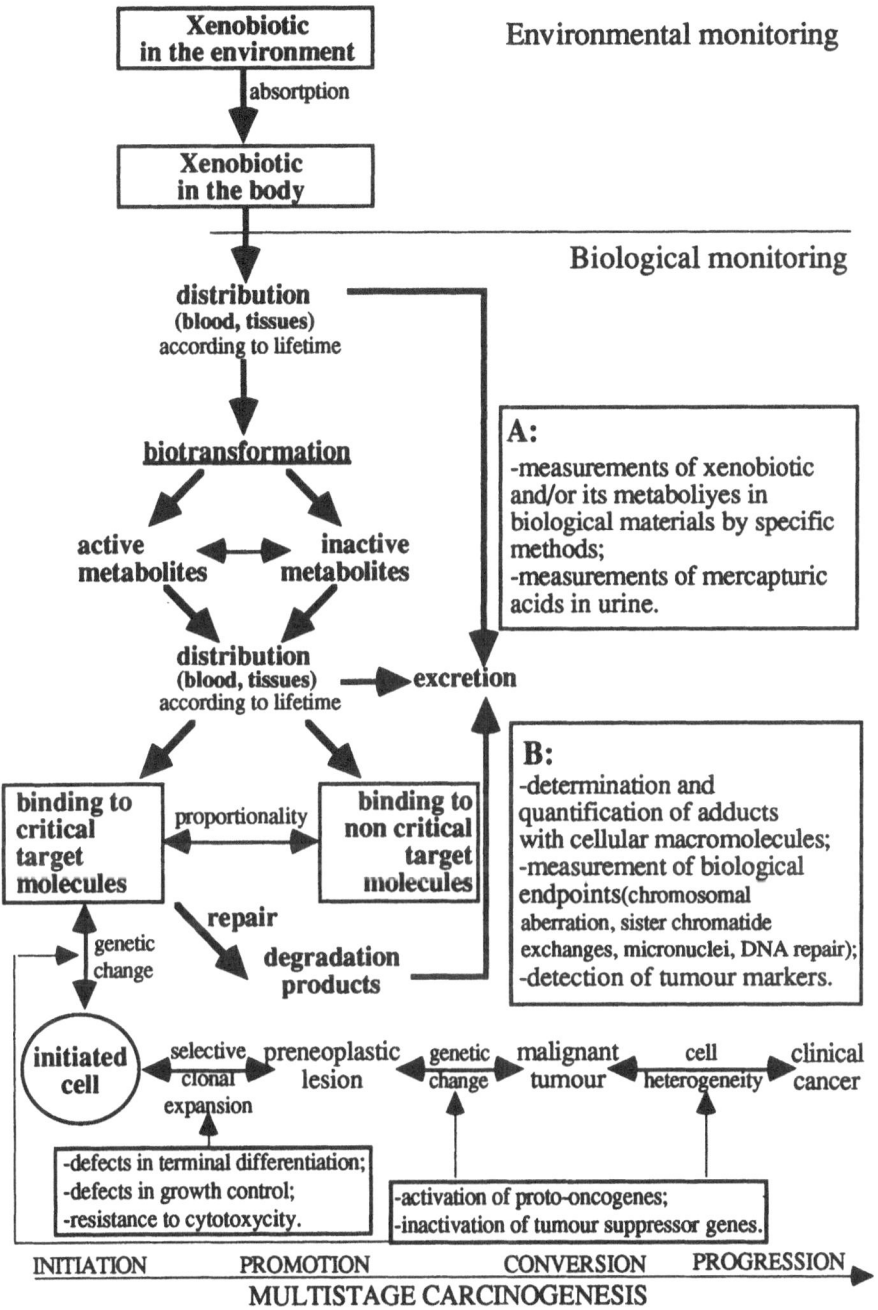

**Figure 1.** Scheme of biological events following human exposure to xenobiotics.

be detected in various biological media, but if the repair mechanisms are inadequate or insufficient, cell lesion may occur. Therefore the **metabolite balance** between activation and deactivation of a pro-carcinogen and the covalent binding of its ultimate carcinogenic metabolite to DNA are considered to be important events in the initial steps of the multistage process of carcinogenesis (10).

## 3. Strategies for identification and control of the genotoxic risk

Traditionally, human exposure to an exogenous chemical (xenobiotic) has been approached by direct repetitive and regular measurements of its concentration, i.e. the amount present in one or several environmental compartments, i.e. environmental monitoring (Fig. 1). For example, in workplaces measurements of xenobiotic in ambient air has long been, and still is, a classical method to evaluate the exposure to xenobiotics. The "external dose" of a xenobiotic, i.e. the amount offered to the body per unit of time over a specified time-interval, is assessed by the measurement of the concentration of chemical in environmental compartments (air, water, food, etc.), duration of exposure, pulmonary ventilation, food consumption, etc..
However, measurement of concentration in air does not provide quantitative information on the body burden in exposed workers. For this reason biological monitoring, i.e. the measurement and assessment of toxic agents or their metabolites in biological samples, is carried out to evaluate exposure and health risk compared to appropriate biological indices.
Environmental and biological monitoring are truly complementary methods to prevent damages to the human health.

The methods currently available for biological monitoring of exposure to genotoxic agents can be distributed between two broad categories:
A-those based on monitoring of uptake;
B-those based on the detection and, possibly, the quantification of those biological changes that result at the target tissue.
The first category (panel A in Fig. 1) includes measurements of xenobiotic and/or its metabolites in biological media by specific methods and, less specific, measurements of mercapturic acid in urine (Fig. 2). Most mercapturic acids have been identified as urinary metabolites originated from xenobiotics, but some endogenous electrophiles are also excreted as mercapturic acids. In this respect more work is necessary to validate the analysis of urinary mercapturic acids as marker of exposure to electrophilic compounds. These methods attempt to detect the absorption of the genotoxic agent and, possibly, to estimate the internal dose. The second category (panel B in Fig. 1) includes: i) determination and quantification of reaction products (adducts) formed between the active genotoxic agent and cellular macromolecules; ii) measurements of biological endpoints, such as chromosomal aberration, sister chromatide exchanges, micronuclei in maturing erythrocytes or in lymphocytes, DNA repair in lymphocytes; iii) detection of tumor markers as indicator of the presence of transformed cells. It is worth noting that biological monitoring methods based on the determination of adducts between genotoxic species and proteins or nucleic acids attempt to estimate the biologically effective dose in various body compartments.

By using the above markers, in conjunction with data obtained from animal experiments, it might be possible to propose preventive measures able to make negligible the cancer

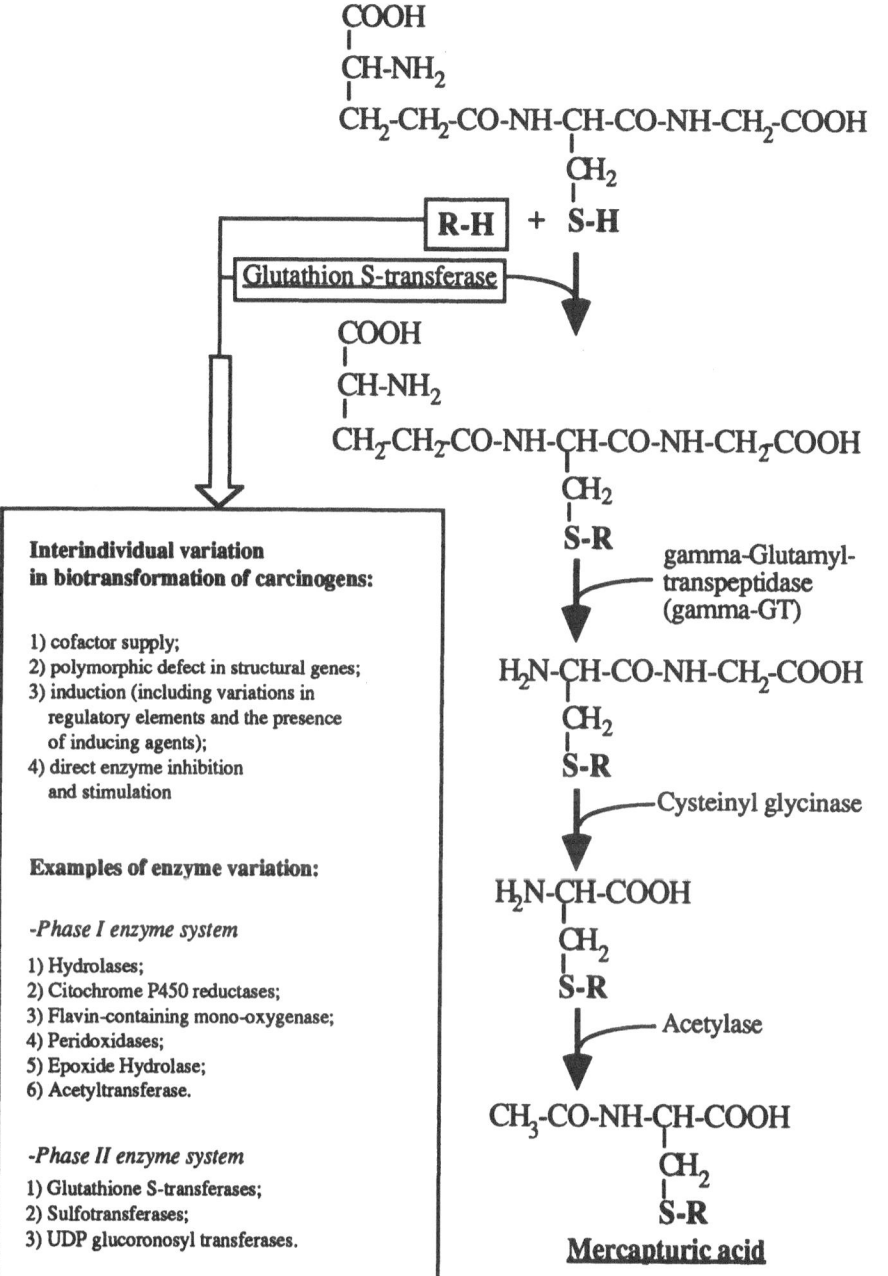

**Figure 2.** Mercapturic acid formation by interaction of electrophylic agents with glutathion depending on interindividual susceptibility.

risk for a single genotoxic agent.

## 4. Molecular Dosimetry

The initiation of toxic effects due to the exposure to electrophilic substance depends on several factors. Not all the electrophilic metabolites formed during biotransformation will bind covalently to cellular macromolecules. In the cell, in fact, there are several detoxification mechanisms that are able to deactivate reactive electrophiles. In this respect binding to glutathion is of particular importance (Fig. 2).

Glutathion is a tripeptide, $\gamma$-glutamyl-cysteinyl-glycine, present in high concentrations in many cells, including red cells. An important property of glutathion is that it is able to conjugate electrophilic agents with the nucleophilic thiol group. The extent of this conjugation process depends on the catalysing action of glutathion S-transferase (11). Glutathion conjugates are converted to cysteine conjugate in two enzymatic reactions: removal of the $\gamma$-glutamyl residue by $\gamma$-glutamyl-transferase (or $\gamma$-glutamyl-transpeptidase) and removal of glycyl moiety by cysteinylglycinase. Often, the cysteine conjugates are N-acetylated by a N-acetyltransferase and the resulting mercapturic acid is excreted subsequently in urine. So, the reaction of electrophilic species with glutathion prevents covalent binding of reactive species with cellular macromolecules and, in this way, glutathion conjugation represents a protective mechanism although in some cases the glutathion conjugates have shown mutagenic properties resulting more toxic than the parent compound. This is the case of 1,2-dihaloethanes (ethylene dibromide/dichloride); they can react directly with proteins but can react with DNA only after conjugation with glutathion through the reactive thiiranium ion formed as intermediate (12,13). So this is one of the cases in which detoxification through conjugation leads new opportunities for genotoxicity. Glutathion S-transferase belongs to a set of enzymes involved in the so called inter individual variation in biotransformation of carcinogens (14) representing a possible variable risk factor since the activation of enzymes responsible for carcinogen metabolism in various human tissues and cells can vary more then 100-fold as in the case of isoenzymes cytocrome P450. The glutathion S-transferase can be grouped into four gene families, which are expressed both in experimental animals and humans. The $\alpha$–$\varepsilon$ group has basic proteins, $\mu$ is a neutral class and $\pi$ is acidic enzyme. They are all cytosolic enzymes; the microsomal glutathion S-transferase is a quite different enzyme. The $\alpha$–$\varepsilon$ and $\mu$ groups appear to be coded by multigene families and the complexity in humans is not yet known. The $\alpha$–$\varepsilon$, $\mu$ and microsomal enzymes are found in liver; the $\alpha$–$\varepsilon$ group in skin and $\mu$ is expressed in leukocytes. The $\pi$ enzyme is normally found in placenta and in several extraepatic tissues; its presence in liver is associated with tumor development and its presence in neoplastic tissues can induce resistance to alkylating chemotherapeutic agents (15). Although the individual glutathion S-transferases are known to vary among people, the hypotheses linking enzyme activity to cancer risk have not been as common as with the P-450 enzymes and acetyltransferases because of the lack of specific assays. Only very recently a specific non invasive assay useful for phenotype characterisation of glutathion S-transferase has been reported (16).

On the basis of the above considerations it is clear that many processes govern the fate of xenobiotics and that the levels of DNA and protein adducts depend on the exposure, on the ratio activation/deactivation processes and, in the case of DNA-adducts, on DNA-repair efficiency.

The dose dependence of macromolecular binding in liver has been studied by several authors by using radioactive labelled carcinogens. For 4-dimethylaminostilbene, administered by single oral dose, a linear relationship over an extremely wide range of doses was observed in 1980 by Neumann (17). The lower end of this relationship corresponds to dose of $5 \times 10^{-10}$ mol/kg, a value three orders of magnitude lower than the lowest daily dose occurring to produce a tumor in rats (18). Within the experimental error, a constant ratio in liver between binding to proteins, DNA and RNA has been demonstrated (17).

Also in the case of aflatoxin $B_1$, the most active natural liver carcinogen, linear relationship has been demonstrated with sub carcinogenic doses administered to rats by single intraperitoneal injection. Again, the ratio of binding to different liver macromolecules remains constant with dose, but binding to proteins is lower than to DNA (19).

Linear relationship has been finally demonstrated in mice in the case on benzo(a)pyrene after oral administration (20). Although reactive electrophiles can be formed from the biotransformation of benzo(a)pyrene by the intestinal flora, the liver acts with an effective detoxifying function, so the ratio of binding to DNA of different organs (liver and stomach) remains constant with dose, but binding to stomach DNA is lower than to liver DNA.

At present it is well known that a 'random' reactivity of electrophilic agents with all nucleophiles takes place with probabilities determined by reaction-kinetic rules.

Since adduct formation is 'random' over cellular nucleophiles, protein adducts will be formed in the body concurrently with DNA adducts. Because modified amino acids are more easily determined than the corresponding changes in DNA and proteins - i.e. haemoglobin and other blood proteins - are easily available in much larger quantities than DNA ( a few mg in a 20 ml blood sample), haemoglobin adducts were proposed in 1976 by the Ehrenberg group in Stockholm for indirect monitoring of DNA adducts and the risk associated (21). They were able to demonstrate that several compounds, both directly alkylating agent and compounds requiring bioactivation, gave a ratio

> | degree of haemoglobin histidine adduct / degree of DNA guanine N-7 adduct |

that is approximately constant as the ratio of reaction rates with these macromolecules determined *in vitro*. This means that alkylated amino acids in haemoglobin are linearly related to dose, as demonstrated in the case of exposure of rats to ethylene oxide in which a linear relationship was observed between the production of N-3'-(2-hydroxyethyl)histidine (HOEt HIS) in haemoglobin and dose of ethylene oxide administered by inhalation (22). Nevertheless, in control rats, in absence of exposure to ethylene oxide, surprisedly a level of HOEt HIS was also measured; this level represents the background alkylation level that reflects the exposure to mutagens or carcinogens of exogenous or endogenous origin, with a possible co-causative role in carcinogenesis. The amounts of these "background" adducts range from 16.4 nmol/g for S-methylcysteine to 0.04 pmol/g for 2-aminonaphthalene sulfinamide (23). The source of the background adducts is in general unknown and several explanations of their formation are possible.

It is worth noting that binding to proteins is far more common than DNA alkylation because generally only a single metabolite binds to DNA. This is the case of vinyl chloride: the binding to DNA is possible only for the epoxide and not for other concurrent or successive metabolites (24). In fact, the role of chlorooxirane and chloroacetaldehyde in the carcinogenicity of vinyl chloride has been studied by comparing the biological effects of vinyl chloride exposure with those of 2,2'-dichloroethyl ether as metabolic precursor of chloroacetaldeyde formed in both cases in liver cell. Following the exposure to vinyl chloride, alkylation of guanine-N-7 in DNA has been detected, but not after exposure to 2,2'-dichloroethyl ether. On the contrary alkylation of proteins in liver and other organs has been observed after vinyl chloride as well as after 2,2'-dichloroethyl ether exposure. This shows that in some circumstances reactive intermediates metabolically derived are capable of alkylating protein structures but not DNA.

Current dogma direct our effort to measure adducts formed by the direct action of the same activated carcinogen metabolite on both DNA and protein nucleophilic sites. However, xenobiotics may also exert their oncogenic effect via indirect damage to macromolecules such as carcinogen-induced formation of oxy radicals, which can cause a further DNA damage, and enzyme/coenzyme modifications, which can cause alteration in metabolic pathways. In particular this induction of the pro-oxidant state, i.e. increased concentration of active oxygen, organic peroxides and oxygen radicals, may also be of importance in tumor promotion.

Having reaffirmed that an ideal biomonitoring procedure would measure the adducts formed by the electrophilic species with critical target molecules, i.e. DNA, at target tissue, because the target tissue is in most cases unknown and the acquisition of sufficient DNA-adduct for the analysis from such target tissue has proved difficult to achieve, it becomes evident from the above discussion that peripheral blood red cells are a reasonable surrogate for target tissue and haemoglobin a valid surrogate for DNA.

Why haemoglobin among other blood proteins? Because:
- this protein is readily available from blood in gram quantities,
-its lifetime is about 120 days in humans and 63 days in rats thus permitting accumulation of exposures over this period,
-the stability of adducts up to now experimentally studied generally approaches that of the protein,
-both human and animals globins have been extensively studied also by fast atom bombardment (FAB) and electrospray (ES) mass spectrometry (MS).
On the contrary albumin is less abundant in blood ( about 42 mg/ml plasma), it has a more rapid turnover than haemoglobin ( about 19 days) and it has not been so extensively studied by mass spectrometry as haemoglobin.

Why mass spectrometry for the analysis of the adducts?
The first justification for the use of mass spectrometry for the analysis of adducts relies on the chemical specificity, i.e. its ability to distinguish and characterise adducts from different carcinogens if they have close related structures. The structural specificity of mass spectrometry is of particular advantage over other methods used in adduct monitoring (i.e. immunoassay, [32]P-post-labelling and fluorescence measurements) in which a possible cross-reactivity between different chemicals is always possible.
The second justification relies on the sensitivity, although the sensitivity requirement for human carcinogen biomonitoring cannot precisely be defined at present. However, in animals DNA modification of 1 in $10^5$-$10^7$ nucleotides is caused by tumorougenic doses of chemical carcinogens. Therefore, one might assume that the sensitivity of the analytical

procedure should range from 1 in $10^7$ nucleotides and 1 in $10^{10}$ nucleotides, since the nucleotides contained in the human diploid genome is in the order of $10^{10}$. The analytical procedures based on mass spectrometry have sensitivities for detecting modified nucleotides within this range. In the case of protein adducts the sensitivity depends on the protein to DNA binding ratio for the electrophile considered. It has been reported that in the case of ethylene oxide the formation of 1 N-7-guanine adduct per $10^{10}$ bases in DNA corresponds to 10 pmoles of N-terminal valine adduct per gram of haemoglobin. Mass spectrometry reaches such sensitivity both using GC-MS and FAB-MS or/and ES-MS; so mass spectrometry has the capability to monitor the equivalent of 1 DNA adduct per cell and this sensitivity is very satisfactory for monitoring procedures (25).

## 5. Identification and quantification of modified amino acids from haemoglobin adducts

According to their chemical nature the carcinogen-protein adducts can be classified in two categories: those that are stable and those that are unstable under the standard acidic conditions of protein hydrolysis. Accordingly two procedures, both based on GC-MS, have been developed for the analysis of haemoglobin-carcinogen adducts.

A) In the case of adducts unstable under the standard acidic conditions of protein hydrolysis the haemoglobin is submitted to a mild basic or acidic hydrolysis; the carcinogen relised is extracted and purified from the haemoglobin and then analysed by GC-MS as such or after derivatization.

The procedure used by Tannenbaum and co-workers (26) for the analysis of adducts formed with haemoglobin following exposure in smokers and non smokers to 4-aminobiphenyl represents an example of analysis of unstable adducts. In fact, 4-aminobiphenyl belongs to the class of carcinogens producing adducts unable to survive not only to strong acid hydrolysis used for degradation of peptide bonds but also to the standard acidic acetone procedure for the precipitation of globins. The chemical nature of such adduct is reported to be of sulphinamidic type. In fact, N-hydroxylation by the action of cytocrome P450 in the liver is the critical step in the metabolic activation of 4-aminobiphenyl; the arylhydroxylamine reacts with oxyhaemoglobin to form nitrosobiphenyl that would be expected to react with nucleophiles, in this case a cysteine residue in haemoglobin, with a rearrangement to sulfinamidic compound. Approximately 5% of a single dose administered to rats is bonded to haemoglobin and the level of adduct formed is proportional to the dose administered (27). The first evidence of the involvement of a cysteine residue has been derived by Tannenbaum and co-workers from the different reactivity *in vitro* of biphenyl hydroxylamine toward human haemoglobin and sperm whale myoglobin, which does not contain any cysteine residue, and from the titration experiment of free SH group with mercuribenzoate. Then, the adduct with human haemoglobin has been crystallised and analysed by X-ray crystallography determining the structure to 4 Å resolution from which the attachment of biphenyl ring system to the cysteine residue in position 93 of the β globin has been deduced (26). The experimental quantitative procedure (26) starts with a mild basic hydrolysis of a known quantity of haemoglobin dialysate after the addition of the internal standard 4'-fluoro-4-aminobiphenyl. Then the hydrolysed amine is extracted into hexane, purified and dried and, finally, redissolved in hexane and derivatised with pentafluoropropionic anhydride, 10 min at room temperature. The sample is evaporated to dryness, the residue redissolved in 20 μl of hexane and analysed by capillary/GC-NICI/MS-SIM following the ions at m/z 295.2 for the pentafluoropropyonil-aminobiphenyl and m/z 313.2 for the standard, both corresponding to the fragment M-H-F]⁻. The application of this method to blood sample from cigarette smokers and non-smokers permitted the evaluation of the adduct levels in both populations; the mean value of the aminobiphenyl for smokers was 154 pg/g of haemoglobin compared to 28 pg/g for non-smokers with no overlap of adduct levels between the two groups (26).

The same method has been applied also to human dosimetry of other aromatic amines in smokers and non-smokers (28). Several amines can be quantified in the same run following specific ions and differentiating isomers by their retention times. The results obtained confirm that in general for aromatic amines the adduct levels are higher in smokers than in non-smokers.

It is interesting to underline that the aromatic amines are potent human bladder carcinogens; the detoxification process of the hydroxylamine metabolites through the conjugation with glucuronic acid ensures the transport of this conjugate into the bladder where the acidic environment is responsible of the formation of the reactive specie involved in critical macromolecular binding reactions.

B) In the case of adducts stable under the standard conditions of protein hydrolysis the haemoglobin is submitted to total hydrolysis with HCl 6N and the carcinogen-modified aminoacid is analysed after a proper derivatization procedure.

The procedure used by Farmer and co-workers (29) is a typical example of analysis used to study this kind of adducts. It involves the hydrolysis of globins *in vacuum* in HCl 6N at 110°C for 24 hours in the presence of an appropriate labelled internal standard. The adduct formed at the nucleophilic site level on the side chain of the target aminoacid is released into the hydrolysate as modified aminoacid that is purified by a chromatographic step using ion exchange chromatography, derivatised by esterification and acylation and analysed by GC-EI/MS-SIM. Typical results obtainable with this procedure are illustrated in the case of determination of hydroxypropylhistidine (30) in haemoglobin as measure of exposure to propylene oxide. Because the EI mass spectrum of N-heptafluorobutyryl methyl ester of N-3'-(2-hydroxypropyl)hystidine show a molecular ion at m/z 619 and a base peak at m/z 560 (624 and 565 in the deuterated labelled reference compound derivative), corresponding to the loss of $CH_3COO^-$, these masses are used in quantitative measurements obtaining clean traces because of the high value of selected masses reducing very much any interference.

By this method it is possible to measure only cysteine and histidine adducts that permit a satisfactory monitoring of high level exposure to the appropriate carcinogens, such as methylating agents (29, 31), ethylene oxide (32) and styrene oxide (33). The main disadvantage of this procedure relies on the difficulties in separating the modified amino acid from the normal aminoacids present in up to $10^6$-fold greater quantities. To overcome this difficulty a significant development has been made by Tornqvist and co-workers who showed that carcinogen adduct with N-terminal valine of haemoglobin could be readily extracted from the remaining of the protein following a modified Edman degradation procedure (34), as reported in separate chapter in this book.

## 6. Characterisation of haemoglobin adducts

In the study of interactions between genotoxic compounds and cellular macromolecules, the best and desirable methods should be based on procedures able to detect any modification in the intact protein identifying both the nature and site(s) of such modification. Strategies, based on FAB-MS, ES-MS and MS/MS, for structural characterisation of proteins have started very recently to be applied for the identification of covalent xenobiotic modifications in haemoglobin (35-38). Although Desiderio and co-workers (39) have already demonstrated the feasibility of quantification of neuropeptides at level of femtomole per mg of tissue by using FAB-MS/MS and stable isotope-incorporated peptide internal standards, more work is necessary to develop methods able to quantify at high level of molecular specificity the intact modified haemoglobin, or peptides obtained by specific enzymatic hydrolysis, carrying such modification. At present the difficulties associated with the use of such advanced mass spectrometric techniques in quantitative determination of covalent adducts formed with haemoglobin following exposure to genotoxic chemical deal only with the problems associated to the preparation of a suitable internal standard.

Based on experiences we developed in structural characterisation of proteins (40-51), in particular haemoglobin, we have designed a general strategy to study the *in vitro* and *in vivo* chemical interaction of xenobiotics with haemoglobin. Mass spectrometry has been extensively applied for the characterisation and quantification of modified amino acid within the polypeptide chain of haemoglobin, combining the use of FAB/MS, GC/MS and ES/MS.

In fact, the modification degree was evaluated by ES/MS analysis through the accurate measurement of molecular weights of modified globins, whereas the reactive amino acid within the polypeptide globin chains were identified by FAB/MS analysis. Because FAB-MS allows the direct analysis of peptide mixtures, the FAB-mapping procedure (41), applied in these studies to peptic hydrolysate of both $\alpha$ and $\beta$ human globin (hG) and rat globin (rG), represents a very powerful tool for the inspection of the modifications induced by xenobiotics on the amino acid residues in hHb and rHb. Such modifications are confirmed and quantified by EI-MS performing the GC-MS analysis of the amino acids obtained by means of enzymatic hydrolysis, after their proper derivatization into volatile derivatives (52).

We applied the above structural-analytical approach to study the molecular interaction between haemoglobins and several xenobiotics, such as styrene oxide, methyl bromide and carmustine.

In the case of styrene oxide the interaction *in vitro* of $\alpha_1$rG and $\alpha$hG was studied and the results obtained suggest that some adducts are formed in correspondence with several nucleophilic amino acids in the case of $\alpha_1$rG ; however N-terminal valine results unreactive unless the styrene oxide concentration reaches 1000 times the globin concentration. On the contrary, $\alpha$hG results unreactive at low concentration of styrene oxide and N-terminal valine results reactive (Tab. 1-2) when adducts start to be formed. The interaction *in vitro* of human red cells (hRC) and human whole blood (hWB) showed a surprising, but expected, results. In fact, the analysis of globin fraction showed the absence of any modification. On the contrary, the analysis of the precipitates showed the occurrence of modified globins as in the case of model peptides (53). In particular, a marked reactivity was observed at level of the N-terminal Val as well as of a particular Histidine residue within the globin chain.

In the case of methyl bromide the interaction with hRC was studied both *in vitro* and *in vivo* by monitoring a group of workers employed in the soil fumigation. After *in vitro* incubation under different conditions the reaction products were analysed. Acid denatured

| | Peptide | Expected MH⁺ | Observed MH⁺ | Number of Adducts |
|---|---|---|---|---|
| 1:10 | 3-23 | 2186 | 2306 | 1 |
| 1:100 | 3-23 | 2186 | 2306 | 1 |
| | 37-66 | 3152 | 3272 | 1 |
| | 67-80 | 1406 | 1526 | 1 |
| | 81-98 | 2048 | 2168 | 1 |
| | 91-105 | 847 | 967 | 1 |
| 1:1000 | **1-4** | **389** | **509** | 1 |
| | 42-53 | 1279 | 1399 | 1 |
| | 42-53 | 1279 | 1519 | 2 |
| | 63-70 | 730 | 850 | 1 |
| | 66-80 | 1519 | 1639 | 1 |
| | 66-80 | 1519 | 1759 | 2 |
| | 66-80 | 15199 | 1879 | 3 |
| | 81-98 | 2048 | 2168 | 1 |
| | 81-98 | 2048 | 2408 | 3 |
| | 99-105 | 847 | 967 | 1 |
| | 99-105 | 847 | 1087 | 2 |
| | 99-105 | 847 | 1207 | 3 |
| | 99-105 | 847 | 1327 | 4 |
| | 99-105 | 847 | 1447 | 5 |
| | 108-116 | 950 | 1190 | 2 |
| | 118-125 | 827 | 1067 | 2 |

**Table 1.** Mass values and identification of alkylated peptide observed in the peptic maps of the α1 rat-globin

globin chains were first separated by RP-HPLC; fractions corresponding to modified α- and β-chains were collected and analysed by ES-MS to figure out the average methyl
group number for each globin. A molecular mass of 16,010.5 ± 2.6 Da was measured for the modified β-globin corresponding to the introduction of 10 methyl groups in the native protein chain (molecular weight 15,867.2). As for the α-chain, a molecular mass of 15,238.9 ± 4.1 Da was recorded (fig. 3) with an increase of 112.5 mass units over the unmodified globin chain (15,126.4 Da), corresponding to the addition of 8 methyl groups.
In order to identify those residues within the polypeptide chains involved in the formation of the adducts, the amino acid analysis was performed by GC-MS, as TBDMS-derivatives, after the acid hydrolysis of incubated protein. The results showed the presence of modifications on the imidazolic ring of histidine (both 1- and 3-methyl isomers), the thiol group of cysteine and the α-amino group of N-terminal valine in both

globins. The relative abundance for each adduct was evaluated by using synthetic methylated amino acids as reference external standard that allowed detection of 1,9 mmole of adduct/mmol of Hb. A better sensitivity was achieved by using the single ion monitoring technique (SIM). Both α and β globins showed the same reactivity either for the amino acids residues involved in the adduct formation and for the extent of modification. It must be observed, however, that for adducts that are labile in the conditions of acidic hydrolysis, the above GC-MS approach cannot be used for the identification of modified amino acids, such as methyl esters of carboxylic amino acids. These labile adducts can be analysed by using enzymatic hydrolysis (53).

|        | Peptide | Expected MH+ | Observed MH+ | Number of Adducts |
|--------|---------|--------------|--------------|-------------------|
| 1:100  | 1-29    | 2909         | 3029         | 1                 |
|        | 33-46   | 1732         | 1852         | 1                 |
|        | 47-69   | 2303         | 2423         | 1                 |
|        | 47-69   | 2303         | 2543         | 2                 |
|        | 70-80   | 1181         | 1301         | 1                 |
|        | 81-98   | 2018         | 2138         | 1                 |
|        | 99-109  | 1239         | 1359         | 1                 |
|        | 110-117 | 855          | 975          | 1                 |
|        | 118-125 | 795          | 915          | 1                 |
| 1:1000 | 1-29    | 2909         | 3149         | 2                 |
|        | 1-29    | 2909         | 3269         | 3                 |
|        | 1-29    | 2909         | 3389         | 4                 |
|        | 34-46   | 1585         | 1705         | 1                 |
|        | 34-46   | 1585         | 1825         | 2                 |
|        | 47-69   | 2303         | 2423         | 1                 |
|        | 47-69   | 2303         | 2543         | 2                 |
|        | 47-69   | 2303         | 2663         | 3                 |
|        | 47-69   | 2303         | 2783         | 4                 |
|        | 47-69   | 2303         | 2903         | 5                 |
|        | 47-69   | 2303         | 3023         | 6                 |
|        | 70-82   | 1339         | 1459         | 1                 |
|        | 70-82   | 1339         | 1579         | 2                 |
|        | 81-98   | 2018         | 2138         | 1                 |
|        | 81-98   | 2018         | 2258         | 2                 |
|        | 81-98   | 2018         | 2378         | 3                 |
|        | 110-125 | 1631         | 1751         | 1                 |
|        | 110-125 | 1631         | 1871         | 2                 |

**Table 2.** Mass values and identification of alkylated peptide observed in the peptic maps of the α human-globin

410

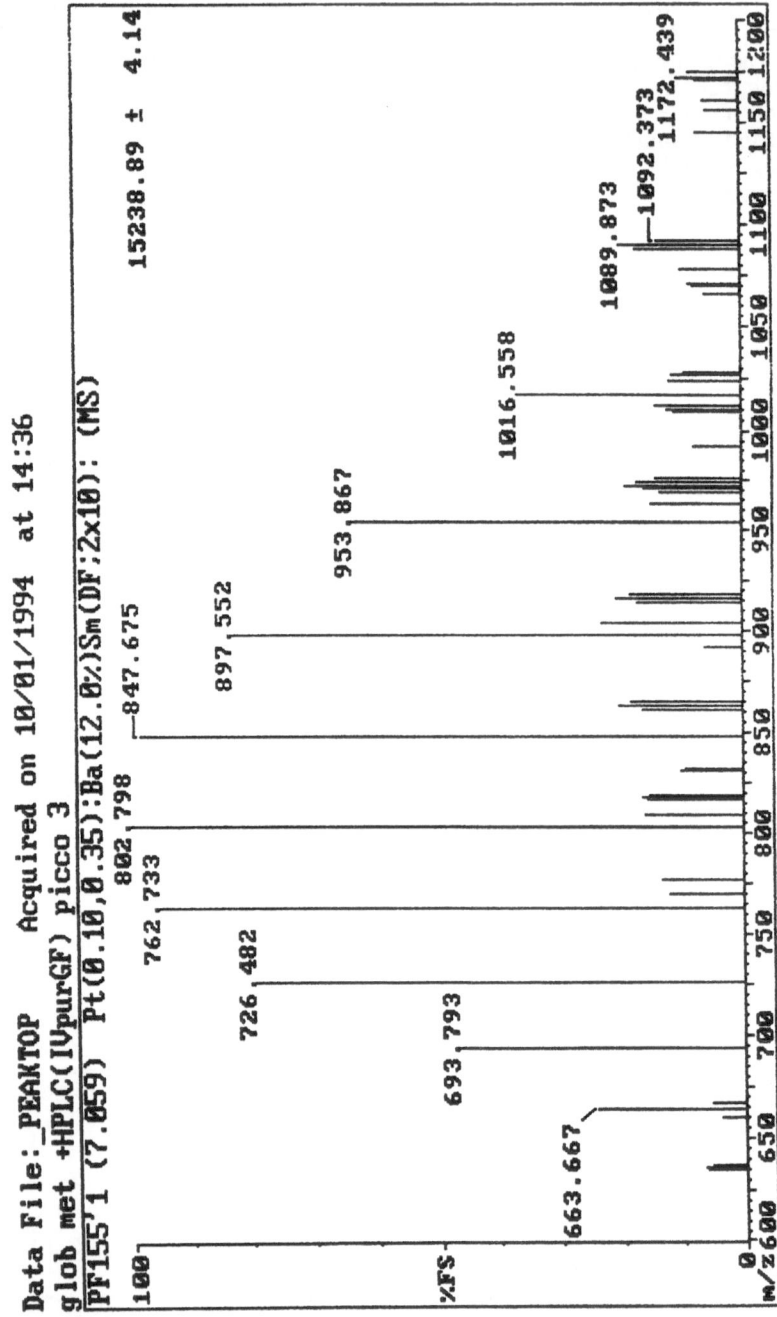

**Figure 3.** Electrospray mass spectrum of HPLC fraction 3 corresponding to α-globin carring eight methyl groups.

By using the same analytical procedure the analysis of blood samples of workers professionally exposed to MeBr was performed; the obtained results confirmed the pattern of adducts formation at level of the N-terminal valine as well as of other nucleophilic sites.

## 7. Conclusions

Mass spectrometry has played an unreplaceable role in the studies of interaction of xenobiotics with cellular macromolecules in order to assess the genotoxic risk.
By using GC-MS-SIM it was possible to quantify modified aminoacids in both animal and human haemoglobins due to exposure to well known genotoxic agents.
By applying in this field the above stated new mass spectrometry methodological approach, based on FAB-mapping, ES, MALDI and MS/MS, more opportunities can be seized to contribute to the understanding of the basic mechanisms of chemical carcinogenesis, a problem of very wide interest not only for the risk assessment in workplace, but also to develop measures against tumour diseases in the general population. In fact, structural-analytical mass spectrometry based advanced procedures offer the possibility of identifying reactive sites upon the exposure to xenobiotic in the environment and represent important tools to understand the mechanisms of interaction of the toxic agent with bio-macromolecules.

## 8. Acknowledgements

These studies have been supported by European Commission (grant CNR/CEE STEP CT91-0161) and by C.N.R., P.S. Aree Metropolitane ed Ambiente.

## 9. References

1. B. Ramazzini, "De morbis artificum diatriba", In: W.C. Write (ed.) The Latin text of 1713, University of Chicago Press, Chicago, 1940; translated by W. C. Write History of medicine, Ser 23, New York Academy of Medicine, New York, 1964.
2. P. Pott, "Chirurgical obsevations relative to the cataract, the polypus of the nose, the cancer of the scrotum, the different kinds of ruptures, and the mortification of the toes and feet", Hawes, Clarke and Collins, London, 1775.
3. J. W. Cook, C. L. Hewett, I. Hieger,, "The isolation of a cancer-producing hydrocarbon from coal tar", Part I, II and III, *J. Chem. Soc.* 395-405 (1933).
4. L. Rehn, "Blasengeschwulste bei Fucsin-Arbreitern", *Arch. Clin. Chir.* 50, 588-600 (1895).
5. R. A. M. Case, M. E. Hosker, "Tumours of the urinary bladder as an occupational disease in the rubber industry in England and Wales", *Br. J. Prev. Soc. Med.* 8, 39-50 (1954).
6. R. A. M. Case, M. E. Hosker, D. B. McDonald, J. T. Pearson, "Tumours of the urinary bladder in workmen engaged in the manufacture of certain dyestuff intermediates in the British chemical industries. Part I. The role of aniline, benzidine, alpha-naphthylamine and beta-naphthylamine", *Br. J. Ind. Med.* 11, 75-104 (1954).
7. R. A. M. Case, J. T. Pearson, "Tumours of the urinary bladder in workmen engaged in the manufacture of certain dyestuff intermediates in the British chemical industries. Part II. Further consideration of the role of aniline and of the manufacture of auramine and magenta (fuchsin) as possible causative agents", *Br. I. Ind. Med.* 11, 213-216 (1954).

412

8. B. N. Ames and L. S. Gold, " Chemical carcinogenesis: to many rodent carcinogens", *Proc. Natl. Acad. Sci.* (U.S.A.) **87**, 7772-7776 (1990a).

9. B. N. Ames and L. S. Gold, "Endogenous mutagens and the causes of aging and cancer", *Mutat. Res.* **250**, 3-16 (1991).

10. C. C. Harris, A. Weston, J. Willey, G. Trives and D. Mann, "Biochemical and molecular epidemiology of human cancer: Indicators of carcinogen exposure, DNA damage and genetic predisposition", *Env. Heatlth Pers.* **75**, 109-119 (1987).

11. L. F. Chasseaud, "The role of glutathione and glutathione S-transferase in the metabolism of chemical carcinogens and other electrophilic agents", *Adv. Cancer Res.* **29**, 175-275 (1979) and references therein.

12. P. J. van Bladeren, A. van der Gen, D. D. Breimer and G. R. Mohn, "Stereoselective activation of vicinal dihalogen to mutagens by glutathione conjugation", *Biochem. Pharmacol.* **28**, 2521-2524 (1979).

13. C. Cmaric, P. B. Inskeep, M. J. Meredith, D.J . Meyer, B. Ketterer anf F. P. Guenguerich, "Selectivity of rat and human glutathione S-transferases in activation of ethylene dibromide by glutathione conjugation and DNA binding and induction of unscheduled DNA synthesis in human hepatocytes", *Cancer Res.* **50**, 2747-2752 (1990).

14. F. P. Guenguerich, "Interindividual variation in biotranformation of carcinogens: basis and relevance", In: J. D. Groopman and P. L. Skipper (eds.) Molecular dosimetry and human cancer: Analytical, Epidemiological, and Social Considerations, pp. 27-51, CRC Press, Boca Raton, 1991, and references therein.

15. G. Batist, A. Turpule, B. K. Sinha, A.G. Katki, C. E. Myers and K. H. Cowan, "Overexpression of a novel anionic glutathione transferase in multidrug-resistant human breast cancer cells", *J. Biol. Chem.* **261**, 15544-15549 (1986).

16. E. Hallier, T. Langhof, D. Dannappel, M. Leutbecher, K. Schöder, H. W. Goergens, A. Müller and H. M. Bolt, "Polymorphism of glutathione conjugation of methyl bromide, ethylene oxide and dichloromethane in human blud: influence of the induction of sister chromatid exchanges (SCE) in lymphocytes", *Arch. Toxicol.* **67**, 173-178 (1993).

17. H.-G. Neumann, "Biochemical effects and early lesions in regard to dose-response studies", *Oncology* **37**, 255-258 (1980).

18. H. Druckrey, D. Schmähl and W. Dischler, " Dosage-effect relations in cancer induction with 4-dimethylamino-stilbene in rats", *Z. Krebsforsch.* **65**, 272-288 (1963).

19. B. Scott Appleton, M. P. Goetchius and T.C. Campbell, " Linear dose-response curve for the hepatic macromolecular binding of aflatoxin $B_1$ in rats at very low exposures", *Cancer Res.* **42**, 3659-3662 (1982).

20. P. B. Dunn, "Wide range linear dose-response curve for DNA binding of orally administered benzo[a]pyrene in mice", *Cancer Res.* **43**, 2654-2658 (1983).

21. S. Osterman-Golkar, L. Ehrenberg, D. Sagerbäck and I. Hällström, "Evaluation of genetic risk of alkylating agents. II. Haemoglobin as a dose monitor", *Mutat. Res.* **34**, 1-10 (1976).

22. S. Osterman-Golkar, P. B. Farmer, D. Segerbäck, E. Bailey, C. J. Calleman, K. Svensson and L. Ehrenberg, "Dosimetry of ethylene oxide in the rat by quantitation of alkylated histidine in haemoglobin", *Teratogen. Carcinogen. Mutagen.* **3**, 395-405 (1983).

23. P. B. Farmer, E. Bailey, S. Naylor, D. Anderson, A. Brooks, J. Cushnir, J.H. Lamb, O. Sepai and Y-S. Tang, "Identification of endogenous electrophiles by means of mass spectrometry determination of protein and DNA adducts", *Environ. Health Perspec.* **99**, 19-24 (1993).

24. L. M. Gwinner, R. J. Laib, J. G. Filser and H. M. Bolt, "Identification of chloroethylene oxide as the reactive metabolite of vinyl chloride towards DNA: comparative studies with 2,2'-dichlorodiethyl ether", *Carcinogenesis* **4**, 1483-1486 (1983).

25. P. Farmer, "Analytical approaches for the determination of protein-carcinogen adducts using mass spectrometry", In: J. D. Groopman and P. L. Skipper (eds), Molecular dosimetry and human cancer: Analytical, Epidemiological, and Social Considerations, pp. 189-210, CRC Press, Boca Raton, 1991, and references therein.

26. M. S. Bryant, P. L. Skipper, S. R. Tannenbaum and M. Maclure, "Hemoglobin adducts of 4-aminobiphenyl in smokers and nonsmokers", *Cancer Res.* **47**, 602-608 (1987).

27. L. C. Green, P.L. Skipper, R.J. Turesky, M. S. Briant and S. R. Tannenbaum, "*In vivo* dosimetry of 4-aminobiphenyl in rats via cysteyne adduct in hemoglobin", *Cancer Res.* **44**, 4254-4289 (1984).

28. W.C. Stillwell, M. S. Bryant and J. S. Wishnok, "GC/MS analysis of biologically important aromatic amines. Application to Human dosimetry", *Biomed. Environ. Mass Spectrom.***14**, 221-227 (1987).

29. P. B. Farmer, E. Bailey, J. H. Lamb and T. A. Connors, "Approach to the quantitation of alkylated amino acids in haemoglobin by gas chromatography mass spectrometry", *Biomed. Environ. Mass Spectrom.* **7**, 41-46 (1980).

30. P. B. Farmer, S. M. Gorf and E. Balley, "Determination of hydroxy-propylhistidine in haemoglobin as a measure of exposure to propylene oxide using high resolution gas chromatography mass spectrometry", *Biomed. Environ. Masss Spectrom.* **9**, 69-71 (1982).

31. M. Törnqvist, S. Osterman-Golkar, A. Kautiainen, M. Näslund, C. J. Calleman and L. Ehrenberg, "Methylation in human hemoglobin", *Mutat. Res.* **204**, 521-529 (1988a).

32. C. J. Calleman, L. Ehrenberg, B. Jansson, S. Osterman-Golkar, D. Seberbäck, K. Svensson and C.A. Wachtmeister, "Monitoring and risk assessment by means of alkyl group in hemoglobin in persons occupationally exposed to ethylene oxide", *J. Environ. Pathol. Toxicol.* **2**, 427-442 (1978).

33. M. B. Nördqvist, A. Löf, S. Osterman-Golkar and S. A. S. Walles, "Covalent binding of styrene and styrene 7,8-oxide to plasma proteins, hemoglobin and DNA in mouse", *Chem. Biol. Interact.* **55**, 63-73 (1985).

34. M. Törnqvist, J. Mowrer, S. Jensen and L. Ehrenberg, "Monitoring of environmental cancer initiators through hemoglobin adducts by Edman modified degradation method", *Anal. Biochem.* **154**, 255-266 (1986a).

35. P. Ferranti, I. Fiume, V. Carbone, N. Sannolo, M. Gallo, C. Magno, A. Milone and A. Malorni, "Incubation of hemoglobin and model peptides with xenobiotics: identification of reaction products by mass spectrometry", 12th International Mass Spectrometry Conference Amsterdam, August 26-30, 1991.

36. N.Sannolo, V.Carbone, P.Ferranti, I.Fiume, A.Milone and A.Malorni, Characterization by mass spectrometry of reaction products of hemoglobin and model peptides with xenobiotics", CEC Contact Group Meeting (Biomonitoring), York (U.K.) 14-16 September 1992.

37. C. Romano, M. Esposito and A. Malorni, Accertamenti medico-legali del danno alla persona da inquinamento atmosferico", In: D. Lauria (Ed.), Patologia Ambientale-Vol. III, pp.103-120, Idelson, Napoli, 1993.

38. S. Kaur, D. Hollander, R. Haas and A. L Burlingame,"Characterization of structural xenobiotic modifications in proteins by high sensitivity tandem mass

414

spectrometry. (Human hemoglobin treated *in vitro* with styrene 7,8-oxide)", *J. Biol. Chem.* **264**, 16981-16984 (1989).

39. D. M Desiderio, "Mass spectrometry of biologically important neuropeptides", In: D.M. Desiderio (Ed.) Mass Spectrometry of Peptides, pp 367-400, CRC Press, Boca Raton, U.S.A., 1991, and references therein.

40. R. De Biasi, D. Spiteri, M. Caldora, R. Iodice, P. Pucci, A. Malorni, P. Ferranti and G. Marino, "Identification by Fast Atom Bombardment Mass Spectrometry of [β112 (G14) CYS -> Arg] hemoglobin (Hemoglobin Indianapolis) in a family from Naples (Italy)", *Hemoglobin* **12**, 323-336 (1988).

41. G. Marino, P. Pucci, A. Malorni and H.R. Morris, "Analysis of post-translational modifications by FAB Mass Spectrometry", In: V. Zappia, P. Galletti, R. Porta and F. Wold (Eds.) Advances in post-translational modidications of proteins and ageing, pp 651-657, Plenum Press, N.Y., 1988 and references therein.

42. P. Pucci, P. Ferranti, G. Marino and A. Malorni, "Characterization of abnormal human haemoglobins by Fast Atom Bombardment Mass Spectrometry", *Biomed. Env. Mass Spectrom.* **18**, 20-26 (1989).

43. R. Porta, C. Esposito, S. Metafora, A. Malorni, P. Pucci, R. Siciliano and G. Marino, "Mass spectrometric identification of the amino donor and acceptor sites in a transglutaminase protein substrate secreted from rat seminal vescicles", *Biochemistry* **30**, 3114-3120 (1991).

44. G. Marsh, G. Marino, P. Pucci, P. Ferranti, A. Malorni, J. Marsh, J. Kaeda and L. Luzzatto, "A third instance of the high-oxygen affinity variant, Hb Heathrow [β103(G5) Phe->Leu]: identification of mutation by mass spectrometry and by DNA analysis", *Hemoglobin* **15**, 43-51 (1991).

45. P. Pucci, R. Siciliano, A. Malorni, G. Marino, M. F. Tecce, C. Ceccarini and B. Terrana, "Human α-Fetoprotein primary structure: a mass spectrometric study" *Biochemistry* **30**, 5061-5066 (1991).

46. A. Malorni, P. Pucci, P.Ferranti and G. Marino, "Characterization of human hemoglobin variants by mass spectrometry", In: M. L. Gross (ed.), Mass spectrometry in the biological sciences: a tutorial, pp. 325-332, Kluwer Academic Publishers (The Netherlands), (1992) and references therein.

47. G. Marino, R. Siciliano, P. Pucci, P. Ferranti and A. Malorni, "Detection of post-translational modification of proteins by mass spectrometry", In: M. L. Gross (ed.), Mass spectrometry in the biological sciences: a tutorial, pp. 333-342, Kluwer Academic Publishers (The Netherlands), (1992) and references therein.

48. P. Ferranti, A. Malorni, G. Marino, P. Pucci, G.H. Goodwin, G. Manfioletti and V. Giancotti, "Mass Spectrometric Analysis of the HMGY Protein from Lewis Lung Carcinoma", *J. Biol. Chem.* **267**, 22486-22489 (1992).

49. A.Parente, C.Verde, A.Malorni, P.C.Montecucchi, F.Aniello and G.Geraci, "Aminoacid sequence of the dimeric myoglobin from radular muscles of the marine gastropod Nassa mutabilis", *Biochim. Biophys. Acta* **1162**, 1-9 (1993).

50. P. Ferranti, A. Malorni, G. Marino, P. Pucci, A. Di Luccia and L. Ferrara, "FAB-OVERLAPPING: a strategy for sequencing homologous proteins", *Int. J. Mass Spectrom. Ion Proc.* **111**, 287-300 (1991).

51. P. Ferranti, V. Carbone, N. Sannolo, I. Fiume and A. Malorni, "Mass spectrometric analysis of rat hemoglobin by FAB-overlapping. Primary structure of the α-major and of four β constitutive chains", *Int. J. Biochem.* **25**, 1943-1950 (1993).

52. H. J. Chaves Das Neves, A. M. P. Vasconcelos, "Capillary gas chromatography of amino acids, including Asparagine and Glutamine: sensitive gas chromatographis-mass spectrometric and selected ion monitoring gas

chromatographic-mass spectrometric detection of the N,O(S)-*tert*-butyldimethylsilyl derivatives", *J. Chromatog.* **392**, 249-258 (1987).

53. P. Ferranti, V. Carbone, N. Sannolo, I. Fiume, A. Milone, M. Ruoppolo, M. Gallo and A. Malorni, "Study of interaction of styrene oxyde with Angiotensin by mass spectrometry", *Carcinogenesis* **13**, 1397-1401 (1992).

# CHARACTERIZATION OF CERAMIDE MIXTURES BY FAST ATOM BOMBARDMENT AND TANDEM MASS SPECTROMETRY

FEDERICO MARIA RUBINO
*ITBA-CNR, v.Ampere 56, I-20131 MILANO (Italy)*
SANDRO SONNINO
*Study Center for the Functional Biochemistry of Brain Lipids; Department of Medical Chemistry and Biochemistry, The School of Medicine, University of Milan, Italy*

ABSTRACT. Fast Atom Bombardment, precursor and fragment ion analysis mass spectrometry were employed to characterize the fatty acid and long chain base composition of a complex ceramide mixture prepared from beef brain lipids. Precursor ion analysis of a sphingenine-specific fragment allowed to selectively recognize ceramides with different fatty acids. Characterization of the *N*-linked fatty acids was performed by analysis of the fragment ions following high-energy collision-induced decomposition of the FAB-desorbed $(M+Li)^+$ species by constant-B/E scanning on both a two and a four sector machine. The latter yielded much better precursor ion resolution and transmission of low-mass fragments and allowed to assign structures even to minor components. The composition of the mixture of linked fatty acid was derived from the intensity of the molecular species both in the source and in the precursor spectra, and was compared with that obtained by capillary gas chromatography of the released fatty acid methyl esters (FAME).

## 1) Introduction.

Ceramides (**SCHEME 1**), which are the amides of long-chain bases such as sphingenine with fatty acids, constitute the lipophilic "tail" imparting amphiphilic properties to glycosphingolipids and sphingomyelin and are also found in the waxy secretions of higher animals [1], as well as in plants and lower organisms [2]. These molecules, either naturally occurring as such, or deriving from the catabolism of sphingolipids, recently raised a great interest, being postulated as modulators of cell growth and differentiation [3].

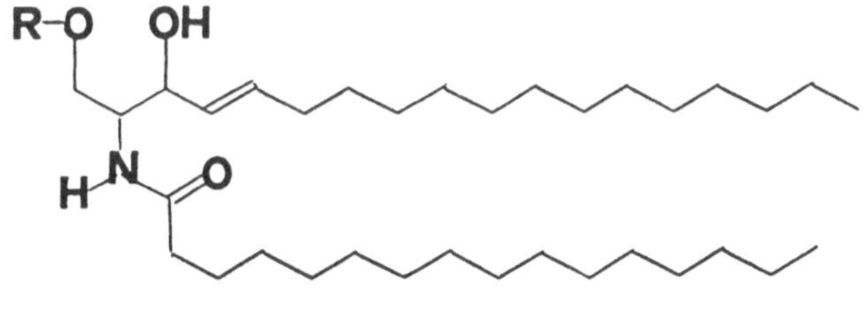

R = H    ceramides
R = $-PO_3^-$-$CH_2$-$CH_2$-$N^+(CH_3)_3$    sphingomyelin
R= oligosaccharide (reducing end)    glycosphingolipids

SCHEME 1

417

*R. M. Caprioli et al. (eds.), Mass Spectrometry in Biomolecular Sciences, 417–428.*
© *1996 Kluwer Academic Publishers.*

Their characterization is usually performed by hydrolysis and separate identification of the fatty acid and long chain base components. In the past, structural analysis by GC-MS of the intact molecules, derivatized as TMS ethers, was reported [1f,4] and recently also LC-MS has been applied to natural ceramides and other N-acyl sphingosines, underivatized [5] and as their TMS ethers [6].

Tandem mass spectrometric analysis of the molecular species of ceramides, sphingomyelin, phosphononosphingolipids and of some glycoconjugates such as cerebrosides, phosphoinositides and sulphatides has been recently reported and reviewed [7]. High-energy CID MS/MS of protonated, deprotonated and alkali-cationized molecules show distinctive fragmentation. SCHEME 2 reports the most structurally characteristic fragments in the spectra. The structure of the N-linked fatty acid can be derived from analysis of the array of ions generated by collision-activated charge-remote fragmentation (CRF) of its carbon backbone. This technique [8], recently introduced by Gross and coworkers as a major structural tool in lipid chemistry, allows to determine the position of double bonds and of other structural motifs such as methyl branching and cyclopropane rings on polymethylene chains, such as those of fatty acids and alcohols.

Although structural variations on the straight C-18 or C-20 carbon chain of the sphingosine base are generally restricted to products of non-mammalian origin [9], characterization of its hydrocarbon moiety would also be desirable, and to this purpose the occurrence of CRF on the protonated long chain base has been demonstrated [10].

This work reports on the characterization of a complex ceramide mixture prepared from beef brain lipids. The required information includes the assessment of the nature of the sphingosine component, of the fatty acid composition and the structural determination of the most abundant molecular species present. Among the employed techniques are precursor and fragment ion analysis of species desorbed by FAB and collision-activated decomposition at high and low energy. Comparisons are made between *a)* the fatty acid composition derived from FAB-MS, from precursor ion analysis MS/MS and from conventional GC and GC-MS of the released FAME; *b)* between the precursor analysis results obtained with different scanning techniques on sector and quadrupole based instruments with respect to mass resolution and *c)* between protonated and cationized molecular species as precursors for fragment ion analysis. One major conclusion is that sector and quadrupole tandem mass spectrometric techniques yield complementary results and are both necessary for the characterization of the analyzed compounds.

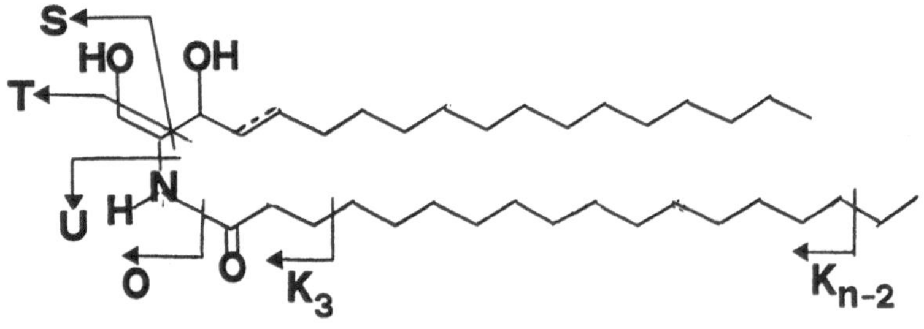

SCHEME 2

Fragment ions and their letter coding for the decomposition of protonated and cationized ceramides are assigned with partial modification according to Adams *et al.* (see text)

## 2) Experimental.

A ceramide mixture was prepared by means of a reported procedure [11] starting from the organic phase of the total lipid extract of beef brain [12] and analyzed by FAB as solutions containing 50 to 0.5ug/uL of mixture. Thioglycerol, m-nitrobenzyl alcohol (mNBA) and mNBA saturated with lithium iodide and with sodium acetate were employed as FAB matrices and were of standard availability in the laboratory.

Most mass spectrometric measurements were performed on a Finnigan MAT90 reverse-geometry mass spectrometer equipped with its own FAB source and an IonTech atom gun (Xe, $5*10^{-5}$ torr, 2uA). Resolution was kept at better than 1500 throughout the work and calibration was performed with cesium fluoride [13]. Precursor and fragment ion scanning was performed with computer-controlled functions at constant $B^2/E$, $B^2*E$ and $B/E$. Collisional activation in high-energy fragment ion spectra was accomplished with Ar in the collision cell located between the source and the magnet, at 50 to 80% attenuation of the main beam. The identification of minor components was achieved on a CONCEPT-II-HH four-sector tandem mass spectrometer (Kratos Analytical, Manchester, UK) fitted with a scanning array detector [14] (data by courtesy of Dr. Su Chen, University of Warwick, UK). Precursor ion spectra were also recorded on a Finnigan TSQ70 triple quadrupole, under comparable FAB conditions. Collisional activation at -30eV collision energy was accomplished with 0.5 mtorr Ar, as measured on the Pirani ion gauge of the collision cell. Precursor ion spectra were recorded as separate 5s scans across the m/z 400-900 range.

All measurements besides source spectra were acquired in the profile mode and the presented spectra are the result of several scans acquired throughout the lifetime of a single sample introduction, software-summed with a 100mmu merge window.

The linked fatty acids of the ceramides were also analyzed by conventional methanolysis and capillary GC and GC-MS.

## 3) Results and Discussion.

### 3.1) SOURCE SPECTRA.

FIGURE 1 shows the source spectra of a ceramide mixture (50ug/uL) extracted from beef brain lipids, recorded from neat mNBA (b) and mNBA/LiI (c). Under a wide range of concentrations (0.5-50ug/uL), the spectra (a) do not show the expected $MH^+$ species, but only prominent $(MH-H_2O)^+$ ions, while that recorded from a mNBA/LiI matrix shows intense $(M+Li)^+$ species. This is not in agreement with the results recently reported by Adams and Ann, who obtained $MH^+$ species from mNBA [20] and reported the CID MS/MS spectra of these species. We have nevertheless noticed that sphinganine-containing ceramides almost exclusively generate $MH^+$ species from mNBA and thioglycerol, while only weak (5-10% of MH-H$_2$O) $MH^+$ signals are indeed formed from sphingenine-containing ceramides when large amounts of pure samples are analyzed (data not shown) We previously observed that also the free sphingosines show the same behaviour, i.e. the unsaturated compound featuring a major $(MH-H_2O)^+$ peak in the source spectrum as a consequence of the presence of an allylic hydroxyl which is extremely prone to racemization, dehydration and hydroxyl 3-->5 shift [10].

The spectrum recorded from mNBA/LiI (c) shows at least 20-fold increase in sensitivity and a more persistent ion current even with the least concentrated sample (0.5ug/uL), therefore we now consider this as the conditions of choice for both structural and analytic measurements.

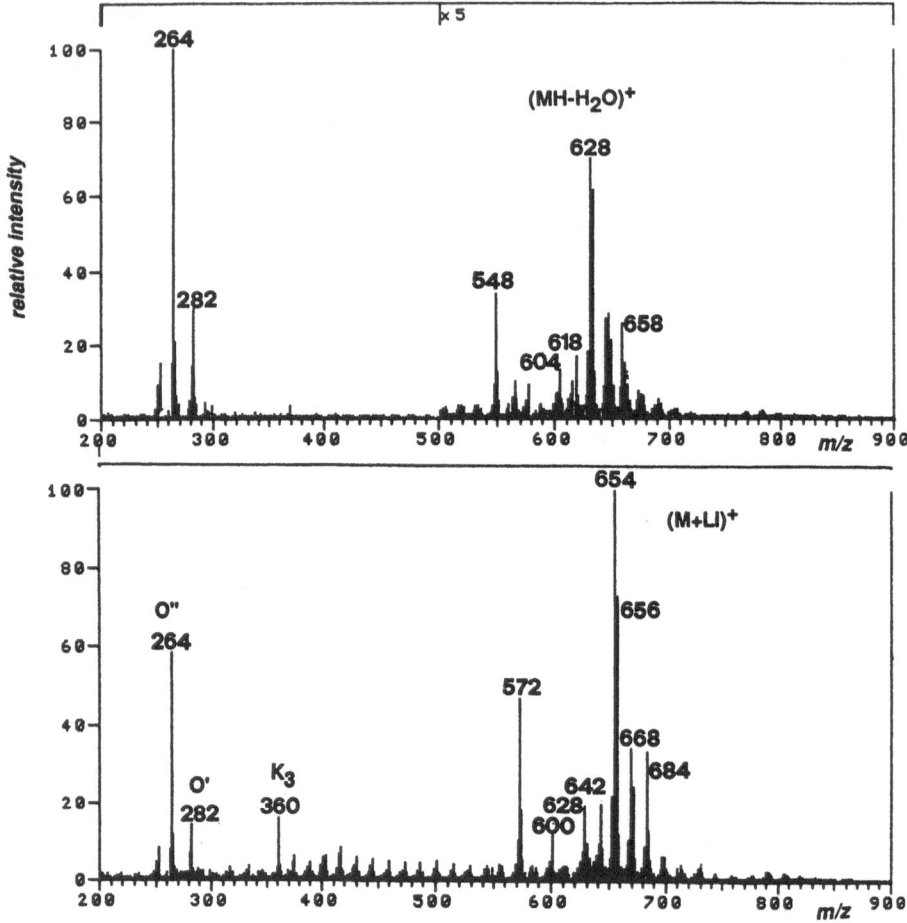

**FIGURE 1** Positive-ion FAB spectra of a beef brain ceramide mixture (50ug/ul) recorded from m-nitrobenzyl alcohol (a) and m-nitrobenzyl alcohol saturated with LiI (b).

The source spectra recorded from both mNBA and mNBA/LiI also feature intense fragment peaks at m/z 264 (O'; Sph+H-2 $H_2O)^+$ and 282 (O"; Sph+H-$H_2O)^+$, due to the C-18 long chain base, while lack of the corresponding signals 28u higher (m/z 292 and 210, respectively) rules out occurrence of the C-20 homologous sphingosine long-chain base. This was an expected result, since it is long known that the neutral glycolipids from which the ceramide mixture was prepared only contain the C-18 long chain base. The source spectrum recorded from mNBA/LiI also shows a signal at m/z 360, which can be interpreted as a $Li^+$-cationized N-acryloyl-C-18-sphingenine generated by a partly charge-driven fragmentation process [15] of the fatty acid backbone.

### 3.2) PRECURSOR ION ANALYSIS.

The "precursor ion" MS/MS approach has been usefully applied for the determination of several phospholipid classes in crude extracts of microorganisms [16], of glycerophosphocholines in neutrophils [17] and to the elucidation of the biosynthesis of Platelet Activating Factor (PAF) and of their 1-O-acyl analogues by endothelial cells [18]. In all these cases, low-energy collision-induced dissociation (CID) in a triple quadrupole instrument was employed.

The occurrence in the source spectra of fragment ions specific of the sphingosine chain suggested that the determination of the fatty acid composition of the ceramide mixture by precursor ion analysis might be feasible. This would in principle allow to discriminate between isomeric compounds (e.g. N-stearoyl-C-18-sphingenine, N-palmitoyl-C-20 sphingenine, N-oleoyl-C-18 sphinganine) present in a mixture, since these fragments are specific for the long chain base; this approach would also be helpful in analyzing crude ceramide preparations extracted from biological materials.

Precursor ion scanning in the reverse-geometry mass spectrometer was accomplished with a constant-$B^2*E$ linked scan which selects decompositions occurring from metastable precursors in the drift region between the magnetic and electric sector; it is described as yielding a better resolution of the precursor ions than that obtainable by probing decompositions in the drift region between the ion source and the magnet (constant-$B^2/E$ scan) [19]. In our case the constant-$B^2*E$ experiment gave rise to broad peaks (FIGURE 2a), the centroids of which match the masses of the (MH-$H_2O)^+$ ceramide species in the mixture , but the attained resolution is much too poor to distinguish fatty acids in mixture differing by one unsaturation, thus making the measurement of little, if any, applicative utility. To overcome this limitation, the detection of individual molecular species was performed with a precursor ion scan on a triple quadrupole instrument, because this instrument yields unit mass resolution on both precursor and fragment ions. FIGURE 2b shows the precursor ions of m/z 264 obtained by 30eV collision-induced decomposition of ceramide species against an Ar target in the collision cell of a triple quadrupole instrument. From the point of view of detection of individual components and of mixture characterization, new signals at m/z 520 (C-16:0), 546 (C-18:1) and 564 (C-18:0-OH) appear in the precursor spectra, which could not be confidently assigned in the source spectra of the mixture, due to the high background "chemical noise" of the FAB matrix. In both cases, the signals representing the ceramides are only due to the (MH-$H_2O)^+$ species, irrespective of the ceramide molecular species (MH-$H_2O$, M+Li, M+Na) present in the source spectra. This unexpected behaviour does not hamper recognition of the individual ceramide species, once taken into consideration and points to the coexistence of more than one ionization process leading to ion production in the source and in the drift regions.

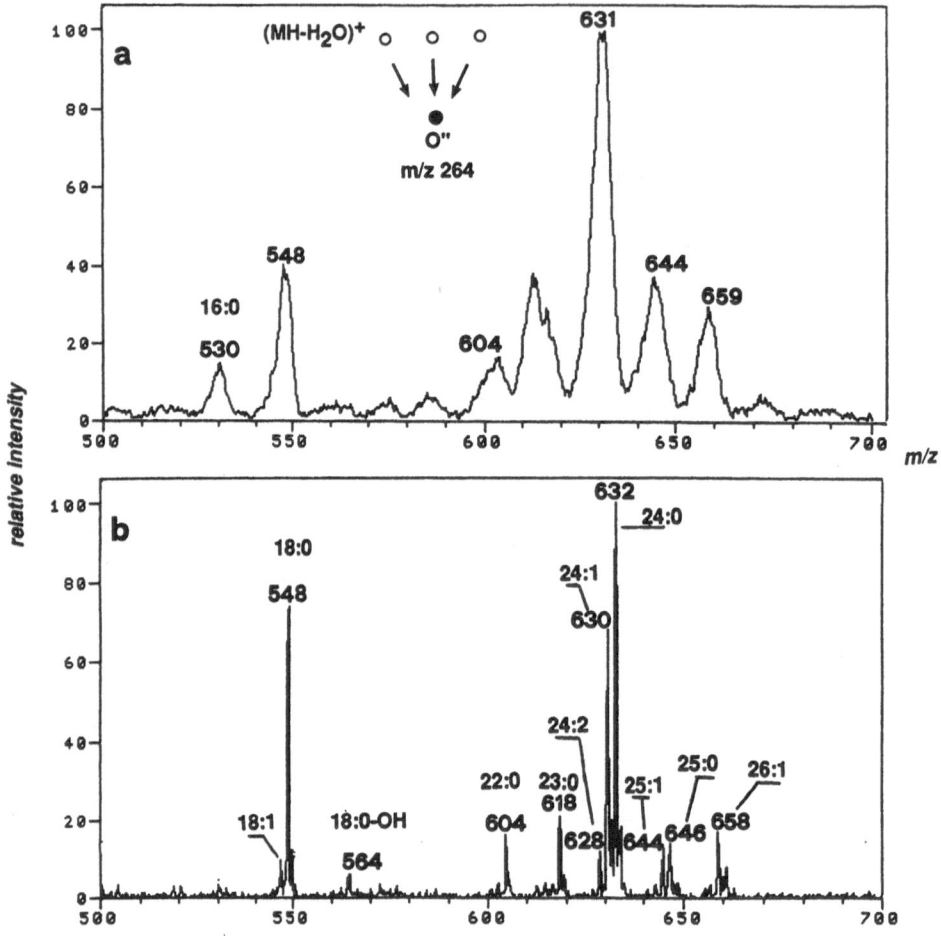

**FIGURE 2** Precursor ion spectra of m/z 264: a) "2nd FFR" metastable scan at constant $B^2*E$; b) triple quadrupole scan (collision energy -30eV; Ar, 0.5 mtorr). Sample as in Figure 1b.

## 3.3) FATTY ACID COMPOSITION.

One major purpose of the work was to start developing alternative analytical procedures for the determination of the fatty acid composition of ceramide mixtures based on FAB-MS and MS/MS. For comparative purpose, the released FAME were analyzed by capillary GC-FID and GC-EIMS. The quantitative results obtained with the employed methods are reported in TABLE 1.

TABLE 1.    Fatty acid composition (moles %) of the ceramide mixture, determined by various analytical techniques.    Relative abundances in each column are only for those fatty acid residues for which presence and identification could be reached on a stand-alone basis with the employed technique.

| Acid | FAB (mNBA) (m/z)[a] | | FAB (mNBA+LiI) (m/z)[b] | | FAB (mNBA+NaOAc) (m/z)[c] | | FAB (Par 264) [a] | FAME GC-FID | FAME GC-EIMS |
|------|------|------|------|------|------|------|------|------|------|
| 16:0 | - | (520) | - | (544) | - | (574) | [d] | 13.7 | 10.4 |
| 18:1 | 2.9 | (546) | - | (570) | 2.4 | (586) | 3.5 | 1.3 | 1.0 |
| 18:0 | 9.9 | (548) | 10.7 | (572) | 8.5 | (588) | 20.5 | 13.3 | 16.9 |
| 18:0-OH | 3.1 | (564) | - | (588) | 1.7 | (604) | 1.7 | _[e] | _[e] |
| 20:0 | 2.8 | (576) | 3.4 | (600) | 3.4 | (616) | - | 2.5 | 1.0 |
| 22:0 | 3.8 | (604) | 5.1 | (628) | 6.0 | (644) | 4.4 | 5.3 | 7.1 |
| 23:0 | 4.9 | (618) | 5.3 | (642) | 11.8 | (658) | 5.5 | 6.4 | 7.5 |
| 24:2 | 5.4 | (628) | 5.3 | (652) | 3.8 | (668) | 3.3 | _[e] | _[e] |
| 24:1 | 20.6 | (630) | 24.3 | (654) | 16.1 | (670) | 18.9 | 8.7 | 8.5 |
| 24:0 | 18.1 | (632) | 20.7 | (656) | 20.6 | (672) | 27.7 | 25.9 | 31.0 |
| 25:1 | 8.0 | (644) | 8.7 | (668) | 6.4 | (684) | 3.3 | 3.3 | 2.1 |
| 25:0 | 8.3 | (646) | 7.5 | (670) | 7.6 | (686) | 3.9 | 7.4 | 7.6 |
| 26:1 | 7.4 | (658) | 9.0 | (682) | 6.5 | (698) | 4.7 | 3.1 | 1.7 |
| 26:0 | 4.5 | (660) | - | (684) | 4.1 | (700) | 2.2 | 4.6 | 5.1 |

[a] $(MH-H_2O)^+$ species
[b] $(M+Li)^+$ species
[c] $(M+Na)^+$ species
[d] C-16/C-18 ratio = 1:5, as determined from the spectrum of Figure 2a
[e] FAME GC-FID and GC-MS analyses were not optimized for the search of hydroxy- and polyunsaturated acids, which may coelute with other compounds.

The relative proportions of the identified ceramides were first assigned by measuring the peak heights of the $(MH-H_2O)^+$ and $(M+cat)^+$ ions (cat= Li, Na) in the source spectra. This simpler approach can be applied in this particular case, since we could demonstrate from the source spectra and from the collision-activated decomposition of the major $(M+Li)^+$ species that the mixture only contains N-acyl-C-18-sphingenines. Although this is the more common case when ceramides are prepared from a total lipid extract, in several biological situations ceramides may contain both C-18 and C-20 sphingosines, thus both a hydrolysis-based fatty acid assay and measurement of the source spectrum would not be sufficiently specific.

To verify if precursor ion analysis can be employed in the more general case of a mixture of homologous long chain bases, the fatty acid composition of the mixture was also derived from the relative abundances of the $(MH-H_2O)^+$ species in the spectra of the precursor ions. In the absence of compound-by-compound standardization, this calls for the assumption that *a)* the relative abundances of the FAB-desorbed ceramide ions closely reflect that of the sample and *b)* the rate constants for the $(MH-H_2O)^+$ ---> O" transition employed for detection do not strongly depend on the nature of the fatty acyl chain.

As for the first point, very similar ratios in the composition of the ceramide mixture are obtained from the source spectra in various matrices and at different sample concentrations. Therefore, the considerable discrepancy in the relative proportions of the fatty acids with respect to the values obtained by gas chromatography (indeed no C-16 containing ceramide was detected) is related to the much different desorption properties of the various ceramide species, whose fatty acyl chains range from 16 to 26 carbon atoms. The complexity of this natural mixture (at least 14 different compounds) and the abundance of long-chain fatty acids makes the problem particularly severe.

As for the second point, the fatty acid composition of the mixture obtained from the precursor ion spectrum of m/z 264 is close to that derived from the source spectra, although some discrepancies are found, compared to the relative proportions found by gas chromatography. From this finding we may conclude that the difference in rate constants for the sampled transition between ceramides containing homologous fatty acids is in first instance negligible. Calibration factors can nevertheless be calculated for each compound and detection mode upon availability of pure standards, if very precise compositional values are needed.

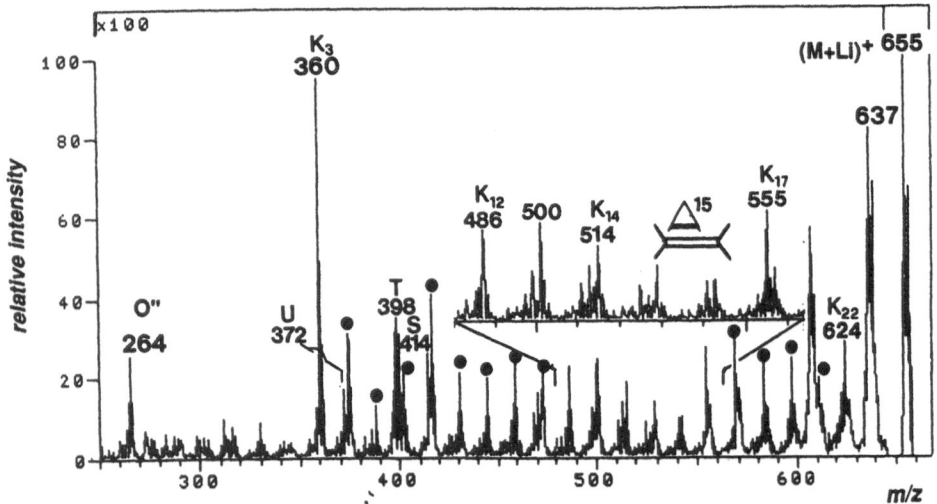

**FIGURE 3** CID Fragment ion spectrum of the $(M+Li)^+$ species at m/z 654 (sample as in Figure 1b). Closed circles label the charge-remote fragments of the fatty acyl chain. The nomenclature for cleavages is slightly modified from that of Adams and coworkers (see Figure 2 and text).

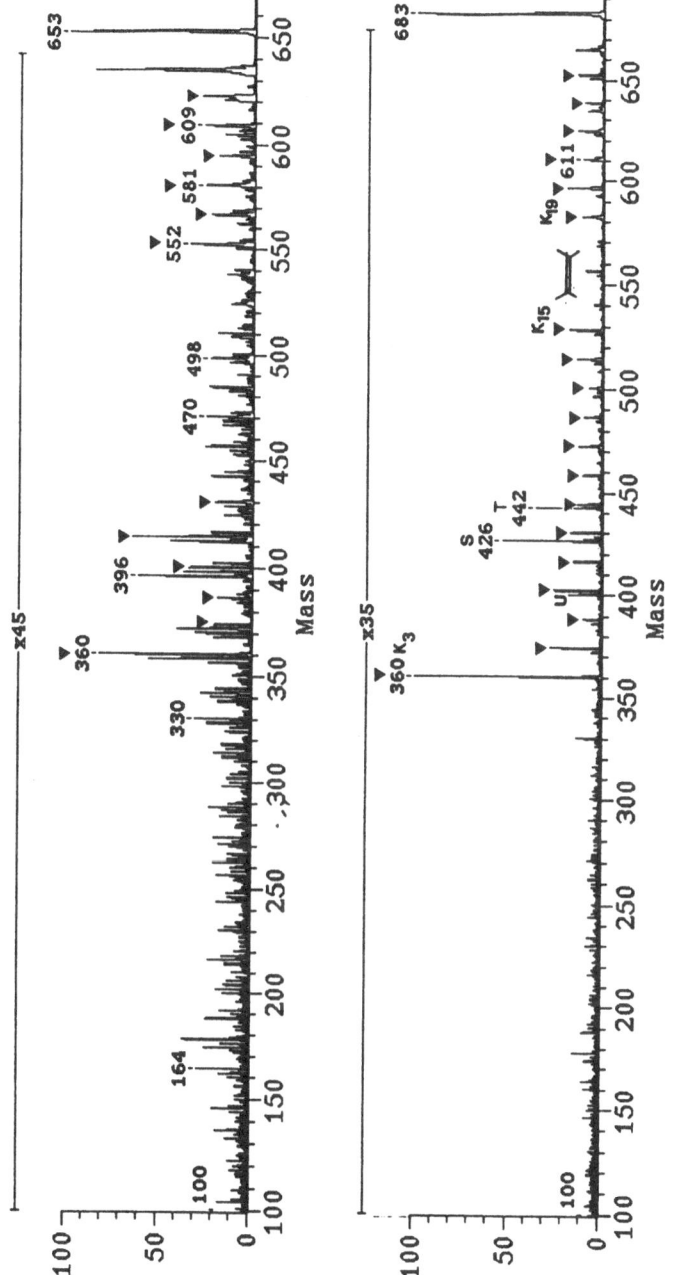

**FIGURE 4** Four-sector CID fragment ion spectra of the (M+Li)⁺ species at m/z 652 (C24Δ12,15) and 682 (C26:Δ17). Sample as in Fig. 1b.

**3.4) FRAGMENT ION ANALYSIS.** The most abundant components of the ceramide mixture could be characterized by collisional spectroscopy of their $(M+Li)^+$ species by constant-B/E linked scanning on a two-sector magnetic mass spectrometer. The nomenclature employed to describe fragmentation (SCHEME 2) is slightly modified from that proposed by Adams and coworkers [15], in that the charge-remote fragments involving the fatty acid chain are explicitly correlated to the lowest homologue, fragment K (now $K_3$) with an index which counts the number of carbon atoms. Moreover the indexed letter L now indicates the corresponding fragments of the long chain base, which also occur by electrocyclic decomposition of $K_3$. With this modification, the fragments which define the occurrence and position of the double bond, labelled L and M in the spectra reported by Adams and Ann, become $K_{17}$ (m/z 555) and $K_{14}$ (m/z 514) in the fragment spectrum of the C24:1 ceramide (m/z 654 $(M+Li)^+$ signal) and allow to define the position of the double bond as $\Delta15$ (nervonic acid) (FIGURE 3). Minor components could be characterized by recording their fragment specta on a four-sector tandem mass spectrometer. This allowed to assign the position of the double bonds in the C24:2 acid as $\Delta12,15$ and that of the double bond of the C-25:1 and C26:1 as $\Delta15$ and $\Delta17$, respectively (FIGURE 4).

Also the characterization of the long-chain base could be achieved by fragment ion analysis of the signals at m/z 360, 282 and 264. The CID spectra of m/z 264 and 360 yielded charge-remote fragments and in particular the former was very similar to that recorded from the corresponding, weak fragment in the source spectrum of authentic C-18 sphingenine ((2R,3S,4E)-2-amino-4-octadecene-1,3-diol) [10], while that of m/z 360 was superomposable to that of authentic, synthesized lithiated N-acryloyl-sphingenine (FIGURE 5).

## 4) Conclusions.

In this work, we have aimed to characterize a complex ceramide mixture with a number of different mass spectrometric experiments, rather than explicitly pursuing the highest sensitivity in the structural determination on a single, purified compound.

Precursor-ion analysis of fragments related to the long-chain base is a viable technique for the detection of ceramides in mixture, expecially when a mixture of homologous sphingosine bases giving rise to isobaric compounds could be suspected. The best resolution of the individual molecular species in the precursor spectra is achieved with the quadrupole-based precursor scan, while high-energy collision-induced decomposition of the cationized ceramides in a magnetic mass spectrometer enables to characterize the hydrocarbon chain of the N-linked fatty acid by charge-remote fragmentation, which is not feasible by low-energy collision in the triple quadrupole instrument. A merge of the performance obtainable with each instrumental configuration is indeed obtainable on hybrid sector-quadrupole machines, the utility of which in the solution of a wide range of problems in the biochemistry field has been demonstrated [20,21].

Measurement of the fragment ion spectra from the less abundant species and at the lower sample concentrations is affected by chemical background and by limited resolution of the precursor ion in the constant-B/E scan on a two-sector machine. Both problems have been overcome by use of true tandem mass spectrometry on a modern, "biochemically-oriented" instrument [22], and both the "chemical noise" and discrimination effects between individual homologues can possibly be further reduced with the use of continuous-flow FAB [23], thus allowing to improve reliability of the fatty acid composition of the ceramide mixture by source spectra and precursor ion analysis. This approach may further evolve into an analytical method for the quantification of ceramide species in biological specimens [24], since the variations in fatty acid composition have possibly

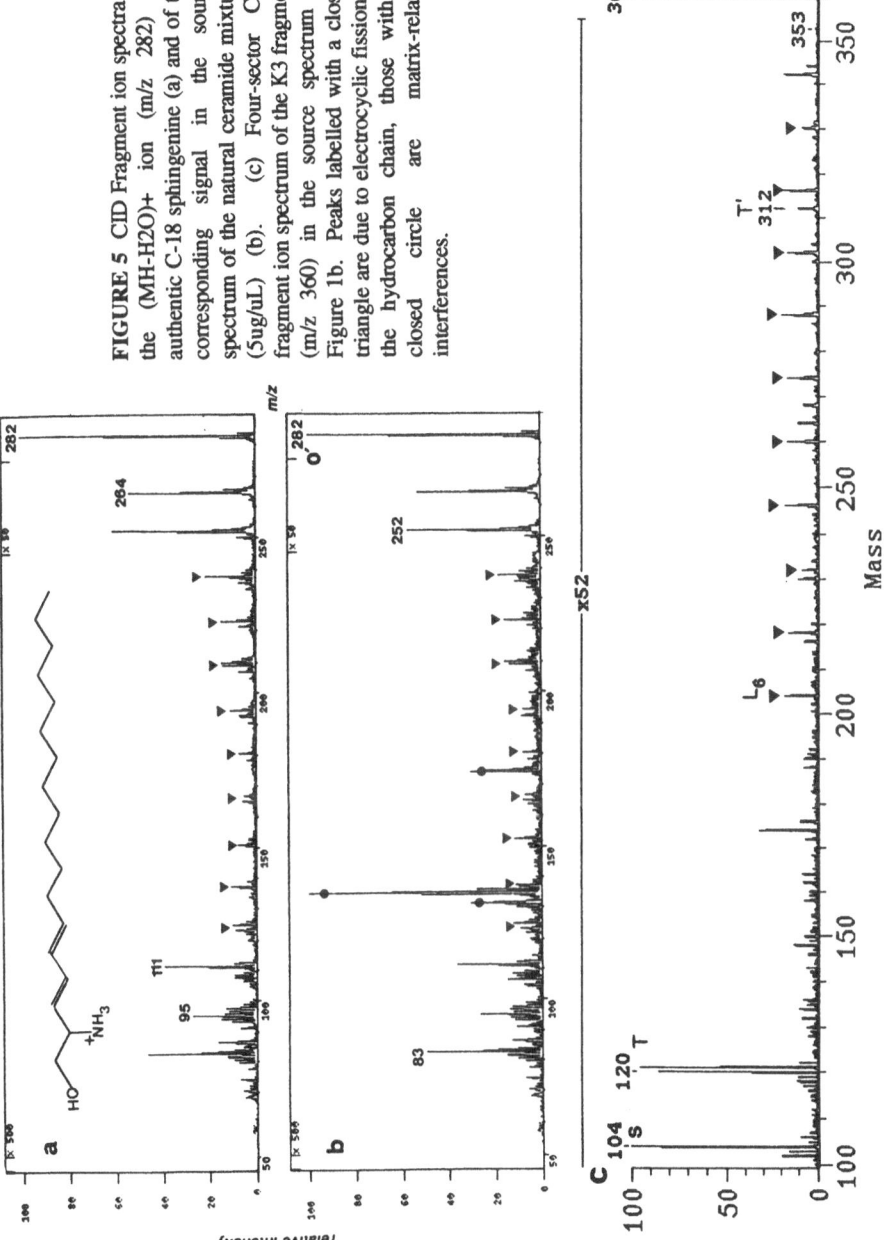

**FIGURE 5** CID Fragment ion spectra of the (MH-H2O)+ ion (m/z 282) of authentic C-18 sphingenine (a) and of the corresponding signal in the source spectrum of the natural ceramide mixture (5ug/uL) (b).    (c) Four-sector CID fragment ion spectrum of the K3 fragment (m/z 360) in the source spectrum of Figure 1b. Peaks labelled with a closed triangle are due to electrocyclic fission of the hydrocarbon chain, those with a closed circle are matrix-related interferences.

428

been related to a wide range of biochemically significant situations, such as cell-cell interactions and second-messenger signalling [25].

## 5) References.

1) *a:* Yasugi,E., Kasama,T., Seyama,Y.: *J. Biochem.*, *102*, 1477-82 (1987); *b:* Yasugi,E., Kasama,T., Seyama,Y.: *J. Biochem.*, *103*, 889-93 (1987); *c:* Yasugi,E., Kasama,T. Shibahara,M., Seyama,Y.: *Biochem. Cell Biol.*, *68*, 154-60 (1990); *d:* Yasugi,E., Kasama,T., Seyama,Y.: *J. Biochem.*, *110*, 202-6 (1991).

2) *a:* Shibuya,H., Kawashima,K., Sakagami,M., Kawanishi,H., Shimomura,M., Ohashi,K., Kitagawa,I.: *Chem. Pharm. Bull*, *38*, 2933-8; *b:* Ishida,R., Shirahama,H., Matsumoto,T.: *Chem. Lett.*, *1993*, 9-12; *c:* Matsubara,T., Hayashi,A., Banno,Y., Morita,T., Nozawa,Y.: *Chem. Phys. Lipids*, *43*, 1-12 (1987); *d:* Cardellina,J.H., Moore,R.E.: *Phytochemistry*, *17*, 554-5 (1978); *e:* Garg,H.S., Sharma-Pandey, M., Bhakuni,D.S., Pramanik,B.N., Bose,A.K.: *Tetrah. Lett.*, *33*, 1641-44 (1992); *f:* Gaver,R.C., Sweeley,C.C.: *J. Am. Chem. Soc.*, *88*, 3643-47 (1966).

3) *a:* Okazaki,T., Bielawska,A., Bell,R.M., Hannun,Y.A.: *J.Biol. Chem.* *265*, 15823-31 (1990); *b:* Bielawska,A., Livardic,C.M., Hannun,Y.A.: *J.Biol. Chem.*, *267*, 18493-97 (1992).

4) Hammarstroem,S.: *Eur. J. Biochem.*, *15*, 581-91 (1970).

5) Jungalwala,F.B., Evans,J.E., Kadowaki,I., McCluer,R.: *J. Lipid Res.*, *25*, 209-16 (1984).

6) Kuksis,A., Marai,L., Myher,J.J.: *Lipids*, *26*, 240-46 (1991).

7) Adams, J.: *Mass Spectrom. Rev.*, *12*, 51-85 (1993).

8) Adams, J.: *Mass Spectrom. Rev.*, *9*, 141-86 (1990).

9) Karlsson, K-A.: *Lipids*, *5*, 878-91 (1971).

10) Rubino,F.M., Zecca,L., Sonnino,S.: *Org. Mass Spectrom.*, *27*, 1357-64 (1992).

11) Carter,H.E., Rothfuss,J.A., Gigg,R.: *J. Lipid Res.*, *2*, 228-34 (1961).

12) Tettamanti,G., Bonali,F., Marchesini,S., Zambotti,V.: *Biochim. Biophys. Acta*, *296*, 160-170 (1973).

13) Rubino,F.M.: *Org. Mass Spectrom.*, *26*, 718-20 (1991).

14) Evans,S., Buchanan,R., Hoffman,A., Mellon,F.A., Price,K.R., Hall,S., Walls,F.C., Burlingame,A.L., Chen,S., Derrick,P.J.: *Org. Mass Spectrom.*, *28*, 289.90 (1993).

15) *a:* Qinghong,A., Adams,J.: *J. Am. Soc. Mass Spectrom.*, *3*, 260-3 (1992); *b:* Qinghong,A., Adams,J.: *Anal. Chem.*, *65*, 7-13 (1993).

16) Cole,M.J., Enke,C.G.: *Anal. Chem.*, *63*, 1032-38 (1991).

17) *a:* Kayganich,K., Murphy,R.C.: *J. Am. Soc. Mass Spectrom.*, *2*, 45-54 (1991); *b:* Kayganich,K., Murphy,R.C.: *Anal. Chem.*, *64*, 2965-71 (1992).

18) Clay,K.L., Johnson,C., Worthen,G.S.: *Biochim. Biophys. Acta*, *1094*, 43-50 (1991).

19) Boyd,R.K., Porter,C.J., Beynon,J.H.: *Org. Mass Spectrom.*, *16*, 490-95 (1981).

20) Gaskell,S.J., Reilly,M.H: *Rapid Commun. Mass Spectrom.*, *2*, 139-42 (1988).

21) Gaskell,S.J.: *Biol. Mass Spectrom.*, *21*, 413-19 (1992).

22) *a:* Gross,M.L. in *Methods in Enzymology*; McCloskey,J.A., Ed.; Academic Press: San Diego, CA, 1990; Vol. 193, pp 131-153; *b:* Yost,R.A., Boyd,R.K. in *Methods in Enzymology*; McCloskey,J.A., Ed.; Academic Press: San Diego, CA, 1990; Vol. 193, pp 738-768.

23) Caprioli,R.M., Moore,W.T. in *Methods in Enzymology*; McCloskey,J.A., Ed.; Academic Press: San Diego, CA, 1990; Vol. 193, pp 738-768.

24) Chen,S., Pieraccini,G., Moneti,G.: *Rapid Commun. Mass Spectrom.*, *5*, 618-21 (1991).

25) Olivera,A., Spiegel,S.: *Glycoconj. Res.*, *9*, 109-17 (1992).

# BIOORGANIC STUDIES OF A NEW PHOTORECEPTOR STRUCTURE

M. ORLANDO and M. L. GROSS
*MidWest Center for Mass Spectrometry*
*Department of Chemistry*
*University of Nebraska-Lincoln*
*Lincoln, Nebraska 68588*

ABSTRACT. The aim of this work is to show the importance of using different instrumental techniques in the field of bioorganic research to determine the structure of unknown compounds present at trace level in biological systems. FAB MS and MS/MS were employed to elucidate structural features of a new type of photoreceptor chromophore. Moreover, a new approach for establishing the positions of OH groups in polyhydroxylated molecules has been developed, and the underlying ion chemistry understood.

## 1. Introduction

*Stentor coeruleus*, a heterotrich protozoan, is a unicellular and blue-green colored ciliate that exhibits a step-up photophobic response and a negative phototactic response [1,2]. Stentorin is the photoreceptor , and its chromophore structure is significantly different from those of rhodopsin, bacteriorhodopsin, careotinoids, chlorophylls, phytochrome, phycocyanins, and flavins. Although stentorin has been proposed to contain a hypericin-like chromophore [3,4], its exact structure is not known. This report describes the structure. The structure of stentorin is not only photobiologically significant but also phototherapeutically relevant because it is related to hypericin, which has highly specific anti-HIV viral activity [5].

## 2. Structural Characterization

The isolation and the purification of this new type of chromophore was carried out at a sample level that places great demands on the analytical methods used for structural characterization. Stentorin which was isolated by sonicating *Stentor* cells in acetone and purified by reverse-phase HPLC, shows an absorption spectrum very similar to that of hypericin with $\lambda_{max}$ at 595 nm (588 nm for hypericin) (Figure 1). Under acidic conditions, the two absorption spectra match more closely, with 587 and 588 nm for stentorin and hypericin, respectively, suggesting that stentorin has a naphthodianthrone skeleton [3,4]. When acidified, stentorin exhibits its a weak hyperchromic effect whereas hypericin shows hypochromism, indicating that stentorin is not structurally identical to hypericin.

### 2.1.    FAB MS AND MS/MS.

At this time attempts to obtain NMR data from the molecule were unsuccessful for the small amount of material and the impurity of the sample. Mass spectrometry was applied to determine the size of the unknown structure and to define the best method to purify the

*R. M. Caprioli et al. (eds.), Mass Spectrometry in Biomolecular Sciences, 429–434.*
© 1996 Kluwer Academic Publishers.

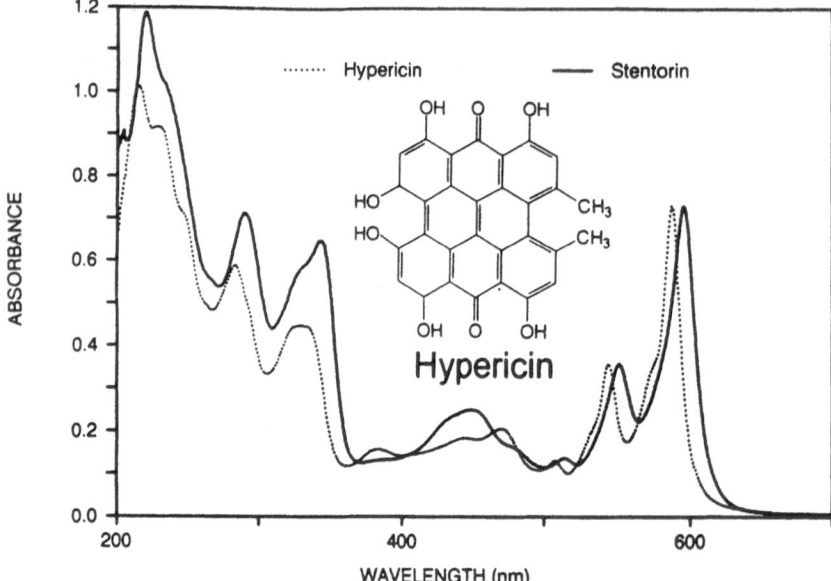

Fig 1 Absorption spectra of hypericin and stentorin in methanol

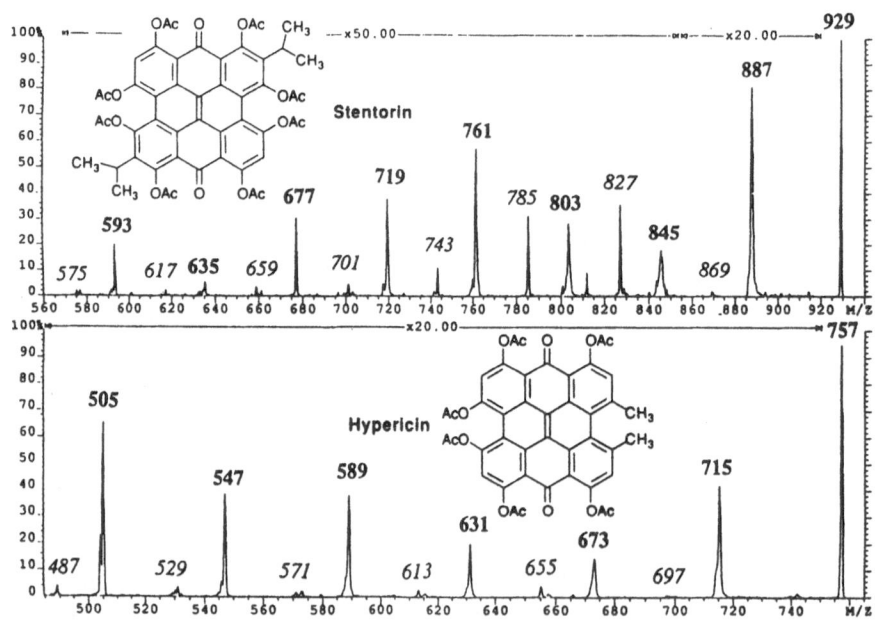

Fig 2: Collisionally activated decomposition (CAD) spectra of (a) acetylated stentorin (b) acetylated hypericin.

sample. FAB MS was carried out using the first two sectors (MS-1) of the VG ZAB-T a four sector tandem mass spectrometer of BEBE design. Stentorin in negative ion mode fast atom bombardment mass spectrometry showed a weak signal for molecular ion at m/z 591.1304, which is in accord with the molecular formula $C_{34}H_{23}O_{10}$ (calculated molecular weight 591.1291). The first attempt to obtain structural information from the CAD spectra of the ion at m/z 591 did not succeed because the structure is rigid. To improve the analyte characterization we acetylated the stentorin molecule. The peracetylated molecule, FAB-desorbed as $(M+H)^+$, shows a series of ions at m/z 593, 635, 677, 719, 761, 803, 845, 887 and 929, with the most abundant ion at 929, whereas acetylated hypericin shows an ion series at m/z 505, 547, 589, 631, 673, 715 and 757, with the most abundant ion at 757. A similar series of ions occurs upon EI of hypericin hexa-acetate.6 The ion series for stentorin establishes that there are eight hydroxyl groups.

Additional confirmation is provided by a collisionally activated decomposition (CAD) spectrum of the $(M+H)^+$ of the octa-acetate (Figure 2a). The CAD spectra were also conducted on the ZAB-T activating the precursor ion in the third field-free region (between MS1 and MS2) by collision with He. Sufficient He had been added to the collision cell to decrease the main beam intensity by 50%. As expected, the parent ion undergoes eight consecutive losses of 42 u (apparently ketene), one for each acetate. More surprising, however, is the loss of 60 u (acetic acid), which is not expected for the acetylated phenol-like functionalities. We have attributed the loss of 60 u to the interaction of the acetoxy groups situated at the 2,2' (or 7,7') positions [7]. The CAD spectrum of the $(M+H)^+$ of hypericin (Figure 2b) also exhibits the same phenomenon and the smaller contribution of this fragmentation pathway has to be related to the absence of the OHs at position 2,2'.

A FAB MS/MS study of diacetates of 4,5-dihydroxy phenanthrene and 2,2'-dihydroxy biphenol has proven that this new fragmentation process can serve as an analytical tool for the establishing the presence of the two OH groups with similar relationship [7]

## 2.2.    FRAGMENTATION MECHANISM STUDY

The FAB mass spectrum, of 4,5-dihydroxyphenanthrene diacetate, 1, shows the protonated molecule of m/z 295 and an abundant fragment ion of m/z 235 (exact mass measured 235.0752, calculated 235.0759 for $C_{16}H_{11}O_2$) which apparently produced by the loss of a molecule of acetic acid. The metastable ion (MI) mass spectrum of the $(M+H)^+$ ion shows only two product ions at m/z 235 and m/z 253 in the ratio 100:15 (Table 1), formed by the losses of $CH_3COOH$ and the ketene, respectively. Although the loss of the ketene from the radical cation of a phenol acetate is an established fragmentation, and the same loss from an $(M+H)^+$ is not surprising, the loss of the acetic acid is a novel and more unexpected. The collisionally-activated dissociations (CAD) of the $(M+H)^+$ of 1 lead to abundant product ions of m/z 253 (15%), m/z 235 (100%), m/z 211 (34%) and a minor ion of m/z 43 (3%). It is proposed that the elimination of a molecule of acetic acid from the protonated molecule of the compound 1 is due to the anchimeric assistance by the second acetoxy group. When the two acetoxy groups are far from each other as in, for example, 1,9-dihydroxy phenanthrene diacetate, 2, no expulsion of acetic acid occurs from the protonated molecule; instead only ketene is eliminated.

The FAB-produced, protonated 4,5-dihydroxyphenanthrene-$d_6$-diacetate dissociates metastably to yield an abundant ion of m/z 238 (formed by elimination of $CD_3COOH$) and of, m/z 257 (loss of $CD_2CO$) in the ratio 100:13. There is no deuterium scrambling in either dissociation. Furthermore, collisional activation of the m/z 238 fragment ion affords abundant ions of m/z 46, $CD_3CO^+$, and m/z 192, from loss of an acetyl radical. These reactions are similar to those of the unlabeled compound 1.

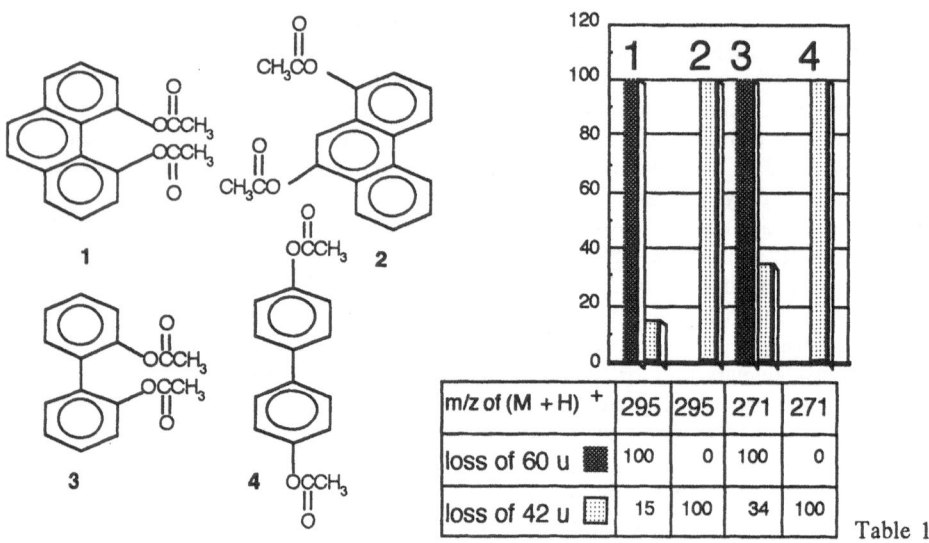

Table 1

| m/z of (M + H) $^+$ | 295 | 295 | 271 | 271 |
|---|---|---|---|---|
| loss of 60 u ■ | 100 | 0 | 100 | 0 |
| loss of 42 u ▦ | 15 | 100 | 34 | 100 |

The diacetate of 2,2'-biphenol, 3, is another molecule with the two acetoxy groups similarly situated as for compound 1. The FAB mass spectrum of 3 shows $(M+H)^+$, m/z 271, and abundant fragment ions of m/z 211 (exact mass measured 211.0765, calculated 211.0759 for $C_{14}H_{11}O_2$) and m/z 229 formed by losses of acetic acid and ketene, respectively. The metastable dissociation of the $(M+H)^+$ of 3 yields fragment ions of m/z 211 (100%), due to the elimination of $CH_3COOH$, and of m/z 229 (34%), produced by the loss of ketene, Table 1. The CAD spectrum of protonated 3, however, shows that the ion of m/z 229 is most abundant and that of m/z 211 is only 25%. This is probably because the $(M+H)^+$ of 3 has to reorient itself such that the two phenyl rings are nearly planar and the acetoxy groups are in close proximity prior to the expulsion of the acetic acid. It is noted, however, that for the $(M+H)^+$ ion of 1, a rigid planar molecule, the elimination of $CH_3COOH$ is a more important process than loss of ketene both for metastable-ion as well as collision-induced dissociation. Our hypothesis was confirmed by a computer modeling study. We defined the three-dimensional structures of the two compounds using Macromodel, software for computer modeling . From the three-dimensional structure of the two molecule is clear that the phenanthrene molecule is quite planar and that the two acetyl groups are closer to each other than in the 2, 2' diacetoxy biphenol molecule (Figure 3) [8].

## 2.3.   NMR

Mass spectrometry is most effectively used for the solution of biological problems when it is a part of an integrated approach that utilizes complementary techniques as well.

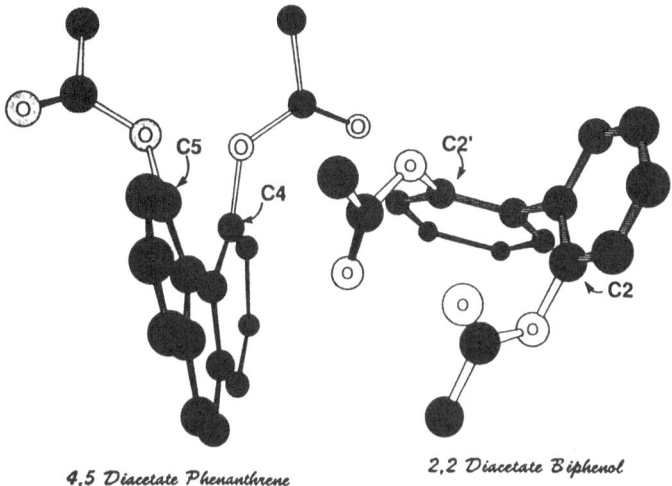

4,5 Diacetate Phenanthrene      2,2 Diacetate Biphenol

Fig. 3: Computer calculated three-dimensional structures

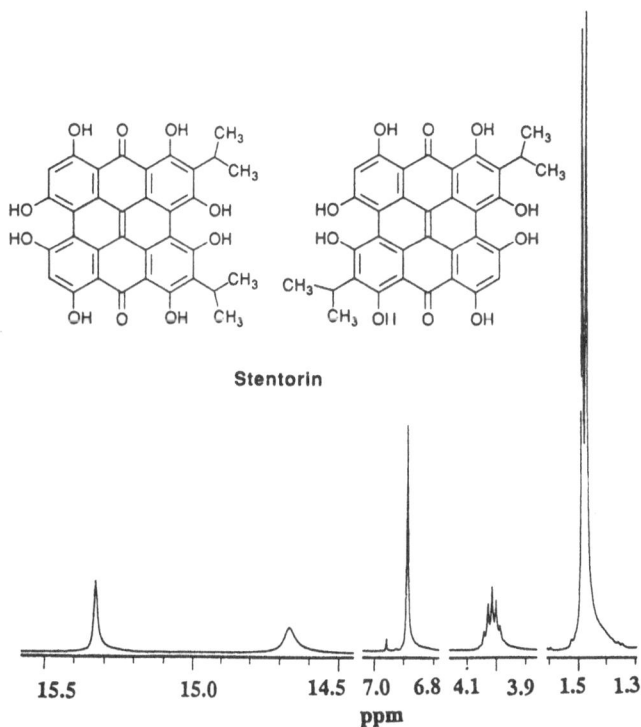

Fig. 4: NMR spectrum of stentorin in DMSO-$d_6$ acquired on GE Omega -500 NMR spectrometer

The [1]H-NMR spectrum of Stentorin (Figure 4) reveals the presence of isopropyl groups, which are characterized by a septet signal centered at δ 4.0 ($J$= 7 Hz, 1 H) and a doublet at δ 1.5 ($J$= 7 Hz, 6 H). Unlike hypericin, which has two aromatic proton signals at δ 7.4 (H–3, H–3', which are the adjoining the methyl groups) and δ 6.6 (H–6, H–6', which are the adjoining the OH groups) in its NMR spectrum, stentorin has only a single peak at δ 6.9, suggesting that only two out of four positions at 3, 3' 6, or 6' are occupied by protons. The other two positions are occupied by isopropyl groups. Thus the structure of stentorin involves one of the three possible symmetrical arrangements of those protons and isopropyl groups.(Figure 4)

The last choice (3) may be ruled out by considering the usual routes of biosynthesis, because hypericin is produced *in vivo* through the polyketide pathway. A study to define the final structure is being done.

## 3. Conclusion

The losses, of acetic acid from protonated 4,5 diacetoxyphenanthrene and 2,2'-diacetoxybyphenyl are examples of a new proximity effect that occurs for closed shell ions desorbed by FAB. Moreover, this novel fragmentation process can be used as a bench mark for identifying the presence of 4,5-dihydroxyphenanthrene moieties in a polyhydroxy aromatic compound, as was the case in the structure proof of the stentorin chromophore.

REFERENCES:

1.  Song, P.S.: Hader, D.P.; Poff, K.L. *Arch. Microbiol.* **126**, 181-186, (1980).
2.  Song, P.S.: Hader, D.P.; Poff, K.L. *Photochem. Photobiol.* **32**, 781-786, (1980).
3.  Moller, K.M. *C. R. Trav. Lab. Carlsberg, Ser. Chim.* **32**, 472-497, (1962).
4.  Walker, E.B.; Lee, T. Y.; Song, P.-S. *Biochim. Biophys. Acta* **587**, 129-44, (1979).
5.  Lavie, G.; Valentine, F.; Levine, B.; Mazur, Y.; Gallo, G.; Lavie, D.; Weiner, D. Meruelo, D. *Proc. Natl. Acad. Sci.* U.S.A. **86**, 5693-5697, (1988).
6.  Brockmann, h.; Spitzner, D. *Tetrahedron Lett.* **1**, 37-40, (1975).
7.  Orlando, M.; George, M.; Gross, M. *Organic Mass Spec.* **28**, 1184-1188 (1993).
8.  MacroModel version 3.5, MM2 force field, PCMS minimization algorithm

# Part IV
# Chromatography Coupled to
# Mass Spectrometry

# GAS CHROMATOGRAPHY AND MASS SPECTROMETRY COUPLING

J. Abián and E. Gelpí
C.I.D. - C.S.I.C.,
Barcelona, Spain

ABSTRACT. Some practical considerations on GC-MS coupling are discussed with special emphasis on vacuum requirements. Coupling devices such as molecular separators and direct connection systems are compared. Finally, the application of GC-MS to quantitative analysis is introduced.

## 1. Introduction.

Until around 1957 gas chromatography (GC) and mass spectrometry (MS) progressed through different avenues in the field of organic analysis. However, the significant analytical potential shown by the first GC-MS coupling attempts (1-3) and their rapid development into a practical and useful new technique have lead to a situation in which GC-MS is recognized as one of the most efficient and versatile systems for the study and identification of organic compounds in complex sample matrices. For a number of years, it was the technique of choice even though with the advent of the more recent LC-MS systems, it has been relegated to a more silent background. This has been due to the introduction of a wide array of new liquid phase ionization techniques which have opened the field of biomolecular polymers to mass spectrometry. Nevertheless, GC-MS is still very much alive and healthy. A recent survey (4) of four years worth of MS publications in the MEDLINE data base has shown that GC-MS is by far the MS technique most widely used throughout the world today. However, it has become so standard and routine that it is not any more fashionable to talk about it in the literature as opposed to the newer techniques of thermospray-, electrospray- or continuous flow-FAB-MS which evidently have stolen the spotlight from GC-MS. As a result recent reviews about GC-MS coupling techniques are scarce (5-7).

When the outlet of a gas chromatograph is connected to the ion source of a mass spectrometer one can obtain structural information for each one of the components in the original sample as they elute sequentially from the GC column. In this fashion the inherent limitations of conventional GC detectors for qualitative analysis are considerably minimized. Such limitations stem from the fact that, whereas gas chromatography can separate the individual components of a sample with a high resolving power, it cannot provide -in a rigorous sense- more than preliminary information on their structure. With the exception of certain specific detectors, GC detectors respond to the amount of sample eluted from the column. Thus, the use of data based on the absolute or relative retention time can be useful for the tentative identification of given compounds in relatively simple mixtures for which standards are available. However, when these methods are taken to extremes such as in complex natural product analyses, results thus obtained lack the necessary qualitative precision. In this regard one of the most rewarding areas of the GC-MS coupling lies in the identification of unknown compounds.

*R. M. Caprioli et al. (eds.), Mass Spectrometry in Biomolecular Sciences, 437–460.*
© 1996 *Kluwer Academic Publishers.*

## 2. Practical Considerations in GC-MS Coupling.

The mass spectrometer is an analytical instrument of extremely high sensitivity potential. In principle, it could detect a signal of less than 30 ions equivalent to $\simeq 10^{-20}$ g for a medium size molecule (MW 300-600). However, due to operational factors such as ionization efficiency, mass resolution, losses upon sample introduction, ion dispersion and transmission efficiency, the minimum required flow of sample into the MS should be of the order of $10^{-8}$ - $10^{-12}$ g/sec. Since the instrument is capable of providing detailed structural information on samples of which we may only have amounts available of the order of $10^{-6}$ - $10^{-11}$ g, it becomes particularly adequate for the direct analysis of compounds eluted from a GC column. Nevertheless, the GC and the MS are basically incompatible in regards to operating pressures. Typically, a GC system operates at atmospheric pressure (760 torr). On the other hand, the MS requires a vacuum of $10^{-5}$ - $10^{-7}$ torr in order to avoid collisions of ions with background molecules in the system, which would result in a considerable loss in sensitivity and resolution (8). Notwithstanding, the two systems show two important common characteristics; both operate in the vapour phase and with sample amounts of the order of the microgram or less.

Fortunately, the lack of compatibility as regards to their different working pressures can be efficiently resolved by various coupling techniques of which the direct capillary connection has become nowadays the method of choice. However, this selection in favour of the direct coupling of capillary GC columns took more than 20 years, mostly on account of basic MS instrument limitations as regards to insufficient pumping speed of vacuum systems, low scan speed and lack of appropriate software and hardware for data processing of the vast amounts of information generated in a high resolution GC-MS run. Various of these limitations were solved over the years with the introduction of differentially pumped mass spectrometers, the improvement of resolution and mass range of quadrupole instruments and with the introduction of low inductance laminated magnets and high currents as well as the advances in electronics and data processing systems. These developments ended two decades of experimentation and use of different interfacing systems designed for the selective removal of the GC carrier gas. These interfaces, with the exception of the molecular jet separator, which is still commercial available and in use, are for the most part presently obsolete. However, from a tutorial point of view it may be instructive to briefly consider them in some detail. In this regard, it is interesting to note that some of this interface technology has given rise today to new applications of mass spectrometry such as the use of skimmers for spray plume sampling and analyte enrichment in modern LC-MS interfaces(9), the development of particle beam interfaces also for LC-MS (10), and the use of permeable polymer membranes for the selective introduction of volatile organic compounds (11, 12).

## 3. Coupling Techniques for GC-MS

Any coupling device must be capable of carrying the analyte in vapour form into the ion source of the mass spectrometer by the helium flow from the GC column while maintaining vacuum in the mass spectrometer. Likewise, samples must be introduced without affecting the chromatographic separation provided by the GC column. These two basic prerequisites establish the only technical incompatibility between the two instruments as regards total gas flow and pressure limitations. As indicated above, the mass spectrometer imposes a low operating pressure ($10^{-6}$ - $10^{-7}$ torr) which necessarily restricts the flow of gas from the GC entering the ion source. This high vacuum in the MS is dictated by the necessity of securing the survival of the newly formed ions for a time long enough for them to reach the detector end of the instrument. Accordingly, a molecule´s mean free path (L), which is defined as the average distance a molecule travels before it collides with another molecule, should be larger than the ion flight path within the various zones of the instrument. In GC-MS, the maximum tolerable pressure in the ion source is determined by that mean free path value allowing for good ion transmission out of the source and through the analyzer.

Other practical considerations calling for high vacuum in the MS are that high pressures in GC-MS ion sources interfere with the ionizing electron beam and its regulation, can lead to electrical discharges, reduce the life of the filament, increase background noise and facilitate ion molecule reactions (at $P > 10^{-1}$ torr). However, most of these are not critical when using Helium as GC carrier gas at $P < 10^{-1}$ torr.

Pressure and volume in vacuum systems are related by Charles Law ($PV = nRT$) which shows that the number of molecules in 1 ml of gas at various pressures is very high even when vacuum can be considered as very good. Molecular density is inversely related to the mean free path of the molecules. For practical purposes the mean free path of a molecule of air at room temperature is related to pressure by the expression:

$$L = 5 \times 10^{-3}/P$$

where P is the pressure in the vacuum system in torrs and L is in cm.

For instance, if we consider the different flight path regions in conventional mass spectrometers, to maintain an $L \simeq 0.5$ cm in the ion source, the maximum pressure in this zone should be $10^{-2}$ torr but to maintain an $L = 100$ cm in the analyzer region the maximum pressure there should not be higher than $10^{-5}$ torr. Under these conditions, we could approximate the maximum allowable gas flow from the GC, considering for this purpose the terms of pumping speed, gas flow rate and conductance.

The pumping speed (S) in liters per second of vacuum pumps is given by the term.

$$S = Q/P \qquad [\text{eq. 1}]$$

where Q is the gas flow rate to the pump in torr - l/s and P the pressure at the pump entrance in torr.

In a vacuum system any plane sustains a flow of gas across it and the pressure at the plane remains constant. The gas flow rate across this plane is given by the expression:

$$Q = P \times S$$

Thus, to maintain a vacuum level of $2 \times 10^{-5}$ torr and assuming a pumping speed in the system of 100 l/s, the gas flow rate evacuated per second would be

$$Q = 2 \times 10^{-5} \text{ torr} \times 100 \text{ l/s} = 2 \times 10^{-3} \text{ torr - l/s}$$

In terms of GC carrier gas flow through the corresponding entrance restrictor or valve leading to the MS this value represents a conductance (C) at this point of:

$$C = Q/\Delta P \quad \text{[eq. 2]}$$

Thus

$$C = 2 \times 10^{-3} \text{ torr - l/s} / 760 \text{ torr} = 2.6 \times 10^{-6} \text{ l/s} = 0.16 \text{ ml/min}$$

where $\Delta P$ is the pressure difference between the two ends of the restrictor or valve. To simplify calculations, we consider the pressure at the outlet under vacuum to be negligible, so that $\Delta P \simeq P_{inlet} = 760$ torr.

The conductance has the same units (volume/sec) as the pumping speed and is the term used in vacuum technology to define the flow of gases through pipes, orifices or other system components. In any vacuum system, the flow of gas is defined by the system conductance C. The three main terms in vacuum technology, Q (gas flow rate, torr-l/s), P (pressure, torr) and C (conductance, l/s) are equivalent to the terms I (current intensity, Coulomb/s), V (voltage, volts) and 1/R (conductance, ohms$^{-1}$) in electrical technology.

Increasing this gas flow into the MS by a factor of 10 to 1-2 ml/min so that:

$$C = 2.6 \times 10^{-5} \text{ l/s}$$

then,

$$Q = 760 \text{ torr} \times 2.6 \times 10^{-5} \text{ l/s} = 2 \times 10^{-2} \text{ torr - l/s}$$

and

$$P = Q/S = 2 \times 10^{-2} \text{ torr - l/s} / 100 \text{ l/s} = 2 \times 10^{-4} \text{ torr}$$

Thus, if we were to allow a flow of 1-2 ml/min the pressure would increase to $10^{-4}$ torr so that with a pumping speed of 100 l/s, in order to maintain a vacuum level not worse than $2 \times 10^{-5}$ torr the mass spectrometer could only accept about 1% of the flow from a GC column operating at 20 ml/min or a 10% of the flow from a 0.01" i.d. capillary. Typical flows through GC capillary columns are as follows: 0.01" i.d., 1-2 ml/min; 0.02" i.d., 2-6 ml/min and 0.03" i.d., 8-15 ml/min.

In the example above we determined that increasing the GC gas flow into a single stage pumped mass spectrometer to 1-2 ml/min with a pumping speed of 100 liters/sec would result in an increase of pressure to the $10^{-4}$ torr range. This would decrease the mean free path in the analyzer from 250 cm to 50 cm.

We can now consider a single pumped mass spectrometer with the pump placed in the analyzer region. For this example, let's assume that the ion source exit slit gives a conductance of 0.2 l/s. For an amount of gas evacuated per second of $Q = 2 \times 10^{-3}$ torr. l/s, as calculated in the first example, the resulting pressure in the ion source would be:

$$P = Q/C = 2 \times 10^{-3} \text{ torr - l/s} / 0.2 \text{ l/s} = 10^{-2} \text{ torr} \quad \text{so that L} \simeq 0.5 \text{ cm}$$

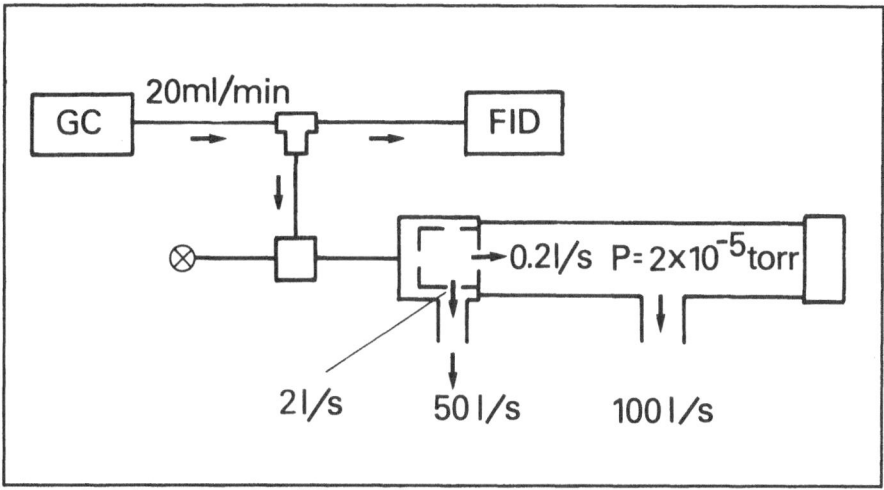

Figure 1. Schematics of a differentially pumped GC-MS system showing GC flow rates, conductance values out of the source, analyzer pressure and pumping speeds. X, variable split valve.

However, in a differentially pumped system (Figure 1), if C at the entrance of the ion source pump is about 2 l/s, then the gas flow rate to be coped with by this pump will be:

$$Q = P \times C = 10^{-2} \text{ torr} \times 2 \text{ l/s} = 2 \times 10^{-2} \text{ torr - l/s}$$

Under these conditions the total amount of gas evacuated from the ion source via the source pump and the analyzer slit would be:

$$Q = (2 \text{ l/s} + 0.2 \text{ l/s}) \times 10^{-2} \text{ torr} = 2.2 \times 10^{-2} \text{ torr - l/s}$$

and

$$C = 2.2 \times 10^{-2} \text{ torr - l/s} / 760 \times 10^2 \text{ torr} = 2.8 \times 10^{-5} \text{ l/s}.$$

equivalent to 1.7 ml/min, that is about 10% of the total GC flow of the same packed column operating at 20 ml/min whereas for 0.01" and 0.02" i.d. capillaries this would represent up to 90-100% of the carrier gas flow.

Nevertheless, in this case the overall sensitivity may be lower due to the lower injection loads in capillary GC columns and to the "dilution effect" of sample in the carrier gas reducing the sample ionization efficiency. Thus, in principle and considering the MS technology of the late 60's and early 70's, these considerations led to the development of devices for the selective removal of GC carrier gas in order to achieve the highest possible sample enrichment factor. GC-MS coupling through flow splitting systems may result in sample losses of up to 99% against the use of enriching devices or molecular separators, capable of changing the ratio carrier gas/sample in favour of the latter.

## 4. Molecular separators

Molecular separators were designed to eliminate most of the carrier gas at the GC outlet before it reaches the ion source of the MS. While acting as a direct coupling device between the MS and the organic compounds eluted from the GC, the molecular separator must also act as an efficient restrictor but only for the carrier gas. In this double role, the ratio of analyte to eluent undergoes an enrichment process so that these systems were often described as "molecular enriching devices" (5).

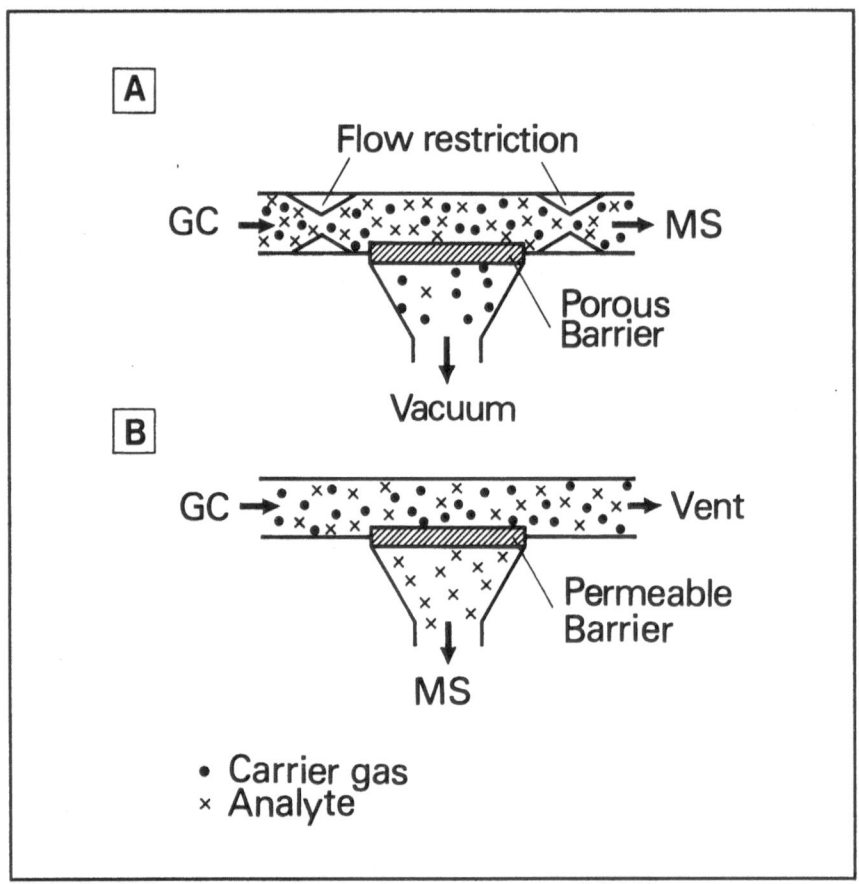

**Figure 2.** Different types of molecular separators for GC-MS

With the exception of Ryhage's jet separator (13), all molecular separators use a physical separation barrier according to two basic designs: **A**) the GC carrier gas preferentially crosses the separation barrier (Figure 2A) and **B**) the eluted compounds preferentially cross the barrier (Figure 2B).

## 4.1. MOLECULAR SEPARATORS OF TYPE A.

In this type of interfaces the chromatographic carrier gas flow goes over a porous barrier on its way to the ion source of the mass spectrometer. At the other side of the barrier there is a vacuum system to remove the carrier gas that preferentially effuses through the pores in the barrier (Fig 2A). Separation barriers are made of materials that are readily penetrated by the carrier gas, such as fritted tubes, porous metal etc., taking as a principal selection criterium that the average pore size should be 1 μm.

### 4.1.1. Watson-Biemann (14).

This interface was the first of its type and was described by Watson and Biemann in 1964. Sample enriching is based in the process of gas effusion through orifices under molecular flow conditions. The carrier gas helium diffuses out of the pores more rapidly than the heavier and bulkier analytes, with the resulting analyte enrichment. The original system used a fritted glass tube, 20cm long by 0.8cm diameter inside a glass chamber connected to a vacuum pump. The average pressure inside the fritted tube must be less than atmospheric but higher than the pressure in the ion source and the surrounding glass chamber (1 torr). In order to maintain molecular flow conditions through the tube pores, the molecules mean free path inside these pores should be about 10 μm (10 x pore length).

Other systems that were based in this design were those of Krueger and McCloskey (15), Cree (16), Blumer (17) and Lipsky (18), in which the fritted glass tube was substituted for one of stainless steel with a mean pore size of 0.1 μm, silver porous membranes (pore size, 2-3 μm) or teflon tubes.

Along these lines, a group at the Jet Propulsion Laboratory in Pasadena (19) designed a similar system fitted with silver-palladium cylinder heated to 250°C, specific for hydrogen as carrier gas. The hydrogen would rapidly diffuse through the silver-palladium wall whereas the organic compounds would go through the cylinder towards the mass spectrometer. Calculations indicate that with a 325 mm long tube, all hydrogen would diffuse out of the system and in fact this process can be so efficient that there would be no hydrogen left as carrier gas to push the analytes on to the ion source. Under these conditions it becomes necessary to add a small flow of inert gas ($N_2$) to carry the compounds into the MS.

### 4.1.2. Brunnée et al. (20)

This separator also operates on the principle of molecular effusion although instead of a porous surface the barrier here is an adjustable annular slit. By its design, based on a variable area of effusion, it is the only separator that was claimed to be able to accept the flow from any GC column (e. g. 1-100 ml/min.). The aperture of the slit could be adjusted by vertical displacement of a plate over a circular edge forming a continuous annular slit. The distance between this plate and the circular edges was continuously variable between 0 and 50 micron.

## 4.2. MOLECULAR SEPARATORS OF TYPE B.

In this case, the carrier gas flow from the GC also goes over the barrier but the organic analytes are the ones that preferentially cross it by absorption and molecular diffusion processes and enter the ion source while the carrier gas is vented to the atmosphere (Fig. 2B). For this purpose, very fine dimethylsilicone copolymer membranes deposited over a mechanical support are used. These membranes preferentially select organic compounds so that they are absorbed and diffused through. Membrane permeability is a function of the solubility (s) and the diffusion rate (D) of the organic matter going through and is defined by:

$$k = sDA_m (P_1 - P_2) / d_m t_m$$

Where $A_m$ is the area of the membrane (2-3 $cm^2$), $P_1$ -$P_2$ is the pressure gradient across, $d_m$ is the membrane thickness and $t_m$ is the time that a molecule needs to go through.

Although the carrier gasses normally used in GC have a high diffusivity, their solubility is practically nil and this results in the selective transport of organic matter from the GC to the MS with concomitant retention of the carrier gas. One of the main problems of this type of silicone membranes is the loss of chromatographic resolution by peak widening. For instance, amines and other highly polar compounds show pronounced tailing effects when going through one of these membranes. The main causes are: 1) mass transport resistance, 2) contact time between membrane and organic compound and 3) separator dead volume.

The time needed by an organic compound to go through the membrane is a function of $d_m$ and D:

$$t_m = d_m / 6D$$

For instance, for $d_m$ = 2.5 x $10^{-3}$ cm and D = $10^{-6}$ $cm^2$/s, $t_m$ will be 1 second (21).

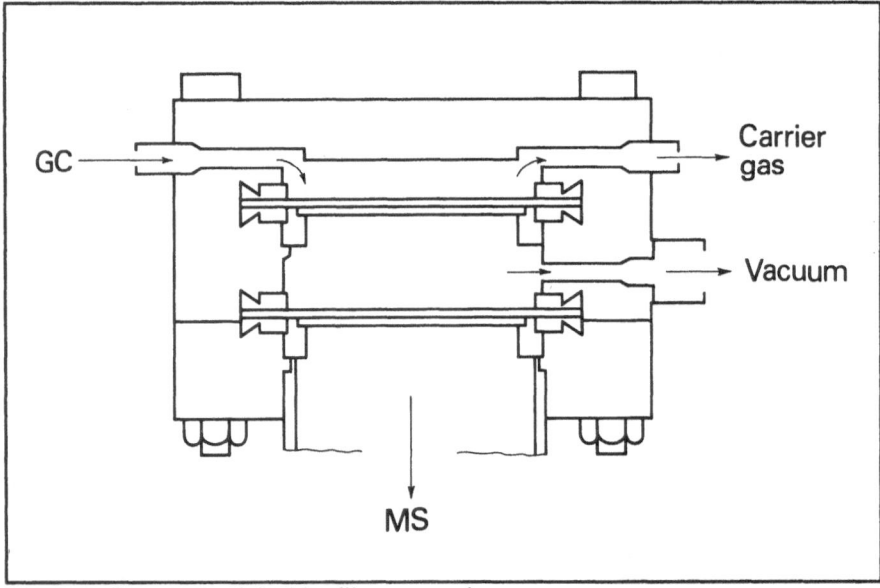

**Figure 3** Separator of double silicone membrane.

A scheme of the first of these separators with two enriching stages is illustrated in Fig. 3. With a flow rate of 30-50 ml/min over the first membrane, this will absorb about a 90% of the organic matter and after the second stage it becomes possible to maintain a helium partial pressure of $10^{-7}$ torr.

A one stage version has been also described (22). This system used a porous silver membrane covered with a phenylmethylsiloxane copolymer, which is considerably less permeable to organics and to helium that the methylsilicone rubbers. It was specifically designed for its use with capillary column flow rates.

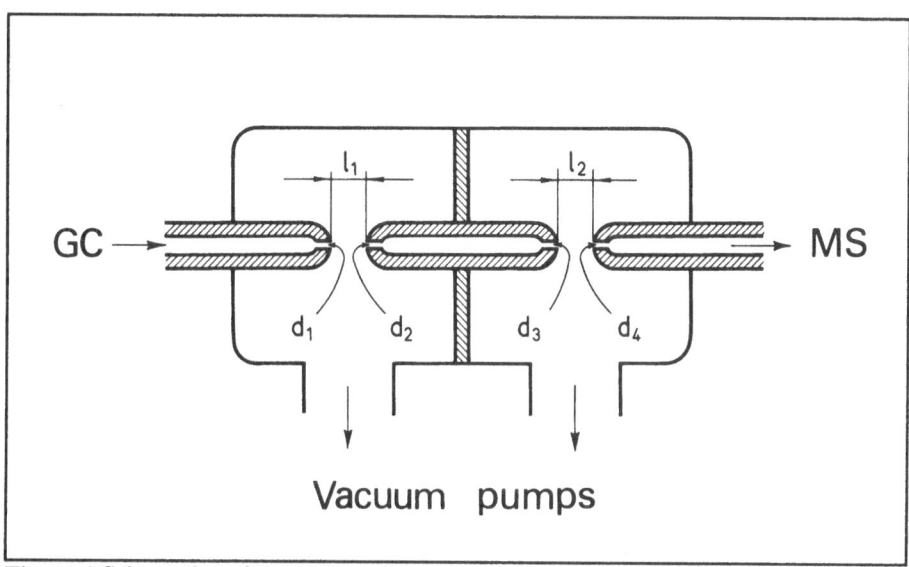

**Figure 4** Schematics of the Ryhage jet separator

### 4.3. OTHER TYPES OF MOLECULAR SEPARATORS.

One of the first and most popular separators designed for the direct connection of a GC to a MS is the so called "jet separator", initially constructed of stainless steel and described by Ryhage (7). This separator, as shown in Fig. 4, is based on the principles advanced for isotope separation (23). The analytes eluted from the GC column together with the carrier gas (helium) go through a small orifice (first jet diameter: $d_1 = 0.1$ mm), from where they emerge at supersonic velocity. Helium due to its low mass and thus small momentum and high diffusibility is thrown off the optical axis aligning the two jets and is evacuated with a pump speed (2.6 l/s) sufficient to reduce pressure in the first chamber to 0.1 torr. The heavier organic material, which for its higher momentum maintains its original trajectory, goes through the second jet ($d_2 = 0.3$ mm; $d_3 = 0.24$ mm) and helium diffuses into the second chamber which is evacuated with a 150 l/s pump to a pressure of $10^{-3}$ torr. The organic compounds pass directly to the last orifice ($d_4 = 0.3$ mm) leading to the ion source of the MS.

To maximize the enriching yield of this separator, distances $l_1$ (0.15 mm) and $l_2$ (0.5 mm) as well as orifice diameters $d_1$ through $d_4$ and pumping speed can be changed. Also, in order to obtain an optimal separation of helium it is required to establish viscous flow conditions through the jets. In other words the mean free path (L) must be smaller than orifice dimensions. In the first stage L is of the order of $10^{-3}$ - $10^{-4}$ cm. At the second stage the pressure is lower so that L is higher. Consequently, to maintain viscous flow in the second jet, this must have a larger diameter.

## 5. Evaluation of the Different Molecular Separators

The yield of a molecular separator device (Y), sometimes referred to also as efficiency, is defined by the ratio of the amount of sample (A) entering the MS ($Q_{MS}$) and that entering the separator ($Q_{GC}$)

$$Y = \frac{Q_{MS}}{Q_{GC}}$$

and it is often expressed in percentage values

The sample enrichment factor is defined by

$$E = \frac{[A]_{MS}}{[A]_{MS}} = \frac{Q_{MS} / He_{MS}}{Q_{GC} / He_{GC}}$$

where $He_{MS}$ or $He_{GC}$ are the flows of helium carrier gas introduced into the MS or into the separator, respectively and [A] is the analyte concentration.

Thus,

$$E = Y \frac{He_{GC}}{He_{MS}}$$

Although E is an important factor, in practice it would be meaningless if Y is too small (e.g. 1%) to obtain a good mass spectrum. For instance, if Y = 1 (no sample loss through separator) an E of 50 would represent an elimination of carrier gas of 98%.

## 6. Practical Considerations Leading to the Development of Molecular Separators

In its most simple conception the GC-MS connection would be a temperature controlled small diameter tube acting as a direct connection between the outlet of the GC and the ion source inlet. However, as we discussed above, on account of the necessary low pressure to be maintained within the MS side of the coupling, in this case total flow would be restricted to about 0.2 to 2 ml/min, depending on whether the system is differentially pumped (high evacuation capacity) or not.

In the first days of GC-MS work, flow splitting was considered as a valid alternative to cope with the GC flow restrictions as imposed by the MS. Evidently, splitter systems were cheap, simple to use and furthermore, the previously optimized GC conditions were not affected. However, in this fashion only 0.1 - 4 ml of helium carrier gas, depending on the type of GC column used, could be diverted into the ion source of the MS which was equivalent to a sample transfer efficacy of 1 - 10% without any enrichment. This, in some cases resulted in undetectable sample amounts ultimately reaching the mass spectrometer. For instance, when only a 1% of the carrier gas flow from a 0.6 mm i.d. packed column is diverted to the MS, this represents that even considering a theoretically very large 100 μg injection of an analyte, only 1 μg would eventually reach the ion source in a period of 30-200 seconds (depending on peak elution time). This is equivalent to about 5-30 ng/sec, a sufficient amount to obtain a good mass spectrum. However, if the same analyte is present at a more realistic level of 0.1 μg then, after the same flow splitting step, only 0.001 μg would reach the ion source at a rate of 0.005 - 0.03 ng/sec, an amount clearly insufficient to obtain a mass spectrum or in some cases even detect the analyte at all.

By contrast, if a molecular separator is used and considering a not too efficient performance with a yield (Y) of 40% and an enrichment of only 20, the amount of analyte reaching the ion source from a packed column operated at a flow of 30 ml/min would be:

$$Q_{MS} = Y \times Q_{GC} = 0.4 \times 0.1 \ \mu g = 0.04 \ \mu g$$

In this case, within a peak elution time period of 30-200 seconds the amount of analyte entering the ion source would be about 1.3 - 0.2 ng/sec or 40 times the amount entering after a direct 1% split.

These 0.04 μg reaching the ion source through the molecular separator would be carried by a total flow at the outlet of the separator of:

$$H_{eMS} = \frac{Y \ H_{eGC}}{E} = \frac{0.4 \times 30 \ ml/min}{20} = 0.6 \ ml/min$$

So the concentration of analyte introduced per minute into the ion source would be:  0.04 μg/ 0.6 ml/min = 0.066 μg x min / ml.
Evidently, considerations such as this represented the driving force for the research and development of molecular separator systems.

## 7. Open Split Coupling Systems

The open split connection, as first described in 1975 by Henneberg et al (24), represented a practical response to the drawbacks and limitations of both molecular separators and the direct connection with no splitting (see below). These drawbacks were centered in the first case on the need for proper maintenance and conditioning to prevent sample losses by adsorptive surface processes as well as on their preferential use with packed columns The later was due to the fact that since all of these devices add dead volume they reduce the resolving power of very efficient open tubular capillary columns. On the other hand, the direct connection was very much restricted to low inlet flows due to lack of the necessary pumping speed in the early commercial MS systems. Furthermore, from the point of view of chromatographic performance, comparison of GC and GC-MS profiles when using the direct coupling was presumably difficulted by the changes in separating power of the columns when the column outlet is operated at reduced pressure.

The open-split is an atmospheric open system consisting of a narrow gap ($\cong$ 2 mm) between the GC capillary column outlet, which is under pressure, and the inlet tubing of the MS, which is under vacuum. The gap between both ends of tubing is swept by helium gas so that if the outflow from the GC exceeds the inflow capacity of the MS inlet restrictor, the excess effluent is swept away by the helium flow to waste and only that fraction of GC carrier gas flow controlled by the split ratio enters the ion source. On the other hand, if the GC outer flow rate is lower, then analytes are diluted with helium from the sweeping line. The later results in a decrease in sample concentration at the MS inlet.

Although, platinum capillaries were used in the early days as the inlet line restrictor to the mass spectrometer (25) later work showed the quality of the chromatogram was less affected by using glass restrictors and modern GC-MS instrumentation is equipped with restrictors made of deactivated fused silica capillary tubes. For instance, one of such systems the GC/ITD from Finnigan Mat Ltd. uses a 1.2 m long fused silica capillary of 0.15 mm i.d. coated with DB-5 phase in order to prevent sample adsorption to the walls of the tube.

There is no enrichment of the sample in this system and the yield depends on the split ratio of the column eluate and MS inlet flows. The open split is perfectly compatible with very high efficiency GC separations, it does not cause band broadening and allows direct comparison of GC chromatograms and MS total ion profiles. Also, column change is rapid and simple and there is no need to break vacuum to replace a GC column.

### 8. Direct Coupling Devices

If the flow restrictor at the MS inlet of an open split is removed and the GC column is inserted through the open split gap past the transfer line into the ion source then direct coupling is achieved.

As indicated above, the major limitation with this type of coupling, as encountered in the early days of GC-MS, was the limited range of carrier gas flow rates that could be accommodated by ion source pumping systems. Thus, usually the end of the capillary column (for instance a 50 m long by 0.25 mm i.d.) was connected through leak tight fittings to capillary restrictors allowing a maximum flow of 1 ml/min into the MS. However, in 1971 the replacement of the 150 l/s oil diffusion pump of a Varian MAT CH5 with a 600 l/s pump allowed helium flows into the ion source of up to 8ml/min at a pressure of $7x10^{-5}$ torr (26).

Later, using a high speed pump at 1500 l/s it was shown that the He inlet flow could be increased up to 20 ml/min (27) although optimum performance was found at 0.3 - 7.2 ml/min. Nowadays, this is the usual range of carrier gas flow rates that can be directly introduced in a differentially pumped MS system so that no intermediate restrictors are necessary any more and the outlet end of the capillary column can be inserted right into the ion source. As reported, in such a configuration the maximum column efficiency suffers only a 12.5 % reduction at most, compared with atmospheric outlet conditions (28). Nevertheless, direct comparison of chromatograms from the GC with those from the GC-MS is not always easy but the connection provides quantitative sample transfer into the MS so that the yield in this case is 100% with no enrichment.

It must be considered that in addition to the improved pumping capacity of modern MS equipment, the advent around 1979 of the flexible fused silica capillary tubing used for the manufacturing of GC capillary columns further facilitated the coupling since it allowed the insertion through a guiding tube of the end of the capillary column into the ion source.

## 9. Basic Operating Parameters for a GC-MS System

After a detailed consideration of the various interfaces available for sample introduction, their different yields and operational efficiencies, it becomes mandatory to study the peculiarities and characteristics of the integrated GC-MS coupling. Although simple in principle, once the coupling has been effectively taken care of, the operation of a GC-MS system presents practical difficulties, many of then of an additive nature in terms of the quality of the final results.

### 9.1. THE CHROMATOGRAPHIC COLUMN

Efficiency, flow rate, sample capacities as well as thermal stability and low bleed are important parameters to consider when selecting a column for GC-MS work. Nowadays most GC-MS separations are carried out on narrow (0.22 mm i.d.) or widebore (0.32 mm i.d.) fused silica columns. A summary of typical parameters of various types of capillary columns is shown in Table 1.

In terms of the GC-MS coupling, the 0.15, 0.22 and 0.32 mm i.d. columns would be best suited for instruments with low pumping capacities (e.g. benchtops) whereas differentially pumped systems would even allow the use of 0.53 mm i.d. capillaries at low flow rates ($<$ 7-8 ml/min). The sample capacity of these various types of columns can go from the very low ng to $\mu$g range, as shown in Table 1. For a particular component the sample capacity is defined by the phase ratio ($\beta$) which describes the relationship between the volume of the stationary phase (film thickness) and the volume of the mobile phase (carrier gas). The higher the value of $\beta$ the lower the solute retention and loading capacity (28).

$$\beta = d/4 \; \mu f$$

where d indicates column diameter and $\mu f$ is the film thickness.

As regards sample loading, it must taken into account that, except for thick film wide bore columns, most capillary columns are restricted in the amount of sample load they can tolerate before a given peak is overloaded. This requires split injection (29). Alternatively a very useful system in GC-MS work is the falling needle injector which greatly facilitates solvent removal and sample preconcentration for very diluted samples (30).

Stationary phase bleed into the MS may seriously affect the overall sensitivity of the analysis as well as conceal the actual MS analyte fragmentation pattern. As it is known that the MS responds linearly to the input mass flow of sample, at a fixed input flow rate the response will be proportional to the concentration of sample so that if the analyte competes with bleeding components continuously emerging from the GC column, the ionization efficiency will decrease. However, since the modern capillary column technology of chemically bonded phases has contributed to a significant reduction in bleed levels, this is not as important as it used to be in quantitative terms. Nevertheless, bleeding can still become a severe limitation in cases where there is direct mass interference of background bleeding ions with ions arising from the analyte. despite all these considerations, to the practising GC-MS operator the well known peaks of bleeding from silicone phases have always been useful internal markers of the mass scale, specially in early days when there were no data systems available.

Table I.-    Characteristics of GC capillary column types[a].

| Diameter (mm) | Film Thickness | Phase Ratio | Efficiency Plates/m | Sample[b] Capacity (ng) | He Flow ml/min |
|---|---|---|---|---|---|
| 0.15 | 0.10 | 375 | 2.600-4.000 | 10 | 0.2-0.6 |
|  | 0.25 | 150 |  | 35 |  |
|  | 0.40 | 95 |  | 55 |  |
| 0.22 | 0.10 | 550 | 2.000-5.000 | 20 | 0.5-1.2 |
|  | 0.25 | 220 |  | 50 |  |
|  | 0.50 | 110 |  | 80 |  |
|  | 1.00 | 55 |  | 190 |  |
| 0.32 | 0.25 | 320 | 1.250-2.100 | 70 | 1.0-2.3 |
|  | 0.50 | 160 |  | 140 |  |
|  | 1.00 | 80 |  | 280 |  |
|  | 4.00 | 20 |  | 1100 |  |
|  | 5.00 | 10 |  | 1200 |  |
| 0.53 | 0.5 | 265 | 650-1.100 | 460 | 1.7-17 |
|  | 1.0 | 132 |  | 2.300 |  |
|  | 3.0 | 44 |  | 3.500 |  |
|  | 5.0 | 26 |  | 4.500 |  |

(a) Data extracted from an SGE catalog
(b) Based on analysis of alkanes on nonpolar silicone stationary phase.

In summary, the effects of GC column parameters on GC-MS optimization have been described in some detail in the literature (31).

## 9.2. TRANSFER LINES

This is a critical part of the system where significant sample losses may occur by condensation on cold spots or by adsorption or degradation due to chemical reactivity of the connecting line surfaces. The degree of adsorption on a given glass or metal surface in general depends on the sample polarity and, as shown many times in the literature, losses of sensitivity of polar compounds due to this effect can be effectively prevented by surface treatment with a silanizing agent. In any GC-MS system, when working with very labile and polar compounds it is advisable to silanize all surfaces in contact with the sample from the injector to the inlet of the ion source. As a rule it has been reported that a truly inert system should give a relative ion abundance of about 3:1 for the $M^{+\cdot}$ ion of cholesterol at m/z 386 and the $[M-18]^{+\cdot}$ ion due to loss of water (6).

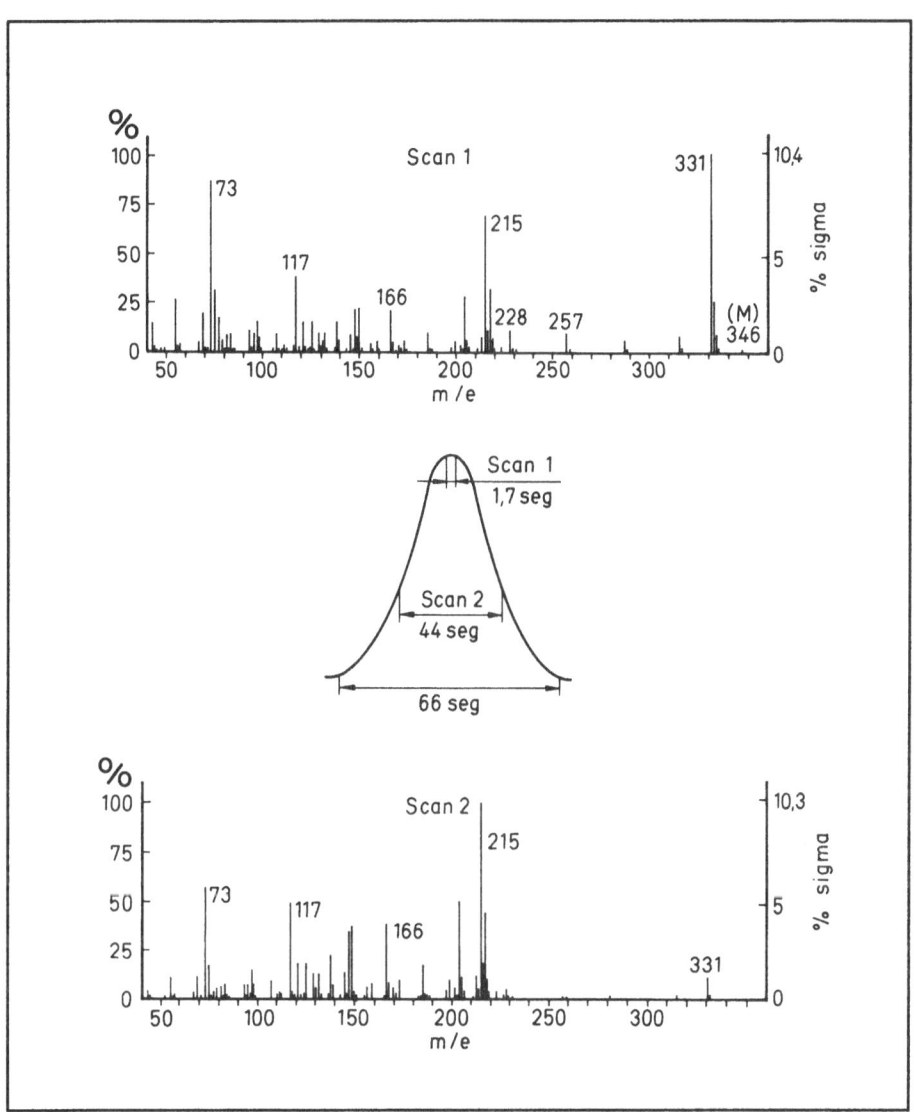

**Figure 5** Distorsion effect upon relative abundances due to slow scanning speed across GC peak profile.

Of course direct insertion of the outlet of the capillary column into the MS ion source avoids many of these problems although the use of deactivated capillaries between the column outlet and the ion source facilitates column changes (32).

## 9.3. MS SCANNING SPEED

Another very critical parameter in the GC-MS combination lies in the scanning speed necessary to sequentially focalize all ions over the collector slit. Since this has to be accomplished in the time period of the chromatographic peak band width the system needs to be able to perform very rapid scans over the complete mass range. In fact, due to the continuous change in analyte concentration over the peak profile, the mass spectrum must be recorded in a fraction of that time. If the scan is slow the mass spectral pattern will be distorted in favour of the central area corresponding to the moment of maximum analyte concentration within the ion source. This was illustrated many years ago in the literature (8) with the example shown in Figure 5. Nowadays, due to the possibility of much faster chromatographic separations the situation is even more complicated. Peak profiles are not usually 66 seconds wide but in extreme eases can be only fractions of a second wide. To avoid a significant distortion of peak intensities, the minimum requirements for the scanning rate in GC-MS would call for scan speeds of 1-2 s/decade or 0.2-0.5 s/decade for packed and capillary columns, respectively. In other words, the higher the frequency of spectrum collection the better the quality of the reconstructed ion chromatogram since we will have more points available to define a given GC profile.

This problem has been addressed recently in the literature (33) regarding the requirements of high resolution GC-MS where in a extreme case of a column with $10^6$ plates, capable of generating 10 peaks per second, individual peak detection would require a frequency rate of 100 Hz (100 scans/s). Given the relatively slow scan rates of magnetic sector instruments (around 3-4 Hz) this would not be possible at all in such a system. Nevertheless, even if scan rates of 0.1 s/decade were feasible, as for instance on quadrupole instruments, dwell times would be too small resulting in impaired detectability due to poor ion statistics. For instance, at 0.1 s/decade, the dwell time per mass unit in the range 50-500 u would be only 0.22 ms so that both detectability and signal reproducibility would be seriously compromised.

A very effective way of increasing dwell times has been in use since the very first days of the GC-MS coupling. The technique, developed at the Karolinska Institutet in 1967 and nowadays known as selected ion monitoring (SIM)(7), is based on the use of the mass spectrometer not as a scanning device but as a selective mass detector. Instead of spending a few milliseconds on each mass unit during a conventional MS scan, the system is set to focus only on a few ions characteristic of the compound of interest eluting from the GC column. Sensitivity is thus dramatically increased and what is even more important, quantitative precision is vastly improved. This is specially true when using isotopically labelled analogues of the compounds to be analyzed, as first described by Sweeley and coworkers in 1966 (34). In fact, the combination of GC-MS/SIM techniques for high sensitivity detection with isotope dilution methods for quantitation of selected compounds stands at present as the most powerful and reliable analytical technique in the Life Sciences (4). The basis of quantitative isotope dilution mass spectrometry (IDMS) on account of its major significance in GC-MS assays is described in some detail below.

**10. Quantitative GC-MS**

The common basis of quantitative analysis is the comparison of the signal generated by the analyte in the sample with the signal from known amounts of the analyte in calibration samples. Low detection limits, precision and accuracy are critical requirements to develop "definitive" quantitative methods.

Accuracy and precision are optimized by the use of an internal standard (IS) that diminishes the effect of instrumental variations and controls recovery. Obviously, the optimum IS should have identical physicochemical characteristics than the analyte. At the same time, it should be naturally absent from the sample and must be resolved from the signal of the analyte. A good approximation to this point is the use of an isotopically labelled form of the analyte as IS. The intrinsic characteristics of the MS analysis allows the application of this method by parallel monitorization of selected ions of the analyte and the IS. Isotope dilution was firstly applied in quantitative MS in combination with GC and electron impact (EI) or chemical ionization (CI) modes. Since then, IDMS has extended its use to practically all the developing MS techniques and has take advantage of the technical improvements on selectivity and sensitivity obtained during this time.

Isotope dilution mass spectrometry is considered today the method of reference for accurate quantitative analysis and is used for the validation of other assays in the development of reference methods (35). The theory and practice of IDMS has been extensively reviewer over the years (36). More recent work in Clinical Chemistry has been reviewed by DeLeenher (36); IDMS of the elements has also been covered by Heumann (38); the use of internal and external standards in trace analysis and the theoretical treatment of the calibration data has been covered by Boyd (39) and the mathematical treatment of data from isotopic enrichment methods in metabolic studies by Rosenblatt et al.(40).

As a stable isotope we understand a non radioactive isotope different from the more abundant one in the nature. Most commonly used isotopes in IDMS are presented in Table II.

TABLE II. Natural abundance of some stable isotopes of the elements.

| Element | Isotope | Relative Abundance |
|---------|---------|--------------------|
| Hydrogen | $^2$H | 0.0148 |
| Carbon | $^{13}$C | 1.11 |
| Nitrogen | $^{15}$N | 0.365 |
| Oxygen | $^{17}$O | 0.037 |
| | $^{18}$O | 0.204 |
| Sulphur | $^{33}$S | 0.760 |
| | $^{34}$S | 4.22 |

Quantitation by IDMS is achieved by the addition of a known amount of labelled standard to the sample. After the separation procedure is completed, the ratio (R) of intensities of a selected ion of the analyte to the corresponding ion of the standard is calculated. This ratio corresponds with the molar ratio of the compounds in the sample. Thus, the amount of analyte can be directly calculated from this value. A hypothetical IDMS analytical process is depicted in scheme I.

Scheme I. Example of a IDMS procedure

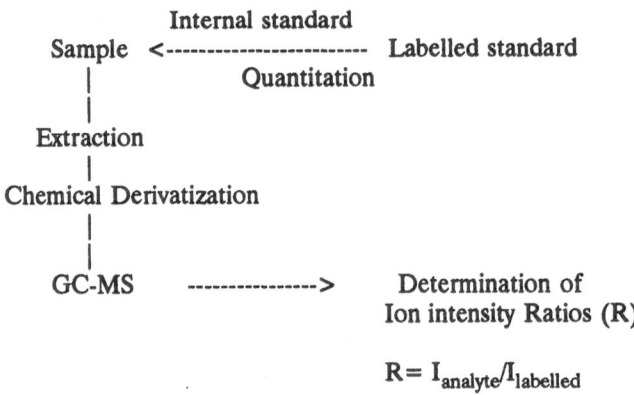

$$R = I_{analyte}/I_{labelled}$$

Addition of the IS at the first steps of the process eliminates procedural errors and makes the result independent of the recovery. In quantitative IDMS the internal standard is added to the sample to be extracted. In this case the IS is sometimes called a surrogate IS to differentiate it from the volumetric IS which is added after the extraction process. A volumetric IS is used to eliminate the dependence of the result on volume changes and errors during analysis. A surrogate IS is, in fact, also a volumetric IS which, in addition, eliminates the effect of variable extraction recoveries in the final results. This is only true under conditions of adequate equilibration of the standard and the sample. Even in this case, irreversible absorption or occlusion of the analyte in the matrix could result in different extraction recoveries for the analyte and the IS.

Quantitative results are obtained taking the added amount of IS as a reference. Thus, accuracy in the IS addition is the more critical step in the analytical procedure.

## 10.1. MATHEMATICAL TREATMENT IN IDMS

In theory, representation of the intensity ratio versus the concentration ratio of the analyte to the IS at a constant IS concentration should generate a straight line with a slope of 1.

In real work, deviations from this situation are frequently encountered. When the ions monitored differ in less than three mass units, the contribution of natural isotopes in the analyte to the labelled standard becomes important and linearity of the response is lost. Also a general problem is the presence of some amounts of unlabelled material in the labelled IS which contribute to the signal of the analyte. Another case of contribution of the signal of the IS to the analyte is related to the presence of IS fragment ions at the mass measured for the analyte. For example, relatively intense signals at $[M-1]^+$ are characteristic of the EI spectra of aldehydes and losses of 1-3 mass units (H, $H_2$) are common in the chemical ionization of several families of compounds.

In order to solve the problem of isotopic overlap, Pickup and McPherson (41) derived the following equation for the Response Ratio R:

$$R = \frac{af_{aa} + sf_{sa}}{sf_{ss} + af_{as}} \qquad [eq.3]$$

were a and s are the mols of analyte and standard respectively; $f_{aa}$ and $f_{as}$ the fraction of analyte contributing to the signal monitored for the analyte and to the one for the standard, respectively; $f_{ss}$ and $f_{sa}$ are the fraction of standard contributing to its monitored signal and to the signal of the analyte, respectively. During the calibration, the amount of standard, s, and the corresponding signal isotopic abundances, $f_{ss}$, $f_{sa}$, $f_{aa}$ and $f_{as}$, are known or can be calculated and remain constant. The equation represents then a hyperbola that is reduced to a linear form when $f_{as}$ is zero. In other words, when contribution of the analyte to the signal of the internal standard is negligible, a linear approximation could be used for calibration. It should be noted that, in practical terms, reliable results imply that linearity must be demonstrated and not assumed. In this respect, the use of polynomials to fit the calibration curves has been considered a more accurate method to describe the situation (42).

Equation 3 reflects the case of two elements overlapping. The more complex situation of three overlapping components was mathematically derived by Bush and Trager (43) and reduced to a linear relationship.

"Bracketing" is, however, the more accurate method for data interpolation (44). In this method unknown samples are determined between measurement of two calibration samples with their isotope ratios confining, as close as possible, that of the unknown. Measurements can then be made by linear interpolation provided that the calibration standards are close one another and that $f_{as}$ is small (<10%)(45). Under these conditions, measurements show high precision. For example, in the determination of serum uric acid by electron impact GC-MS, Elleber et al. use a 0.5% upper limit for the coefficient of variation (CV) encountered in the duplicate determinations of one sample (45). Common values of the CVs for single measurements are reported in the range 0.1%-0.4% (44,46-48), when the measured intensity ratios are near 1.

Application of bracketing implies that the concentration of the sample is approximatively known in order to select the calibration samples. That means that the sample must be analyzed first by other method or by conventional interpolation procedures, making the whole process inappropriate for routine analyses. Thus, the bracketing technique is used only in cases where high precision is required (validation of reference methods or certification of reference materials).

All the cases discussed until now use only one ion from the analyte and one from the IS for the calculation. In some cases, response ratios are calculated using the sum of several ions in the spectra (49). In general, the ratio is calculated from the sum of the intensities of 2 or more ions of the analyte and the sum of the corresponding ion intensities for the labelled analogue. The use of one ion of the analyte versus two ions of the IS has also been reported (49). These procedures are applied in order to counteract variations in the ionization process due to punctual changes in the source conditions that could affect precision.

## 10.2. CHARACTERISTIC OF A LABELLED INTERNAL STANDARD.

Development of reliable IDMS methods implies careful selection of the stable isotope and the position of labelling. In addition, the isotopic purity and number of substituted atoms are important factors in simplifying the calculation processes. It is generally accepted that the standard should contain 3 or more labelled atoms to effectively reduce the effect of $f_{as}$ on the signal from the labelled standard. For example, in the analysis of caffeine by particle beam MS, a second order response ratio is observed when using $[3-^{13}C_1]$ caffeine as IS and no overlap correction is applied (50). The use of $[1,3,7-^{13}C_3]$ caffeine affords linear plots with $r_2$ =0.998 without any mathematical correction. Contributions to the ion at m/z m+3 from compounds consisting on C, O and H atoms and masses in the order of 300 u are around 1% of the signal at mass m. This should be taken into account when measuring kinetic parameters where the maximum tracer to tracee ratios are around 0.1. Under these conditions the contribution of the unlabelled compound to the total signal observed at m+2 and m+3 can be at least the 60% and 50%, respectively.

Critical characteristics to take in account are the stability of the IS and the lack of isotopic effects.

### 10.2.1. *Stability*
Labelled IS should maintain its labelling throughout the analytical procedure. Losses of the label can take place as result of metabolic processes, during sample treatment, derivatization and during MS analysis.
For example, Haroldsen et al (51) have recently reported loss of label from the IS during derivatization of platelet activating factor (PAF). In this work, labelled PAF is hydrolysed to 1-hexadecyl-2-$[d_3]$-acetylglycerol and then derivatized with pentafluorobenzoyl chloride for its analysis by NICI GC/MS. The problem was greatly diminished using lower reaction temperatures (0 °C instead 120 °C) and the presence of base. These authors also suggest the use of 1-$[^{13}C_2]$-hexadecyl-PAF as internal standard.

### 10.2.2. *Isotope effects*
The isotope effect is the result of the different physicochemical characteristic of the labelled molecule and the natural one due to the isotopic substitution. This effect is detected as differences in the behaviour of the labelled and unlabelled compounds during the analytical process and leads to a loss of precision and accuracy.
The isotope effect can be detected in any of the steps of the analytical procedure such as the extraction process, chromatography and mass spectrometric analysis, as discussed below.

Isotopic effects can be produced in the analytical sample through differential adsorption processes to other components in the matrix and/or through selective enzymatic or chemical processes. The analyte and IS can show different affinities for this adsorption into proteins or other macromolecules in the sample. This could be reflected in several ways such as incomplete equilibration or different distribution of the IS in the analytical pool, different metabolization rates or unequal extraction recoveries. Isotope effects on the protein binding of drugs have been studied by Cherrah et al.(52) using ID GCMS. The intensity of the effect is related to the number of labelled atoms and the position of the label.

Also, when the target compounds are exposed to metabolization or degradation processes, differential reaction rates will afford erroneous data. For example, very different half life values were calculated for 25-hydroxy-vitamin $D_3$ in human plasma depending on the position of the isotopic labelling (49). A half life of 15-27 days have been reported for 25-hydroxy-Vitamin $D_3$ using tritiated tracer standards labelled at positions 23,24 or 26,27, or when calculated using $[1,2-^2H_3]$-vitamin $D_3$ as precursor. However, the half life was calculated to be 10 days by IDMS using $[6,19,19-^2H_3]$-25-hydroxy-vitamin $D_3$. These differences are due to the existence of metabolic oxidation processes implicating the labelled position in the tritiated standard indicated above. Tritium and deuterium atoms in these positions slow down the metabolization rate in relation to hydrogen. The isotopic effect is thus reflected as an apparent longer half life of the analyte.

In chromatographic separation, the isotope labelling causes differences in the retention times of the labelled and unlabelled molecule. Recently, this isotope effect has been examined by Matucha et al. for deuterium and $^{14}C$ labelled alkanes, methylpalmitate and hexane (53). The isotope effect is proportional to the number of heavy atoms in the molecule and is much larger for deuterium than for $^{14}C$-labelled compounds. Mathematical treatment of isotope ratio data based on eq 3 assumes coelution of the labelled and unlabelled analyte. The parameters $f_{sa}$ and $f_{as}$ on eq 3, calculated by separate injection of pure standard and pure labelled compound, are not correct when only a fraction of the labelled and unlabelled compounds contribute to the area of each other.

Mass spectral patterns are highly reproducible using electron impact ionization. Although the change of one or more atoms in a molecule could produce changes in the relative abundances of fragment ions, precision is expected to be maintained. This could not be the case in ionization techniques such as chemical ionization (CI) where ionization and fragmentation can be strongly affected by experimental parameters such as source pressure and temperature. In these cases, the analyte and the labelled standard can behave differently and in an unexpected way leading to the loss of reproducibility in the measure of isotope ratios. For example, in the analysis of Lormazepan by negative chemical ionization using $[1,1,1-d_3]$-lormazepan a large coefficient of variation was noticed and the calibration curve deviated from the expected theoretical line (54). The response ratio was calculated for the base peak in the spectra that was produced by loss of trimethylsilylalcohol from lormazepan. This fragmentation implied loss of one of the deuterium atoms in the labelled molecule and an isotope effect was suspected. Using positive CI or electron impact, were no loss of TMSiOH is observed, no isotope effect was detected. Substitution of the IS for the $[3,4,5,6-d_3]$-lormazepan also provided precise quantification.

## 11. Applications

Applications of GC-MS in its various modes of operation are wide and diverse, comprising still at present the bulk of publications in the Health Sciences (4). A detailed account of the vast number of possible applications is certainly beyond the scope of this chapter. However, the interested reader is advised to consult the monthly "Current Awareness" section in the Journal of Biological Mass Spectrometry, where one may find the latest in applications in the fields of aminoacids, peptides and proteins, carbohydrates, lipids, nucleic acid constituents, pharmacology and toxicology, natural products, environmental chemistry and trace elements. Similar data can also be found in the annual reviews of Mass Spectrometry which appear every two years in Analytical Chemistry (55).

## REFERENCES

1. Brunnée, C., Jenkel, L. and Kronenberger, K.; Anal. Chem., 189, 50 (1962).

2. Ryhage, R., Arkiv Kemi, 20, 185 (1962).

3. McFadden, W.H., Teranishi, R., Black, D.R. and Day, J.C., J. Food Sci., 28, 316 (1963).

4. E. Gelpí, Int. J. Mass Spectrom., 118/119, 683 (1992).

5. Ten Noever de Brauw, J. Chromatogr., 165, 207 (1979).

6. Greenway A.M. and Simpson, C.F., J. Phys. E. Sci. Instrum., 13, 1131 (1980).

7. Ryhage R., Mass Spectrom. Reviews., 12, 1 (1993).

8. Leemans, F.A.J.M. and McCloskey, J.A., J. Am. Oil Chemists Soc., 44, 11 (1967).

9. Bruins, A.P., Mass Spectrom. Rev., 10, 53 (1991).

10. Willoughby R.C. and Browner R.F., Anal. Chem., 56, 2626 (1984).

11. Kotlaho, J., Lauritsen, R.R., Chodhury, T.K., Cooks, R.G. and Tsao, G.T., Anal. Chem., 63, 675A (1991).

12. Tsai, G.. -J., Austin, G.D., Syu, M.J., Tsao, G.T., Hayward, M.J., Kothaho, T. and Cooks, R.G., Anal. Chem., 63, 2460 (1991).

13. Ryhage, R., Anal. Chem., 36, 759 (1964).

14. Watson, J.T. and Biemann, K., Anal. Chem., 36, 759 (1964).

15. Krueger, P.M. and McCloskey, J. A., Anal. Chem., 41, 1930 (1969).

16. Cree, R. F., General Electric Publications, Nov. 1957. General Electric, Schenectady, New York.

17. Blumer, M., Anal. Chem., 40, 1591 (1968).

18. Lipsky, S.R., Horvath, C.G. and McMurray, W.J., Anal. Chem., 38, 1585 (1966).

19. Simmonds, P.G., Schoemake, G.R. and Lovelock, J.E., Anal. Chem., 42, 881 (1970).

20. Brunnée, C., Bültemann, H.J. and Kappus, G., A.S.T.M. Meeting on Mass Spectrometry, 20.5.1969, Dallas, Texas.

21. Watson, J.T. in "Ancillary Techniques of Gas Chromatography", Chapter 5. Ettre, L.S. and McFadden, W.H., eds. Wiley Interscience (1969).

22. Black, D.R., Flath, R.A. and Teranishi, R., J. Chromatog. Sci., 7, 284 (1969).

23. Becker, E.W. "Separation of Isotopes", p 360, Newness, London, 1961.

24. Henneberg, D., Henrichs, V. and Shomburg, G., Chromatographia, 8, 449 (1975).

25. N. Neuner-Jehle, Etzweiler, F. and Zarske, G., Chromatographia, 7, 323 (1974).

26. Schulze, P. and Kaiser, K.N., Chromatographi., 4, 381 (1971).

27. Henderson, W. and Steel, G., Anal. Chem., 44, 2302 (1972).

28. Cramers, C.A., Scherpenzel, G.J. and Leclerq, P.A., J. Chromatogr., 203, 207 (1981).

29. Hinshaw, J.V., J. Chromatogr. Sci., 26, 142 (1988).

30. Ros A., J. Chromatogr., 7, 252 (1965).

31. Huber, J.F.K., Matisova, E. and Kenndler, E., Anal. Chem., 54, 1297 (1982).

32. Peltonen, K. Lakklsto, V. and Rosenberg, C., LC.GC Int., 1, 48 (1988).

33. Holland J.F., Anal. Chem., 55, 997A (1983).

34. Sweley, C., Elliot W., Fries I. and Ryhage, R., Anal Chem., 38, 1549 (1966)

35. Tietz N.W., Clin.Chem.,25, 833 (1979).

36. Gardland W.A. and Powell M.L., J.Chrom.Sc.,19, 392 (1981).

37. De Leenheer A.P. and Thienpont L.M., Mass Spectrom. Reviews 11, 249 (1992).

38. Heumann K.G., Mass Spectrom.Reviews., 11, 41 (1992).

39. Boyd R.K., Rapid Comm Mass Spectrom., 7, 257 (1993).

40. Rosenblatt J., Chinkes D., Wolfe M. and Wolfe R.R., Am. J.Physiol 263, 584 (1992).

41. Pickup J.F. and McPherson K., Anal.Chem., 48, 1885 (1976).

42. Jonckheere J.A., DeLeenheer A.P. and Steyaert H.L., Anal.Chem., 55, 153 (1983).

43. Bush E.D. and Trager W.F., Biomed.Mass Spectrom. 8, 211 (1981)

44. Cohen A., Hertz H.S., Mandel J., Paule R.C., Schaffer R., Sniegoski L.T., Sun T., Welch M.J. and White V E., Clin.Chem., 26, 854 (1980).

45. Yap W.T., Schaffer R., Hertz H.S., White V E. and Welch M.J., Biomed.Mass Spectrom. 4, 262 (1983)

46. Ellerbe P., Cohen A., Welch M.J. and White V E., Anal Chem., 62, 2173 (1990).

47. Welch M.J., Cohen A., Hertz H.S., Ruegg F.C., Schaffer R., Sniegoski L.T. and White V E., Anal.Chem., 56, 713 (1984).

48. Welch M.J., Cohen A., Hertz H.S., Ng K.J., Schaffer R., Van Der Lijn P. and White V E., Anal.Chem., 58, 1681 (1986).

49. Vicchio D., Yergey A., O'Brien K., Allen L., Ray R., Holick M., Biological Mass Spectrometry, 22, 53 (1993).

50. Doerge D.R., Burger M.W., Bajic S., Anal.Chem., 64, 1212 (1992).

51. Haroldsen, P.E., Gaskell S.J., Weintraub S.T. and Pinckard R.N., J.Lipid Res., 32, 723 (1991).

52. Cherrah Y., Falconnet J.B., Desage M., Brazier J.L., Zini R. and Tillement, Biomed.Environ.Mass Spectrom., 17, 245 (1988).

53. Matucha M., Jockish W., Verner P. and Anders G., J.Chromatogr., 588, 251 (1991).

54. Takahashi S., Biomed.Environ.Mass Spectrom., 14, 257 (1987).

55. Burlingame, A.L., Baillie, T.A. and Russell, D.H., Anal Chem., 64, 467R (1992)

# RECENT DEVELOPMENTS IN COMBINED SUPERCRITICAL FLUID CHROMATOGRAPHY/MASS SPECTROMETRY (SFC/MS)

P. ARPINO [1], F. SADOUN [2], H. VIRELIZIER [2]

1 Laboratoire de Chimie Analytique, Institut National Agronomique
16 rue Claude Bernard,
75231 Paris-05 (France)

2 Commissariat à l'Energie Atomique
Centre d'Etude de Saclay
SPEA/SAIS
91190 Gif-sur-Yvette (France)

ABSTRACT. During the three year period since the last review [1], the technique of SFC/MS has not progressed significantly, and it can be considered to be still in a steady state. An excellent critical analysis of the capabilities, advantages, and shortcomings of the various types of SFC/MS interfaces can also be found in a recent review by Pinkston [2]. The reasons for this situation are examined, and future directions that could give a new start are discussed, and in particular, the use of packed rather than capillary SFC columns, interfaced to an atmospheric pressure ionization source rather than to a conventional EI/CI source under a vacuum.

## 1. Introduction

When SFC was rediscovered in the early 1980s, it was frequently estimated that a strong driving force to its development would be the ease of devising a simple SFC/MS interface. This was believed to be easily achieved if analytical conditions were limited to capillary SFC columns as a general separation tool, and to the choice of neat $CO_2$ as the unique supercritical fluid. The low flow rate of mobile phase delivered by capillary columns was easy to accommodate by the vacuum equipment of standard mass spectrometers, and the specific physical properties of $CO_2$ made possible solute ionization by different ion-molecule reactions, especially charge exchange ionization. This approach has not lived up to all of its promises [3]. The major causes of the observed mismatch are the large variations of the MS source pressure as a result of the $CO_2$ pressure gradient at the SFC column inlet, the low sensitivity of charge exchange ionization at these high MS source pressures, and the inability to handle polar and nonvolatile molecules. Adaptation of LC/MS interfaces, such as the thermospray interface or the particle beam interface, to SFC/MS conditions was a step forward, but these devices have their own limitations. Alternative methods to direct SFC/MS coupling have been investigated recently. They are based on the use of packed columns rather than capillaries, and on solute ionization at atmospheric pressure rather than under a vacuum, by means of either gas-phase corona discharge ionization or liquid-phase electrospray ionization. These new developments may revive research into the design of reproducible and sensitive SFC/MS systems where the number of recent studies is still low compared with other chromatography/mass spectrometry coupling studies.

R. M. Caprioli et al. (eds.), Mass Spectrometry in Biomolecular Sciences, 461–474.
© 1996 Kluwer Academic Publishers.

The first successful method of interfacing supercritical fluid chromatography to mass spectrometry (SFC/MS) was reported in 1978 and used a packed column and a molecular beam separator [4]. Nevertheless, the most common method of SFC/MS coupling during the subsequent years was to directly insert the restrictor, attached to the end of a capillary SFC column, into the ion source of a mass spectrometer of the type normally designed for chemical ionization (CI). The only critical requirements of the interface design were to provide sufficient heating of the restrictor tip, to compensate for the severe cooling of the rapidly expanding supercritical fluid; to allow for a rapid restrictor interchange in case of plugging; to minimize dead volumes along the interface; and to provide convenient addition of a reagent gas for chemical ionization. This direct approach, sometimes referred to as direct capillary SFC/MS coupling, or the direct fluid interface (DFI), was pioneered by Smith and co-workers, and was adopted by other authors, in particular by Arpino and co-workers [5, 6]. Earlier work has been reviewed [2, 7, 8, 9] and various forms of imaginative adaptation to practice, and applications to various classes of organic compounds have been reported more recently [10, 11, 12]. One may quote in particular: the analysis of low molecular-weight fatty acids and triglycerides in hamster feces [13], the determination of tocopherol in a deodorizer distillate and commercial antioxidant formulations [14].

The direct fluid interface has also been used in a number of fundamental studies for the direct sampling of supercritical fluid solutions to the mass spectrometer with no chromatographic separation. For example, it was used for the study of gas-phase ion-molecule reactions at high pressures [15], and for the determination of high-pressure phase distribution isotherms for supercritical fluid chromatographic systems [16, 17].

## 2. Direct SFC/MS at low flow rates used with capillary column SFC

Reasons in favor of direct capillary SFC/MS have been frequently discussed and need only to be briefly recalled here: the interface is a mechanically simple low-cost device that can be easily assembled in the laboratory; it is possible to use commercial unmodified SFC and MS instruments, including benchtops [18]; the operator may select the recording of library searchable electron impact-like mass spectra, resulting from charge exchange ionization [18, 19, 20], or of protonated molecule mass spectra upon CI with a proton donor reacting gas, for example, methane, isobutane or ammonia.

Negative aspects were unfortunately soon evident, some due to the choice of capillary SFC with $CO_2$ as the exclusive mobile phase, others to specific MS limitations, and a last group that are common to both techniques.

Among SFC reasons is the low sample capacity of typical SFC capillary columns with internal diameter of 50 $\mu$m and a coating layer of a chemically bonded stationary phase with an average thickness of 0.3 $\mu$m, at maximum. Conservation of the column resolving power imposes sample loads of a few nanograms per peak and maximum sample volumes of 100 nL. Consequently, sample solution concentrations in real case studies are typically 50-100 ppm. Minor constituents in complex mixtures that are hardly detected by a flame ionization detector (FID) are totally missed by a mass spectrometer with a DFI interface because it generally exhibits a lower basic sensitivity. Despite recent reports on improved injection techniques for capillary column SFC [21], many analysts have chosen to shift to packed columns which, although they have fewer plates, provide much easier injection conditions, in addition to other specific advantages including faster analyses, and reduced restrictor plugging problems.

Packed column SFC was also reconsidered because the solvating power of pure $CO_2$ was found not sufficient for chromatographic elution of polar solutes. It could be increased by the addition of a few percent of a polar liquid solvent such as methanol, but addition of such polar modifiers was found to be less effective with capillary columns compared to packed columns.

Erratic flow transmission through the necessary restrictor at the exit of the SFC system was another serious nuisance, both affecting chromatographic retention parameters and ion recording by the mass spectrometer. Despite a variety of available restrictor types [7, 22] those that are suitable for use with capillary SFC/MS comprise pinhole orifices or narrow channels with diameters of a few microns or less. Modification of their flow permeability, not to mention fatal clogging, depends on the investigated sample and this is an unacceptable random risk in a control laboratory.

In the above limitations of capillary SFC, mass spectrometry played little or no part, but generally speaking, capillary SFC/MS made no significant contribution which could be said to make capillary SFC any better. This and the fact that a mass spectrometer is still a complex and expensive detector for chromatography would have been sufficiently convincing reasons to explore other SFC/MS routes, but the situation was made even more unfavourable by specific MS limitations.

Significant differences have been occasionally reported between true EI mass spectra and $CO_2$ charge exchange mass spectra, thus making the identification of some chlorinated pesticides not feasible using a library of EI mass spectra [23].

Under the usual capillary SFC condition of fluid pressure programming, the performance of the mass spectrometer was repeatedly reported to be affected by the introduction of the large quantities of the mobile phase, especially at the end of the analysis, when unfortunately the solutes of greatest interest in SFC are eluted. The situation is diametrically opposed to that in direct capillary column GC/MS coupling, where high mass and low volatile samples are ionized under a minimum MS source pressure, as a result of the mobile phase flow rate decreasing with increasing GC column temperatures.

Reductions in important MS parameters such as sensitivity, mass resolving power and dynamic range were reported for all kinds of mass spectrometers directly coupled to SFC, including quadrupoles [5, 18], ion traps [24, 25], magnetic sectors [26, 27, 28] and Fourier transform mass spectrometers [29]. The effect was particularly severe when the gas admitted into the ion source was pure $CO_2$ and solute ionization was by charge exchange with $CO_2^{+\cdot}$ reactant ions. The magnitude of the drop may depend on different mass spectrometer models (for example, high mass discrimination at high source pressures is more or less pronounced for some quadrupole mass spectrometers), but the reduction in sensitivity appears to be always unavoidable. This is the reason why, in general, authors have attempted to limit the input of $CO_2$ into the ion source to a minimum by using very narrow flow restrictors. In one study, a hybrid instrument combining an external differentially pumped ion source quadrupole and an ion trap for final mass determination has been assembled with a view to protecting the pressure sensitive ion trap from excess of $CO_2$ [25]. The loss of sensitivity is reduced when using significantly higher ionizing electron energy of ca. 200-500 eV, instead of the conventional 70 eV, so that electrons can penetrate further into the ion source [27], but it was observed that this also increases sample fragmentation resulting in a decrease of the signal-to-noise ratio [23]. A substantial background signal in the mass range 180-270 was also observed when using a rhenium filament for producing the ionizing electrons. Thermo-chemical reactions were evidenced by the presence of several rhenium oxide derived species. Replacement by a thoriated iridium filament eliminated this inconvenient [23].

On the other hand, better sensitivity and good resolution was observed when sample ionization was by proton transfer CI, even for high mass solutes, after addition of a suitable reactant gas into the ion source. Sensitivity at high mass for the protonated sample molecule was observed to increase with increasing ammonia partial pressure [20]. Ammonia negative chemical ionization was found to be particularly sensitive for the SFC/MS investigation of diphenylmethane polyisocyanate mixtures with masses up to 774 Da [30]. A high partial pressure of the CI reagent was also required if mixed charge exchange/proton transfer ionization was not wanted, a condition sometimes difficult to control because of the continuously changing input flow rate of $CO_2$ during pressure programmed SFC [5]. Nevertheless, these large amounts of reactant gas were additional to the already high gas load to be evacuated by the vacuum equipment. Under such CI conditions, structure elucidation by library search is not possible, thus eliminating one of the claimed advantage of direct coupling, although molecular mass determination, target compound detection, and MS/MS collision studies remained as compensating advantages.

## 3. Direct SFC/MS at high flow rates used with packed column SFC

A further development of direct coupling was the investigation of systems accepting higher mobile phase flow rates, while retaining the basic concepts of direct coupling, and was achieved by the addition of one stage of rotary vacuum pumping.

Smith and co-workers [31] preferred to introduce the extra vacuum line in front of the ion source entrance, using a purpose-built interface. Their so-called "high flow rate interface" was successfully used in a number of applications [31, 32, 33] but the work was not pursued recently by this research group. A similar SFC/MS coupling to a high resolution magnetic sector instrument was realized in Germany using a modified GC-transfer line and an additional vacuum pump connected to a prevacuum chamber [34]. The system was applied to the identification of polymer additives from the supercritical fluid extraction of a polyethylene film.

The particle beam interface discussed in the next section similarly removes part or all of the mobile phase before sample introduction into the MS source and can be considered as an extension of the model of Smith and co-workers.

Chapman [35], and Games and co-workers [36] preferred to utilize the line placed past the ion source block of an unmodified commercial thermospray source, originally designed for LC/MS coupling. In both cases, highest sensitivity was obtained by proton transfer CI from the addition of a suitable reagent gas. Filament-on or discharge mode of operation was required for the thermospray source. Coupling to packed SFC columns was demonstrated, and signal enhancement upon the addition of less frequently used CI reagents, such as methylene chloride or hexane, was noteworthy [35]. Applications using the thermospray source for SFC/MS work have been reported [37-42].

Nevertheless, a thermospray source is designed to operate at an optimum constant vapour throughput, which corresponds to an optimized constant flow rate of mobile phase entering the MS source [43]. For example, the distance between the ionizing electron entrance, or the discharge needle, and the ion exit aperture is fixed and was optimized for typical LC conditions of ca. 1 mL/min of water. Consequently the thermospray source is more suitable to SFC with mobile phase gradient composition at constant flow rate, than to fluid pressure gradient (although MS source pressure variation may be reduced by the addition of an excess of a make-up gas). This would be favourable to packed column SFC, yet the major disadvantage of the thermospray source remains and is independent of the connected chromatographic system, be it LC or SFC. Thermospray sources suffer from an inherent low sensitivity, often requiring injection of microgram amounts of sample, because many of generated solute ions are not adequately directed toward the mass analyser, but are pumped away along with the mobile phase vapour.

## 4. SFC/MS with mobile phase separators

In order to operate the mass spectrometer in the absence of high levels of neutrals and ions derived from the mobile phase, classical LC/MS interfaces with a high enrichment, such as the moving belt transport interface [44] and the two-stage momentum separator of the particle beam interface [45], were transposed to SFC/MS conditions. Both interfaces can decouple the operations of the SFC and the MS instruments by selectively eliminating most of the mobile phase, making possible a choice between true EI or CI, whilst coupling with packed SFC columns. Interesting applications have been reported using the moving belt interface [38, 39, 46]; on the other hand, the particle beam interface has received much less attention and is only described in one report [45]. The inherent limitations of these LC/MS interfaces [47, 48] were, however, not improved by SFC/MS conditions, thus a low overall sensitivity had to be often accepted, and the investigation of nonvolatile and polar molecules was in general not possible.

## 5. SFC/MS with fast-atom bombardment ionization

This combination is described in a single report [49] and was developed in an attempt to couple SFC with an ionization method capable of producing ions from high mass solutes. It is a transposition to SFC/MS conditions of a continuous flow FAB interface originally designed for LC/MS coupling. Although it was used with packed column, only 1% of the sample was used effectively for mass analysis. The eluent (200 μl/min) from a narrow-bore packed column (1mm i.d.) was mixed with an equal amount of a liquid matrix made of 1% glycerol in methanol containing 10% tetrahydrofuran. A splitter served the purpose to vent $CO_2$ and to sample 5 μl/min of the glycerolic solution to a FAB source, via a so-called "frit/FAB" LC/MS interface. The system was applied to the characterisation of oligomers with n=2-21 in Triton X-100, a synthetic mixture of non-ionic surfactant molecules with 1-alkyl-4-polyethoxybenzene structures.

## 6. SFC/MS with atmospheric pressure ionization sources

Previous SFC studies have demonstrated the advantages of detectors located beyond the restrictor and working at constant pressure, especially at atmospheric pressure. Within some practical limits, these detectors are independent of flow rate variations when the mobile phase is pressure or temperature programmed. In addition to the well documented flame ionization detector, other examples of such a detector type include the flame photometric detector (PID) [50], the light scattering detector [51] and the ion mobility detector [52, 53]. Their potential contribution to SFC is less important when the detector response is affected by the supercritical fluid composition: for example, the FID is not compatible with mobile phases containing polar organic modifiers, thus precluding its use for packed-column SFC.

The recent growing development of atmospheric pressure ionization (API) techniques in mass spectrometry, especially the electrospray (ES), which has already deeply influenced the conception of new efficient LC/MS interfaces, has also promoted studies on possible application to SFC/MS. Of course, reports on the use of API in SFC/MS are far fewer than those on its use in LC/MS. Note that the moving belt transport interface discussed above also operates at atmospheric pressure and can accommodate variable flow rates of SFC mobile phases, but it is distinct from true API sources because with a moving belt transport interface, sample ionization is produced under a vacuum into a conventional MS source.

Two fundamentally different processes can be utilized in an API source to ionize solutes from a SFC separation:

## 6.1. GAS-PHASE IONIZATION IN A CORONA DISCHARGE

If samples exist in the gas phase with a minimum lifetime of a few milliseconds, after adequate heating if necessary, they can be ionized at atmospheric pressure using a variety of methods. These include the interaction with keV electrons emitted from a $^{63}$Ni foil, UV-photons, alkaline cations, and organic clusters formed in a corona discharge. The first two reactions have been used in non MS-based SFC detectors, e.g. the ion mobility detector [52, 53] and the photoionization detector [50]. Mass spectra of lithium ion adducts formed at atmospheric pressure have been reported for some volatile substances, and possible coupling to SFC was claimed but not demonstrated [54]. Much more convincing results have been reported using corona discharge ionization in an SFC/API/MS system [55-59] probably because this API source is more rugged and often exhibits a wider detection dynamic range than other API sources. Corona discharge ionization requires restrictor temperatures well above 200°C for total solvent and solute vaporization, and gas-phase ionization (Fig. 1). It has been used, together with packed column SFC and organic solvent modifiers, for the analysis of steroids in biological matrices [55], aromatic hydrocarbons in coal tar and sand oil extracts [56], various synthetic mixtures including polyethylene glycol and polystyrene oligomers, and vitamins [57], and a synthetic mixture of polypropylene glycols [58,59].

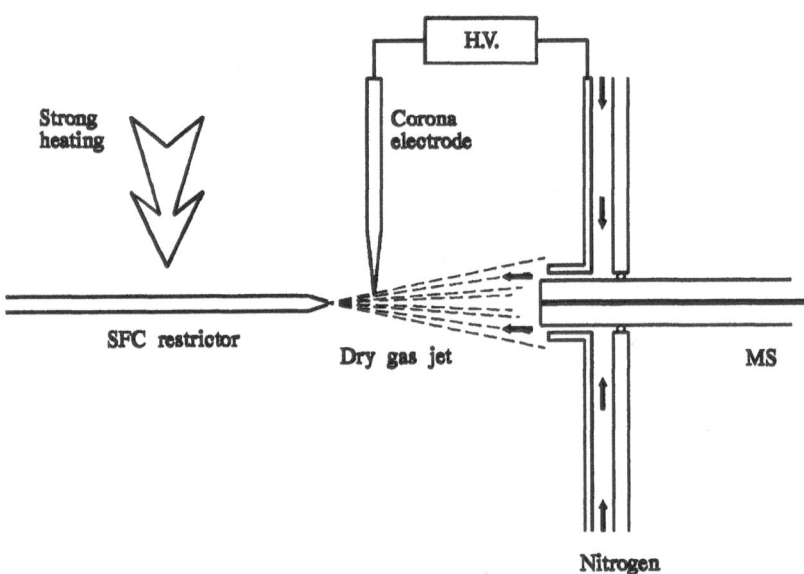

Figure 1. Schematic principle of SFC/API/MS with gas-phase ionization.

Positive ions were recorded in all cases. Under the experimental conditions of positive-ion API, a series of complex gas-phase chemical ionization reactions end in the formation of intact solute-derived ions. Anacleto and co-workers [56] assume that primary reactant ions such as $N_2^+$, $O_2^+$, and $H_2O^+$ are generated via the impact of corona-created electrons on ambient atmosphere constituents into the API chamber. These cations trigger a complex series of ion-molecule reactions culminating in the formation of an equilibrium set of hydrated protons $(H_2O)_nH_3O^+$. The carbon dioxide mobile phase does not appear to play a major role, as no $CO_2$ derived ions are observed among primary reactant ions.

Two concurrent ionization mechanisms were observed when sample molecules react with $(H_2O)_nH_3O^+$: charge exchange leading to $M^+$, and proton transfer leading to $MH^+$. The relative importance of these two mechanisms are governed by the relative electron recombination energy (RE) and proton affinity (PA) of the sample molecules. As proton affinities usually increase with molecular masses, this factor becomes predominant in determining the ratio of $[M+H]^+$ to $M^+$ ions for high-molecular weight solutes. On the other hand, charge exchange reactions leading to $M^+$ can be enhanced if traces of a substance with a low RE, e.g. benzene (RE=9.3 eV), are introduced into the ionization chamber. The overall ionization phenomenon is thus complex, and the authors believe that the composition of the surrounding gas at atmospheric pressure into the API chamber should be carefully controlled, in particular its humidity, in order to keep constant the ionization conditions. This observation has been confirmed recently [59].

On the positive side, it has been claimed that corona discharge API system is simple and trouble free, with a near universal response to most analytes [55]. At first glance, the limitation to volatile compounds, because ionization occurs in a gas phase, would seem to be somewhat restricting. Nevertheless, SFC in fact shows an upper limit of sample masses and polarities that may not be far from that of the corona discharge ionization.

An accessory for SFC/MS coupling using this ionization mode is commercially available from Sciex, but very few results using this system have been reported so far and they have been obtained after some laboratory-built modifications [58, 59]. More generally, the limited number of reports on this SFC/API/MS coupling can be accounted for by the rather small number of SFC systems in use, and by the presently preferred dedication of API/MS sources to the investigation of high mass biological compounds in liquid solutions with electrospray ionization.

6.2. LIQUID PHASE IONIZATION BY PURE ELECTROSPRAY IONIZATION

Solute ions can also be formed at atmospheric pressure, directly from liquid solution aerosols in electrical fields of a few volts/Ångstrom, by electrospray ionization (ES) [60]. Electrospray ionization coupled with mass spectrometry (ES/MS) has the potential to analyse large polar molecules with masses over 50,000 Da [61] as well as molecules with masses below 1000 Da [62]. Ionspray is a pneumatically assisted variant of electrospray ionization sharing an identical ion formation mechanism [62, 63].

ES/MS is well documented in case of direct solution sampling and LC/MS coupling [64], but it has not received attention in case of supercritical fluid solutions. This approach was explored recently using an SFC/MS instrument that could record ions produced from electrospray ionization of solutes in liquid droplets that are briefly formed in the jet at the restrictor exit (Fig. 2) [65]. The instrument is rather similar with those described in references [55-59], with the noticeable exception that the corona needle electrode was omitted. It was used at low temperatures at the nebulizer exit in order to produce droplets in the expanding gas jet. These conditions are believed to be more suitable to the investigation of thermally labile samples. The light scattering detector adapted to SFC monitoring [51] is another example of an atmospheric pressure system that similarly utilizes the liquid droplets in suspension in the subcritical gas phase which results from mobile phase decompression and cooling.

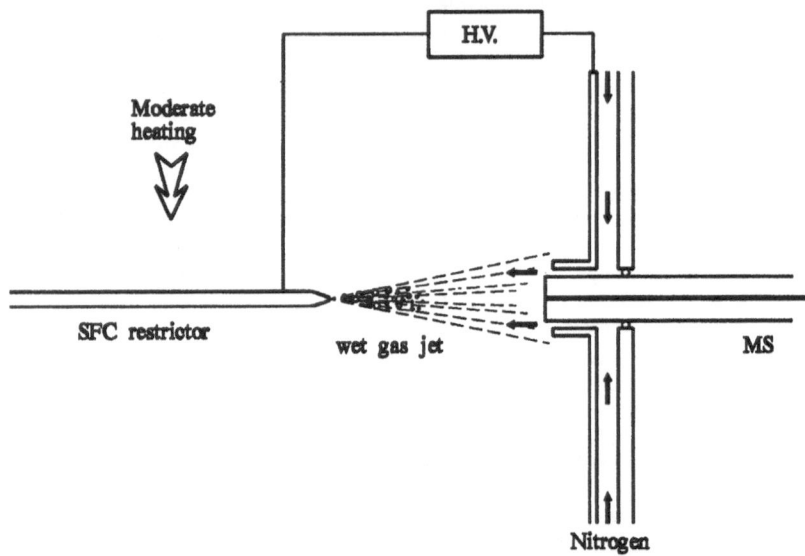

Figure 2. Schematic principle of SFC/API/MS with liquid-phase ionization.

Supercritical solutions of polar basic solutes in $CO_2$ modified with organic solvents were introduced at mobile phase flow rates compatible with the use of packed columns, i.e. 150-1000 µL/min (Fig.3). The mass spectrometer was a single quadrupole equipped with an Analytica electrospray source of first generation. With the exception of the SFC/MS interface, the Analytica electrospray source was used without modification (Fig. 4). A stream of dry nitrogen, at flow rate and temperature of 9 L/min and 80°C, respectively, built a small positive pressure into the ES source that prevented the introduction of contaminants from the laboratory atmosphere, assisted solution vaporization, and broke heavy ion clusters prior to MS analysis.

The SFC/MS interface was very simple and consisted of a restrictor with the end tip located at ca. 5 cm and in direct line of sight with the ion sampling aperture. Linear restrictors were generally used and were made of 25-30 cm long fused silica tubing, with 25 µm i.d. Proper electrospray ionization requires that the nebulizer be electrically conducting, to avoid disturbing charge build-up. Consequently, the polyimide coating of the fused silica tube was covered with an electrically conducting layer of a nickel doped polyurethane painting over a length of ca. 10 cm. In general, the restrictor end was kept at ground potential, and was only warmed by the surrounding nitrogen bath at 80°C in the atmospheric pressure ion source. Another restrictor type was used occasionally and consisted of a stainless steel tube (30 cm x 1 mm i.d.) crimped at the terminal end, such that a pressure drop of 100 bars is measured when the mobile phase flow rate is 1 mL/min of $CO_2$.

The column oven was used to adjust the mobile phase temperature. By keeping this parameter at a sufficiently low value, and by establishing an electrical potential of a few kilovolts between the restrictor end and the MS sampling orifice, both solvent and sample ions were produced. The fluid temperature must be carefully controlled as excessive heating, above 150°C, vaporizes the SFC solution completely and stops the ionization process, whilst insufficient heating, below 50°C, generates solid microparticles that also give no ions. The conditions appeared to be identical with those existing in conventional electrospray ionization, a liquid phase ionization process that is known to require the presence of charged liquid droplets. Such charged liquid droplet can be transiently obtained during decompression and cooling of the expanding subcritical jet at the restrictor outlet, thus allowing solvent derived cluster ions and preformed sample ions to escape from the charged droplets, into the gas phase, and be mass analysed.

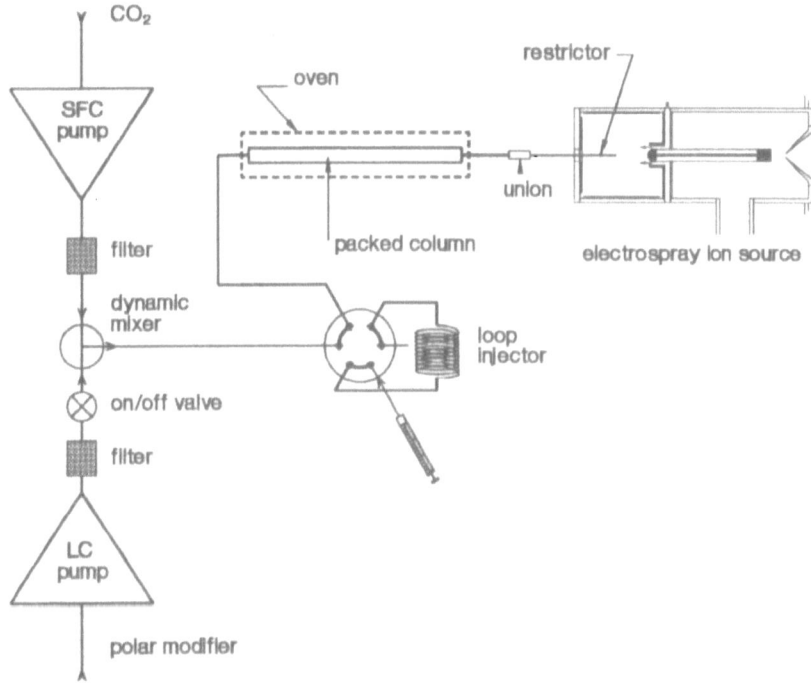

Figure 3. Schematic diagram of the SFC/API/MS setup at Saclay.

Solvent and sample ion generation strongly depends on the mobile phase composition. Pure $CO_2$ produced no ion for any investigated fluid temperature and flow rate, and API source voltage conditions. The similarity with the absence of ions when non polar solvents are used in thermospray experiments is also noticeable [43]. On the other hand, solvent and solute derived ions were present when a minimum amount of a polar solvent was added to the mobile phase.

When $CO_2$ modified by a few percents of a methanol/water mixture is introduced into the source, the observed solvent derived ions and clusters correspond to $H_3O^+$ (m/z = 19), $(CH_3OH)_nH^+$ with n= 1,2 (m/z = 33, 65), and a mixed cluster $(H_2O)(CH_3OH)H^+$ (m/z=51). The presence of an ion at m/z=61 is more difficult to explain, unless one postulates a reaction involving $CO_2$ and MeOH that would lead to $CH_3COOH_2^+$; this would be the only visible ion arising from $CO_2$. Other frequently observed ions in SFC/MS studies using pure $CO_2$ and a CI source, e.g. $CO_2^+$ and higher cluster ions [5], are totally absent.

If $CO_2$ does not appear to play any direct role in the ion formation, because of the lack of any abundant $CO_2$ derived primary ions (with the possible exception of the $m/z=61$ ion), the methanol and solute derived ion currents remain a function of the total mobile phase flow rate, and is therefore a function of $CO_2$, although its role is still not understood.

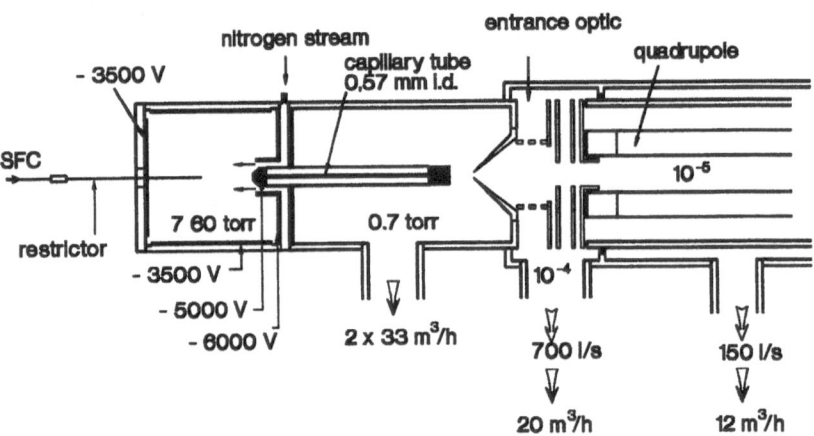

Figure 4. Schematic presentation of the SFC/MS interface and ion source.

As with corona discharge ionization, a problem will arise when planning to achieve real SFC/ES/MS separations with a gradient of mobile phase composition, because constant conditions of ion formation appears to be not possible. Methods to overcome this limitation, probably by the use of post-column addition of a suitable solvent additive, will be needed. Nevertheless, there is at least one very important positive influence of the presence of $CO_2$, and this is the increase in the tolerable input of mobile phase flow rate into the API source. The basic Analytica ES source is limited to very small liquid flow rate inputs, below $5\,\mu L/min$ [66]. Much higher flow rates were possible in our experiments, because the decompressing $CO_2$ acts as a nebulizer gas. We believe that $CO_2$ has the same beneficial influence on the tolerable fluid input as the added gas in an ionspray experiment [63].

An example of the SFC/ES/MS analysis of a synthetic herbicide mixture (Fig. 5) was accomplished in ca. 2 min, using a 2.1 mm i.d. column, packed with 5 μm Zorbax Rx-C18, and eluted with a 0-4% step gradient of methanol in supercritical $CO_2$ (in 4 equal steps of 30 s width), at a flow rate of 1 mL/min. Peaks labelled A,C,D correspond to 1 ng of each sample deposited on the column, peak B is an impurity in sample A, and corresponds to 100 pg. The lower limit of detection, using selective ion recording was 10 pg (with a S/N =3). The restrictor temperature was kept at 80-100°C. Positive ion mass spectra are extremely simple, showing singly charged $MH^+$ species with no or low abundance fragments. Adduct ions at the low mass end consist exclusively of methanol clusters, with no visible $CO_2$ derived ions.

Although intact molecule derived ions were obtained for many organic compounds, including mixtures of polyethylene oligomers with masses above 1000 Da, poor chromatographic peak shapes and excessive peak tailing were frequently observed for heavier compounds, because of cold trapping at the nebulizer tip. Peak asymmetry was reduced when the mobile phase was heated, but this also decreased ion currents, thus the operating temperature was often a compromise between conservation of chromatographic peak integrity and good MS sensitivity. In reality, most of the trapped sample appear to be retained on the outer surface of the capillary nebulizer end, and a liquid sheet of methanol, or another suitable polar solvent, could continuously wash this surface. Post SFC column addition of the polar additive would also provide ES ionization more independent of the chromatographic requirements of the mobile phase composition.

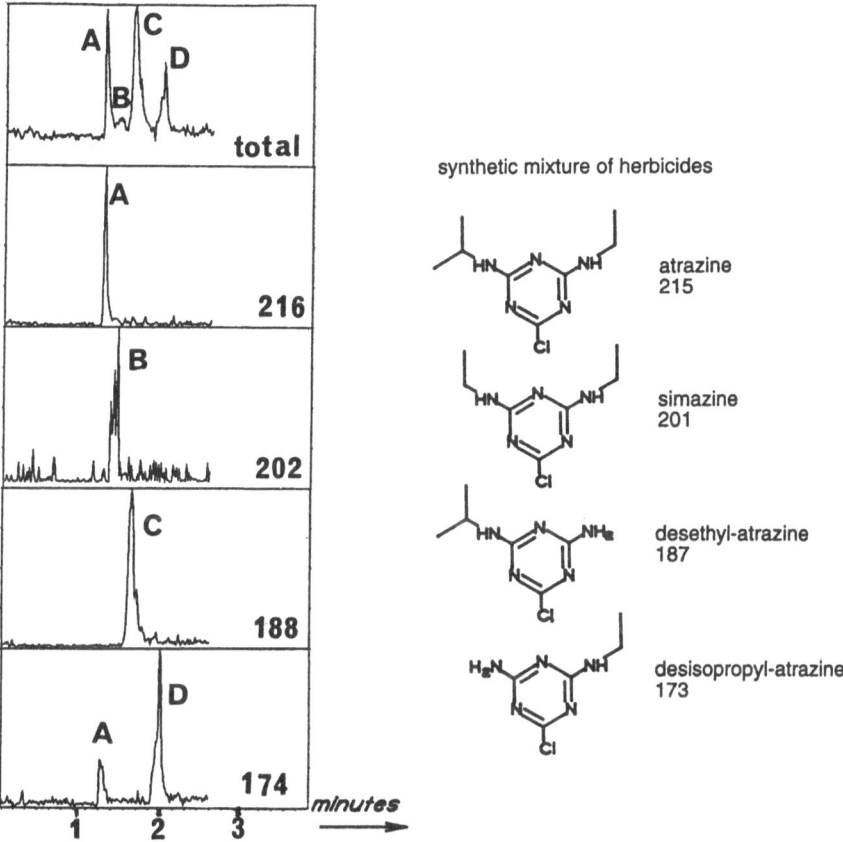

Figure 5. SFC/API/MS analysis of a synthetic mixture of herbicides. The signal was obtained by atmospheric pressure electrospray ionization from charged liquid droplets in the decompressing solution.

Pure electrospray ionization for SFC/MS coupling has both advantages and disadvantages that remain to be solved, in particular, the influence of the low temperature restrictor on chromatographic resolution, and the yet unproven capability to ionize very heavy solutes. A detailed comparison between the results from corona discharge ionization and pure electrospray ionization must also be done.

## 7. Conclusion

Atmospheric ionization sources, with ionization in either the gas phase or liquid phase, could be quite attractive for SFC/MS coupling because of their independence from the fluid flow rate and their high sensitivity to many solutes, on the other hand, both ionization methods are not independent from the SFC mobile phase and ion source atmosphere compositions. Nevertheless, as the two ionization methods can be easily adapted on the same general instrument, SFC/API/MS offers new promising openings for SFC/MS coupling.

At one time, SFC/MS was often assumed to be a substitute for LC/MS because of difficulties faced by established LC/MS systems such as the direct liquid introduction, the moving belt interface [47], the thermospray interface [43], – and because of the apparent simplicity of direct capillary column SFC/MS coupling. Today, the situation is reversed, and it is the success of recent LC/MS techniques that could give a new start to SFC/MS coupling.

## 8. References

1. Arpino, P.J. (1992) *in* Mass Spectrometry in the Biological Sciences: a tutorial (Gross, M.L. Ed), pp. 269-280, Kluwer Academic Publishers, Dordrecht.
2. Pinkston, J.D. (1992) *in* Analysis with supercritical fluids: extraction and chromatography (Wenclawiak, B. Ed.) pp. 151-177, Springer-Verlag, Heidelberg.
3. Arpino, P.J., Sadoun, F., Virelizier, H. (1993) *Chromatographia,* **36,** 283.
4. Randal, L.G., Wahrhaftig, A.L. (1978) *Anal. Chem.,* **50,** 1705.
5. Cousin J., Arpino, P.J. (1987) *J. Chromatogr.,* **398,** 125.
6. Arpino, P.J., Dilettato, D., Nguyen, K., Bruchet, A. (1990) *J. High Resolut. Chromatogr. Chromatogr. Commun.,* **13,** 5.
7. Smith, R.D., Kalinoski, H.T., Udseth, H.R. (1987) *Mass Spectrom. Rev.,* **6,** 445.
8. Olesik, S.V. (1991) *J. High Resolut. Chromatogr. Chromatogr. Commun.,* **14,** 5.
9. Arpino, P. (1989) *Advances Mass Spectrom.,* 11B, 1170.
10. Blum, W., Grolimund, K., Jordi, P.E., Ramstein, P. (1988) *J. High Resolut. Chromatogr. Chromatogr. Commun.,* **11,** 441.
11. Kalinoski, H.T., Hargiss, L.O. (1989) *J. Chromatogr.,* **474,** 69.
12. Pinkston, J.D., Delaney, T.E., Bowling, D.J. (1990) *J. Microcolumn Sep.,* **2,** 181.
13. Pinkston, J.D., Delaney, T.E., Bowling, D.J., Chester, T. (1991) *J. High Resolut. Chromatogr. Chromatogr. Commun.,* **14,** 401.
14. Snyder, J.M., Taylor, S.L., King, J.W. (1993) *J. Am. Oil Chem. Soc.,* **70,** 349.
15. Arpino, P.J., Cousin, J. (1987) *Rapid. Commun. Mass Spectrom.,* **1,** 29.
16. Strubinger, J.R., Song, H., Parcher, J.F. (1991) *Anal. Chem.,* **63,** 98.
17. Strubinger, J.R., Song, H., Parcher, J.F. (1991) *Anal. Chem.,* **63,** 104.
18. Murugaverl, B., Voorhees, K.J., DeLuca, S.J. (1993) *J. Chromatogr.,* **633,** 195.
19. Lee, E.D., Hsu, S.H., Henion, J.D. (1988) *Anal. Chem.,* **60,** 1990.
20. Houben, R.J., Leclercq, P.A., Cramers, C.A. (1991) *J. Chromatogr.,* **554,** 351.
21. Koski, I.J., Markides, K.E., Richter, B.E., Lee, M.L.(1992) *Anal. Chem.,* **64,** 1669.
22. Pinkston, J.D., Hentschel, R.T. (1993) *J. High Resolut. Chromatogr. Chromatogr. Commun.,* **16,** 271.
23. Jablonska, A., Hansen, M., Ekeberg, D., Lundanes, E. (1993) *J. Chromatogr.,* **647,** 341.
24. Todd, J.F.J., Mylchreest, I.C., Berry, A.J., Games, D.E., Smith, R.D. (1988) *Rapid Commun. Mass Spectrom.,* **2,** 55.

25. Pinkston, J.D., Delaney, T. E., Morand, K.L., Cooks, R.G. (1992) *Anal. Chem.*, **64**, 1571.
26. MacKay, G.A., Reed, G.D. (1991) *J. High Resolut. Chromatogr. Chromatogr. Commun.*, **14**, 537.
27. Huang, E.C., Jackson, B.J., Markides, K.E., Lee, M.L. (1988) *Anal. Chem.*, **60**, 2715.
28. Huang, E.C., Jackson, B.J., Markides, K.E., Lee, M.L. (1990) *J. Microcol. Sep.*, **2**, 88.
29. Baumeister, R., West, C.D., Ijames, C.F., Wilkins, C.L. (1991) *Anal. Chem.*, **63**, 251.
30. Blum, W., Ramstein, P., Grether, H.J. (1990) *J. High Resolut. Chromatogr. Chromatogr. Commun.*, **13**, 290.
31. Smith, R.D., Udseth, H.R. (1987) *Anal. Chem.*, **59**, 13.
32. Kalinoski, H.T., Smith, R.D. (1988) *Anal. Chem.*, **60**, 529.
33. Kalinoski, H.T., Wright, B.W., Smith, R.D. (1988) *Biomed. Environment. Mass Spectrom.*, **15**, 239.
34. Bücherl, T., Gruner, A., Palibroda, N., Wolff, E., Mapelli, G.P. (1993) *in* Proceedings of the Second European Symposium on Analytical Supercritical Fluid Chromatography and Extraction, Riva del Garda, May 27-28, 1993, P. Sandra and K. Markides, Eds., pp 87-93, Huethig, Heidelberg.
35. Chapman, J.R. (1988) *Rapid Commun. Mass Spectrom.*, **2**, 6.
36. Berry, A.J., Games, D.E., Mylchreest, I.C., Perkins, J.R., Pleasance, S. (1988) *Biomed. Environ. Mass Spectrom.*, **15**, 105.
37. Raynor, M.K., Kithinji, J.P., Bartle, K.D., Games, D.E., Mylchreest, I.C., Lafont, R., Morgan, E.D., Wilson, I.D. (1989) *J. Chromatogr.*, **467**, 292.
38. Perkins, J.R., Games, D.E., Startin, J. R., Gilbert, J. (1991) *J. Chromatogr.*, **540**, 239.
39. Perkins, J.R., Games, D.E., Startin, J. R., Gilbert, J. (1991) *J. Chromatogr.*, **540**, 257.
40. Morgan, E.D., Murphy, S.J., Games, D.E., Mylchreest, I.C. (1988) *J. Chromatogr.*, **441**, 165.
41. Musser, S.M., Callery, P.S. (1990) *Biomed. Environ. Mass Spectrom.*, **19**, 348.
42. Scalia, S., Games, D.E. (1992) *Org. Mass Spectrom.*, **27**, 1266.
43. Arpino, P. (1990) *Mass Spectrom. Rev.*, **9**, 631.
44. Berry, A.J., Games, D.E., Perkins, J.R. (1986) *J. Chromatogr.*, **363**, 147.
45. Edlund, P.E., Henion, J.D. (1989) *J. Chromatogr. Sci.*, **27**, 274.
46. Ramsey, E.D., Perkins, J.R., Games, D.E., Startin, J.R. (1989) *J. Chromatogr.*, **464**, 353.
47. Arpino, P. (1989) *Mass Spectrom. Rev.*, **8**, 35.
48. Arpino, P. (1990) *Fresenius J. Anal. Chem.*, **337**, 667.
49. Matsuura, K., Takeuchi, M., Nojima, K., Kobayashi, T., Saito, T. (1990) *Rapid. Commun. Mass Spectrom.*, **4**, 381.
50. Sim, P.G., Elson, C.M., Quilliam, M.A. (1988) *J. Chromatogr.*, **445**, 239.
51. Nizery, D., Thiebaut, D., Caude, M., Rosset, R., Lafosse, M., Dreux, M. (1989) *J. Chromatogr.*, **467**, 49.
52. Rokushika, S., Hatano, H., Hill, H.H.Jr. (1987) *Anal. Chem.*, **59**, 8.
53. Huang, M.X., Markides, K.E., Lee, M.L. (1991) *Chromatographia*, **31**, 163.
54. Fujii, T. (1992) *Anal. Chem.*, **64**, 775.
55. Huang, E., Henion, J., Covey, T. (1990) *J. Chromatogr.*, **511**, 257.
56. Anacleto, J.F., Ramaley, L., Boyd, R.K., Pleasance, S., Quilliam, M.A., Sim, P.G., Benoit, F.M. (1991) *Rapid. Commun. Mass Spectrom.*, **5**, 149.
57. Matsumoto, K., Nagata, S., Hattori, H., Tsuge, S. (1992) *J. Chromatogr.*, **605**, 87.
58. Tyrefors, L.N., Moulder, R.X., Markides, K.E. (1993) *in* Proceedings of the Second European Symposium on Analytical Supercritical Fluid Chromatography and

Extraction, Riva del Garda, May 27-28, 1993, P. Sandra and K. Markides, Eds., pp 203-205, Huethig, Heidelberg.

59. Tyrefors, N.L., Moulder, R.X., Markides, K.E. (1993) *Anal. Chem.,* accepted for publication.

60. Fenn, J.B., Mann, M., Meng, C.K., Wong, S.F., Whitehouse, C.M. (1990) *Mass. Spectrom. Rev., 9,* 37.

61. Henry, K.D., Quinn, J.P., McLafferty, F.W. (1991) *J. Am. Chem. Soc.,* 113, 5447.

62. Pleasance, S., Quilliam, M.A., De Freitas, A.S.W., Marr, J.C., Cembella, A.D. (1990) *Rapid. Commun. Mass. Spectrom., 4,* 206.

63. Covey, T.R., Bonner, R.F., Shushan, B.I., Henion, J.D. (1988) *Rapid Commun. Mass Spectrom., 2,* 249.

64. Ikonomou, M.G., Blades, A.T., Kebarle, A.T.P. (1990) *Anal. Chem., 62,* 957.

65. Sadoun, F., Virelizier, H., Arpino, P.J. (1993) *J. Chromatogr., 647,* 351.

66. Whitehouse, C.M.; Dreyers, R.N.; Fenn, J.B. (1985) *Anal. Chem., 57,* 675.

# GC-MS ANALYSIS OF VOLATILE COMPOUNDS OF MESIR

Yaşar HIŞIL, Neriman BAĞDATLIOĞLU, Semih ÖTLEŞ
Department of Food Engineering
University of Ege
Bornova, Izmir 35100
Turkey

ABSTRACT. Gas chromatography - mass spectrometry (GC-MS quadrupole, EI) with multiple ion detection analysis was performed in order to obtain information on the volatile compounds prepared from Mesir, an unknown special blend of forty-one spices. Analysis by gas chromatography coupled with mass spectrometry led to the identification (mass spectra and retention characteristics) of the various volatile substances which have not yet been reported in Mesir.

## 1. Introduction

Mesir is, in fact, considered a natural blend containing forty-one different spices, delicious to taste and exotic in flavour. It is no trading commodity and can be found producing principally in the Western Turkey (Manisa), only in April. Mesir, a traditional special mixture for Turkish people since hundred-years, is being used in human dietary, that possess different physiological and medicinal properties (1). Table 1 contains the summing up of the spices used in the preparation of Mesir.

Volatile compounds produced during growth of plant are responsible for the characteristic aroma of spices. Those compounds are end-products of metabolic pathways which vary due to variety and species, and are influenced by the factors such as ecological and agronomic conditions. Although the volatile compounds of spices have been extensively studied, all these results on the composition of volatile compounds are difficult to compare. This can be explained by the different extraction procedures or by the varietal aspects (2 - 6).

The combination of gas chromatography with mass spectrometry (GC-MS) is one of today's most powerful analitical techniques owing to its ability to provide sensivite and selective analysis of volatile compounds on the basis of mass separation. Gas chromatography is usually used to identify the volatile substances responsible for their characteristic taste and odor; the identities are then confirmed by mass spectrometry (7 - 18).

## 2. Experimental Section

*Materials.* The samples of Mesir were obtained from wholesale distributors in Manisa.

*Extraction of volatile constituents.* 60 g of Mesir was dissolved in 500 ml distilled water and placed in a 1 L round-bottomed flask. The sample flask was connected to the Clevenger distillation apparatus and 1 ml n-pentane was placed in graduated section. The distillation was run for exactly 1 hour. The volatile

R. M. Caprioli et al. (eds.), Mass Spectrometry in Biomolecular Sciences, 475–482.
© 1996 Kluwer Academic Publishers.

*Table 1.* The spices in Mesir

--------------------------------------------------------------

| | |
|---|---|
| Anason | Fructus Anisi vulgaris, Umbelli |
| Hindistan Cevizi ve | |
| Besbase | Macis, Myristicaceae |
| Hindistan Çiçeği | Clove, Cassia |
| Çivit | Indigo, Leguminosae |
| Çöpçini | Rhizoma chinae, Liliceae |
| Çörek Otu | Semen nigellae, Ranunculaceae |
| Dar-ì Fülfül | Piper longum, Piperaceae |
| Hardal Tohumu | Semen sinapis, Crociferae |
| Havlican | Rhizoma galangae, Zingiberaceae |
| Hìyarşenbe | Fructus cassiae, Leguminosae |
| Kakule | Fructus cardomoni, Zingiberaceae |
| Kal Barda | Brigth Scarlet Color, Galibarda |
| Karabiber | Fructus piperis nigri, Piperaceae |
| Karanfil | Caryophyllus, Myrtaceae |
| Kebabe | Fructus cubebae, Piperaceae |
| Kimyon | Fructus cumuni, Umbelliferae |
| Kirìm Tartar | Potassi Tartaras |
| Kişniş | Fructus coriandri, Umbelliferae |
| Limon Tuzu | Citric acid |
| Iksir | Elixir |
| Mai-Leziz | Eau dauce, Eav fraiche |
| Meyan Balì | Succus liguiritiae, Leguminosea |
| Portakal Kabuğu | Ecorce d'orange |
| Ravend Kökü | Rhizoma rhei, Polygonaceae |
| Resene | Fructus fueniculi, Unbelliferae |
| Safran | Terminalia citrine, Combretaceae |
| Sinameki | Follium sennae, Leguminosae |
| Şamlì (Şaşlì) | Dame d'onze heures |
| Şeker | Sucros |
| Tarçìn | Cortex cinnamoni, Lauraceae |
| Tarçìn Çiçeği | Flores cinnamoni |
| Teke Mersini (Tohumu) | Airelle, raisin debois, uyrtille |
| Tiryak | Theriaque |
| Ud-ül Kahar | Radix pyrethri, Compositae |
| Vanilya | Fructus vanillae, Orchidaceae |
| Yeni Bahar | Fructus pimentae, Myrtaceae |
| Zencefil | Rhizoma zingiberis, Zingiberaceae |
| Zerde Çöp | Rhizoma curcumea, Zingiberaceae |
| Zulùmba | Rhizoma zedoariae, Zingiberaceae |

substances were collected in 1 ml n-pentane. 1 µl aroma fraction was injected to GC and GC-MS.

*Gas chromatography.* Gas chromatographic analyses were performed using a Shimadzu GC-14A equipped with a flame ionization detector (FID) and a fused silica capillary column DB-1701 FSC (30 m x 0.324 mm i.d., $d_f$ = 1 µm). The helium was used as the carrier gas (0.8 atm). A splitter was used at a split ratio of 1 : 30. The column temperature was programmed from 30°C (10-min hold) to 50°C at 2°C / min and after 5 min hold in 50°C to 225°C at 3°C / min and then 7-min hold in 225°C. The temperatures of injection port and detector (FID) were 200°C and 250°C, respectively.

*Gas chromatography - Mass spectrometry (GC-MS).* Gas chromatographic peaks were identified by gas chromatography-mass spectrometry (GC-MS) using a Shimadzu GC/MS - QP2000A coupled with a Shimadzu GC-14A equipped with the same capillary column, directly connected to the ion source, as in the GC analyses. Mass spectra were recorded with an ion source energy of 70 eV. The temperature of the ion source was 250°C. Sample components were tentatively identified by mass spectrum matching with a mass spectral library collection (NIST, NIH and EPA).

## 3. Principles of MS-Quadrupole

The quadrupole instruments built for general use were introduced about 20 years ago. A quadrupole type mass spectrometer consist of an ion source ionizes the sample and has a lens system to focus the ions into the quadrupole mass filter confines ions in two dimensions that ions should be accelerated through the length of the device by external electric fields (ions travel at a slow speed along the x axis of the electrodes, and they are forced by the r.f. and d.c. field to undergo transverse motion in the x or y direction. Here, an external ion source is necessary and the length of time is fixed that ions reside within the quadrupole field.

If the conditions are stable, a and q can be defined by the following relations :

$$a = 8 z u / m r_0 2 w^2 \quad (1)$$

$$q = 4 z v / m r_0 2 w^2 \quad (2) \quad (w = 2 \Pi f)$$

The subscripts z and r represent motion between and perpendicular to the andcaps, respectively, u is the d.c. bias on the endcap electrodes, v is amplitude of r.f. voltage, $r_0$ is the radius of the ring electrode, and m is the mass number (19 - 20).

Mass analyzing quadrupole is set to pass an ion of a single m / z ratio to monitor a particular reaction.

$$m / z = k \text{ (constant) } v / r_0 2 f^2$$

Quadrupole mass spectrometer have a great potential for use, especially by scientist interested in biomolecules because of their unique configuration.

*Table 2.* Volatile compounds identified by GC-MS in Mesir

| $R_t 1$ | $P_n 2$ | Compound |
|-----|-----|----------|
| 30.13 | 32 | 3-isothiocyanato 1-propene |
| 30.73 | 33 | 1,6-octadien-3-ol, 3,7-dimethyl-propanoate |
| 35.76 | 42 | D-limonene |
| 37.23 | 46 | cineole (van) |
| 45.60 | 62 | 1,6-octadien-3-ol,3,7-dimethyl |
| 49.56 | 71 | 3-cyclohexan-1-ol,4-methyl-1-(1- methylethyl) |
| 51.46 | 78 | 3-cyclohexene-1-methanol,.alpha.,.alpha.4-trimethyl |
| 55.06 | 91 | (2-propynyloxy)-benzen |
| 55.23 | 92 | 4-(1-methylethyl)-benzaldehyde |
| 56.23 | 95 | 1-methoxy-4-(2-propenyl),benzen |
| 59.40 | 108 | 3-phenyl,2-propenal |
| 61.56 | 117 | 2-metoxy-4-(2-propenyl),phenol |
| 62.43 | 119 | 1,2-dimetoxy-4-(1-propenyl),benzen |
| 62.70 | 121 | alpha cubebene |
| 67.86 | 143 | 4-metoxy-6-(2-propenyl)-1,3-benzodioxole |
| 68.96 | 148 | 2H-1-benzopyran-2-one |
| 72.10 | 161 | 1-naphthalenol,1,2,3,4,4a,7,8,8a- octohydro-1,6-dimethyl-4-(1-methyl-ethyl) |
| 78.86 | 187 | tetradecanoik acid |
| 85.83 | 208 | hexadecanoik acid |

[1] retention times of peaks in ion chromatogram
[2] peak numbers in GC chromatogram

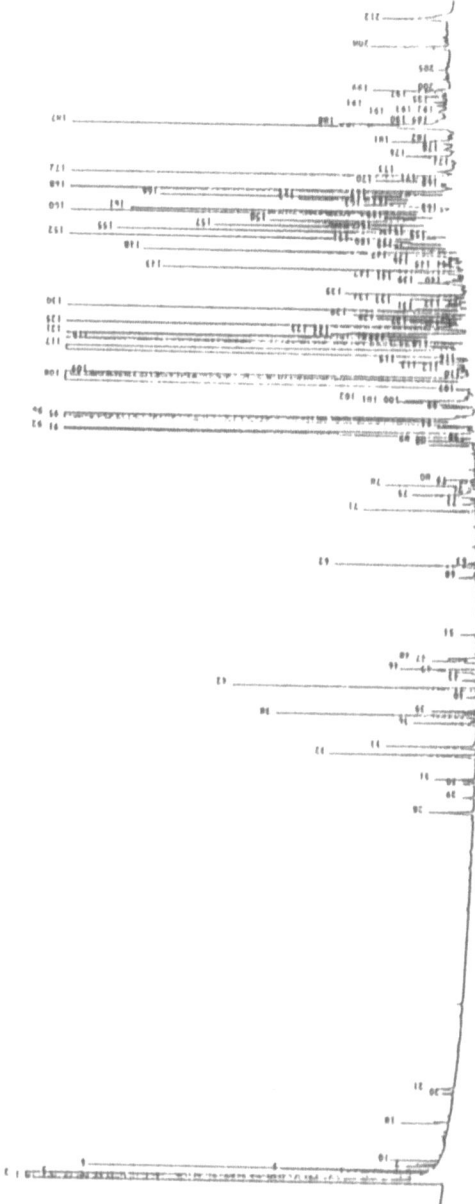

Figure 1. Gas chromatogram of volatile compounds in Mesir.

480

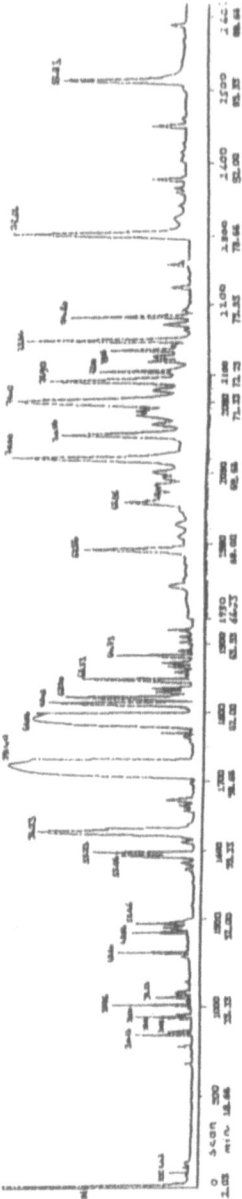

Figure 2. Reconstructed ion chromatogram from GC-MS analysis of Mesir volatile compounds.

## 4. Conclusions

The qualitative examination by GC-MS (Quadrupole) of the volatile compounds in Mesir was realized by steam distillation sampling method with Clevenger apparatus. Those compounds concentrated in 1 ml of n-pentane was analysed by capillary gas chromatography using DB-1701 column, FID detector, and split mode (split ratio : 1/30). Figure 1 shows a GC/FID chromatogram of volatile compounds isolated by steam distillation. About two hundred peaks, including very small ones, were observed on GC/FID chromatogram. The total ion current chromatograms of volatile substituents of Mesir obtained by GC-MS (EI) are shown in Figure 2. The constituents identified by the reconstructed chromatograms on the basis of the total ionization current of the volatile compounds in Mesir are summarized in Table 2. 19 peaks found in significant amounts were selected for identification. We have found terpens, aromatic hydrocarbons, aldehydes, ketones, alcohols, esters, phenols, and even carboxylic acids. The peak 108 (3-phenyl, 2-propenal) was a very large peak followed by peaks 117 [2-methoxy-4-(2-propenyl), benzen] and 95 [1-methoxy-4-(2-propenyl), benzen]. The volatile compounds of various spices have been investigated by many scientist. Naturally some spices show very characteristic volatile compound profiles, e.g. cineol, linalool, methyl eugenol in star anise (I. verum); carvone in caraway; thujone, camphor in sage oil; cineol, α-terpinyl acetate in cardamom; l-pinene, myrcene in J. rigida; α and ß-pinene, limonene, ß-caryophyllene in black pepper. However, even such spices contain a lot of other components. The amounts of these components also change to the harvesting season, the maturation and the conditions of cultivation area. No published data in the literature has been found for the volatile compounds in Mesir to compare with our data. Therefore, it must be done additional studies on the compounds of Mesir.

## 5. Acknowledgements

We gratefully acknowledge the assistance of many colleques in Health Protection Institute with the measurement of the mass spectra.

## 6. References

1. Bayat, A.H. 1980. Man. Tur. Der. Yay., 2, 1-41.
2. Aishima, T. The shelf life of foods, 755-774.
3. Fröhlich, O.; Schreier, P. 1990. J. Food Sci., 55, 176-180.
4. Pruthi, J.S. 1980. Spices and condiments, New York.
5. Quere, J.L.L.; Latrasse, A. 1990. J. Agric. Food Chem., 38, 3-10.
6. Whitfield, F.B.; Mottram, D.S.; Brock, S.; Puckey, D.J.; Salter, L.J. 1988. J. Sci. Food Agric., 42, 261-272.
7. Arpino, P.J. 1990. NATO-ASI, 253-268.
8. Burlingame, A.L.; Maltby, D.; Russell, D.H.; Holland, P.T. 1988. Anal. Chem., 60, 294-342.
9. Cerny, R.L.; Gross, M.L. 1990. NATO-ASI, 407-412.
10. Douillard, C.; Guichard, E. 1989. Sci. Alim., 9, 53-76.

11. Geno, P.W. 1990. NATO-ASI, 133-144.

12. Marino, G.; Siciliano, R.; Pucci, P.; Ferranti, P.; Malorni, A. 1990. NATO-ASI, 333-342.

13. Mihara, S.; Tateba, H.; Nishimura, O.; Machii, Y.; Katsumi, K. 1987. J. Agric. Fd. Chem., 35, 532-537.

14. Moore, W.T.; Caprioli, R.M. 1990. NATO-ASI, 229-252.

15. Paschke, A.; Herbel, W.; Steinhardt, H.; Franke, S.; Francke, W. 1992. J. High Res. Chrom., 827-833.

16. Sindona, G. 1990. NATO-ASI, 383-406.

17. Trainor, J.R.; Derrick, P.J. 1990. NATO-ASI, 3-28.

18. Vernin, G.; Metzger, J.; Suon, K.N.; Fraisse, D.; Ghiglione, C.; Hamoud, A.; Parkanyi, C. 1990. Lebensm. Wiss. u. Technol., 23, 25-33.

19. Dorey, R.C. 1990. NATO-ASI, 79-92.

20. Wysocki, V.H. 1990. NATO-ASI, 59-78.

# THE APPLICATION OF MASS SPECTROMETRY TO FOOD AND NUTRITION RESEARCH

R. SELF
*School of Chemical Sciences*
*University of East Anglia*
*Norwich*
*NR4 7TJ*
*UK*

F.A. MELLON
*Institute of Food Research*
*Norwich Laboratory*
*Norwich Research Park*
*Colney Lane*
*Norwich*
*NR4 7UA*
*UK*

B.A. McGAW
*School of Applied Sciences*
*The Robert Gordon University*
*Aberdeen*
*AB1 1HG*
*UK*

A.G. CALDER, G.F. LOBLEY and E. MILNE
*Rowett Research Institute*
*Greenburn Road*
*Bucksburn*
*Aberdeen*
*AB2 9SB*
*UK*

ABSTRACT. A selection of topics, illustrating the use of recently-developed mass spectrometric methods in food, agriculture and nutrition research, has been made from our own research programmes. No attempt has been made to present a critical review of the state of the art; instead a wide range of applications is intended to stimulate ideas and applications of these powerful analytical procedures to the solution of problems in the food and nutritional science area. The dialogue is restricted to research rather than routine analysis, although several of the methods described are gaining acceptance as a consequence of the success of recent inter-laboratory collaborative trials.

*R. M. Caprioli et al. (eds.), Mass Spectrometry in Biomolecular Sciences*, 483–515.
© 1996 *Kluwer Academic Publishers.*

## 1. Introduction

The main impetus behind the development of organic mass spectrometry was probably founded in the petroleum industry`s requirement, in the late 1940s and early 1950s, to elucidate the structure of apolar and low molecular weight (by today's standards, <200 daltons),volatile fractions from oil and gas products ideal for gas phase ionisation by electron bombardment. It was soon realised that food volatiles, i.e. those substances comprising aroma and contributing to flavour, were also low molecular weight, but unlikely to be apolar. In fact many aroma components are alcohols, acids, aldehydes, esters, etc and early experiments showed that the electron ionisation (EI) mass spectra of these compounds contained informative, structurally significant fragment ions. The atmosphere at the time was caught by participants in the 3rd NATO Advanced Study Course in Food Research held in Cambridge in 1962 [1]. GC/MS, first described by Holmes and Morrell as early as 1957 [2], was expensive and only available in one or two laboratories, but flavour chemists were quick to recognise the enormous advantages of the on-line separation and identification of extremely complex mixtures of both synthetic and natural volatile substances. Consequently, mass spectrometry became an essential tool in the natural products laboratory. The limitations imposed by low volatility were addressed, by forming chemical derivatives of many of the less volatile compounds. This was especially evident in the design of quantitative assays for naturally-occurring constituents or for man-made additives and contaminants, spreading the technology into the food quality control and safety areas. But it was the advent of practical, solid phase, particle desorption ionisation in the 1980s, e.g. fast atom bombardment mass spectrometry (FABMS) [3], which opened up mass spectrometry to wide ranging studies of the composition of food and agricultural products [4] with work on the verification of the molecular structures of lipids, peptides, carbohydrates, vitamins, antinutrients and other biologically-active components.

The more recent applications in food science, of tandem mass spectrometry (MS/MS), liquid chromatography/mass spectrometry (LC/MS), pyrolysis mass spectrometry, inorganic trace analysis as well as the established EI and chemical ionisation (CI) methods for flavour research, were also reviewed in 1987 [5]. And capillary electrophoresis/mass spectrometry (CE/MS) is pushing into new and exciting areas in molecular biology and microphysiology. However, tandem mass spectrometry was probably the most significant development at this time, allowing the return to fundamental studies of the relationship between fragmentation and structure during controlled, collision-induced dissociation of structurally significant primary ions. A detailed example is the elucidation of the tandem mass spectra of pyridinoline and related compounds, the inter-molecular cross-linking substances formed during the maturation of extracellular collagen fibres.

In common with all other biosciences, food and agricultural research calls into play at one time or another many different forms of mass spectrometry. Applications of stable isotope ratio mass spectrometry (SIRMS) and inorganic nutrient studies using stable isotope tracers are now essential techniques for the measurement of protein turnover and energy expenditure as well as for the quantitative analysis of inorganic and organic food components, additives and contaminants.

## 1.1 BIOLOGICAL STUDIES

Food and agricultural sciences depend upon fundamental explorations into the structure of matter and the relationship this has with the function of the food (or feed) in the supply of nutrients to the human or animal recipient. These relationships are invariably complex and simpler models are used to aid research. In a way, an isolated component is a model because it is removed from the complex interactions occurring in the cell or similar local environment, and therefore its role in isolation must be viewed with caution. Nevertheless, structural studies have moved apace in the last 10 years, especially now that x-ray crystallography and NMR techniques are more readily available. However, their requirement for quite large quantities of material (milligram amounts) allows mass spectrometry to maintain its vanguard role in recognising novel compounds and confirming the general (primary) structures of substances of interest at the sub-nanogram level. For specific molecules, mass spectrometry is exquisitely sensitive and qualitative and quantitative probes have been designed at the femtomole level of detection.

The continuing interest in high instrumental sensitivity and specificity in the nutritional sciences relates to the need to identify compounds, from highly specific locations, which control the mechanisms of food intake, absorption and transport, at the physiological level. This, for contemporary analysts, means reaching down to sub-femtomole levels.

## 1.2 INSTRUMENTAL APPROACH

In studies of nutrients and their interactions, it is apparent that only the highest resolving power in pre-analytical extraction and purification procedures will permit unimpeded examination of the target analytes, and therefore, the biochemistry and physiology of food and agricultural science relies on rigorous protocols to separate the compounds of interest from their natural matrices. While GC/MS continues to play a major role in all areas where stable volatile products are of interest, LC/MS is being used in the detection of involatile metabolites down to the picogram level. As the mass of the analyte increases, so does the number of isobaric alternatives possible at that mass. Thus the continuing improvement in resolving power of HPLC columns is mandatory if meaningful interpretations of mass spectra are to be made. And even so, the presence of isomers in the mixture can only be detected if the chromatographic stationary phases are capable of resolving them prior to mass spectral analysis. Nevertheless, the attainment of higher mass ranges is outstripping the increase in resolving power and, for biomolecules of greater than 10,000 daltons mass, it is often necessary to pre-purify the compound of interest to avoid confusion caused by interfering co-chromatographed contaminants. Single pure compound studies using fast atom bombardment (FAB) and other desorption ionisation methods, including matrix-assisted laser desorption ionisation (MALDI), are being used to great effect in nutritional studies on novel metabolic systems containing complex mixtures of protein isoforms of unknown biological activity. A detailed example is given later of a study of the metallothionein isoforms. Ultimately, the need for rapid analyses has overtaken the use of labour-intensive pre-purification processes and effort is switching to the combination of capillary electrophoresis (CE) - with its inherent $10^6$ theoretical plate resolving power and rapid analysis times - with high molecular weight mass spectrometry. Alternatively, the creation of multiply-charged species using a combination of electrospray ionisation and medium mass range instruments allows high mass molecular weight determination with 0.1-0.01% accuracy of mass measurement.

Both inorganic and organic nutrients are of interest to growers, food processors and nutritionists. To provide a rounded approach to the analytical biochemistry of micronutrients essential to the well-being of animals and humans, the development of methods for the measurement of inorganic trace elements has taken place over the past decade. The accuracy and precision of thermal ionisation mass spectrometry (TIMS) and the complementary, semi-quantitative, broad survey capabilities of inductively-coupled plasma mass spectrometry (ICPMS) will be discussed in detail here. Organic mass spectral techniques have been fully described in other chapters.

Perhaps one of the most significant developments of recent times has been the improvement in the range and sensitivity of focal plane detectors and the latest electro-optical array detectors based on a charge-coupled device provide an increase of two orders in sensitivity over that of a conventional point detector while covering the whole mass range rapidly. When these arrays are fitted to tandem mass spectrometers, the overall sensitivity is improved to the picomole level, facilitating structural studies on trace metabolites such as the glycoalkaloids, a group of nitrogenous, steroidal glycosides with potential for human toxicity. This, and other specific examples from our own laboratories will be used to illustrate applications based on the latest developments in instrumentation.

## 2. Food and Agricultural Research

### 2.1 FLAVOURS - INSTRUMENT AND SENSORY EVALUATION

The instrumental opportunities made available by GC/MS were quickly exploited in the 1960s by groups at the USDA laboratories and elsewhere. At a Symposium on Foods at Oregon State University in 1966 [6] contributors described their research aimed at identifying various flavour components. By the end of the meeting it was obvious that similar chemicals were contributing to flavours as diverse as fish, milk, vegetables and beef, albeit in different proportions. Nevertheless, GC/MS studies on food flavour components accelerated and diversified for another twenty years e.g. over 700 components of coffee flavour have been tabulated [7], but even today the function of these volatile substances is not well understood, and the research emphasis has moved toward the elucidation of interactions between volatiles and receptors in the olfactory epithelium.

The analysis of interstitial cellular components using microdialysis [8] and the detection of components in single cells using micropipettes with CE [9], might eventually lead to a technique sensitive enough to measure physiological changes in response to the adsorption of the stimulant molecules at the receptor surface. If so, mass spectrometry in the form of CE/MS will feature once again in the fundamental understanding of the phenomenon of olfaction. Already microdialysis has been combined with MS/MS [10] and continuous flow FAB [11] and the sensitivity of CE-ESI/MS has been reported to be in the attomole range [12].

Before leaving the area of flavour research it is apposite to mention the problem of stoichiometric perturbation caused by the act of measurement itself. The possibility of rearrangement or at least disturbance of the stoichiometry of the natural system has to be considered whenever relatively energetic processes, like adsorption chromatography, are involved in the analysis. Although direct analysis of headspace volatiles [13] has been found to be reliable for many years, the conventional methods of concentrating flavours involving steam distillation, solvent extraction etc, have been supplemented recently by supercritical fluid extraction (SFE) in readiness for GC/MS analysis [14]. Nevertheless SFE imposes severe limits on the classes of chemical compounds which can be extracted successfully. The use of combined sensory and instrumental assays based on split flow techniques help to confirm the olfactory quality of the composition of the mixture being analysed [15].

The combination of gas chromatography with SIRMS [16] to measure $^{13}C/^{12}C$ ratios with high precision has potential for use in validating the origin of flavours as natural or artificial based on minor changes in $^{13}C/^{12}C$ ratio caused by the different mass discrimination introduced by the $C_3$ and $C_4$ metabolic cycles in plants. Other methods of validation are based on the GC/MS recognition of impurity peaks associated with the synthetic process.

## 2.2 PEPTIDES, METALLOPROTEINS AND COLLAGEN CROSSLINKS

2.2.1 *Peptides.* The application of mass spectrometry to peptide and small protein analysis has progressed rapidly since the introduction of ionisation methods which enabled large biopolymer molecules to be desorbed from the surfaces of polar liquid matrices with relatively little excess energy. An intimate knowledge of the chemistry and the immediate past history of the isolation and purification processes has allowed many peptide sequences to be elucidated by mass spectrometry first to determine the mass and then establish the amino acid order from the series of low abundance fragment ions usually produced. Hydrophobic peptides are more likely to give good spectra than hydrophilic peptides, which are susceptible to suppression by surface-active competitors. When isolated in high purity, hydrophilic peptides also give good mass spectra. In either case, the efficacy of the mass spectral approach is enhanced by the use of MS/MS methods, when, collision-induced decomposition of the molecular ion often yields sufficient information to sequence the whole amino acid chain using well-established fragmentation rules. Several reviews summarise progress in this field: the article by Biemann [18] covers most of the applications relevant to nutritional science. Experience in working with physiological fluids has been reported for the recognition of peptides from the pituitary gland [19]. With so many signalling compounds known to be peptidic in nature, nutritionists are beginning to use mass spectrometric techniques to unravel the extremely complex mixtures emanating from target organs/glands of the body [20]. Problems can occur when peptides react with liquid matrices to form adduct ions with masses greater than the molecular ion [21]. Such ions might help to confirm the molecular weight of a known peptide, but equally might cause serious confusion if unknown species are being studied.

Specific applications of mass spectrometry to food science were reviewed in 1987 [5]. Since then, the techniques have developed substantially and MS/MS with charge-coupled devices and microchannel plate scanning-array detection systems [22] promises to produce femtomole sensitivity for amino acid sequence determinations. When complete sequence information is not available in the mass spectrum, it is possible to hydrolyse samples in time course studies in order to acquire additional information through the formation of families of sub-peptides, each capable of entering the gas phase as a protonated or cationised molecular ion under desorption ionisation conditions, thus "filling in" the gaps often present in single compound mass spectra [23]. A knowledge of the kinetics of the 400 or so peptide bonds constituting the amino acid chains of proteins is helpful. The unusual 34 amino acid residue peptide nisin has been studied extensively by mass spectrometry [24-27]. This antibiotic, which is widely used as a food preservative, is isolated from strains of *Streptococcus lactis*. The presence of two minor analogues was observed by both teams and one component was assigned the structure C-des Dha-Lys nisin by PDMS [25] and confirmed by FABMS/MS [24].

2.2.2 *Metalloproteins*. The metallothioneins (MT) [28] are a unique class of proteins capable of occluding seven bivalent metal ions by thiolate coordination in two separate clusters [29]. They are thought to be involved in the regulation of the flow of essential trace elements Zn and Cu through the cell. Over 50 metallothionein proteins are known among animal, plant and microorganism species all of which have conserved over 55% of the 61 or so amino acid residues, including the 20 cysteine residues, during evolution. Thus many members of the class have molecular weights between 6000 and 6500 daltons, and even when several changes in the sequence of amino acids occur among isoforms, the net difference in molecular mass may be small. There are two major isoforms in mammalian tissue, MT-1 and MT-2, but some species are known to have several sub-isoforms in each of these groups. While isoforms MT-1 and MT-2 differ by a single negative charge at neutral pH, which allows electrophoretic methods to be used in their separation, the sub-isoforms are more difficult to separate. It is important to be able to resolve these mixtures by mass spectrometry. Ideally, nominal mass resolution at 6000 daltons would be required since there are only 4 daltons difference between two of the compounds, but an accuracy of $\pm 5$ mass units will solve most problems. A particularly useful role for mass spectrometry will be to establish the nature of the N-terminus [30] and determine if it is a free $NH_2$ group, or is blocked by acetyl or another group.

The molecular weights of several isoforms of metallothionein from different species have been determined by MALDI mass spectrometry in collaboration with Kratos Analytical using their Kompact II instrument. Each well on the 20-well sample carrier was loaded with 1-10 femtomoles of sample in approximately 1 μl aliquots of nearly saturated matrix solution to which about 1 femtomole of myoglobin (M = 16950.5) was added to act as internal mass marker. The well-mixed solution was dried and the carrier placed on the robotic arm for transmission to the ionisation chamber. Doubly- and triply-charged protonated molecular ions of myoglobin appeared in the MALDI spectrum at m/z 8475.25 and 5653.5 daltons, bracketing the region of interest on the mass scale. Interpolation afforded mass accuracies of $\pm 2$ daltons. Matrices used included sinapinic acid and α-cyano-4-hydroxy cinnamic acid (CHCA).

In general, the MALDI mass spectra of the metallothioneins are matrix-dependent. With sinapinic acid, the major peak is due to the protonated molecular ion of the apoprotein, but sometimes a minor adduct ion $[M + A]^+$ appears. With CHCA, it would seem that not all metal ions are ejected and a family of minor peaks of increasing metal ion content occurs on the tail of the major apoprotein peak. In this case, the adduct ion is less noticeable. Bovine insulin was used to check the performance of the instrument. because it is available as a pure standard. Chicken MT was thought to be a single organometallic compound isolated in high purity. The molecular weight of the major peak was determined by interpolation as 6500 daltons, which suggests that the apoprotein is acetylated (±2 daltons; Table 1). Rat MT was separated into two isoforms, MT-1 and MT-2, and their mass spectra fixed the molecular weights within two daltons of those expected from the amino acid composition, if the proteins were acetylated (Table 1).

Table 1. Possible pseudomolecular ions of metallothionein apoproteins

| Species | Isoform | Mwt | $MH^+$ | $(M + Na)^+$ | $(AcMH)^+$ | Measured Mass 1 | 2 | Average mass |
|---------|---------|-----|--------|--------------|------------|-----------------|-----|--------------|
| Chicken | MT | 6461 | 6462 | 6484 | 6504 | 6499 | 6500 | 6499.5 |
| Rat (RRI) | MT-1 | 6006 | 6007 | 6029 | 6049 | 6049 | 6047 | 6048 |
| | MT-2 | 6145 | 6146 | 6168 | 6188 | 6186 | 6189 | 6187.5 |
| Rat (Abdul) | MT-1 | 6006 | 6007 | 6029 | 6049 | 6048 | 6047 | 6047.5 |
| | MT-2 | 6145 | 6146 | 6168 | 6188 | 6186 | 6190 | 6188 |
| Sheep | MT-1a | 5951 | 5952 | 5974 | 5994 ) | | | |
| | MT-1b | 5981 | 5982 | 6004 | 6024 ) | 5986 | 5986 | 5987 |
| | MT-1c | 5985 | 5986 | 6008 | 6028 ) | | | |
| | MT-2 | 6028 | 6029 | 6051 | 6071 | 6052 | 6053 | 6052.5 |

However, the metallothioneins of sheep are thought to comprise three sub-isoforms of MT-1 and a single pure compound MT-2 (Table 1). The mass spectrum of the MT-1 isoforms shows two peaks at 5986 and 6017 daltons, with an adduct ion at 6165 daltons. The major isoform is known to be MT-1a, with much less of MT-1c and only a trace of MT-1b. The spectrum was interpreted knowing that the accuracy of mass measurement was likely to be ±2 daltons only for pure compounds, and the resolving power available was likely to mean that peaks of up to 15 daltons mass difference may not be completely separated. Therefore it was the adduct peak at 6165 which suggested a molecular mass of 5976 (i.e. 6165-189). If the major isoform was sodiated, not protonated, then the molecular weight would be 5974 (i.e. within the ±2 dalton expected accuracy, but the molecule would not be acetylated. Thus the mass spectrum suggests that sheep MT-1a is different from other MTs in being unacetylated at the N-terminus. If this is correct, would it be reasonable for the unacetylated molecule to be more likely to occlude sodium? The peak at 5986 could be interpreted as the unresolved composite of sodiated MT-1a and MT-1c, but until higher mass spectral resolution is available it will not be possible to validate these tentative interpretations. In the meantime, these compounds have been studied by CE and preliminary results have been reported [31, 32].

### 2.2.3 Collagen Crosslinks.

2.2.3 *Collagen Crosslinks.* Growth and certain disease conditions are related to high turnover rates of collagens, the fibrous scleroproteins of connective tissue and bones. Tissue metabolism results in the release into blood and urine of fragments, some of which contain the collagen-specific 3-hydroxypyridinium crosslinks e.g. pyridinoline. FABMS was used soon after its introduction to determine the structure of pyridinoline [33, 34] Preliminary studies, using these components to monitor collagen degradation rates [35] have indicated that a large proportion comprise small peptides. As the crosslinks are found in a variety of tissues containing different genetic types of collagen, identification and sequencing of these peptides would facilitate the development of tissue-specific markers of collagen degradation. As a prelude to this development, the tandem mass spectra of pyridinoline (Pyd), the major trifunctional crosslink of cartilage and deoxypyridinoline (Dpd) which is primarily located in bone collagen and glucosylgalactosylpyridinoline (Glc-Gal-Pyd) have been studied. A 1:1 mixture of glycerol/thioglycerol was found to be superior to glycerol alone, and nitrobenzyl alcohol was found to reduce the sensitivity of the secondary ion mass spectrum of Pyd and Glc-Gal-Pyd to impractical levels [36]. The collision induced decomposition (CID) mass spectra of the molecular cation, the predominant species of the three compounds, produced fragment-rich mass spectra. In the timescale of collision cell transit, at least three successive generations of fragments can form, collide and partially decompose. It can be argued that hydrogen rearrangement processes involving, e.g. 1:4 elimination, occur in any of the three side chains liberating molecules of ammonia $(M - 17)^+$, formic acid $(M - 46)^+$ and glycine $(M - 75)^+$, and that simple ß-fission also occurs leading to loss of an alanyl radical leaving a radical cation capable of rearrangement to the stable azatropylium ion $(M - 88)^+$. However, once any of these primary processes has occurred, the subsequent losses e.g. of 131 Daltons from $m/z$ 354, can only occur from the $R_3$ sidechain. This indicates that an ion at $m/z$ 223 is of particular structural significance. Similarly, there is some evidence that the initial cleavage of the N-C bond of $R_3$ leads to $(M - R_3 + H)^+$ at $m/z$ 284 which decomposes further by loss of 87 daltons from $R_1$ or $R_2$ sidechain to $m/z$ 197, a relatively stable ion. Dehydration via the ring hydroxyl would give $m/z$ 179.

A rational interpretation of the spectra suggests that the first loss of ammonia or formic acid or glycine occurs largely from $R_1$ because subsequent losses of greater than 75 daltons, which seem to be part of the cascade, could only be generated from $R_2$ or $R_3$. The exception to this assumption would be the loss of 88 daltons (alanyl radical) from $R_1$. Similarly the loss of masses greater than 102 daltons is likely to originate only from $R_3$. Thus, it is hoped that these processes might be recognisable in the fragmentation of crosslinking compounds having peptide chains substituted at any of the terminal amino or carboxyl groups. Work is beginning on the isolation and identification of these compounds.

## 2.3 GLYCOBIOLOGY AND THE GLYCOSIDES

Mass spectrometry has not been so effective in solving general structural problems in oligosaccharide chemistry as it has in peptide chemistry because many monosaccharide units are isomeric, and carbohydrates are less polar than most oligopeptides, but nevertheless there have been notable successes. The location of oligosaccharide linkages using partially-methylated alditol derivatives and electron ionisation mass spectrometry [37, 38] has been widely used. Acidic groups are recognised by using deuterated reducing agents during the second reduction step. These methods have been applied to the analysis of cell wall polysaccharides in wheat bran [39], cabbage [40], runner beans [41] and apples [42]. Desorption ionisation techniques have aided the structure determination of polysaccharides, especially if permethylation is feasible [43] where the spectral improvement is due to the reduction in hydrogen bonding and increase in hydrophobicity. FABMS has been used routinely for checking the mass of oligosaccharides and has provided structural information on occasions [44]. A general knowledge of carbohydrate fragmentation is valuable when working with glycosylated structures such as phenolic polysaccharides, glycoalkaloids and pyridinolines and tandem mass spectrometry is producing more informative fragmentation patterns [45].

## 2.4 VITAMINS

The need to assay vitamins rapidly and accurately for improved legal food labelling requirements is focusing attention on HPLC methods, but the extraction and storage of vitamin samples has to be rigorously controlled to achieve the desired reproducibility. Mass spectrometry may have a role where incomplete separation of the target vitamin from analogues etc causes difficulty for simple HPLC-based methods. LC/MS, but more likely CE/MS, should provide the required assurance if stricter control of vitamin content of foods is made compulsory. So far, mass spectrometric methods using isotope dilution techniques have been developed for vitamins $D_2$ and $D_3$ and their metabolites [46-49] and for sub-nanomolar quantities of vitamin $B_6$ in liver, milk, urine and faeces [50]. Also nicotinamide has been determined in meat and meat products as 3-cyano pyridine [51]. Interest is being shown in the HPLC separation of eighteen carotenoids [52], along with related antioxidant vitamins such as A and E, and in their roles in providing protection against oxidative damage to membranes and nuclear material; a process which has been linked to the development of cancer [53]. Tandem mass spectrometry has been used to characterise tocochromanols with vitamin E activity [54] and FABMS was reported for folic acid and the folates [55].

## 2.5 ANTINUTRIENTS - SECONDARY METABOLITES

Not all components of foods are efficacious and toxicants and antinutrients are present to some extent in all our natural sources. Some plants prioritise the repulsion of predators through the production of bitter tasting constituents and poisons. Refined cultivars contain reduced concentrations of these repellents and the developed countries' supplies are usually free from high levels of toxicants. In countries where famine and drought prevail, the population resort to less nutritious plant species which are invariably higher in tannins and other polyphenols, and in glucosinolates, and glycoalkaloids, which can reduce even further the nutritional value of the food. Furthermore, attack from bacteria, moulds, fungi, insects etc can introduce toxins, such as the aflatoxins, which are harmful if present in high concentrations. The inhibition of enzymes and binding of mineral nutrients are two ways in which these chemicals reduce the benefit derived from food. Mass spectrometry has been used to characterise members of these chemical classes and particular success has been obtained with FAB in both positive and negative ion modes [56].

2.5.1 *The Glucosinolates*. Over 100 compounds have been identified, which have both beneficial and detrimental properties [57]. Their FAB mass spectra are usually intense and informative [58]. Thermospray LC/MS has been used effectively for both glucosinolate and desulphoglucosinolate analysis [59, 60]. The intense protonated molecular ions produced under FAB are ideal for tandem mass spectrometry as demonstrated by Heeremans *et al.* [60] for glucosinolates in sprout extracts. The same group used continuous flow FAB LC/MS to identify intact glucosinolates [61]. Saponins also respond to fingerprinting by FABMS and the composition in crude extracts of different varieties of legume seeds has been reported [62].

2.5.2 *The Glycoalkaloids*. These compounds give rise to intense FAB mass spectra first reported in 1981 [63]. The localisation of the positive charge on the steroidal nitrogen and the polarisation of the negative charge on the carbohydrate moiety produce complementary positive and negative ion FAB mass spectra [4]. The method has been applied to the qualitative identification of solanum glycoalkaloids in crude extracts from potatoes [64] and the quantitative estimation of a-tomatine in tomato fruit [65]. Recently, the intense protonated molecular ions of several glycoalkaloids have been subjected to tandem mass spectrometry under the auspices of the Agricultural and Food Research Council's Linked Research Group at the Institute of Mass Spectrometry and Department of Chemistry at the University of Warwick, UK. The positive and negative ion CID spectra of glycoalkaloids from potato shoots and tomatoes contain a wealth of fragment ions from the sugar chain but very little fragmentation of the steroidal ring system was observed [66]. The method has been shown to be sensitive down to the 200 femtomole level using a novel scanning-array detection system based on a charge coupled device [67].

The application of mass spectrometry to the analysis of mycotoxins has been reviewed [68, 69, 70]. Tandem mass spectrometry has been shown to be effective in determining crude aflatoxins and sterigmatocystin-related compounds [71] and macrocyclic tricothecenes [72]. The trifluoroacetyl ester of deoxynivalenol was prepared for quantitative GC/MS analysis down to picogram levels [73].

2.6 POLYPHENOLS (VEGETABLE TANNINS)

Considerable interest has been generated in plant phenolics because of their diverse roles in food processing and preservation, agriculture, nutrition, and general areas of food acceptability and safety. The chemical class is also diverse, making rules of analysis difficult to define. The analysis of the condensed tannins is particularly intractable due to their propensity for polymerisation, and consequently, energetic analytical processes tend to be unsuccessful. FABMS opened up the analysis of many different phenolic classes [4, 56] and work has continued since then, routinely recognising anthocyanin variants from a host of different plant sources [e.g. 74-80]. Thermospray LC/MS has been applied to the analysis of the procyanidin B2 dimer [59] and hopefully the growing availability of electrospray mass spectrometry will enable more of this class of compounds to be studied (in the LC/MS mode) by this approach. FABMS has continued to be used in the recognition of novel phenol-containing cell wall polysaccharides [81]. Phenolic acids such as p-coumaric and ferulic are covalently bound to cell-wall polysaccharides as esters, and they can be released on treatment with alkali. Interest in these compounds stems from the idea that when cell walls are treated with alkali, the amount of phenols released is correlated with the amount of forage digested by ruminants. Other glucosylated phenolic compounds from sorghum have been studied by negative ion FABMS [82]. Novel procyanidins having the basic formula epicatechin-(epicatechin)-catechin and monomeric flavonoids eriodictyol 5-glucoside and (+)-taxifolin 7-glucoside and their related aglycones were found.

## 3. Nutrition Research

Investigations are required at the systemic, cellular and molecular levels in nutrition research. The major systemic constituents are being studied in greater detail now that rapid HPLC methods allow their routine measurement. The reliability of assays is still somewhat questionable when extraction from complex matrices (foods and physiological fluids) is involved, and many vitamins and other metabolites are subject to collaborative inter-laboratory tests to improve the accuracy and precision of assays required to give information on the nutritional value of foods, or the nutritional status of the individual. So far, the measurement of the absorption of nutrients from food via the gut has been a difficult area for biochemical analysis because the microbiological status of the gastrointestinal tract is variable and many physiological and psychological parameters change simultaneously in response to food intake and nutrient absorption. Stable isotopes are now seen as ethical, rigorous tracers for both organic and inorganic nutritional studies on all subjects, even the most vulnerable( infants and pregnant women, for example), and many reports have appeared where stable isotope ratio mass spectrometry has been used to measure the amount and origin of nutrients and their absorption, metabolism and bioavailability. However, the physiological system can be perturbed by the addition of stable isotopes and excessive amounts can lead to toxic overload. If too little is added however, the accuracy of measurement of isotopic ratio will be low. Thus, factors such as body pool size and excretion rates have to be considered along with the instrumental limitations of precision of isotope ratio measurement, upon which the calculation of absorption and metabolism of nutrients (from foods) is based.

In general, the measurement of isotope ratios can be performed with acceptable accuracy in biological experiments, where the errors associated with the sampling of complex matrices normally obscure the lesser instrumental errors. Therefore, it is essential to define the sampling procedures in great detail if reproducible results are to be achieved on an inter-laboratory basis. The rules for the sampling of body fluids and solids are well known in clinical laboratories as are those required as a prelude to mass spectrometry. Uncontaminated homogenisation and long-life storage in inert containers are basic requirements. Care has to be taken to avoid contamination from the laboratory air, water and apparatus. Clean procedures are also well established for handling vulnerable samples. The sample size should be carefully considered, especially as the sensitivity of mass spectral methods increases. Too small a sample, although measurable, might not be representative of the whole, leading to inaccurate information. Too large a sample may result in inefficient extraction of the analyte with again inaccurate results. Above all, the accurate use of SIRMS requires the whole procedure to be calibrated, preferably by "bracketing" with certified reference material (CRM). Some nutrients require sophisticated chemical extraction protocols and only highly trained assayists can achieve the desired level of reliability. The latest practical approaches relating to mass spectrometric methods in human nutrition research have been reviewed [83].

The ratios of the stable isotopes of the pure elements are in general, known with high accuracy. However, in nature these ratios can be altered by processes which discriminate among isotopes, largely due to their difference in mass - the mass fractionation or mass discrimination effect. Thus, if a known amount of a pure isotope of an element is added to a biological system, it may not pass through the various metabolic stages without perturbation, and these chemical, enzymatic and mass fractionation effects are being studied. From the analytical viewpoint, the measurement of isotope abundance in a sample of biological origin is only meaningful if all these natural phenomena are understood and precautions are taken to avoid fractionation during the sample preparation stages. Only then can the appropriate corrections be made to the observed value to arrive at an absolute value for the incorporation of the label, and hence the quantity of the labelled element or compound in the sample. One particularly dramatic natural effect is due to the ability of all terrestrial plants to discriminate against $^{13}CO_2$. Two photosynthetic groups, the $C_3$ and $C_4$ plants fix $CO_2$ by different pathways resulting in different $^{13}C/^{12}C$ ratios in all subsequent metabolic products. Therefore, extracts from the $C_3$ and $C_4$ plants can be used as natural labels (tracers), and since the differences in isotope concentration can be monitored by isotope ratio mass spectrometry, experiments have been designed to take advantage of this natural phenomenon. Animal metabolism based on food plant sources will perpetuate this variation and add to it other changes based on 'higher order' processes. The adage that "you are what you eat" is a fundamental law when stable isotope ratio mass spectrometry is used in animal and human nutrition research.

With the exception of Na, Al, P, Mn, Co and As, the other elements with nutritional connotations are isotopic and most of them are commercially available in high purity at a cost that permits their use as internal standards for quantitative analysis. If the rules for standard addition are observed, it is possible to design accurate assays for organic and inorganic nutrients based on various mass spectrometric techniques for measuring isotope ratios. The use of GC/MS, gas isotope ratio (GIRMS), GC/IRMS, inductively-coupled plasma (ICPMS), thermal ionisation (TIMS) and FABMS techniques will be described, giving examples of the types of biological assays which can be undertaken and the accuracy and precision that can be expected from them.

Whilst instrument sensitivity limits many applications, there are other issues such as the cost and toxicity of the isotope, the amount of sample and the concentration of the analyte, and the perturbation of the system under examination, which must also be addressed. The ultimate combination of these factors and the accuracy and precision available from the measuring instrument determine the viability of any assay [84].

As with all internal standardisation methods the incorporation of a stable isotope as tracer assumes that the tracer mixes intimately with the analyte, is chemically indistinguishable from it, and does not separate from it during the time scale of the experiment. The accuracy of the method is usually determined by comparing the measured value of the unknown to that of a reliable standard under identical experimental conditions. Calibrated sub-standards may be used and related to absolute standards, and preference is given to the principle of bracketing with standards to produce increased confidence in the result. The accuracy is measured in units relative to an agreed international standard. The precision required of the measuring instrument is different depending on the particular isotope ratio to be measured. Usually the variability in sample preparation procedures exceeds the instrumental variability and therefore particular attention is paid to the extraction, derivatisation or combustion processes used.

## 3.1 BIOAVAILABILITY OF MINERAL NUTRIENTS

Any mass spectral technique which will produce inorganic ions or transfer existing inorganic ions from sample matrices to the gas phase for mass analysis is a candidate for SIRMS. FABMS [4], TIMS [85] and (ICPMS) [85] are in regular use. In our laboratories, studies of element metabolism were first performed by FABMS [86], simply because it was the only technique available [87, 88]. It was necessary to validate results against classical methods such as neutron activation and atomic absorption spectroscopy (AAS) and with CRMs. Later on, the FAB method was validated against the other mass spectral techniques [89, 90] and eventually it was phased out as new and better instrumentation became available. The studies on the quantitative analysis of iron by FABMS were reviewed in 1987 [4]. The laboratory procedures adopted for this work were ultimately extended to TIMS methodology, and to the investigation of other elements. Recent studies on zinc and calcium have been reviewed [91]. Enriched stable isotope compositions were checked mass spectrometrically before use, and the concentrations of solutions of minerals were determined by AAS. Isotopes were administered orally and/or by intravenous injection. Carmine and/or radio-opaque markers were also given with labelled meals when total faecal samples were required. Blood samples were centrifuged and the plasma or deproteinised supernatant fraction stored. Complete urine samples were made for each 24 hour period, and saliva samples were collected in plastic tubes at regular intervals, and stored. Sample purification methods have been updated since the programme started and the latest methods have been described in detail [91].

Faecal samples for Zn analysis were ashed, ion-exchanged, eluted with HCl and dried. Blood and urine samples were digested in $HNO_3$ in a teflon bomb in a microwave oven, evaporated to dryness and ion-exchanged as above. Urine and saliva samples for calcium were treated similarly (but not the same) as for zinc except a two-stage, ion-exchange purification was found to be necessary.

A Finnigan MAT GmbH ThermoQuad mass spectrometer (THQ) was used for TIMS analyses and calibrated using isotopic standards (Technical and Optical Equipment, London, UK; Medgenix, Ratingen, Germany or Oak Ridge National Laboratory, Oak Ridge, USA) yielding precisions of 0.1-0.5% relative standard deviation (RSD) for isotope ratios of samples of more than 1 μg and <1% RSD with between 100 ng and 1 μg. The plasmaquad (VG Elemental, Winsford, UK) was used for ICPMS analyses [92]. Samples were introduced into the plasma by either flow injection (FI) or continuous nebulisation (CN). The ICP technique is described in detail in section 3.2.

The bioavailability of calcium (Ca) was measured by FABMS using $^{44}$Ca to label the food, while $^{42}$Ca was injected intravenously to label plasma and urine, allowing for the excretion of endogenous Ca. Assuming that the two isotopes are metabolised at the same rate, when equilibrium has been reached, absorption from the oral dose can be calculated in plasma, urine and saliva from the following equation:-

$$ABS = \frac{(na^{44}Ca)(^{42}Ca\ i.v.) \times D\%CS^{44}Ca \times 100}{(na^{42}Ca)(^{44}Ca\ oral) \times D\%CS^{42}Ca}$$

where ABS is the percentage absorption, na is the natural abundance of the isotope relative to $^{46}$Ca, i.v. and oral are the exact doses of $^{42}$Ca and $^{44}$Ca given and Δ%XS is the % difference between the enriched ratio and the natural abundance. Results showed that the absorption of Ca from skimmed milk was significantly higher (P<0.001) than from watercress soup. Precision of 0.5% RSD was in agreement with earlier experiments [86] and is adequate for measuring changes of isotope ratio of 5-15% above normal - the recommended level. The FABMS method was well suited to the measurement of calcium bioavailability but was too slow for modern laboratory programmes.

Changes in the rate of absorption of zinc as a result of food processing were measured using FABMS, quadrupole TIMS and ICPMS and the three instrumental methods compared. Evidence has been reported [93] that extrusion cooking of a high fibre cereal product, compared with the conventional cooking, reduces the bioavailability of zinc in ileostomy patients. Here 5.51 mg of 93.11 atom% (see section 3.3.4 for definition) $^{67}$Zn was administered in 170 ml of a cola drink, markers were administered with the meals which included 40 g extrusion-cooked wheat bran/white wheat flour (50:50 mix) or a conventionally-cooked equivalent meal, and faecal samples were collected. The $^{64}$Zn/$^{67}$Zn ratios were determined by the three instrument methods. Total Zn content was measured by AAS. In this study, using normal healthy subjects, no significant difference was found in the absorption of zinc from either meal. The detailed comparison of the results from the three techniques [90, 91] showed that there were no significant differences between ratios determined by the different MS methods. A precision of better than 10% RSD was achieved for FAB, quadrupole TIMS and continuous nebulisation ICPMS, which is considered to be adequate for human mineral studies [94]. However, in spite of its extensive availability, FABMS has not been widely applied, possibly because of its poorer sensitivity and lower reproducibility compared to dedicated techniques of inorganic mass spectrometry.

Kinetic modelling based on $^{66}Zn/^{70}Zn$ isotope ratios measured in plasma, urine and faeces has been used to measure the periodic total $^{70}Zn$ in different pools and the rate of turnover of the pools [95].

## 3.2 QUANTITATIVE ANALYSIS OF METALS BY ICPMS

The relatively high energy volatilisation and dissociation of inorganic molecules has been accomplished using ICPMS technology, a comparatively recent development. The first commercial instruments were manufactured by Fisons (PlasmaQuad system) and Sciex (Elan system) and became available in 1983. The interest in ICP as an analytical tool began with the use of inert gas plasmas as sources for atomic emission spectroscopy (AES). Despite problems associated with sensitivity and with interferences in the line rich spectra of complex matrices, ICP/AES remains an important analytical technique, especially for multi-element applications.

Inert gas plasmas operate at high temperatures (5000-8000°K) and allow efficient transfer of energy to the sample, thereby serving as an excellent source of ions. When coupled to a mass spectrometer, greater sensitivity and specificity could be achieved than that offered by optical systems, thus paving the way for a more powerful multi-element instrument which also offered the capability of isotope analysis. Historical aspects of the development of ICPMS have been covered in detail by Gray [96, 97] and are beyond the scope of this review. For detailed overviews of ICPMS systems the reader is referred to reference [98].

Samples are usually analysed as solutions which are delivered to the instrument by a peristaltic pump and nebulised into the spray chamber in a stream of argon gas. A variety of nebuliser systems have been employed in ICPMS; the most common being the Meinhard concentric and V-groove types. The latter is especially useful for analysing solutions containing high dissolved solids (e.g. urine, plasma, blood etc). Other methods of sample introduction are also available for ICPMS, namely: direct insertion, laser ablation and electrothermal vaporisation for solids; and flow injection, hydride generation and slurry nebulisation for liquid samples. These techniques are extending the applications of ICPMS, but for most biological materials it is a simple matter to mineralise the sample (see Section 3.2.1) and present it as a solution in dilute mineral acid or, if a biological fluid, to analyse the sample directly with minimal sample pretreatment.

The aerosol produced by the nebuliser is carried to the ICP torch via the spray chamber which can be constructed of teflon, borosilicate glass or quartz. The spray chamber acts as particle size separator allowing particles that can be supported by the gas stream to be carried into the torch. Particles with a diameter of > 4 μm fall out of the gas stream and condense on the walls of the chamber before being pumped away via a drain. Although only a small proportion of the analyte is carried into the torch, this process is important for two reasons: firstly, water cools the plasma thereby reducing its efficiency as an ion source and secondly, water is the main source of oxygen in the formation of interfering oxide species (see table 2 below). Further reductions in the amount of water transported to the plasma are made by using a jacketed spray chamber cooled by water at 5-10°C circulated through a thermostatically controlled reservoir.

From the spray chamber, the analyte is swept into the quartz plasma torch, the tip of which is surrounded by a water cooled RF induction coil. The argon gas supply to the torch has three components: the auxiliary supply (0-1 Lmin$^{-1}$), the cooling supply (12-15 Lmin$^{-1}$) and the nebuliser supply (0.5-1.5 Lmin$^{-1}$). Initially, the argon gas is seeded with electrons from a tesla electrode connected to an ignition coil, but thereafter the plasma is self-sustaining as the RF energy of the load coil is coupled to free electrons generated by the plasma itself. The ionised gases from the plasma are extracted through a small aperture (ca 1.0 mm) in the water cooled sampling cone (usually of Ni) and thence into the expansion stage which is held at an operating pressure of approximately 2 mbar. As a result of expanding into this region a supersonic jet forms behind the aperture and the ionised gases cool rapidly, freezing the composition of the gases and preventing further reactions. A shock front, known as the Mach disc, is established behind the aperture of the sampling cone. In this region the gas flow is subsonic so it is essential to place the next cone (the skimmer cone) upstream of this disc. The skimmer cone, again usually of Ni, is acutely angled to prevent formation of further shock waves and has a small aperture (ca. 0.75 mm) through which the gas passes into the intermediate stage (operating pressure < 10$^{-4}$ mbar). Positively charged ions that enter this stage are extracted, focused around a photon stop, and then passed through a differentially pumped aperture into the third (analyser) stage, which contains the quadrupole mass spectrometer held at an operating pressure of around 3 x 10$^{-6}$ mbar. Transmitted ions are detected with an off-axis continuous dynode channel electron multiplier which is generally used in the pulse counting mode (linear range = 5-6 orders of magnitude). Analogue counting is also possible on some ICPMS systems giving an extended dynamic range (up to 8 orders of magnitude). Faraday cups, discrete dynode electron multipliers and Daly detectors have also been used.

The above description is based on the Fisons PlasmaQuad system which is similar to that of the Sciex Elan. Other instrument manufacturers, namely Finnigan-MAT, Spectro Analytical, Varian and Jeol also market ICPMS systems. There is now a considerable range of options available, including: high resolution (magnetic sector) ICPMS and combined glow discharge sources [99]. The former is a logical development because a resolution of > 10,000 is capable of eliminating most of the polyatomic interferences (see next section) and gives ICPMS a truly multi-element capability. Unfortunately high resolution ICPMS is extremely expensive and to our knowledge has not yet been applied to biological problems.

3.2.1 *Sample Preparation.* As stated above, it is usually a simple matter to mineralise biological samples, and therefore to analyse the sample in dilute mineral acid, using a peristaltic pump to deliver the solution to the nebuliser. There are a number of approaches that can be made for mineralising samples, such as; wet ashing, dry ashing and microwave digestion. The choice of method is dependant on factors such as volatility of the elements of interest, potential sources of contamination, sample throughput, cost, availability of equipment etc. In essence, the same careful approaches that are adopted for AAS must be rigorously enforced in sample preparation for ICPMS. Plastic or teflonware is preferable to glass, but all laboratoryware needs to be meticulously cleaned and stored in a dust-free environment. Reagents need to be of the highest specifications and above all the ICPMS laboratory requires an adequate supply of good quality water (18 M$\Omega$) and mineral acids and a clean dust-free area in which to perform sample manipulations. It is good practice to use sub-boiling quartz or teflon stills for the production of high quality acids that can then be stored in teflon bottles prior to use. For a comprehensive review of sample preparation for the analysis of trace elements in biological matrices the reader should consult the work of Bock [100] and Mizuike [101] (for discussions of classical techniques) and Kingston and Jassie [102] (for microwave techniques).

Apart from the care that needs to be taken to avoid contamination and loss of sample during work-up there are some considerations that are specific to ICPMS analysis. Firstly, it is sometimes possible to avoid extensive sample pre-treatment. Biological fluids like blood, plasma, milk and urine can often be analysed whole or after minimal pre-treatment. Trace metals can be quantified in whole blood by ICPMS by diluting EDTA-treated blood, adding a small amount of Triton 100X and adjusting the pH to 8.5 with ammonium hydroxide [103]. This procedure gives a clear solution that remains stable at room temperature over several weeks as well as lowering sample viscosity and salt concentration. The latter are important because they prevent nebuliser blockage and excessive build-up of salt on the sample cone, allowing the rapid sample throughput potential of ICPMS to be exploited. A similar approach was used to obtain precise measurements of iron isotope ratios from whole blood [104]. Other groups have obtained quantitative and/or isotope ratio data by ICPMS using whole blood [105-107] and urine [110] diluted with mineral acids. Interferences have been categorised and grouped under the following headings: oxides, doubly charged ions, polyatomic ions, isobaric overlaps, matrix suppression and physical effects [98]. Some examples of these effects for elements of biological importance are given in Table 2.

Quadrupole ICPMS systems are not able to resolve the overlapping peaks produced by these interferences and unless steps are taken to eliminate them it may not be possible to obtain accurate measurements of the elemental concentration or its isotope ratio. Where the element in question has several isotopes, it is usually possible to find one where the interferences are insignificant, but this is not always the case, for example some elements are mono-isotopic (e.g. Al, Mn) and others are in "difficult" mass regions where polyatomic interferences are abundant (e.g. Ca, Fe, Se). One outcome of this is that the limits of detection (LODs) of ICPMS for Ca, Fe and Se are uncharacteristically high (ca. 1 ng mL$^{-1}$). For elements which are substantially interference-free, LODs can be as low as 1 pg mL$^{-1}$; although this is also dependant upon such factors as ionisation energies and isotope abundance.

Table 2 covers only a fraction of known potential interferences and the analyst must be vigilant. A detailed knowledge of the elemental composition of the sample will give indications of the kinds of interference problems that are likely to be encountered but only when good agreement is obtained between unknown quantitative data and those certified for Standard Reference Materials (SRMs) will confidence in the data be gained for a particular matrix.

Table 2. Examples of interference effects for some elements of importance in human nutrition

| Mass | Analyte | Interference | Source of interference |
|------|---------|--------------|------------------------|
| 24 | $^{24}Mg$ | $^{12}C_2^+$ | Atmospheric $CO_2$, organic and inorganic carbon in sample |
| 40 | $^{40}Ca$ | $^{40}Ar^+$ | Plasma argon |
| 54 | $^{54}Fe$ | $^{40}Ar^{14}N^+$ | Plasma argon, atmospheric/sample nitrogen, nitric acid |
| 56 | $^{56}Fe$ | $^{40}Ar^{16}O^+$ | Plasma argon, $H_2O$, atmospheric oxygen |
| 57 | $^{57}Fe$ | $^{40}Ar^{17}O^+$ $^{40}Ar^{16}O^1H^+$ | Plasma argon, $H_2O$, atmospheric oxygen Plasma argon, $H_2O$, atmospheric oxygen |
| 63 | $^{63}Cu$ | $^{40}Ar^{23}Na^+$ | Plasma argon, sample sodium |
| 65 | $^{65}Cu$ | $^{32}S^{16}O_2^+$ | Sulphuric acid, $H_2O$, atmospheric oxygen |
| 70 | $^{70}Zn$ | $^{35}Cl_2^+$ | Hydrochloric acid, chloride salts |
| 76 | $^{76}Se$ | $^{40}Ar^{36}Ar^+$ | Plasma argon |
| 78 | $^{78}Se$ | $^{40}Ar^{38}Ar^+$ | Plasma argon |
| 80 | $^{80}Se$ | $^{40}Ar_2^+$ $^{40}Ar^{40}Ca^+$ | Plasma argon Plasma argon, sample calcium |

Most interference problems can be overcome, although this often involves lengthy sample pre-treatment procedures that reduce the analytical productivity of the ICPMS system. The perceived advantages of ICPMS over TIMS for isotope ratio work (around 4x greater sample throughput potential and minimal sample pre-treatment before analysis) are often nullified by the requirement to remove interferences. Generalised approaches aimed at reducing the population of all polyatomic species (such as use of $N_2$ or small amounts of Kr and Xe in the plasma gases and organic solvents in the sample) have been promising, but never wholly satisfactory [111-113]. Background subtraction (for argon oxides, or nitrides, etc) can only be a realistic option if the analyte signal is considerably stronger than that derived from the interference. The best solution is to use high resolution ICPMS, but for reasons of cost the usual option is to use chromatography, hydride generation or precipitation to selectively remove the interfering species or the analyte of interest. In this way Lyon et al. [114] used gel-filtration to de-salt human serum thereby removing interferences at m/z 77 (due to $^{40}Ar^{37}Cl$) and enabling them to quantify Se.

Similar approaches have been adopted [115-121] to measure isotope ratios of Cu, Fe, Li, Mg, Se and Zn for both metabolic studies and for quantification of these elements by isotope dilution ICPMS.

Another area that has seen considerable development in recent years is speciation studies using liquid chromatography coupled to ICPMS [122]. In this way, a variety of separation systems (ion exchange, reverse-phase, normal phase and size-exclusion chromatography) have been used to separate different elements, organometallics and species of different oxidation states.

*3.2.2 Standardisation, Calibration, Accuracy and Precision.* With ICPMS there are three approaches that can be made to quantification. Firstly, there is the semi-quantitative approach that utilises a curve of response (area counts s$^{-1}$) versus mass, generated by a mixture containing several elements (usually around 8) of known concentration. The use of internal standards ($^{115}$In is commonly used in such analyses) allows changes in the performance of the system to be corrected during a run. In this way, Amarasiriwardena *et al.* [123] were able to quantify 23 elements in a variety of biological SRMs within 30% of certified values. Semi-quantitative measurements are therefore useful for multi-element surveys and are often used as a preliminary analysis and to help identify possible interference problems. The second approach, the so-called "fully quantitative" method, uses calibration curves that have been generated for each element of interest. There are many publications showing excellent comparisons between ICPMS data and those certified in SRMs of biological origin [108, 109, 124-126]. The final approach to quantification uses stable isotope dilution (ID) techniques which are widely accepted as the "gold standard". There are several reports in the literature showing the power of ID ICPMS. Schuette *et al.* [120], for instance, obtained the following values for Mg in National Bureaux of Standards (NBS) Bovine Liver and International Atomic Energy Agency (IAEA) Animal Bone (certified values in parentheses): $617.0 \pm 44.0$ µg g$^{-1}$ ($600.0 \pm 15.0$ µg g$^{-1}$) and $3585.0 \pm 16.0$ µg g$^{-1}$ ($3550.0 \pm 90.0$ µg g$^{-1}$) respectively. Similar agreements have been obtained for Li, Se, Fe, Cu and Zn [127]. For quantitative studies, matrix-matched SRMs enable analysts to have confidence in their procedures, but they are not always available. Although the number of SRMs of biological origin has increased in recent years and includes such matrices as; urine, blood, serum, muscle, liver, kidney, milk powder, brown bread, fish tissue, wholemeal flour and hay, there are some obvious omissions (e.g. faeces and "mixed diets") which are of great importance for nutritional studies and which have never existed or have been exhausted. In other cases matrix-matching is possible, but the SRM is not certified for the mineral of interest.

No SRMs exist with certified isotope ratios for nutritionally-important elements (although these do exist for Pb - NBS SRMs 981 and 982 - because this is one of the few metals to exhibit variations in natural isotope ratios). Attempts have been made to compare isotope ratio data obtained by ICPMS with other techniques. Eagles *et al.* [90] for instance, obtained excellent agreement between data on $^{64}$Zn/$^{67}$Zn ratios in artificially enriched human faeces measured by FABMS, TIMS and ICPMS. The mean RSD of the ICPMS data was 0.3%. This level of precision is routinely achieved for isotope ratio measurements by ICPMS for Mg, Cu and the major isotopes of Zn in diverse biological matrices.

## 3.3 PROTEIN TURNOVER AND ENERGY EXPENDITURE BY GIRMS

The technique requires samples to be converted to gaseous products either as oxides e.g. $CO_2$, $SO_2$ or elemental forms e.g. $N_2$, $H_2$ and $O_2$. Thus, if specific compounds are to be monitored they must be isolated in a pure form prior to analysis. The first isotopic tracer study was reported in 1935 by Schoenheimer and Rittenburg who used deuterium to study metabolism in animals and man [128]. It was not, however, until the 1950s when Isotope Ratio Mass Spectrometry (IRMS) was developed for the analysis of geological samples [129] that the use of the technique increased in popularity in physiological and nutritional research. With improvements in instrument specifications, smaller differences in stable isotope ratios can be measured to a higher degree of accuracy and precision. This has led to a gradual increase in their use over the last two decades, with particular applications in end-product measurement of protein turnover, substrate oxidation studies, and more recently, the estimation of energy expenditure in free living animals, including man, using the doubly labelled water method.

Gas isotope ratio analysis mass spectrometers are capable of producing extremely precise and accurate results. The precision obtainable from both dual inlet and continuous flow instruments largely depends on the type of sample and the preparation system being used, but to generalise, an RSD of 0.001% can be obtained in $^{13}C$ and $^{15}N$, 0.3% in H/D and 0.01% in $^{18}O$ studies. This is largely due to the use of the classical Faraday cup dual inlet design, combined with the fixed geometry of the mass spectrometer analyzer. The system can measure small differences in enrichment, typically $\pm$ 0.0004 Atom%, but micromole quantities of material are required. Gas isotope ratio mass spectrometers are based on the Nier - McKinney design [130, 131] and with advances in electronics, vacuum technology and computing, modern instruments are now fully automated. The instrument is made up of four components: the sample preparation system, inlet, ion source and collectors. Instruments can be fitted with many different types of preparation system all of which are designed to purify the sample gas prior to admission to the mass spectrometer inlet. Cryogenic purification is used when the dual inlet system is employed to analyze carbon dioxide (see below). Firstly, water is removed at -80°C, $CO_2$ is then trapped at -197°C while non-condensable contaminant gases are pumped away. The pure $CO_2$ is admitted to the mass spectrometer. When measuring nitrogen and hydrogen, the samples are purified off-line and the resulting gas admitted directly to the inlet of the mass spectrometer. Non-cryogenic on-line sample preparation systems are discussed in greater detail below.

The instrument is fitted with a dual inlet system comprising two identical halves, one side containing reference gas of known isotopic composition, and the other the sample gas. The pressure in both sides of the inlet is equalized using variable volume bellows. The sample and reference gases are then passed through matched capillaries into the changeover valve which allows either the reference or the sample gas into the mass spectrometer while the other goes to waste at exactly the same rate. This rather complicated procedure allows the movement of the gas around the inlet and into the mass spectrometer in such a way that isotopic fractionation effects associated with gas handling and ion beam formation are cross-cancelling. Such factors therefore do not need to be taken into consideration in the calculation of the enrichment values.

Ionisation of the sample gas is effected by electron impact at 80-100 eV and the resulting ions are accelerated, focused and raised to a potential of approximately 2500 V ($CO_2$), 3700 V ($H_2$) or 4000 V ($N_2$). The singly-charged ions pass into a static magnetic field and separation occurs according to their mass to charge ratio.

Modern instruments use three Faraday cups mounted behind fixed slits of different widths, calculated to collect ions of a specified mass as they emerge from the analyser. This so-called universal detector can be used to measure $SO_2$, $CO_2$, $O_2$ and $N_2$. The measurement of hydrogen/deuterium having a 2:3 mass ratio (compared to $^{12}CO_2/^{13}CO_2$ of 44:45) requires a separate collector system.

Instrument control and data management are facilitated by dedicated personal computers. Software is provided by the manufacturers. Unfortunately, in most cases, this is not easily modified to allow new or specialised sample manipulation or information-collection procedures to be inserted.

3.3.1 *Continuous Flow Techniques.* The dual inlet system allows "geological" precision e.g. 0.00004 Atom% for $CO_2$ but it requires large quantities of sample to compensate for the `dead volume' associated with the inlet, and, even with the use of automated cryogenic microanalysis, the analysis time is long (20-30 min per sample). The realisation over the past few years that for many physiological experiments the biological variability far outweighs the precision of the mass spectrometer when coupled with the large number of samples produced in such studies, has led to the introduction of continuous flow isotope ratio mass spectrometry (CF-IRMS). In CF-IRMS the sample gas is carried directly into the mass spectrometer in a stream of helium. The use of on-line, packed-column chromatographic purification procedures increases the sampling rate and permits the use of smaller samples while the precision of isotope ratio measurement under continuous-flow conditions (approximately half that of the dual inlet approach) is adequate for many biological applications. Most elements of the CF mass spectrometer are the same as the dual inlet version; the major difference is the increased range of preparative systems which can be connected directly to increase the sample throughput.

The Automated $^{15}N$ $^{13}C$ Analyser (ANCA) produces nitrogen, nitrogen oxides and carbon dioxide by Dumas combustion [132, 133] of the samples at $1000^{o}C$ which are then passed through a reduction furnace to convert nitrogen oxides to molecular nitrogen. Subsequent chemical traps remove water and, if desired, carbon dioxide. Next, the sample gas is passed through a gas chromatographic packed column to separate nitrogen and carbon dioxide prior to their entry into the mass spectrometer. With the correct software control, isotope ratio data can be obtained for both gases from the same sample. A typical analysis time for nitrogen is 5 min. alone and 8 min. in combination with carbon dioxide, and the sample for nitrogen can be as low as 10 µg (cf. 500 µg for the dual inlet system).

The Automated Breath $^{13}C$ Analyser (ABCA) is used for the analysis of carbon dioxide from samples other than combusted materials. The wet sample is flushed out of the sample vessel with helium, dried during passage through a chemical trap, and the carbon dioxide chromatographically separated from the nitrogen, which is diverted before the gas stream enters the mass spectrometer, where the isotopic ratio of the pure carbon dioxide is measured. In some applications, sample size can be 1 µmole of carbon or less (compares with 3 to 20 μmoles, depending on the use of cryogenic microanalysis, with dual inlet systems).

3.3.2 *GC/IRMS.* A recent innovation uses IRMS coupled to a high resolution gas chromatograph with an intermediate in-line combustion system. Conventional derivatisation produces volatile compounds from organic mixtures which are separated into their constituents by the gas chromatograph. The emerging carrier gas passes through a combustion furnace and the isotopic ratio of the carbon dioxide gas produced, is then measured in the mass spectrometer. The use of high resolution chromatography ensures that several constituents from the one sample (e.g. plasma, urine, tissue extracts) can be measured. The GC/IRMS method permits rapid, on-line metabolite purification which facilitates good accuracy ($\pm$ 0.2 ‰, see section 3.3.4 below) on nanogram amounts of sample. Disadvantages include the addition of extra carbon atoms during the derivatisation step and so far, nitrogen cannot be measured using this system.

3.3.3 *Sample Preparation for IRMS.* There is a range of sample preparation systems available to convert heavy-isotope labelled biological materials into the appropriate isotopic gases for mass spectral analysis. Included among these are:

Breath analysis ($^{13}CO_2$, $C^{18}O_2$); The subject breathes into a bag from which a sub-sample containing expired air at 4-6% $CO_2$ (i.e. 50 μmoles $CO_2$) is taken into a pre-evacuated 20 ml Vacutainer ® [134].

Combustion of organic material ($^{13}CO_2$, $C^{18}O_2$, $^{15}N_2$); A sub-sample of tissue, biological fluid or diet material is combusted off-line in a sealed vial with copper oxide and silver, or in the case of $C^{18}O_2$ [135]. An alternative technique involves the use of the Dumas combustion procedure in the ANCA preparative system, where a weighed amount of the material (solid or liquid) is loaded onto the carousel of the instrument in a small tin capsule for on-line combustion at 1000°C [136].

Liberation of blood bicarbonates ($^{13}CO_2$); A blood sample, usually 0.7-1.0 ml, taken anaerobically, is added to a pre-evacuated Vacutainer ® containing frozen lactic acid and stored frozen if required. For analysis, the sample is thawed (if required) and mixed to initiate the release of $CO_2$ [137] in the reaction between lactic acid and the blood bicarbonate.

Liberation of carbonates ($^{13}CO_2$); Carbon dioxide is liberated from the sample (e.g. $BaCO_3$, prepared by passing exhaled breath or respiration chamber gases through a carbon dioxide trapping solution) or standard carbonate, by reaction with 100% phosphoric acid [135].

Decarboxylation of isolated amino acids ($^{13}CO_2$); The amino acid(s) of interest are isolated using ion-exchange chromatography, desalted and reacted with ninhydrin to liberate the C-1 atom as carbon dioxide [137].

All the above techniques can use Vacutainers ® to contain the sample gas for use with the dual inlet instruments. Inclusion of a carousel system with cryogenic purification allows automated overnight analyses to be performed.

Oxidation of ammonium sulphate($^{15}N_2$); The total nitrogen of the $^{15}N$-labelled biological sample is first converted to $(NH_4)_2SO_4$, by Kjeldahl digestion [133], and transferred to a pyrex reaction bottle (Lowers Happert, Netherlands). Alkaline hypobromite is added and the bottle evacuated and placed in an oven at 80°C. The resultant nitrogen gas is admitted to the mass spectrometer [138]. If the Dumas combustion system is used, raw plant or soil material can be analyzed directly. Similarly, certain compounds e.g. urea, can react directly with hypobromite to produce $N_2$ gas in which both atoms originate from the same urea molecule.

Reduction of water (Deuterium); Water from urine, plasma or other biological fluids can be reduced, either directly or after distillation, to hydrogen gas by reaction with zinc under vacuum at a temperature of 470°C [139]. The deuterium gas is admitted to the mass spectrometer through a manifold.

Equilibration of water with carbon dioxide ($C^{18}O_2$); In samples of biological material, the heavy oxygen isotope in water can be measured following equilibration with carbon dioxide gas of known isotopic content. The sample is placed in a Vacutainer and shaken on an orbital shaker for 12 hours when hydroxyl oxygen exchanges with $CO_2$. The equilibrated $CO_2$ is then admitted to the mass spectrometer from the vacutainer via a water trap [140].

3.3.4 *Standardisation, calibration, accuracy and precision* in GIRMS. All measurements are made relative to a secondary reference material which has been calibrated against an internationally-accepted standard. Because the measurements are made relative to a reference, and the precision of the instrumentation is such that very small differences can be measured, the delta unit is commonly used to express results:-

$$\delta \text{ per mil} = \frac{R_{sample} - R_{reference}}{R_{reference}} \times 1000$$

where R = the ratio of the total ion current at m/z 45 to the total ion current at m/z 44 (for $CO_2$), and m/z 29 to m/z 28 for $N_2$ etc.

This convenient terminology is then converted to atom per cent (Atom%) for the calculation of isotope enrichment (gains or losses):-

$$Atom\% = \frac{100}{(1 + 1/R)}$$

(NB the denominator becomes 2/R for diatomic nitrogen).

The commonly accepted standard material for measuring $^{13}C$ is PDB (a limestone sample taken from the rostrum of a Cretaceous belemnite, *Belemnitella americana*). This material is now exhausted, so although results are still expressed relative to PDB the standardisation is performed using NBS 20 Solenhofen Limestone supplied by the National Bureau of Standards. The quoted values for NBS 20 are $^{13}C_{PDB} = -1.06‰$ and $^{18}O_{PDB} = -4.14‰$. When measuring $^{15}N$, atmospheric nitrogen is used as the acceptable standard [141], with a value of 0.3665 Atom%. Two standards are used when measuring Deuterium: Vienna-Standard Mean Ocean Water (V-SMOW) and Standard Light Antartic Precipitation (SLAP), and to facilitate inter-laboratory comparisons a value has been adopted internationally for SLAP relative to V-SMOW of $^{18}$O = -55.5‰ and D = -428‰ with the standard delta calculation modified to:-

$$\delta \text{ per mil} = \frac{R_{sample} - R_{v\text{-smow}} \times [\delta^{\circ}_{slap} / ((R_{slap} - R_{v\text{-smow}}) / R_{v\text{-smow}})]}{R_{v\text{-smow}}}$$

where R is $^{18}$O/$^{16}$O or D/H and $\delta^{\circ}_{slap}$ is the pre-assigned value of SLAP [142].

For analyses of $^{13}$C and $^{15}$N the reference gas (carbon dioxide or oxygen free nitrogen respectively) is calibrated to the standards mentioned above. A series of 'working standards', carbonates for $^{13}$C and ammonium sulphate for $^{15}$N, are then used for the daily instrument and sample preparation checks. The procedure for Deuterium is similar, however to account for differences in raw delta values which have been observed on sample preparation and analysis both in the zinc reduction and water equilibration techniques, laboratory standards (calibrated on the V-SMOW/SLAP scale) are incorporated into each batch run and used to correct for drift [142].

3.3.5 *Biological procedures* Stable isotopes have been used as tracers to study many metabolic processes in animals and humans. A few important examples include protein turnover measurement, based on either carbon or nitrogen kinetics, and whole body energy expenditure estimated using doubly labelled water, $^2$H$_2$$^{18}$O.

All nutrients including amino acids are in a state of dynamic flux with metabolites released from endogenous pools and the dilution of an ingested dose of labelled isotope provides an estimate of the extent of the turnover of the body substrate pools e.g. protein from amino acid studies. When labelled amino acids enter the body they are rapidly incorporated into various body tissue proteins. The rate of turnover varies markedly between tissues with liver turning over more rapidly than muscle (e.g. a 10 fold difference in fractional terms in both man and farm animals). Different rates of turnover can be measured under various normal and clinical conditions by following the dilution of a tracer e.g. the amino acid [1-$^{13}$C] leucine [143-145]. Because $^{13}$C metabolites have a natural abundance of approximately 1.1 Atom%, and different compounds, e.g. fat, sugar, etc., have different natural abundances, and therefore different protein bound amino acids, a 'background' sample is taken prior to infusion. This can either be a biopsy sample of the tissue, or plasma protein will often suffice [144,145]. At specific times after administration of the tracer, a biopsy sample is taken from the relevant tissue, hydrolysed and the leucine fraction isolated by chromatography. Great care has to be taken to ensure that no other amino acid is isolated and that the complete amino acid peak is collected since fractionation of $^{12}$C and $^{13}$C isotopes occurs during ion exchange, HPLC and GC separations. Then $^{13}$CO$_2$ is released from the carboxyl group of leucine by the ninhydrin reaction (see section 3.3.3 above).

Alternatively, $^{15}$N labelled amino acids are often used to monitor whole body protein synthesis. A single dose or continuous infusion of say, [$^{15}$N] glycine, is given and total urine collected over 12-48 hours. The enrichment in either eliminated ammonia or urea is determined, after conversion to N$_2$, by the Kjeldahl - Rittenberg method. Alternatively, urinary urea, ammonia and plasma urea enrichment can be measured by trapping ammonia from alkaline urine in 1 M sulphuric acid, the residual urea is then treated with urease and the liberated ammonia trapped similarly. Plasma urea is treated as urinary urea. The rates of nitrogen flux, protein synthesis and protein degradation can then be calculated from the proportions of the dose recovered in urine and the amount of urine-N [146].

Energy expenditure is classically measured by either direct or indirect calorimetry [147, 148] involving confinement of the subjects. An alternative technique was developed by Lifson [149] to measure $CO_2$ production by using water enriched with stable isotopes of oxygen and hydrogen. The principle involved is that oxygen will exit the body both as carbon dioxide and water (due to exchange via carbonic anhydrase) but hydrogen will leave only as water. Therefore if daily samples of body fluid are taken (usually urine, plasma or saliva), analysis of the decline in enrichments of the two isotopes yields two different decay curves, the difference in rate constants represent $CO_2$ production. From this, and assuming a value for the respiratory quotient, energy expenditure can be estimated. There are several corrections which have to be taken into account, as discussed by Haggarty [150]. The advantage of this approach is in application to free-living animals and man when appropriate social and environmental factors are thus included in the measurement.

## 3.4 PROTEIN AND CARBOHYDRATE TURNOVER USING STABLE ISOTOPE RATIO GC/MS

Although gas chromatography has been mentioned already for sample purification in GIRMS, the use of conventional organic mass spectrometers in the conventional GC/MS coupled mode for quantitative analysis using SIRMS techniques is discussed in this section. Stable isotope-labelled volatiles or volatile derivatives e.g. amino acids, can be estimated rapidly using GC/MS. Initially, the technique suffered from a lack of precision (around ± 0.1% compared to ± 0.00001% with GIRMS), although its high sensitivity permitted the use of small (picomolar) amounts of sample. The relative merits of the various techniques have been discussed and their clinical applications critically evaluated in 1989 [151]. Stable isotope-labelled amino acids have gained widespread use in human protein kinetics studies [152-154]. To determine the protein synthesis rate in human tissues requires tracer enrichment to be measured in both precursor and product amino acids. While GC/MS traditionally has been used to measure free amino acids (0.5-20 atom percent excess (APE)), the more demanding measurement of protein-bound amino acids at 0.002-0.1 APE has been the domain of GIRMS. The introduction of GC/IRMS using high resolution capillary columns and GIRMS [155] has obvious attractions, but the high enrichments in plasma-free amino acids still require conventional GC/MS.

The incorporation of deuterated lysine into plasma protein [156] and the use of $d_5$-phenylalanine [157] incorporated in tissue protein demonstrate the latest low-level applications of GC/MS, and since higher levels of enrichment in the free amino acids can also be measured by this same technique, it is now considered to be the method of choice for studying tissue protein synthesis [157]. Enrichments of 0.002-0.09 APE have been measured in 2 mg biopsy equivalents of muscle protein and plasma albumin, where it was found necessary to convert phenylalanine to phenylethylamine by enzymatic decarboxylation to avoid background interference with the ions of choice. The heptafluorobutylamine derivative was chromatographed on a 30 m x 0.25 mm x 0.25 μm SE-30 fused silica capillary column and the (P + 5)/(P + 2) ratio at m/z 109/106 was measured under electron impact - selected ion recording conditions with the electron energy set at 22 eV for $d_5$-phenylalanine enrichments of less than 0.5 APE. The (P + 5)/(P + 2) ratio was monitored instead of the conventional (P + 5)/P ratio because, at these low levels of enrichment, obtaining a measurable signal at the (P + 5) ion leads to the P signal overloading the amplifier. Peak areas rather than peak heights were used in the quantification to avoid problems of peak asymmetry caused by column overload. A mean value of 0.00531 ± 0.00032 APE (n = 29) (RSD = 6.0%) was found to be acceptable for biological studies. The satisfactory application of a multiply deuterated tracer for GC - selected ion recording MS using much lower levels of enrichment (0.002-0.1 APE) at a throughput of around 30 samples each day is expected to accelerate progress in protein turnover and other biosynthetic studies.

For example, the same philosophy has been applied to glucose turnover studies in blood using universally-labelled $^{13}$C-glucose as the aldonitrile acetate derivative for GC/MS on a 12.5 m OV-1701 flexil column (Phase Sep Ltd) [158]. The (M + 6)/M ratio was monitored under chemical ionisation conditions at high mass resolution (411.171/405.151) by monitoring each peak for 0.2 s in rotation, allowing a settling time of 0.05 s on each ion.

The (M + 6) concentration in the infusion fluid was calculated from standard additions of glucose to create isotope ratios similar to those found in plasma [159]. The effect of acetate absorbed from the gut on glucose turnover was tested in four healthy human subjects during fasting and during an intravenous glucose infusion using [U-$^{13}$C] glucose. Sodium acetate (15 mmol) was given by mouth. It was concluded that acetate substitutes for lipid oxidation during metabolism and has no direct effect on glucose turnover. In this study, the monitoring of relatively high mass ions ensured that the chemical background was low, allowing the measurement of glucose turnover rates as 1.88 ± 0.1 mg min$^{-1}$ kg$^{-1}$ during fasting and 4.0 ± 0.08 mg min$^{-1}$ kg$^{-1}$ during glucose infusion to be measured.

## 4. Future Prospects

The increasing sensitivity of mass spectral methods, especially the current developments in electrospray ionisation, should increase the rate of application of qualitative and quantitative measurements of food components and nutritional metabolites at the cellular level of investigation. In concert with sampling techniques such as microdialysis it should be possible to study interstitial fluids and localised cellular material to create "high resolution mapping" of selected metabolic pathways. As the combination of capillary electrophoresis and mass spectrometry matures it should be possible to separate the metabolites isolated by microdialysis and identify them using MS/MS methods to produce structural information at the femtomole level of investigation. Success will depend upon the development of selective membranes for microdialysis which will have low and high mass cut-off admitting only those metabolites which can be phoresed and ionised with the chosen instrumentation. Improved methods of removing the inorganic electrolytes used in

CE will allow other variants such as isoelectric focussing and micellar electrokinetic chromatography to be put on line.

Higher resolution in MALDI mass spectrometry is needed to open up mixtures of isoforms of metalloproteins, etc. where separation by chromatographic and electrophoretic means is proving difficult. The new CCD array detectors are providing femtomole sensitivity in MS/MS which is seen as a most valuable adjunct to the elucidation of the stucture of low level biologically-active substances e.g. neuropeptides and growth hormones, hitherto out of reach with our present-generation instruments.

## 5. Acknowledgements

We wish to thank our colleagues in various biological institutions who have supplied the samples for the exploitation of new techniques and our colleagues at Kratos Analytical and the Warwick Institute of Mass Spectrometry for access to MALDI and MS/MS respectively. We acknowledge the financial support given by the Agricultural and Food Research Council, the European Union, MAFF and the Scottish Office Food and Agriculture Department.

## 6. References

1. *Recent advances in food science - 3 Biochemistry and biophysics in food research* (1963) eds J. Muil Leitch and D.N. Rhodes, Butterworths, London.
2. Holmes, J.C. and Morrell, F.A. (1957) *Applied Spectroscopy* 11, 86
3. Barber, M., Bordoli, R.S., Sedgwick, R.D. and Tyler, A.N. (1981) *J. Chem. Soc., Chem. Commun.* 325
4. Self, R. (1987) in *Applications of Mass Spectrometry in Food Science*. ed. J. Gilbert, Elsevier Applied Science, London pp 239-279.
5. *Applications of Mass Spectrometry in Food Science* (1987) ed. J. Gilbert, Elsevier Applied Science, London.
6. *Symposium on Foods : The Chemistry and Physiology of Flavours* (1987) eds H.W. Schultz, E.A. Day and L.M. Libbey, Avi, Westport, Conn.
7. van Straten, S. and Maarse, H. in *Volatile compounds in foods - quantitative data*, 5th Ed. (TNO-CIVO Food Analysis Institute, Zeist, The Netherlands, 1983 - Suppl. 1984, 1985 and 1986)
8. Watford, M. and Fried, S.K. (1991) *TIBS* 16, 201-202
9. Olefirowicz, T.M. and Ewing, A.G. (1990) *Anal. Chem.* 62, 1872-1876
10. Menacherry, S.D. and Justice, J.B. Jr. (1990) *Anal. Chem.* 62, 597-601
11. Caprioli, R.M. and Lin, S-N (1990) *Proc. Natl. Acad. Sci. USA* 87, 240-243
12. Wahl, J.H., Goodlett, D.R., Udseth, H.R. and Smith, R.D. (1992) *Anal. Chem.* 64, 3194-3196
13. Self, R. (1961) *Nature* 189, 223
14. Hawthorne, S.B., Krieger, M.S. and Miller, D.J. (1988) *Anal. Chem.* 60, 472

510

15. Self, R. (1968) in *Mass Spectrometry* eds R. Brymner and J.R. Penney, Butterworths, London, 93-102
16. Matthews, D.E. and Hayes, J.M. (1978) *Anal. Chem.* **50**, 1465
17. Horman, I. (1984) *Gazz. Chim. Ital.* **114**, 297
18. Biemann, K. (1992) *Annu. Rev. Biochem.* **61**, 977-1010
19. Feistner, G.J., Hojrup, P., Evans, C.J., Barofsky, D.F., Faull, K.F. and Roepstorff, P. (1989) *Proc. Natl. Acad. Sci. USA* **86**, 6013-6017
20. Curtis, J.M., Derrick, P.J., Self, R. and Morgan, P.J. *12th. Int. Mass Spectrom. Conf.*. Amsterdam, August 1991
21. Dass, C. and Desiderio, D.M. (1988) *Anal. Chem.* **60**, 2723-2729
22. Hoffman, A.D., Ryan, P.A., Ireland, D., Derrick, P.J., Morgan, P.J. and Self, R. (1992) Proc. Kyoto '92 Int. Conf. Biol. Mass Spectrom, ed. T. Matsuo. SAN-EI Publ. Co. Kyoto, Japan 280-81
23. Self, R. and Parente, A. (1983) *Biomed. Mass Spectrom.* **10**, 78-82
24. Barber, M., Elliot, G.J., Bordoli, R.S., Green, B.N. and Bycroft, B.W. (1988) *Experientia* **44**, 266-270
25. Roepstorff, P., Nielsen, P.F., Kamensky, I., Craig, A.G. and Self, R. (1988) *Biomed. Environ. Mass Spectrom.* **15**, 305-310
26. Roepstorff, P. and Nielsen, P.F. (1988) *Biomed. Environ. Mass Spectrom.* **17**, 137
27. Self, R. and Mellon, F.A. (1988) in *Analytical Applications of Spectroscopy*, eds. C.S. Creaser and A.M.C. Davies, Royal Society of Chemistry, London, pp 294-297
28. *Metallothionein in Biology and Medicine* (1991) Ed. C.D. Klaasen and K.T. Suzuki, CRC Press, Boca Raton, Florida.
29. Kagi, J.H.R. and Hunziker, P. (1989) *Biol. Trace Element Res.* **21**, 111-118
30. Unger, M.E., Chen, T.T., Murphy, C.M., Vestling, M.M., Fenselau, C. and Roesijadi, G. (1991) *Biochim. Biophys Acta* **1074**, 371-377.
31. Beattie, J.H., Richards, M.P. and Self, R. (1993) *J. Chromatogr.* **632**, 127-135
32. Richards, M.P., Beattie, J.H. and Self, R. (1993) *J. Liq. Chromatogr.* **16** (9 and 10), 2113-2128
33. Barber, M., Bordoli, R.S., Elliot, G.J., Fujimoto, D. and Scott, J.E. (1982) *Biochem. Biophys. Res. Commun.* **109**, 1041-1046
34. Robins, S.P. and Duncan, A. (1983) *Biochem. J.* **215**, 167-173
35. Pratt, D.A., Daniloff, Y., Duncan, A. and Robins, S.P. (1992) *Anal. Biochem.* **207**, 168-175
36. Self, R., Pratt, D.A., Robins, S.P. Chen, S. and Derrick, P.J. (1993) 41st ASMS Conf. on Mass Spectrom. Allied Topics, San Francisco, USA
37. Bjorndal, H., Hellerqvist, C.G., Lindberg, B. and Svensson, S. (1970) *Angew, Chem. Internat. Edn.* **9**, 610
38. Selvendran, R.R. (1983) in *Recent Developments in Mass Spectrometry in Biochemistry, Medicine and Environmental Research*, ed. A. Frigerio, Elsevier, Amsterdam, pp 159-203
39. DuPont, M.S. and Selvendran, R.R. (1987) *Carbohydr. Res.* **163**, 99
40. Stevens, B.J.H. and Selvendran, R.R. (1984) *Phytochem.* **23**, 339
41. Selvendran, R.R. and King, S.E. (1989) *Carbohydr. Res.* **195**, 87
42 Stevens, B.J.H. and Selvendran, R.R. (1984) *Carbohydr. Res.* **135**, 155

43. Dell, A. and Panico, M. (1986) in *Mass Spectrometry in Biomedical Research*, ed. S.J. Gaskell, Wiley, Chichester, pp 149-179
44. Nothnagel, E.A., McNeil, M., Albersheim, P. and Dell, A. (1983) *Plant Physiol.* **71**, 916
45. Domon, B. and Costello, C.E. (1988) *Glycoconj. J.* **5**, 397
46. Coldwell, R.D., Trafford, D.J.H., Varley, M.J., Makin, H.L.J. and Kirk, D.N. (1988) *Biomed. Environ. Mass Spectrom.* **16**, 81
47. Coldwell, R.D., Trafford, D.J.H., Varley, M.J., Kirk, D.N. and Makin, H.L.J. (1989) *Clin. Chem. Acta* **180**, 157
48. Zagalak, B. and Borschberg, H.J. (1988) *Spectroscopy (Ottawa)* **6**, 203
49. Reddy, G.S. and Tserng, K.-Y. (1990) *Biochemistry* **29**, 943
50. Hachey, D.L., Coburn, S.P., Brown, L.T., Erbelding, W.F., DeMark, B. and Klein, P.D. (1985) *Anal. Biochem.* **151**, 159
51. Tanaka, A., Iijima, M., Kikuchi, Y., Hoshino, Y. and Nose, N. (1989) *J. Chromatogr.* **466**, 307
52 Khachik, F., Beecher, G.R., Goli, M.B., Lusby, W.R. and Smith, J.C. jnr. (1992) *Anal. Chem.* **64**, 2111-2122
53. Agriculture and Food Research Council, Institute of Food Research, Annual Report for 1992. ISBN 0 7084 0532 0
54. Walton, T.J., Mullins, C.J., Newton, R.P., Brenton, A.G. and Beynon, J.H. (1988) *Biomed. Environ. Mass Spectrom.* **16**, 289
55. Self, R., Taylor, L.C.E., Bradley, C.V., Santikarn, S. and Williams, D.H. (1982) in *Introduction to Modern Mass Spectrometry* Ed. S. Daolio, Consiglio Nationale della Ricerche, Padova
56. Self, R., Eagles, J., Galletti, G.C., Mueller-Harvey, I., Hartley, R.D., Lea, A.G.H., Magnolato, D., Richli, U., Gujer, R. and Haslem, E. (1986) *Biomed. Environ. Mass Spectrom.* **13**, 449-468
57. Heaney, R.K. and Fenwick, G.R. (1989) in *Aspects of Applied Biology 19. Antinutritional Factors, Potentially Toxic Substances in Plants* Eds J.P.F. D'Mello, C.M. Duffus and J.H. Duffus, Association of Applied Biologists, Wellesbourne, UK
58. Fenwick, G.R., Eagles, J. and Self, R. (1982) *Org. Mass Spectrom.* **17**, 544
59. Mellon, F.A., Chapman, J.R. and Pratt, J.A.E. (1987) *J. Chromatogr.* **394**, 209
60. Heeremans, C.E.M., van der Hoeven, R.A.M., Niessen, W.M.A., Vink, J., de Vos, R.H. and van der Greef, J. (1989) *J. Chromatogr.* **472**, 219
61. Kokkonen, P., van der Greef, J., Niessen, W.M.A., Tjaden, U.R., Ten Hove, G.J. and van der Werken, G. (1989) *Rapid Commun. Mass Spectrom.* **3**, 102
62. Price, K.R., Eagles, J. and Fenwick, G.R. (1988) *J. Sci. Food Agric.* **42**, 183
63. Self, R., Coxon, D.T., Taylor, L.C.E. and Evans, S. (1982) 29th ASMS meeting, Minneapolis
64. Price, K.R., Mellon, F.A., Self, R., Fenwick, G.R. and Osman, S.F. (1985) *Biomed. Mass Spectrom.* **12**, 79
65. Price, K.R., Fenwick, G.R. and Self, R. (1986) *Food Addit. Contam.* **3**, 241
66. Chen, S., Derrick, P.J., Mellon, F.A. and Price, K.R. (1994) Analyt. Biochem. (in press)

512

67. Evans, S., Buchanan, R., Hoffman, A., Mellon, F.A., Price, K.R., Hall, S., Walls, F.C., Burlingame, A.L., Chen, S. and Derrick, P.J. (1993) *Org. Mass Spectrom.* **28**, 289-290

68. Gilbert, J. (1987) in *Applications of Mass Spectrometry in Food Science,* Ed. J. Gilbert, Elsevier Applied Sciences, pp 73-140

69. Catlow, D.A. and Rose, M.E. (1989) in *Mass Spectrometry,* Vol 10, ed. M.E. Rose, Royal Society of Chemistry, London.

70. Evershed, R.P. (1989) in *Mass Spectrometry,* Vol 10, ed. M.E. Rose, Royal Society of Chemistry, London.

71. Uyakul, D., Isobe, M. and Goto, T. (1989) *J. Assoc. Off. Anal. Chem.* **72**, 491

72. Krishnamurty, T., Black, D.J. and Isensee, R.K. (1989) *Biomed. Mass Spectrom.* **18** 287

73. Wreford, B.J. and Shaw, K.J. (1987) *Food Addit. Contam.* **5**, 141

74. Takeda, K. Enoki, S., Harborne, J.B. and Eagles, J. (1989) *Phytochemistry* **28**, 499

75. Williams, C.A., Harborne, J.B. and Eagles, J. (1989) *Phytochemistry* **28**, 1891-1896

76. Harborne, J.B., Greenham, J. and Eagles, J. (1990) *Phytochemistry* **29**, 2899

77. Harborne, J.B., Greenham, J., Eagles, J. and Wollenweber, E. (1991) *Phytochemistry* **30**, 1044

78. Harborne, J.B., Greenham, J., Williams, C., Eagles, J. and Markham, K.R. (1992) *Phytochemistry* **31**, 305-308

79. Williams, C.A., Greenham, J., Harborne, J.B., Eagles, J. and Markham, K.R. (1992) *Phytochemistry* **31**, 555

80. Williams, C.A., Harborne, J.B., Greenham, J., Eagles, J. and Markham, K.R. (1993) *Phytochemistry* **32**, 731-735

81. Mueller-Harvey, I. and Hartley, R.D. (1986) *Carbohydr. Res.* **148**, 71-85

82. Gujer, R., Magnolato, D. and Self, R. (1986) *Phytochemistry* **25**, 1431-1436

83. Crews, H.M., Ducros, V., Eagles, J., Mellon, F.A., Kastenmayer, P., Luten, J.B. and McGaw, B.A. Review in preparation

84. Hachey, D.L., Wong, W.W., Boutton, T.W. and Klein, P.D. (1987) *Mass Spectrometry Reviews* **6**, 289-328

85. Ure, A.M. and Bacon, J.R. (1987) in *Applications of mass spectrometry in food Science.* Ed. J. Gilbert, Elsevier Applied Science, London, pp 343-372

86. Smith, D.L. (1983) *Anal. Chem.* **55**, 2391

87. Eagles, J., Fairweather-Tait, S.J. and Self, R. (1985) *Anal. Chem.* **57**, 469-471

88. Gharaibeh, A.A.R., Eagles, J. and Self, R. (1985) *Biomed. Mass Spectrom.* **12**, 344-347

89. Eagles, J., Fairweather-Tait, S.J., Portwood, D.E. and Self, R. (1989) *Anal. Chem.* **61**, 1023-1025

90. Eagles, J., Fairweather-Tait, S.J., Mellon, F.A., Portwood, D.E., Self, R., Götz, A., Heumann, K.G. and Crews, H.M. (1989) *Rapid Commun. Mass Spectrom.* **3**, 203-205

91. Mellon, F.A., Eagles, J., Fairweather-Tait, S.J., and Fox, T.E. (1994) Anal. Chim. Acta (in press)

92. Dean, J.R., Ebdon, L., Crews, H.M. and Massey, R.C. (1988) *J. Anal. At. Spectrom.* **3**, 349

93. Kivisto, B., Anderson, H., Cederblad, A.-S. and Sandström, B (1986) *Brit. J. Nutr.* **55**, 255
94. Ting, B.T.G. and Janghorbani, M. (1986) *Anal. Chem.* **58**, 1334
95. Fairweather-Tait, S.J., Jackson, M.J., Fox, T.E., Wharf, S.G., Eagles, J. and Croghan, P.C. (1993) *Brit. J. Nutr.* **70**, *221-234*
96. Gray, A.L. (1985) *Spectrochim. Acta* **40B**, 1525-1537
97. Gray, A.L. (1986) *J. Anal. Atom. Spectrom.* **1**, 403-405
98. Gray, A.L. (1989)   In: *Applications of Inductively Coupled Plasma Mass Spectrometry* eds A.R. Date and A.L. Gray, pp 1-42 Blackie London.
99. Sargent, M. and Webb, K. (1993) *Spectroscopy Europe* **5**, 21-27
100. Bock, R. (1979) A Handbook of Decomposition Methods in Analytical Chemistry. International Textbook Company Limited, Blackie, London.
101. Mizuike, A. (1983) *Enrichment Techniques for Inorganic Trace Analysis.* Chemical Laboratory Practice, 19. Springer-Verlag, Berlin
102. Kingston, H.M. and Jassie, L.D. (1988) In *American Chemical Society Professional Reference Book*, eds H.M. Kingston and L.B. Jassie, American Chemical Society, Washington DC
103. Lutz, T.M., Nirel, P.M.V. and Schmidt, B. (1991) In *Applications of Plasma Source Mass Spectrometry*, eds G. Holland and A.E. Eaton, pp 96-100, The Royal Society of Chemistry, Cambridge
104. Whitaker, P.G., Barrett, J.F.R. and Williams, J.G. (1992) *J. Anal. Atom. Spectrom.* **7**, 109-113
105. Delves, H.T. and Campbell, M.J. (1988) *J. Anal. Atom. Spectrom.* **3**, 343-348
106. Campbell, M.J. and Delves, H.T. (1989) *J. Anal. Atom. Spectrom.* **4**, 235-236
107. Vicizian, M., Lasztity, A. and Barnes, R.M. (1990) *J. Anal. Atom. Spectrom.* **5**, 293-300
108. Vanhoe, H., Vandecasteele, C., Versieck, J. and Dams, R. (1989a) *Anal. Chem.* **61**, 1851-1857
109. Vanhoe, H. Vandecasteele, C., Versieck, J. and Dams, R. (1989b) *Mikrochim. Acta III*, 373-379
110. Mulligan, K.J., Davidson, T.M. and Caruso, J.A. (1990) *J. Anal. Atom. Spectrom.* **5**, 301-306
111. Evans, E.H. and Ebdon, L. (1990) *J. Anal. Atom. Spectrom.* **5**, 425-430
112. Lam, J.W.H. and Horlick, G. (1990) *Spectrochim. Acta* **45B**, 1313-1325
113. Beauchemin, D. and Craig, J.M. (1991) *Spectrochim. Acta* **46B**, 603-614
114. Lyon, T.B.D., Fell, G.S., Hutton, R.C. and Eaton, A.N. (1988) J. Anal. Atom. Spectrom. 3, 601-603.
115. Ting, B.T.G. and Janghorbani, M. (1987) *Spectrochim. Acta* **42B**, 21-27
116. Ting, B.T.G. and Janghorbani, M. (1988) *J. Anal. Atom. Spectrom.* **3**, 325-336
117. Ting, B.T.G., Mooers, C.S. and Janghorbani, M. (1989) *Analyst* **114**, 667-674
118. Ting, B.T.G., Lee, C.C., Janghorbani, M. and Prohaska, J.R. (1990) *J. Nutr. Biochem.* **1**, 249-255
119. Lyon, T.D.B. and Fell, G.S. (1990) *J. Anal. Atom. Spectrom.* **5**, 135-137
120. Scheutte, S., Vereault, D., Ting, B.T.G. and Janghorbani, M. (1988) *Analyst* **113**, 1837-1842

514

121. Buckley, W.T., Budac, J.J., Godfrey, D.V. and Koenig, K.M. (1992) *Anal. Chem.* **64**, 724-729

122. Dean, J.R., Crews, H.M. and Ebdon, L. (1989) In *Applications of Inductively Coupled Plasma Mass Spectrometry*, eds A.R. Date and A.L. Gray, pp 141-168, Blackie, London.

123. Amarasiriwardena, C.J., Gercken, B., Argentine, M.D. and Barnes, R.M. (1990) *J. Anal. Atom. Spectrom.* **5**, 457-462

124. Ridout, P.S., Jones, H.R. and Williams, J.G. (1988) *Analyst* **113**, 1383-1386

125. Satzger, R.D. (1988) *Anal. Chem.* **60**, 2500-2504

126. Lyon, T.D.B., Fell, G.S., McKay, K. and Scott, R.D. (1991) *J. Anal. Atom. Spectrom.* **6**, 559-564

127. Janghorbani, M. and Ting, B.T.G. (1989) in *Applications of Inductively Coupled Plasma Mass Spectrometry*. eds A.R. Date and A.L. Gray, pp 115-140, Blackie, London

128. Schoenheimer, R. and Rittenberg, D. (1940) *Physiol. Rev.* **20**, 218-248

129. Hoefs, J. (1973) *Stable Isotope Geochemistry*, Springer Verlag, Berlin.

130. Nier, A.O. (1940) *Rev. Sci. Inst.* **11**, 212-216

131. McKinney, C.A., McCrea, J.M., Epstein, S., Allen, H.A. and Urey, H.C. (1950) *Rev. Sci. Inst.* **21** 724-730

132. Dumas, J.B.A. (1831) Procedes de l'analyse organique. *Ann. Chim. Phys.* **247**, 198-213

133. Bremner, J.M. and Mulvaney, C.S. (1982) in *Methods of Soil Analysis Agronomy* **31**, 595-624

134. Milne, E. and McGaw, B.A. (1987) *Biomed. Environ. Mass Spectrom.* **15**, 467-472.

135. McGaw, B.A., Milne, E., Duncan, G.J. (1988) *Biomed. Environ. Mass Spectrom* **16**, 269-273

136. Europa Scientific Application Note 4. Europa Scientific Ltd, Crewe, Cheshire.

137. Read, W.W., Read, M.A., Rennie, M.J. and Griggs, C. (1984) *Biomed. Mass Spectrom.* **11**, 348-352

138. Hauck H.D. (1982) in *Methods of soil Analysis agronomy* **36**, 735-779.

139. Wong, W.W., Lee, L. and Klein, P.D. (1987) *Amer. J. Clin. Nutr.* **45**, 905-913

140. Midwood, A.J., Haggarty, P., Milne, E. and McGaw, B.A. (1992) *Appl. Radiat. Isot.* **43**, 1341-1347

141. Mariotti, A. (1983) *Nature* **303**, 685-687

142. Gonfiantini, R. (1981) *Stable Isotope Hydrology: Deuterium and Oxygen 18 in the water cycle.* International Atomic Energy Agency, 35-85, Vienna

143. Garlick, P.J., Wernerman, J., McNurlan, M.A., Essen, P., Lobley, G.E., Milne, E., Calder, A.G. and Vinnars, E. (1989) *Clin. Sci.* **77**, 329-336

144. Lobley, G.E., Harris, P.M., Skene, P.A., Brown, D., Milne, E., Calder, A.G., Anderson, S.E., Garlick, P.J., Nevison, I. and Connell, A. (1992) *Brit. J. Nutr.* **68**, 373-388

145. Heys, S.D., Park, K.G.M., McNurlan, M.A., Milne, E., Ermin, O., Wernerman, J., Kennan, R.A. and Garlick, P.J. (1991) *Br. J. Surg.* **478**, 483-487

146. McNurlan, M.A., McHardy, K.C., Broom, J., Milne, E., Fearnes, L.M., Reeds, P.J. and Garlick, P.J. (1987) *Clin. Sci.* **73,** 69-75

147. McClean, T.A. and Tobin G. (1987) *Animal and human calorimetry.* Cambridge University Press, London

148. Blaxter, K.L., Graham, N.McC. and Rook, J.A.F. (1954) *J. Agric. Science (Camb.)* **45,** 10-18

149. Lifson, N., Gordon, G.B. and McClintock, R. (1955) *J. Appl. Physiol.* **7,** 704-710

150. Haggarty, P., Franklin, M.F., Fuller, M.F., McGaw, B.A., Milne, E., Duncan, G., Christie, S.L. and Smith, J.S. (1993) *Proc Nutr. Soc.* **51,** 71A.

151. Thompson, G.N., Pacy, P.J., Ford, G.C. and Halliday, D. (1989) *Biomed. Environ. Mass Spectrom.* **18,** 321-327

152. Halliday, D. (1990) in *Paediatric Nutritional and Metabolic Research.* eds T.E. Chapman, R. Berger, D.J. Reingoud and A. Okken, Intercept, Andover, Hampshire

153. McNurlan, M.A. and Garlick, P.J. (1989) *Diab. Metab. Rev.* **5,** 165

154. Smith, K and Rennie, M.J. (1990) *Bailliere's Clin. Endocr. Metab.* **4,** 461

155. Freedman, P.A., Gillyon, E.C.P. and Jumeau, E.J. (1988) *Am. Lab.* **20,** 114

156. Patterson, B.W., Hachey, D.L., Cook, G.L., Amann, J.M. and Klein, P.D. (1991) *J. Lipid Res.* **32,** 1063

157. Calder, A.G., Anderson, S.E., Grant, I., McNurlan, M.A. and Garlick, P.J. (1992) *Rapid Commun. Mass Spectrom.* **6,** 421-424

158. Scheppach, W., Wiggins, H.S., Halliday, D., Self, R., Howard, J., Branch, W.J., Schrezenmeir, J. and Cummings, J.H. (1988) *Clin. Sci.* **75,** 363-370

159. Wooton, R., Ford, G.C., Cheng, K. and Halliday, D. (1985) *Physics in Medicine and Biology* **30,** 1143-1149

# INDEX

## A

Ab initio methods 89
Acetylated hypercin 430
Acetylated stentorin 430
Adsorption 448, 450, 456
Advanced glycation end products (AGE) 318
Alanine 103, 104, 105, 106, 107
AM1 90, 103, 104, 105, 107
Amadori rearrangement 317
$\alpha$-amino acids 103,107
Ammonia negative chemical ionisation 464
Analysis of nucleosides and bases in mixtures 357
Animal globins 404
Anthocyanin variants 493
Antinutrients 491
Array detector 51
Atmospheric pressure ionisation (API) 465
Atmospheric pressure ionisation source 461
Automated $^{15}$N $^{13}$C analyser 503
Automated Breath $^{13}$C analyser 503
Azatropylium ion 490

## B

Background adducts 403
Background subtraction 278, 500
Band broadening 448
Basic features of nucleic acid structure 352
Benzyloxycarbonyl (Z) 26
Bile salts 286
Bioavailability of mineral nutrients 495
Biological fluids 499
Biological matrices 501
Biological monitoring 400
Biological procedures 506
Biological variability 503
Biotransformation 398
Bleeding 450
Blood samples 495
Body tissue proteins 506
Bond orders 93
Bracketing 455
Bracketing with standards 495
Breath analysis 504

## C

C-terminal sequencing 64
$C_3$ and $C_4$ plants 494
Caffeine 456
Calcium 495
Calibration 501, 504
Calibration standards 455
Cancer initiators 381
Cancer risk 391
Capillary columns 440, 445, 447, 448, 449
Capillary tubes 448
Carbohydrate turnover 507
Carbonic anhydrase 64
Carboxypeptidases 64
Carcinogens 397
Carotenoids 3, 491
Carrier gas 438, 442, 443, 445, 446, 447
Carrier gas flow 440, 448
CE-ESI/MS 486
CE/CF-FAB 287
CE/MS 491
Cell wall polysaccharides 491, 493
Centre-of-mass collision energy 241
Ceramide 417
Certified reference material 494
CF-FAB direct injection analysis 280
CF-FAB of tryptic digests of small peptides 281
Charge density 93
Charge exchange ionisation 462
Charge transfer complexation 25
Charles Law 439
Chemical ionisation 484
Chemical noise 4
Chicken MT 489
Chromatographic purification 503
CID mass spectra 490
$\alpha$-Cleavage to a carbonyl bond 94
$CO_2$ charge exchange mass spectra 463
Collagen crosslinks 487, 490
Collector slit 452
Collectors 502
Collision gas pressure effects on translational energy losses 232

517

Column diameter 449
Combination of mass spectrometry with LC
and CZE 282
Combustion of organic material 504
Concentration ratio 454
Condensed tannins 493
Conductance 439, 440, 441
Continuous flow FAB 486
Continuous flow instruments 502
Continuous flow isotope ratio mass
spectrometry 503
Continuous flow techniques 503
Continuous nebulisation 496
Copolymer membranes 443
Cryogenic microanalysis 503
Cryogenic purification 502
Cyclotron frequency 149
Cytochrome C 64

D

Decarboxylation of isolated amino acids 504
Deoxypyridinoline 490
Desorption mechanism 4
Desorption ionisation techniques 3
Determination of kinetic energy release 100
Determination of molecular weight 62
Determination of molecular weight and
oligonucleotide mapping 366
Determination of nucleic acid sequence 367
Differentially pumped GC-MS system 441
Differentially pumped MS system 448
Diffusion rate 444
Dinucleotides phosphodiesters 21
Direct analysis of a sample by flow injection
280
Direct capillary SFC/MS coupling 462
Direct coupling devices 442, 448
Direct fluid interface 462
Direct momentum transfer mechanisms 227
Discharge ionisation 461
Disulphide bonds 71
Dithymidylic acid 20
DNA adducts 403
Doubly labelled water method 502
Drug metabolites 286
Dual inlet 502
Dumas combustion 503
Dwell times 452

Dynamic FAB 274

E

Edman method 383
Efficiency, flow rate, sample capacities 449
(1:1) Electrolytes 17
Electrohydrodynamic decomposition 62
Electron scavengers 24
Electrophilic agents 403
Electrospray (ES) 465
Electrospray ion source 469
Electrospray ionisation mass spectrometry
(ESI MS) 61
Electrospray source coupled to a quadrupole
analyser 62
Elimination of the radial component 162
Elisa 319
Energy expenditure 502, 507
Energy minimum 106
Energy of the primary particles 15
Energy partitioning 93
Energy shifts in collisional activation 201
Enrichment process 442
Entropy contribution 105
Environmental monitoring 400
Enzymatic decarboxylation 507
Equation of motion 181
Equilibration of water with carbon dioxide
505
ESI MS spectra of tuna cytochrome C 69
ESI MS to determine a-helical content 69
Exopeptidases 64

F

FAB of $C_6$-$C_{16}$ alkyl trimethylammonium
bromides 8
FAB matrices 24
FAB response 10
FAB spectrum of carnitine 17
FAB spectra of protected diribonucleotides 4
Faecal samples 495
Falling needle injector 449
Faraday cups 503
Fast atom bombardment mass spectrometry 4
Fast Ion Detection 126
FFI 321
Film thickness 450
Flame ionisation detector (FID) 462, 465
Flame photometric detector 465

Flavour components 486
Flavour research 484
Flavours 486
Flow injection 496
Flow rates used with CF-FAB 275
Flow restrictors 463
Flow splitting 446, 447
Folded TOF Mass Analyzers 126
Food and agricultural research 486
Food contaminants 287
Food preservatives 488
Food volatiles 484
Fourier transform ion cyclotron resonance spectrometers 177
Fourier transform mass spectrometers 463
Fourier transform mass spectrometry (FTMS) 147
Free living animals 502
Frit FAB 274
Fritted glass 443
Furosine 320
Fusapyrone 22
Fused silica columns 449
Fused silica capillary tubing 448

G

Gangliosides 287
Gas flow and pressure limitations 438
Gas isotope ratio analysis 502
Gas-phase basicity 105
Gas-phase ionisation in a corona discharge 466
GC-MS analysis 475
GC-MS coupling 437, 438, 441, 452
GC/IR/MS 504
Genotoxic compounds 381
Genotoxic risk 398
GIRMS 505
Glass restrictors 448
Glucose infusion 508
Glucose turnover 508
Glucosinolates 492
Glucosylated phenolic compounds 493
Glucosylgalactosylpyridinoline 490
Glycated albumin 323
Glycated polylysine 323
Glycated proteins 320
Glycerol-sodiated cluster peaks 13

Glycerol-to-sample ration (G/S) 7
Glycine 103, 104, 105, 106, 107
Glycoalkaloids 491, 492
Glycobiology 491
Glycolipid 51
Glycopeptides 286
Glycosides 491
Glycosphingolipids 287

H

Haemoglobin adducts 405
Haeme globin interaction 77
Haemoglobin (Hb) accumulates adducts 382
Halogenated nucleosides 23
Hartree-Fock 90
Heptafluorobutylamine derivative 508
High energy decompositions of peptides 304
High flow rate interface 464
High mass-to-charge ions 163
High mass-to-charge ions. Improving detection performance 156
High performance detection capabilities of FTMS 150
High resolution GC-MS 452
High resolution ICPMS 500
High vacuum 439
High-mass resolving power of FTMS 155
Horse heart apomyoglobin 63
HSAB principle 16
Human apolipoprotein AI 65
Human nutrition research 494
Human protein kinetics 507
Hybrid instruments of unusual geometry 261
Hybrid ion traps 265
Hybrid sector/quadrupole instruments 299
Hybrid TOF 262
Hydrogen rearrangement processes 490
Hydrogen-deuterium exchange 69
Hydrogen-deuterium exchange and ESI 68
α-hydroxy cinnamic acid 489
Hydroxy-vitamin $D_3$ 457
Hyphenated methods 320
3-hydroxypyridinium crosslinks 490

I

Impulsive collision transfer theory 228
In vivo microdialysis/CF-FAB MS/MS 291
In-line combustion 504
Inductively-coupled plasma (ICPMS) 494

520

Inlet 502
Inner salts 17
Interferences 499
Internal energy distribution of gaseous ions produced by FAB 16
Internal standard 453, 454, 455, 494
Intimate ion-pairs 20
Intrinsic basicity 103, 107
Ion cloud focusing 167
Ion evaporation model 62
Ion mobility detector 465
Ion motion and interactions with the electric field 153
Ion remeasurement 169
Ion structures and reaction pathways 92
Ion transmission 439
Ion trap 188
Ion trapping with static fields 150
Isobaric overlaps 499
Isotope dilution 452, 455
Isotope dilution ICPMS 501
Isotope effects 456, 457
Isotope enrichment 505
Isotope Ratio Mass Spectrometry (IRMS) 502
Isotopic enrichment 453
Isotopic fractionation effects 502
Isotopic standards 496

J

Jet separator 438, 445

K

Kinectis and energetics of sputtering in FAB 5
Kinetic method 13
Knock-on mechanism 5

L

Labelled amino acids 506
Labelled internal standard 456
Labelled standard 453, 454, 457
Laser desorption 51
LC/CF-FAB 282
LC/MS 472, 491
LC/MS analysis of the crude enzyme digest 285
LC/MS interfaces 461
Liberation of blood bicarbonates 504
Light scattering detector 465

Liquid phase ionisation by pure electrospray ionisation 467
Liquid-phase electrospray ionisation 461
Lithium ion adducts 466
Lormazepan 457
Low energy CAD of peptides 299
Low energy decompositions of peptides 301

M

Magnetic sector instruments 452
Magnetic sectors 463
Maillard reactions 317
MALDI-MS 33
Mass determination of biological molecules 33
Mass discrimination 494
Mass fractionation 494
Mass/charge ratio-deconvoluted, energy-selected momentum spectra of the peptide ions 310
Mass spec. of isomers. Theoretical studies 98
Matched capillaries 502
Matrix suppression 499
Matrix-assisted laser desorption/ionisation 33, 488
Matrices in FABMS 7
Md/MS 292
Md/MS methodology for in vivo studies 292
Mean free path 439, 440, 443, 445
Mechanism of ion trapping 148
Membrane permeability 444
Mesir 475
Metabolic processes 506
Metabolite balance 400
Metal-protein interactions 79
Metallothioneins 488
Metallothioneins of sheep 489
Microdialysis / MS 290, 486
MNDO 90
MO methods 89
Mobile phase 449
Modified cells of simple geometries 160
Molar ratio 454
Molecular density 439
Molecular diffusion 443
Molecular dosimetry 402
Molecular effusion 443

Molecular separators 437, 441, 442, 443, 445, 446, 447
Moving belt transport interface 465
MS/MS analysis 300
MS$^n$ experiments 307
Multi-element surveys 501
Multipass TOF Mass Analysers 125

N

N-9-Fluorenylmethoxycarbonyl (Fmoc) 26
(n:p) Electrolytes 19
Nebuliser systems 497
Negative ion chemical ionisation 385
Neutral polar substrates 22
Nisin 488
Nucleic-acids modified nucleosides 353
Nucleobases 105
Nucleosides 105
Nutritional research 502

O

Oligonucleotides 21, 287
Oligosaccharides 51, 286
One-electron capture reactions 19
Open split 447, 448
Optimisation of the electric trapping field 158
Optimum Pressure for Collision-Induced Decomposition 203
Orthogonal TOF mass analyser 123
Oxidation of ammonium sulphate 504

P

PA 105
Packed columns 441, 447, 462
Packed columns SFC 462
Parametric resonances in cells with non-quadrupolar potentials 162
Partially and fully protected oligonucleotides 27
Partially-methylated alditol derivatives 491
Particle beam 456
Particle beam interfaces 438, 465
PDB 505
PDMS and SIMS 4
Peak band width 452
Pentafluorophenylthiohydantoin 383
Pentosidine 320
Peptide ion rearrangement processes 310
Peptide mapping 64
Peptide mapping by LC/MS 285

Peptides, metalloproteins 487
Peptides low energy fragmentations 301
Permethylation 491
Phase ratio 447, 450
Phenolic polysaccharides 491
Phenylalanine-d$_5$ 507
Photoreceptor chromophore 429
Physiological fluids 487
Plant phenolics 492
Plasma torch 498
Plasma-free amino acids 507
Platelet activating factor 456
Platinum capillaries 448
PM3 103, 104, 107
Pneumatically generated ions by gas phase mobility 62
Polar modifiers 463
Polyphenols 492
Pore size 443
Porous barrier 443
Post-column addition of FAB matrix 284
Potential energy profiles 92
Precharged substrates 17
Precision 452, 453, 455, 456, 457, 495, 501, 505
Preformed ions 8
Preparation of oligonucleotides for analysis 365
Pressure programmed SFC 464
Primary structure of proteins 64
Principles of FTMS experiments 148
Principles of TOF-Mass-Analyzers 112
Procedural errors 454
Procyanidins 493
Prostaglandins 287
Protein adducts 403
Protein bound amino acids 506
Protein degradation 507
Protein folding 70
Protein structure 61
Protein turnover 502, 506, 507
Proton affinities 14, 103, 105, 106,
Proton transfer CI 463
Proton-bound dimers 13
Protonation process 103, 107
Pulsed FAB source 9
Pulsed Ion Production 118

Pumping speed 438, 439, 440, 441, 445, 447,
Pyr /GC/MS 317
Pyridinoline 490
Pyrolisis mass spectrometry, inorganic trace
analysis 484

Q

Quadrupole field 179
Quadrupole instruments 452
Quadrupole ion traps 177
Quadrupole mass filter 177, 184
Quadrupoles 463
QuadrupoleTIMS 496
Quantification of nucleic acid constituents
364
Quantitative analysis 437, 494
Quantitative analysis of metals 497
Quantitative GC-MS 453
Quantitative isotope dilution mass
spectrometry 452
Quantum chemical methods 89
Quasi-Continuous Ion Production 119
Quaternary structure of proteins 81

R

Radiation damage 4
Rat MT 489
Reaction intermediate scanning 307
Reduction of disulphide bonds in proteins 76
Reduction of water 505
Reductive power 24
Reference methods 455
Repeller field TOF Mass Analysers 114
Respiratory quotient 507
Response ratio 455, 456, 457
Restrictor 440, 442, 448, 462, 468
Rhenium filament 463
Ring TOF Mass Analysers 125
Ruthenium(II) polypyridine complexes 19
Ryhage jet separator 442

S

Saliva samples 495
Sample capacity 450
Sample enrichment factor 441, 446
Sample loading 449
Sample pre-treatment 499
Sample preparation 498
Sample preparation for IRMS 504
Sample preparation system 502

Sampling cone 498
Scan rates 452
Scanning speed 452
SCF/MS with mobile phase separators
secondary reference material 505
Secondary structure 69
Sector field TOF Mass Analysers 115
Selected ion monitoring 452
Selected ion recording 508
Selective mass detector 452
Selvedge 6
Semiempirical methods 89
Sensitivity 449, 450, 452, 453
Separation barrier 442, 443
Sequential mass spectrometry of peptides 307
Sequential product ion scanning 307
SFC capillary columns 462
SFC/API/MS 472
SFC/ES/MS separations 470
SFC/MS Interfaces 461
SFC/MS with API 465
SFC/MS with FAB ionisation 465
Silicone membranes 444
Silver-palladium 443
Sinapinic acid 489
Skimmer cone 498
Skimmers 438
Solubility 444
Solvent-induced conformational changes in
proteins 74
Space charge effects 155
Spin density 93
Split injection 449
Split ratio 448
Spray chamber 497
Sputtering mechanism of organic molecules 6
Stability 456
Stable isotope 453, 456
Stable isotope as tracer 495
Stable isotope ratio GC/MS 507
Stable isotope ratio mass spectrometry 493
Stable isotope-labelled volatiles 507
Standard addition 494
Standard Light Antarctic Precipitation 505
Standard Reference Materials 499
Standardisation 501, 505
Stationary phase (film thickness) 449

Stationary phase bleed 449
Stentor coeruleus 429
Storage ion source 122
Structure elucidation of peptides 299
Suppression effect 8, 277
Surrogate IS 454

T

t-Butoxycarbonyl (Boc) 26
Tandem instrument of BEqQ configuration 299
Tandem mass spectrometry 68 261, 386, 484, 492
Tandem-in-time technique 148
Tannins 492
Target gas excitation-induced decomposition? 216
Target gas mass 217
Tertiary structure of a single polypeptide 70
Thermal denaturation 72
Thermal ionisation (TIMS) 494
Thermal stability 449
Thermospray 469
Thermospray interface 461
Thermospray source for SFC/MS 464
Thoriated iridium filament 463
Time dependence of 10
Time-of-flight mass spectrometers 33, 111
Tissue protein synthesis 507
TOF Mass Analysers for-Low-Energy Ions 117
Transfer linc 448, 450
Translational energy loss/fragment ion mass relationship 222
Translational energy spectra 251
Trietylammonium salt 20

U

Universally-labelled $^1$3C-glucose 508
Urine samples 495
Use of surfactants 9

V

Vacuum pumps 439
Vacuum systems 438, 439, 440
Vacuum technology 440
Valinomycin 3
Vancomycin 24
Variants of albumin proteins 68
Variation coefficient 455

Vienna-Standard Mean Ocean Water 505
Vitamins 491
Volumetric IS 454

W

Watson and Biemann with capillary column SFC 443
Whole body protein synthesis with packed column SFC 506

Z

Zinc 495
Zwitterionic compounds 17